T0251492

ADVANCES AND TRENDS IN ENGINEERING SCIENCES AND TECHNOLOGIES

PROCEEDINGS OF THE INTERNATIONAL CONFERENCE ON ENGINEERING SCIENCES AND TECHNOLOGIES, TATRANSKÁ ŠTRBA, HIGH TATRAS MOUNTAINS, SLOVAKIA, 27–29 MAY 2015

Advances and Trends in Engineering Sciences and Technologies

Editors

Mohamad Al Ali & Peter Platko

Institute of Structural Engineering, Faculty of Civil Engineering, Technical University of Košice, Košice, Slovakia

CRC Press
Taylor & Francis Group
Boca Raton   London   New York   Leiden

CRC Press is an imprint of the
Taylor & Francis Group, an **informa** business

A BALKEMA BOOK

CRC Press/Balkema is an imprint of the Taylor & Francis Group, an informa business

© 2016 Taylor & Francis Group, London, UK

Typeset by MPS Limited, Chennai, India

Published by: CRC Press/Balkema
                P.O. Box 11320, 2301 EH Leiden, The Netherlands
                e-mail: Pub.NL@taylorandfrancis.com
                www.crcpress.com – www.taylorandfrancis.com

ISBN: 978-1-138-02907-1 (Hardback)
ISBN: 978-1-315-64464-6 (eBook PDF)

Table of contents

Part B: Buildings and structures, Water supply and drainage, Construction technology and management, Materials and technologies, and Environmental engineering

Preface

The International Conference on Engineering Sciences and Technologies (ESaT 2015) was organized under the auspices of the Faculty of Civil Engineering, Technical University in Košice, Slovakia. The conference was held May 27–29, 2015 in the High Tatras, Slovakia.

The purpose of the conference was to facilitate a meeting that would present novel and fundamental advances in the fields of Engineering Sciences and Technologies for scientists, researchers and professionals around the world. Conference participants had the opportunity to exchange and share their experiences, research and results with other colleagues, experts and industrial engineers. Participants also had the possibility to extend their international contacts and relationships for furthering their activities.

The Conference focused on a wide spectrum of topics and areas of Civil engineering sciences as listed below:

- Materials and technologies
- Survey, mapping and planning
- Buildings and structures
- Roads, bridges and geotechnics
- Reliability and durability of structures
- Mechanics and dynamics
- Water supply and drainage
- Heating, ventilation and air condition
- Environmental engineering
- Construction technology and management
- Computer simulation and modeling

73 papers from more than 10 countries around the world have been accepted for publication in the conference proceedings of the Conference. Each of the accepted papers was reviewed by selected reviewers in accordance with the scientific area and orientation of the papers.

The editors would like to express their many thanks to the members of the Organizing and Scientific Committees. The editors would also like to express a special thanks to all reviewers, sponsors and conference participants for their intensive cooperation to make this conference successful.

Mohamad Al Ali   Peter Platko

Committees of ESaT 2015

Scientific Committee

Prof. Vincent Kvočák
Dean of the Faculty of Civil Engineering, Technical University of Košice, Slovakia

Prof. Stanislav Kmeť
Vice-rector of the Technical University, Technical University of Košice, Slovakia

Prof. Mohamad A. E. Al Ali
University of Aleppo, Syria

Prof. Safar Al Hilal
Zirve University, Turkey

Prof. Ivan Baláž
Slovak University of Technology in Bratislava, Slovakia

Prof. Bystrík Bezák
Slovak University of Technology in Bratislava, Slovakia

Prof. Abdelhamid Bouchair
Blaise Pascal University, France

Prof. Ján Brodniansky
Slovak University of Technology in Bratislava, Slovakia

Prof. Ján Bujňák
University of Žilina, Slovakia

Prof. Ján Čelko
University of Žilina, Slovakia

Prof. Kostas Goulias
University of California, Santa Barbara, USA

Prof. Károly Jármai
University of Miskolc, Hungary

Prof. Marcela Karmazínová
Brno University of Technology, Brno, Czech Republic

Prof. Dušan Katunský
Technical University of Košice, Slovakia

Prof. Dušan Knežo
Technical University of Košice, Slovakia

Prof. Mária Kozlovská
Technical University of Košice, Slovakia

Prof. Aleksander Kozlowski
Rzeszow University of Technology, Poland

Prof. Stanislav Krawiec
Silesian University of Technology, Katowice, Poland

Prof. Víťazoslav Krúpa
Institute of Geotechnics of SAS, Košice, Slovakia

Prof. Jindřich Melcher
Brno University of Technology, Brno, Czech Republic

Prof. Abolfazl K. Mohammadian
University of Illinois at Chicago, USA

Prof. Adrian Olaru
University Polytehnica of Bucharest, Romania

Prof. Ján Ravinger
Slovak University of Technology in Bratislava, Slovakia

Prof. Nadežda Števulová
Technical University of Košice, Slovakia

Prof. Nikolai I. Vatin
Saint-Petersburg State Polytechnical University, Russia

Prof. Josef Vičan
University of Žilina, Slovakia

Prof. Zuzana Vranayová
Technical University of Košice, Slovakia

Assoc. Prof. Naqib Daneshjo
University of Economics in Bratislava, Slovakia

Assoc. Prof. Paulo B. Cachim
University of Aveiro, Portugal

Assoc. Prof. Danica Košičanová
Technical University of Košice, Slovakia

Assoc. Prof. Hana Krejčiříková
Czech Technical University in Prague, Czech Republic

Assoc. Prof. Ján Mandula
Technical University of Košice, Slovakia

Assoc. Prof. Vít Motyčka
Brno University of Technology, Brno, Czech Republic

Assoc. Prof. Miloslav Řezáč
Technical University of Ostrava, Czech Republic

Assoc. Prof. Brigita Salaiová
Technical University of Košice, Slovakia

Assoc. Prof. Anna Sedláková
Technical University of Košice, Slovakia

Assoc. Prof. Stanislav Szabo
Czech Technical University in Prague, Czech Republic

Assoc. Prof. Michal Varaus
Brno University of Technology, Brno, Czech Republic

Assoc. Prof. Michał Zielina
Krakow University of Technology, Poland

Dr. Mohamad Al Ali
Technical University of Košice, Slovakia

Dr. Cengiz Erdogan
TRW Automotive, Germany

Dr. Enayat Danishjoo
TRW Automotive, Germany

Dr. Elżbieta Radziszewska–Zielina
Krakow University of Technology, Poland

Organizing Committee

Dr. Mohamad Al Ali, Dr. Peter Platko
esat2015@tuke.sk

Dr. Jozef Junák, Dr. Gabriel Markovič, Dr. Marcela Spišáková, Dr. Clayton Stone

Eng. Rastislav Gruľ', Eng. Marek Spišák, Eng. Martin Štefanco

Reviewers Committee

Prof. Vincent Kvočák, *Head of the Committee*
Prof. Magdaléna Bálintová
Prof. Mária Kozlovská
Prof. Zuzana Vranayová
Assoc. Prof. Ján Mandula

List of Reviewers

Al Ali M., Bajzecerova V., Baláž I., Balintova M., Baskova R., Beke P., Brodniansky J., Burák D., Cachim P., Cápayová S., Čelko J., Demjan I., Dimitriu D., Diniz J. A. V., Ďurove J. B., Eštoková A., Fecko L., Fedorová E., Flimel M., Harabinová S., Harbuľáková V. O., Hirš J., Hudák A., Hudák M., Jármai K., Junák J., Kaposztasova D., Karmazínová M., Kmeť S., Kosecki A., Kozlovska M., Krejčiříková H., Kušnír M., Kvočák V., Lazarová E., Leśniak A., Lis A., Lopušniak M., Lumnitzxer E., Mandula J., Marková J., Markovič G., Melcer J., Melcher J., Mesároš P., Mojdis M., Molčíková S., Motyčka V., Nikolic R., Ondova M., Palou M. T., Panulinová E., Platko P., Prascakova M., Priganc S., Ravinger J., Rudy V., Řezáč M., Sabol P., Salaiová B., Satnko Š., Sedivy J., Sedláková A., Schosser F., Sicakova A., Sincak P., Stanko Š., Stevulova N., Struhárová A., Sudzina F., Sverak T., Šabíková J., Šeminský J., Šimčák M., Škultétyová I., Šlezingr M., Špak M., Terpáková E., Tomko M., Uhmannová H., Vaclavik V., Varga T., Vatin N.I., Vavrek P., Vičan J., Vilčeková S., Vranay F., Vranayova Z., Zelenakova M., Zielina E. R., Zima K., Zuzulová A.

*Part A: Mechanics and dynamics,
Reliability and durability of structures,
Roads, bridges and geotechnics, and
Computer simulation and modeling*

Advances and Trends in Engineering Sciences and Technologies – Al Ali & Platko (Eds)
© 2016 Taylor & Francis Group, London, ISBN: 978-1-138-02907-1

Resistance of closed compressed cold-formed steel cross-sections with intermediate stiffeners

M. Al Ali
Faculty of Civil Engineering, Technical University in Košice, Slovakia

ABSTRACT: The paper presents fundamental information about running experimental-theoretical research oriented to determine the resistance of thin-walled cold-formed compressed steel cross-sections with and without intermediate stiffeners. The investigated members had closed cross-sections made from homogeneous materials. The theoretical analysis in this research is oriented to determine the resistance of mentioned members according to relevant standards, while the experimental investigation is to verify the theoretical results and to investigate the real behavior of mentioned members during the loading process.

1 INTRODUCTION

Many researches and studies have been devoted to the field of steel cold-formed members and profiles, Rendek and Baláž (2004), Rasmussen (2006), Young et al. (2007 and 2008). Some of the researches were focused on the closed cross sections, Yuan-Qi et al. (2007), Gardner et al. (2010), Al Ali et al. (2013 and 2014). Other authors investigated the resistance of cold-formed members with open profiles, Zhang et al. (2007), Jurgen et al. (2008), Garifullin and Vatin (2014), Vatin et al. (2014).

Theory and design development of cold-formed steel members and profiles creates a certain knowledge base for their practical application in civil engineering. However, this fact does not mean that all complex and challenging processes of the behavior of cold-formed members during the loading process are sufficiently investigated. The local stability requirements related to unfavorable buckling effects of the compressed parts are very significant. Favorable effects, related to membrane stresses and post-critical behavior are also important. In this context, the longitudinal stiffening in the compressed areas is generally effective. In terms of carrying capacity, the question of interaction of individual webs, as well as the question of the effectiveness of longitudinal stiffeners of such members are very important. Different calculation procedures with different results in the relevant standards, STN 73 1402 (1988), EN 1993-1-3 and EN 1993-1-5 (2006), and their confrontation with experimental results indicated the need for further investigation of post-critical behavior of these members.

2 EXPERIMENTAL PROGRAM AND TESTED MEMBERS

Mentioned research included 36 cold-formed test members having closed cross-sections with different dimensions, advisable chosen to eliminate the global stability problems, to reflect the post-critical behavior of the individual thin webs and to present the interaction of the adjacent parts in the loading and failure processes. 18 members had plane webs without stiffeners (cross-sectional groups A and B), Al Ali M. (2014), other 18 members had two webs with longitudinal intermediate stiffeners (cross-sectional groups C and D).

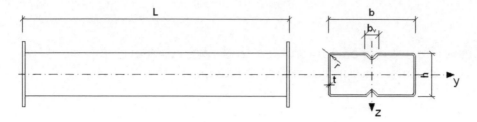

Figure 1. Cross-sections with intermediate stiffeners – scheme of the test members.

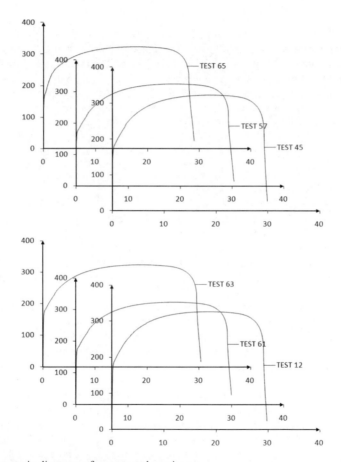

Figure 2. Stress–strain diagrams of some tested specimens.

The obtained results from mentioned research are very extensive; therefore the paper presents only the results of cross-sections with longitudinal intermediate stiffeners, see Figure 1.

All tested members are made from steel sheets with nominal thickness 2 mm. Three material samples were taken from each used steel sheet to make normative shaped test specimens. The test specimens underwent a tension tests to find the stress-strain diagrams and material properties. Some results from the tension tests are illustrated by Figure 2.

Detailed measuring of real dimension of the tested members was done before the loading tests, in order to consistent evaluation of the experimental results, Al Ali et al. (2013). The averages of measured values are presented in Table 1.

4

Table 1. Average values of real measured dimensions and material characteristics.

Tested members Cross-section	Marking	b	h	t	b_v	r	L	f_y	f_u
		[mm]						[MPa]	
C 1	C11	151.42	101.95	2.10	20.0	3.0	455.88	175.33	323.33
	C12	151.10	102.12	2.10		3.0	453.25		
	C13	151.55	102.13	2.10		3.0	460.50		
2	C21	200.42	102.12	2.10		3.0	609.75		
	C22	200.48	101.42	2.10		3.0	610.75		
	C23	200.37	102.07	2.10		3.0	611.25		
3	C31	250.30	102.33	2.10		3.0	760.50		
	C32	248.23	102.50	2.10		3.0	756.50		
	C33	250.65	102.35	2.10		3.0	757.75		
D 1	D11	150.72	102.47	2.10	30.0	3.0	455.50	178.33	324.67
	D12	150.33	102.93	2.10		3.0	456.50		
	D13	151.12	101.60	2.10		3.0	457.75		
2	D21	201.43	102.37	2.10		3.0	609.75		
	D22	201.55	101.60	2.10		3.0	610.25		
	D23	201.32	102.33	2.10		3.0	610.00		
2	D31	251.90	101.97	2.10		3.0	761.75		
	D32	252.12	102.25	2.10		3.0	759.25		
	D33	251.55	102.07	2.10		3.0	753.50		

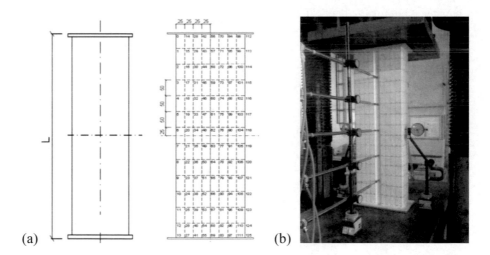

(a)   (b)

Figure 3. Generated raster (a) and general arrangement of the test (b).

3 TESTING METHODOLOGY AND OBTAINED RESULTS

Detailed information about ultimate loads, strains, and failure of the tested members, considering their actual geometrical and material parameters were obtained during the experimental tests. In accordance with the research target, the emphasis has been imposed on the post-critical behavior and interaction of the individual thin webs. The initial buckling shapes of the webs of all tested members were measured using previously generated raster, by means of inductive sensors, before tests initiating. Figure 3 illustrates the concept of the generated raster and general view on the test.

Figure 4. Completion of the test and overall failure, member C21.

During consecutive programmed loading of the tested members, the strains ε were measured using resistive strain-gauges located in their middle cross-sections. Deflections (buckling) of the webs w were measured using inductive sensors located in different places, according to members lengths, see Figure 3. The resistive strain-gauges and inductive sensors were connected to the computer for direct evaluation of obtained results, Tomko and Demjan (2014). Loading process of each member was regulated close to its real behavior, measured strains ε and deflections w. Each test continued until total failure, defined by the beginning of continuous increasing strains ε and deflections of the webs w. The final buckling shapes after test finishing were also revealed, Al Ali et al. (2013). Example of the test completion and overall failure of tested member C21 is given by Figure 4.

4 THEORETICAL AND EXPERIMENTAL LIMIT LOADS

According to research targets, the following characteristic resistances are important:

$N_{c,Rk,EN}$ – The resistance of a cross-section for compression, calculated with considering the European standard EN 1993-1-3.
$N_{c,Rk,STN}$ – The resistance of a cross-section for compression, calculated with considering the previous national standard STN 73 1402.
N_{exp} – The measured experimental force, acted on tested member before the beginning of continuous increasing strains ε and deflections of the webs w.

Taking into account the real measured dimensions and yield stresses, the resistances of all tested members were calculated according to the national standard (given by index STN) and to the European standard (given by index EN). Theoretical and experimental resistances with their comparison are presented in Table 2.

For improved illustration of the relationship between calculated and measured limit loads, graphical representation of their values is given by Figure 5.

From Table 2 and Figure 5 it is evident that the theoretical resistances, calculated according to the relevant standards are different. It is also evident that the resistances $N_{c,Rk,STN}$ are bigger than $N_{c,Rk,EN}$ and than N_{exp} in the all cases.

The comparison in Table 2 and Figure 5 also indicates the cases, where the experimental resistances are slightly smaller than the theoretical resistances $N_{c,Rk,EN}$, but in acceptable rate.

Table 2. Theoretical and experimental limit loads of tested members.

Tested member Cross-section		Marking	$N_{c,Rk,STN}$ [kN]	$N_{c,Rk,EN}$	N_{exp}	$N_{c,Rk,STN}/N_{exp}$	$N_{c,Rk,EN}/N_{exp}$
C	1	C11	186.01	170.76	170.08	1.094	1.004
		C12	185.90	170.70	177.52	1.047	0.962
		C13	186.24	170.87	172.21	1.081	0.992
	2	C21	207.87	183.75	189.21	1.099	0.971
		C22	207.69	183.54	197.72	1.050	0.928
		C23	207.82	183.73	189.21	1.098	0.971
	3	C31	210.01	186.92	191.34	1.098	0.977
		C32	210.03	186.81	195.59	1.074	0.955
		C33	210.02	186.96	193.47	1.086	0.966
D	1	D11	199.67	186.51	192.40	1.038	0.969
		D12	199.72	186.47	181.77	1.099	1.026
		D13	199.32	186.43	183.90	1.084	1.014
	2	D21	237.57	207.71	210.47	1.129	0.987
		D22	237.09	207.50	208.35	1.138	0.996
		D23	237.46	207.66	209.41	1.134	0.992
	3	D31	262.29	213.55	212.60	1.234	1.004
		D32	262.48	213.63	214.73	1.222	0.995
		D33	262.39	213.60	225.36	1.164	0.948

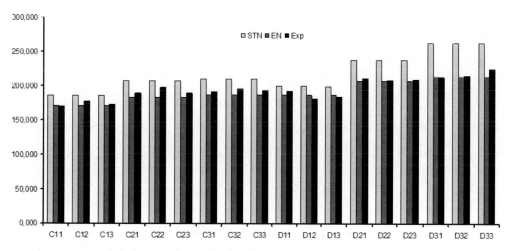

Figure 5. Illustration and comparison calculated and measured limit loads.

5 CONCLUSION

Compared to the National standard, the European standard came up with a completely different view and calculation procedure regarding the calculation of cross-sectional characteristics and the stiffeners effect on the resistance of cold-formed profiles.

The author's previous extensive research focused on the cold formed profiles with flat webs (without stiffeners), Al Ali et al. (2013), raised serious questions about the influence of initial imperfections and how they have been taken into account in the calculation of such cross-sections and members. Based on obtained results it was evident that the resistance of the compressed cold-formed steel members is significantly influenced by the initial imperfections and by the initial buckling shapes of their individual webs. This case also created the presumption that the obtained experimental resistances were influenced by the initial imperfection (cross-sectional groups A and B).

For this reason the second cross-sectional group of tested members had two webs with intermediate stiffeners to investigate their favorable effect, (groups C and D). The results were relatively positive and the resistances calculated according to the European standard are very close to the experimental values, see Table 2 and Figure 5. However, there are a cases where the experimental resistances are a bit smaller (members C11, D12, D13 and D31).

The obtained results expand on the knowledge base regarding the elastic-plastic resistances of cold-formed steel members, as well as the influence of their initial imperfections and the favorable effect of intermediate stiffeners.

ACKNOWLEDGEMENT

This paper is prepared within the research project VEGA 1/0582/13 "The elastic-plastic behaviour of compressed thin-walled cold-formed steel elements and stress-strain analysis of welded steel beams", supported by the Scientific Grant Agency of the Ministry of Education of Slovak Republic and Slovak Academy of Sciences.

REFERENCES

Al Ali, M., et al. 2013. Analysis of the resistance of thin-walled cold-formed compressed steel members with closed cross-sections, Part 1. *Magazine of Civil Engineering* 5(40): 38–43.

Al Ali, M., et al. 2013. Resistance of Compressed Thin-Walled Cold-Formed Steel Members with Regard to the Influence of Initial Imperfections. *Design, Fabrication and Economy of Metal Structures* International Conference Proceedings, Miskolc, Hungary.

Al Ali, M., et al. 2014. Analysis of the resistance of thin-walled cold-formed compressed steel members with closed cross-sections, Part 2. *Magazine of Civil Engineering* 1(45): 53–58.

Al Ali, M. 2014. Compressed Thin-Walled Cold-Formed Steel Members with Closed Cross-Sections. *Advanced Materials Research* Vol. 969: 27–29.

Chen, J., Young, B. 2007. Cold-formed steel lipped channel columns at elevated temperatures. *Engineering Structures* 29: 2445–2456.

EN 1993-1-3:2006: Design of steel structures – Part 1-3: Supplementary rules for cold-formed members and sheeting. CEN, Brussels 2006.

EN 1993-1-5:2006: Design of steel structures – Part 1-5: Plated structural elements. CEN, Brussels 2006.

Gardner, L., Saari, N., Wang, F. 2010. Comparative experimental study of hot-rolled and cold-formed rectangular hollow sections. *Thin-Walled Structures* 48: 495–507.

Garifullin, M., Vatin, N. 2014. Buckling analysis of thin-walled cold-formed beams — short review. *Construction of Unique Buildings and Structures* 6(21): 32–57.

Jurgen, B., Maura, L., Rasmussen, J.R. K. 2008. The direct strength method for stainless steel compression members. *Journal of Constructional Steel Research* 64: 1231–1238.

Nuno, S., Young, B., Camotima, D. 2008. Non-linear behaviour and load-carrying capacity of CFRP-strengthened lipped channel steel columns. *Engineering Structures* 30: 2613–2630.

Rasmussen, J.R. K. 2006. Bifurcation of locally buckled point symmetric columns – Experimental investigations. *Thin-Walled Structures* 44: 1175–1184.

Rendek, S., Baláž, I. 2004. Distortion of thin-walled beams. *Thin-Walled Structures* 2(42): 255–277.

STN 73 1402:1988, equals ÈSN 73 1402:1988, Design of thin-walled profiles in steel structures. UNM, Prague 1988.

Tomko, M., Demjan, I. 2014. Development of plastic zones in the pin connection of the steel roof structure. *Interdisciplinarity in theory and practice* Vol. 4: 61–67.

Vatin, N., Nazmeeva, T., Guslinscky, R. 2014. Problems of Cold-Bent Notched C-Shaped Profile Members. *Advanced Materials Research* 941–944: 1871–1875.

Yuan-Qi, L., et al. 2007. Analysis and design reliability of axially compressed members with high-strength cold-formed thin-walled steel. *Thin-Walled Structures* 45: 473–492.

Zhang, Y., Wang, Ch., Zhang, Z. 2007. Tests and finite element analysis of pin-ended channel columns with inclined simple edge stiffeners. *Journal of Constructional Steel Research* 63: 383–395.

Reparation of the fractured mandrel axle-shaft by welding

D. Arsić, V. Lazić & S. Aleksandrović
Faculty of Engineering, University of Kragujevac, Kragujevac, Serbia

R.R. Nikolić & B. Hadzima
Research Center, University of Žilina, Žilina, Slovakia

ABSTRACT: Problems of reparatory welding of the broken mandrel axle-shaft are considered in this paper. After visual inspecting of the damaged part and analysis of the crack position and static and dynamic loads, which caused the fracture, it was estimated that the broken part could be repaired, but by a very complex welding procedure combined with heat treatment. The prescribed reparation procedure consisted of proposing the welding procedures, selecting the filler metals and way of groove preparation, welding with the so-called EB insert, defining the welding parameters, prescribing the heat treatment before and during the welding, defining the welds deposition order, selecting the additional machining and defining the control type. The axle-shaft was successfully repaired and the spinner was capable of operating. The downtime was significantly decreased and the costs of procuring/manufacturing the new part avoided. The reparation procedure was done in own plant what provided for significant techno-economic benefits.

1 INTRODUCTION

In this paper, a special emphasis was placed on selection of technology for welding by which the broken axle-shaft could be repaired. These authors have already dealt with the similar problems, when the optimal technology for reparation of certain machine parts and devices had to be prescribed, Jovanović and Lazić (2008a), Lazić et al. (2009), Lazić et al. (2015), Lazić et al. (2012a). In addition, certain research has been done, related to steel's weldability, based on which the welding technology should be prescribed, Mutavdžić et al. (2008), Lazić et al. (2012b), Arsić et al. (2015).

After the detailed analysis of weldability of the steel that the axle-shat was made of and of the fracture surface and operating conditions of this, dynamically loaded part, the complete reparation technology was prescribed, Jovanović et al. (2008). It consisted of proposition of the welding procedures (TIG and MMAW), selection of the filler metals, ways of groove preparation at the place of fracture, welding execution with the so-called EB insert, calculation of parameters for the selected welding procedures (TIG – TIG for the root welds and MMAW for the filling and finishing passes), prescribing the heat treatment (prior to and concurrent with welding), weld passes deposition order, selecting the final machining type, current and finishing control, etc. The prescribed technology was executed to an extent that was possible in the production conditions. For the cover passes, the MMAW procedure was selected due to its higher productivity with respect to the TIG method, Chen et al. (2014), Ericsson and Sandström (2003).

2 CHEMICAL COMPOSITION ANALYSIS OF THE AXLE-SHAFT BASE METAL

Chemical analysis has shown that the axle material was the low-alloyed steel (single alloyed by manganese) what approximately corresponds to steel Č3100 – SRPS (EN – E 355N, DIN – ST52.4). Percentage content of individual elements is shown in Table 1, Jovanović and Lazić (2008a), ASM – Metals Handbook (1979), while the mechanical properties of this steel, for thickness s = 16–40 mm are the following: $R_m = 470$–$560\,MPa$, $R_{0.2} = 280\,MPa$, $A_5 = 22\%$. The low-alloyed steel of this class has very low carbon content and it is well weldable. However, due to axle-shaft purpose, the

Table 1. Prescribed and analyzed chemical composition of the mandrel axle-shaft steel (Č3100 – E 355N).

Chemical composition, %

	C	Si	Mn	Cr	S	P
Prescribed	0.14–0.20	0.20–0.40	0.90–1.20	–	0.050	0.050
Analyzed	0.15	–	0.85	0.090	0.020	–

preheating is recommended up to 200°C, Jovanović and Lazić (2008a), ASM – Metals Handbook (1979), Jovanović et al. (1996), Jovanović et al. (2007).

3 THE BASE METAL OF THE AXLE-SHAFT WELDABILITY ESTIMATE

Though the hollow axle-shaft is made of steel that belongs into a class of relatively well weldable materials, due to the shaft's purpose, especially its length and the wall thickness, certain precautionary measures must be taken in order to equalize the metallurgical-mechanical properties of all the zones of the welded joint. Those measures are related to preheating, additional heating and, if necessary, heating through (both the complete welded joint and the zone in its immediate vicinity). Weldability was estimated based on various expressions from references, Lazić et al. (2012a), Mutavdžić et al. (2008), Lazić et al. (2012b), ASM – Metals Handbook (1979), Jovanović et al. (1996). With great simplifications, neglecting a series of influential factors, it could be accepted that the steel is weldable if the final hardness within the heat-affected zone (HAZ) does not exceed 350 HV. It is considered that up to this value martensite would not be created during the welded joint cooling phase. This limiting hardness corresponds to chemically equivalent carbon content of $CE = 0.45\%$. It is thus adopted that steels with $CE < 0.45\%$ are weldable without application of special measures, while the steels with $CE > 0.45\%$ are conditionally weldable or not weldable at all. The obtained values of equivalent carbon for this steel (according to different expressions) were: $CE_{(1)} = 0.31\%$, $CE_{(2)} = 0.31\%$ and $CE_{(3)} = 0.16\%$. Though those obtained values were well below the adopted critical limit ($CE = 0.45\%$), it was necessary to apply prior and concurrent heat treatment, due to structural inhomogeneity of the zone around the circular weld and the operating conditions of the axle-shaft.

Majority of steels, during the welding, are prone to appearance of cold and hot cracks, as well as to numerous interior flaws. The measures to prevent all the flaws consist in application of such welding technologies that would eliminate the transition brittleness. Cold cracks would appear if the martensite or low-bainite structure would be created within the heat-affected zone and if simultaneously appeared large quantities of diffused hydrogen, as well as the internal tensile stresses. The last two factors are caused by the design-technological solutions, so they cannot be influenced. This is why the only possibility left is to prevent the quenching structures to appear during the transformation. That is generally achieved by increase of the driving energy of welding or by preheating. That implies reducing the cooling speed within the area of the least stability of austenite.

The most used method for calculation of the preheating temperature for this class of steel is the method by Seferian, Jovanović et al. (1996), Jovanović and Lazić (2007). By entering data into the Seferian's formula, the preheating temperature of about 70°C was obtained for the thickness of s = 27.5 mm. Taking into account that the preheating temperature must not be higher than the M_s temperature (circa 440°C, CCT diagram), Jovanović and Lazić (2008b), and the mentioned operating conditions of the axle-shaft, the preheating temperature was adopted to be within range $T_p = 150–200 \pm 10°C$. There, it had to be kept in mind that the own thermal stresses (transition and residual ones) must be kept at the minimum level.

4 THE FILLER METAL SELECTION

For filling the groove, after the root pass was deposited by the TIG procedure, it is recommended to apply the MMAW procedure and electrode EVB 50, Jovanović and Lazić (2008a). The

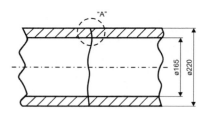

Figure 1. Schematic details for welding the axle-shaft.

Table 2. Recommended electrodes and welding parameters.

Technological parameters of the TIG procedure					Heat treatment	
Electrode mark	d_e [mm]	I [A]	Current type	Electrode drying	Prior T_p [°C]	Current $T_{heat-through}$ [°C]
EVB 50	3.25	110–140 A	= (+)	400°C/1 h	150–200	150–200
	4.00	140–180 A	= (+)	400°C/1 h	150–200	150–200

recommendations are that the few initial passes should be executed by the TIG procedure (Fig. 1), the filling passes close to the root pass by the MMAW procedure and with electrode of diameter $d_e = 3.25$ mm, while the rest of the filling and the cover passes to be executed with the electrode of diameter $d_e = 4.0$ mm. In Table 2 are presented the recommended welding parameters. Selection of the filler metal has the strong influence on final mechanical properties of the welded joint, what was shown by Mazur et al. (2014). The filer metals are chosen from certain catalogues of the filer metals manufacturers, (Catalogues and prospects, 2014).

5 THE WELDING TECHNOLOGY SELECTION

One starts from preparation for making the circular V-groove by mechanical machining. All the noticed cracks must be ground, i.e., the "liberation" of all the flaws is done and the groove is formed for welding. It is recommended that the design and preparation of the groove should be done according to Figure 1 and to make the so-called the EB-insert in order to deposit the root pass by the TIG procedure, Jovanović and Lazić (2008a). The method of the melting insert with the root pass was originally applied for the very responsible structures and was mainly developed for assemblies accessible from one side only, when the smooth interior surface without underfills is needed and when the root pass of unconditionally high quality is required (pipes joints).

It is assumed that the insert is completely melted by the TIG welding procedure. Deposition of the root pass, which is smooth and uniform, is enabled even when welding is done from one side only. The insert provides for complete welding-through of the root in grooves, which are accessible from the top side only. The melting insert must be precisely positioned and fixed and it serves as the underlying ring, what is often specified in welding of pipes. In many cases of execution of responsible and multi-pass welds with the melting insert, the TIG procedure is applied, but only for the joining (staple) welds and the root pass.

The typical procedure for application of the melting inserts is joining of two pipes of different cross-sections, in the horizontal position and it consists of the following activities:

– careful placement of circular melting EB inserts into the previously prepared V-groove;
– careful execution of the stapling welds by joining the EB inserts and the pipe walls by application of the TIG burner;
– after the partial joining of the melting semi rings and the pipes walls, it is necessary to carefully execute the root pass by application of the previously determined welding parameters (tungsten electrode diameter 2.4 to 3.2 mm, welding current 100 to 180 A, working voltage 12 to 17 V,

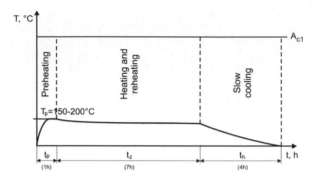

Figure 2.   Complete heat treatment cycle of the axle-shaft reparation.

Table 3.   Welding parameters for the TIG procedure and estimated characteristics of the HAZ.

d_e [mm]	I [A]	U [V]	v_z [mm/s]	q_l [J/mm]	$t_{8/5}$ [s]	HAZ microstructure	HAZ hardness
3.25	110	24.5	≈1.24	1738.7	11.2–14.15	B + F	$210 < HV < 285$
4.0	170	27	≈1.60	2295.0	13.46–17.00	B + F	$210 < HV < 245$

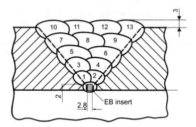

Figure 3.   Deposition of the passes by the MMAW procedure (1 to 4-d_e = 3.25 mm; 5 to 13-d_e = 4.0 mm).

polarity E-, welding speed ≈180 mm/min, protective gas flux ≈10 lit/min, nozzle size 9.5 to 11 mm);
– EB insert should be made of low carbon steel (e.g. Č0146 – EN-DC01) or austenitic stainless steel.

In manufacturing the semi rings one should predict the extensions for easier placing, which would be cut off or melted after placing the semi rings into the groove. Based on the obtained results, one can conclude that the relatively favorable microstructure in the HAZ is obtained (B + F) and relatively low hardness. This is why additional tempering was not necessary, since the unfavorable brittle structures did not appear. In addition, the convenient circumstance is that the following passes relax the previous ones, and the finishing, non-tempered layers are additionally mechanically processed. It is recommended that the fractured zone of the axle should be first preheated. Thus, this zone should be heated slowly until the preheating temperature is reached, then it should be held at that temperature for a certain time – heating through, until the whole cross-section wall does not reach that temperature 150 to 200°C, Figure 2.

After the preheating, one should deposit the root pass by the TIG procedure (Table 3) with the EB insert; then the groove filling should be executed by the MMAW procedure (Fig. 3).

During the whole reparation time, it is necessary to maintain the required preheating temperature, i.e., the part should be constantly reheated by special heaters or the gas burner. It is necessary to provide the most favorable position for the welder, what is achieved by utilities (holders, positioning tools, lunette, turning by the angular velocity adjusted to the prescribed welding speed for the given electrode diameter, etc.). Welding should be done according to the proposed order, Figure 3.

Figure 4.   Appearance of the repaired mandrel axle-shaft.

Order of operations in the reparation procedure should be as follows:

– Prepare the groove;
– Make the two-part EB insert according to the given dimensions from the recommended material – the low carbon steel (e.g. Č0146 (EN – DC01), Č0147 (EN – DC03), ...) or from the austenitic steel;
– Secure the two welding devices: the TIG apparatus for the root pass deposition and the MMAW apparatus with direct current for the filling passes deposition;
– Procure the basic electrodes (EVB 50) of diameters 3.25 mm and 4.0 mm, dry them out according to the prescribed regime to eliminate the moisture from the coating (at least one whole package of each of the given diameters, at least 5 kg each);
– Deposit the root pass by the TIG procedure with help of the EB insert, then the filling passes by the proposed electrodes, first of diameters 3.25 and then 4.0 mm (Fig. 3);
– Apply the adequate technology of block-sequences in welding-filling the groove and during depositing the finishing passes (Fig. 3);
– Immediately after the deposition of each layer, it should be mildly forged at temperature above 480°C in order to preserve the impact toughness;
– The slag should be completely removed by the steel brush after each pass and finally it should be blown away by the compressed air;
– The preheating temperature should be constantly controlled;
– It is necessary to heat additionally, occasionally or constantly, the welded area to maintain the prescribed temperature; (it is the best to do it by the propane burner)
– After the complete reparation procedure was executed, the regenerated part should be heated through for a shorter time and then slowly cooled down to room temperature; (the whole welded area should be covered by asbestos or the hot sand).

For this type of joints, the control is mandatory. The prior control consists of checking of the base and filler metals, evaluation of the welding equipment condition, insight into the welders' qualifications and checking of the groove dimensions and coaxiality of both parts of the broken pipe. The current control consists of: checking whether the prescribed electrodes are properly used, whether the voltage was adequately selected, whether the order of individual passes is being respected, whether the slag is regularly removed after each pass. The final control enhances the visual control, the control with the magnifying glass, the control by magnetic flux, penetrant liquids and ultrasound defectoscopy. Appearance of the repaired part is presented in Figure 4.

6   CONCLUSION

By executing the prescribed technology the mandrel axle-shaft was successfully regenerated and the machine – the spinner successfully repaired for work. The downtime of the machine was significantly reduced, the supply costs lowered as well costs of making the new axle-shaft, the reparation procedure costs were also lower since the whole job was done in own plant, i.e., the significant techno-economic benefits were realized.

The whole reparation procedure of the broken mandrel axle-shat enabled modeling of a complex procedure for some future reparation tasks for parts of the similar shape and dimensions. However, despite this fact, each reparation procedure must be approached in a special way. By application of the new knowledge, modern technologies and filler metals, it is realistic to expect multiple positive effects of the part regeneration, which are reflected primarily in saving of the expensive imported parts – the base metal, shortening the downtimes of machinery, reconstruction of some technical solutions, etc.

ACKNOWLEDGEMENT

This research was partially financially supported by European regional development fund and Slovak state budget by the project "Research Center of the University of Žilina" – ITMS 26220220183 and by the Ministry of Education, Science and Technological Development of Republic of Serbia through grants: ON174004, TR32036, TR35024 and TR33015.

REFERENCES

Arsić, D., Lazić, V., Nikolić, R., Hadzima, B., Aleksandrović, S., Djordjević, M. 2015. The optimal welding technology of high strength steel S690QL. *Materials Engineering – Materiálové inžinierstvo* 22(1): 33–47.
ASM-Metals HandBook Vol 6-Welding-Brazing-Soldering. 1979. Metals Park Ohio, USA: ASM
Catalogues and prospect of electrode manufacturers 2014. SŽ Fiprom-Jesenice, Slovenia; FEP-Plužine, Monte Negro; Elvaco-Bijeljina, Bosnia and Herzegovina; Esab-Goteborg, Sweden; Lincoln Electric- Euclid, Ohio, US.
Chen, J., Wu, C. S., Chen, M. A. 2014. Improvement of welding heat source models for TIG-MIG hybrid welding process. *Journal of Manufacturing Processes* 16(4): 485–493.
Ericsson, M., Sandström, R. 2003. Influence of welding speed on the fatigue of friction stir welds, and comparison with MIG and TIG. *International Journal of Fatigue* 25(12): 1379–1387.
Jovanović, M., Adamović, D., Lazić, V. 1996. *Welding technology – Handbook*. Kragujevac, Serbia: Faculty of Mechanical Engineering.
Jovanović, M., Lazić, V. 2007. *Practicum for MMAW and TIG welding*. Kragujevac, Serbia: Faculty of Mechanical Engineering.
Jovanović, M., Lazić, V. 2008. *Reparation of the damaged mandrel axle-shaft – technical report*. Holding "Kablovi" a.d., Jagodina, Serbia.
Jovanović, M., Lazić, V. 2008. *Practicum of gas and argon welding*. Kragujevac, Serbia: Faculty of Mechanical Engineering.
Lazić, V., Aleksandrović, A., Nikolić, R., Prokić-Cvetković, R., Popović, O., Milosavljević, D. 2012. Estimates of weldability and selection of the optimal procedure and technology for welding of high strength steels. *Procedia Engineering* 40: 310–315.
Lazić, V., Sedmak, A., Aleksandrović, S., Milosavljević, D., Čukić, R., Grabulov, V. 2009. Reparation of damaged mallet for hammer forging by hard facing and weld cladding. *Tehnički vjesnik – Tehnical Gazette* 16(4): 107–113.
Lazić, V., Sedmak, A., Milosavljević, D., Nikolić, I., Aleksandrović, S., Nikolić, R., Mutavdžić, M. 2012. Theoretical and experimental estimation of the working life of machine parts hard faced with austenite-manganese electrodes. *Materiali in tehnologije/Materials and technology* 46(5): 547–554.
Lazić, V., Sedmak, A., Nikolić, R., Mutavdžić, M., Aleksandrović, S., Krstić, B., Milosavljević, D. 2015. Selection of the most appropriate welding technology for hard facing of bucket teeth. *Materiali in tehnologije/Materials and technology* 49(1): 165–172.
Mazur, M., Ulewicz, R., Bokuvka, O. 2014. The impact of welding wire on the mechanical properties of welded joints. *Materials Engineering – Materiálové inžinierstvo* 21 (3): 122–128.
Mutavdžić, M., Čukić, R., Jovanović, M., Milosavljević, D., Lazić, V. 2008. Model investigations of the filler metals for regeneration of the damaged parts of the construction mechanization. *Tribology in Industry* 30(3): 3–9.

Advances and Trends in Engineering Sciences and Technologies – Al Ali & Platko (Eds)
© 2016 Taylor & Francis Group, London, ISBN: 978-1-138-02907-1

The application of geodetic methods for displacement control in the self-balanced pile capacity testing instrument

M. Baca, Z. Muszyński & J. Rybak
Wrocław University of Technology, Wrocław, Poland

T. Żyrek
Silesian University of Technology, Gliwice, Poland

ABSTRACT: Field tests of foundation pile's capacity by means of load tests constitute the fundamental method of verifying computations. Due to the necessity of controlling the displacement of
many points on the head of the pile under load, as well as for the purpose of measuring the deformations of the anchor piles and of the loading system itself, it is indispensable to double the traditional
measurements (using dial indicators) by means of geodetic measurements. In this way, the records
of displacement by means of dial indicators might be supplemented, at the same time eliminating
systematic errors brought about by the system of reference instability. This work presents the initial
results of the geodetic techniques application in the field of innovative self-balanced static tests of
pile axial capacity.

1 INTRODUCTION – PRINCIPLES OF THE PROPOSED PILE TESTING METHOD

Traditional methods of static load testing make it possible to measure the pile head displacement
under physically applied load provided to the pile by a hydraulic jack. Despite its verified reliability,
the static load test has some disadvantages. The most significant include a relatively long time of
investigation and the complications with building a reaction system. To minimize some of these
disadvantages, in 1984 Osterberg proposed a new type of pile investigation method, named "the
Osterberg test" after its originator. In this test, a specially designed device, called the Osterberg Cell
or an O-cell, is placed inside the pile during its construction. The load applied through the O-cell
makes it possible to measure simultaneously the displacement of two sections of a pile – above
and below the O-cell. Therefore, the Osterberg test enables to carry out separate investigations
for the load capacity and the side shear of a pile. Despite some doubts as to the comparability of
results between the Osterberg test and the traditional static load test, there are many works which
confirm the sufficient comparability between these two investigation methods (England 2003,
2009, Osterberg 1998, Schmertmann & Hayes 1997).

A different approach was primarily introduced in the case of slender piles. The work presented
by Hayden (2013) describes a testing appliance which enables to control the bearing capacity of
micropiles, ductile piles and soil nails. The idea of a removable hydraulic jack was than adopted
by the authors of this paper in the case of closed-end steel pipe piles. The force loading the pile
base is transmitted from hydraulic jack at an instrumented pile head by means of a removable
piston. The hydraulic jack can be assembled on top of the piston and disassembled when the test
is completed. The testing system consists of a hydraulic jack with electric pump, reaction beams,
instrumentation for pile displacement control and the reference system for geodetic measurements.
Preliminary results of relative displacement measurements in this new self-balanced pile capacity
testing method seem to be very promising. Contemporary authors' experience proves that this new
idea also allows for repeated static load tests which are necessary for the "pile setup" estimation.

2 TRADITIONAL AND GEODETIC SURVEY METHODS

The basic measurement tools used for measuring vertical displacement are most often dial indicators with the readout precision of 0.01 mm. They are fastened to a stiff base, typically made of steel profiles. Such base should be set outside the zone of influence of the loaded pile settlement and outside the scope of subsoil deformation due to the raising of the anchoring piles. It was signaled in professional literature that when the piles are loaded with considerable forces, the settlement values thus measured are sometimes largely distorted, due to the displacement of the whole base to which the dial indicators were fastened. That base is typically situated in the area of deformation around the pile under test (which spreads as far as several meters). The errors may, in that case, amount to 10%, which undermines the sense of performing the measurements with the maximum precision enabled by the dial indicators' capacity. Such distorted results may lead to the faulty assessment of piles' bearing capacity.

The method of load tests is employed in geotechnics mainly for the purpose of testing load capacity of foundation piles, as well as in the investigation of the subsoil density ratio and of the load capacity of a structure's foundation. The basic task of geodetic survey in this method is to record, at suitable moments in time, vertical and/or horizontal displacement of selected structural elements (Muszyński & Rybak 2010). Geodetic survey activity begins with the stabilization of points of reference (reference marks) in the field, located beyond the zone of impact of the structure under test, at the places ensuring the fixity of the mutual position of those points. The identification of fixed reference points is an important stage of the test result elaboration and has been described e.g. in (Muszyński 2010). The structure under investigation is represented by an appropriately selected number of check points. Displacement of the check point is understood as the change in the position of that point, in relation to the point of reference, occurring at a specific time interval. The subsequent geodetic interpretation of the displacement makes it possible to assess the significance of the displacement values from the point of view of measurement accuracy, and to determine the behaviour of the structure's elements in space. Generally that behaviour may be described by rotation angle values, gradient, translation components and also after checking if the structure under control displaces as a rigid body or it is subject to deformation. A short overview of two modern geodetic methods applied in the field of foundation capacity testing is presented below.

2.1 Structural monitoring

In recent years more and more often the so called structural monitoring is used for the investigation of displacements. That is a modern system combining different absolute measurements of geodetic survey with the relative measurement of displacement, as well as with various types of sensors recording the behavior of the structure under scrutiny (e.g. high-precision electronic inclinometers, piezometers). Structural monitoring makes it possible to record automatically the state of the structure, cyclically at predefined time intervals. Thus collected data are gathered, corrected in relation to the influence of the survey environment and calculated in the appropriate manner. The obtained values of displacement are compared with the permissible values, visualized and sent remotely to relevant supervisory bodies. The main component of the system are the motorized electronic total stations equipped with a function of the automatic target (prism) detection.

2.2 Terrestrial laser scanning

The ground-based laser scanning, used in the survey of displacement and deformation, has recently become an increasingly frequent and rapidly evolving technology which makes it possible to survey a great number of points in an instant (Gordon et al. 2001, Olsen 2015). During the scanning a vertically rotating prism sends the impulses of a laser beam which are reflected from the structure under scrutiny. The scanner rotates simultaneously at a predetermined leap in the horizontal plane. For each point under survey, the reflectorless measurement of the distance (up to 300 m) is taken, and

(a)   (b)

Figure 1.   Instrument (a) and reference system (b) applied to provide an appropriate accuracy of results.

the horizontal and vertical angles are recorded. The accuracy of the measurement at the distance of up to 50 m amounts to 4 mm. The number of the points under survey is defined by way of determining the horizontal and vertical dimensions of the mesh of rectangles which the points under survey constitute on the predefined surface. The scanner is equipped with a high-resolution digital camera with an automatic zoom, which takes the photos of the structures under survey and thus enables the archiving of their condition and their realistic visualization.

In the case of the pile load test, the terrestrial laser scanning is used rather experimentally. Despite its relatively small accuracy of determining the position of a single point, the laser scanning has a great advantage. In a short time we are able to obtain the measurement of the position of millions of control points which form a fully metric, three dimensional model of a pile, the testing structure and the ground in the vicinity of tested pile. The technology is remote and does not require contact with the structure under survey. It makes it possible to register the condition of the whole structure, and not only of some selected characteristic points. Additional HDS targets mounted on checked structure allow to control displacements of selected points with an accuracy of ±2 mm.

3   EXAMPLES OF GEODETIC METHODS APPLICATION ON TEST FIELD

The closed-end steel pipe pile testing program in Bojszowy Nowe (Southern Poland) started on Feb. 3rd 2015. The 400 mm diameter and over 8.0 m long steel pipe was driven to the depth 6.0 m by means of vibrohammer. An internal piston (steel pipe) was placed inside the pipe-pile and a removable reaction system was assembled over it. When only the standard 1700 kN hydraulic jack was inserted between the piston and the reaction system, the pile was ready for load testing. The first attempt to Static Load Testing was done just 2 weeks after the driving. A set of tests (including maintained load test and rapid cyclic loading) was performed under the supervision of the authors. In general, the relative displacement of the internal and external pipe was measured by means of dial gauges (Fig. 1b). The uplift of the external pipe was also controlled by dial gauges attached to an independent reference system situated in appropriate distance from pile shaft. At the same time, geodetic measurements were also applied to provide a parallel and independent displacement control (Fig. 1a). So, in this case, dial gauges formed basic measurement system but geodetic techniques were also applied for the reason of higher result reliability.

The next test started on the 4th week after driving. The two subsequent photographs (Figs 2a, 2b) present the instrumented pile head and geodetic measuring techniques used for the field test, respectively. The application of terrestrial laser scanning and motorized tacheometry enabled us to control the displacement of the total construction, including the deformation of the reaction system. The surveying began with the stabilizing the points that made up the measuring network.

Figure 2. Instrumented pile head (a) and geodetic instruments applied for the displacement control (b).

For the terrestrial laser scanner there were 3 reference points and 4 check points, indicated by HDS target. For the motorized total station there were 3 reference points and 6 check points indicated by prism. The localization and numbering of check points and dial gauges are presented on Figure 2a. The measurements were made independently with motorized electronic total station Trimble S3 and Leica ScanStation C10 laser scanner. Throughout the duration of the measurements, all reference points were mounted on tripods, whereas all check points were mounted to the structure under scrutiny using magnetic grips. Total station measurements were performed in two phases, using the automatic targeting on points. Distance measurements to all measured points were performed twice for each phase. Measurements with terrestrial laser scanner were performed from a distance 13 m. The nominal density of the scanning was identical in both directions and was 1 mm.

4 PRELIMINARY TEST RESULTS

So far, only a few results were gathered by the authors, however they seem to be quite promising. For the first stage of test, when the external pipe is driven only 6 m into the subsoil, the shaft capacity seem to be insufficient. On Feb. 18th (2 weeks after driving), load in the piston reached 170 kN in the first load cycle, the external pipe working in tension (pull off) started to creep in the ground. The further cyclic testing proved that once the shaft capacity is broken, the ability for the mobilization of shaft resistance need a certain time to be regained. After the next 2 weeks, similar shape of load-displacement graph could have been achieved.

The next load tests of the pile were performed on Feb. 28th 2015. The test was split into a number of load cycles. The results of the third cycle are presented below. The values of relative vertical displacements from the dial gauges are juxtaposed in the Table 1. The vertical displacements of check points calculated on the basis of motorized tacheometry are shown on the Figure 3. The comparison of relative displacement values from dial gauges and total station is presented on Figure 4.

It must be highlighted that the last load step is shorter (from 121 kN to 134 kN) due to the observed reduction of pressure in the hydraulic system caused by increasing settlements. As it was impossible to stabilize the constant load of 141 kN, the authors decided to treat the "stabilization load level" (at 134 kN) as the last load step.

Table 1.   The values of relative vertical displacements from dial gauges.

| Load [kN] | Displacements [mm] from dial gauge | | |
	no. 1	no. 2	no. 3
0	0.00	0.00	0.00
36	0.19	0.02	0.22
73	1.97	1.04	1.71
97	2.75	1.58	2.49
121	3.60	2.25	3.37
134	5.14	4.74	5.91
0	2.50	2.09	2.45

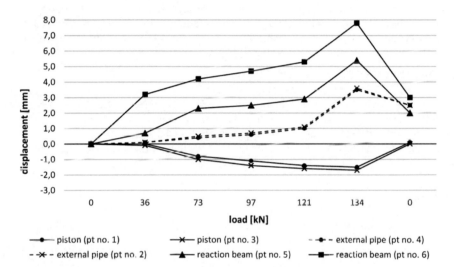

Figure 3.   Vertical displacements of the check points on the basis the total station measurements.

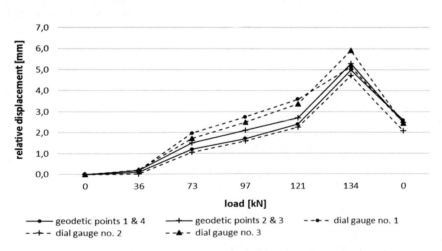

Figure 4.   Comparison of relative displacement values from dial gauges and total station.

19

5 BASIC CONCLUSIONS

However the idea of Osterberg-like "budget" test of closed-end steel pipe piles is still in a phase of a prototype, it seems that it has a market potential, at least compared with the other time and money consuming testing procedures. It can be used as a model test to determine soil capacity. The results of numerical simulations (Le 2004, Baca & Rybak, in press) have shown comparability between the traditional static load test and the Osterberg test.

Modern techniques of surveying may be used for fast and reliable measurement of displacement of many points in terms of pile testing. The role of those techniques becomes crucial when the thermal factors (due to insolation) and the movements of the ground around the pile under test at the terminal phase of the test affect the results of traditional dial gauges measurements.

To sum up, the modified Osterberg test can be an interesting option in the pile testing methods which can help in analysis of pile bearing capacity, especially for closed end steel pipe piles with high capacity. The application of terrestrial laser scanning allows fast control of many points, including the usually neglected deformations of the testing structure and reference system for dial gauges. That seems to be crucial when we want the results to be free from systematic errors related to climate conditions, ground movements and dynamic influence of works in the vicinity of the performed test.

ACKNOWLEDGMENTS

Co-author Tomasz Żyrek recieved scholarship in DoktoRIS project – Scholarship's program for innovating Silesia, co-financed with European Union in European Social Fund.

The authors would like to express their gratitude to PPI CHROBOK S.A. for technical support in the assembling of the Osterberg-like static load test appliance and kind providing of the test field in Bojszowy Nowe.

REFERENCES

Baca, M. & Rybak, J. (in press). Osterberg test as an alternative pile testing method. *55 years of geotechnical engineering in Slovakia, Bratislava 01.–02. June 2015.*

England, M. 2003. Bi-directional static load test – state of art. *Proc. of the 4th International Symposium on Deep Foundation on Bored and Auger Piles*, pp. 309–313.

England, M. 2009. Review of methods of analysis test results from bi-directional static load tests, *Proc. of the 5th Int. Symp. on Deep Foundation on Bored and Auger Piles*, pp. 235–239.

Gordon, S., Lichti, D. & Stewart, M. 2001. Application of a high-resolution, ground-based laser scanner for deformation measurements. *The 10th FIG International Symposium on Deformation Measurements 19–22 March 2001.* Orange, California, USA, pp. 23–32.

Hayden, M. 2013. Pile HAY-Proof-System: New test method for static load tests of slender piles. *Ground Engineering*, December 2013, pp. 25–28, http://www.nce.co.uk/Journals/2014/06/03/e/w/f/GE-Dec-2013-Pile-HAY-Proof-System-New-test-method-for-static-load-tests-of-slender-piles-Hayden.pdf

Le, Yi. 2004. *Finite Element Study on Static Load Testing.* A thesis submitted for degree of master of engineering, National University of Singapore.

Muszyński, Z. 2010. Application of selected robust estimation methods for calculating vertical displacements of hydrotechnical structures. *Studia Geotechnica et Mechanica* 32 (1): pp. 69–80.

Muszyński, Z. & Rybak, J. 2010. Application of geodetic survey methods in load capacity testing of piles *From research to design in European practice*: *proc. of the XIVth Danube-European Conference on Geotechnical Engineering, Bratislava, Slovak Republic, 2nd–4th June 2010*, pp. 1–9.

Olsen, M. 2015. In Situ Change Analysis and Monitoring through Terrestrial Laser Scanning. *Journal of Computing in Civil Engineering*, 29 (2).

Osterberg, J.O. 1998. The Osterberg Load Test Method for Bored and Driven Piles – The First Ten Years, *7th International Conference 7 Exhibition on Piling and Deep Foundations*. Deep Foundation Institute, Vienna, Austria.

Schmertmann, J. & Hayes, J. 1997. The Osterberg Cell and Bored Pile Testing – a Symbiosis. *Proceedings at the Third Annual Geotechnical Engineering Conference*. Cairo University, Cairo, Egypt.

Advances and Trends in Engineering Sciences and Technologies – Al Ali & Platko (Eds)
© 2016 Taylor & Francis Group, London, ISBN: 978-1-138-02907-1

Critical moment of timber cantilever under point end load F

I.J. Baláž & T.J. Živner
Department of Metal and Timber Structures, Faculty of Civil Engineering, STU in Bratislava, Slovakia

Y.P. Koleková
Department of Structural Mechanics, Faculty of Civil Engineering, STU in Bratislava, Slovakia

ABSTRACT: Lateral torsional stability of timber beams with monosymmetric cross-sections. Proposals based on large parametrical studies (Baláž 1999a, Baláž & Koleková 1999a, Koleková 1999a, b) given in (Baláž & Koleková 2000a, b, 2002a, b) for calculation of elastic critical moment M_{cr} of beams under different loadings and various boundary conditions were accepted by prEN 1999-1-1: May 2004 and EN 1999-1-1: May 2007 for design of aluminium structures. They are used also for design of steel structures. This procedure may be used for design of timber structures too. Comparison and evaluation of different procedures for critical moments M_{cr} calculation for timber beams. Examples show advantages of authors procedure which is already used for design of metal structures. Use of Bessel functions for critical moments M_{cr} calculation.

1 PROCEDURES FOR CRITICAL MOMENT CALCULATIONS OF TIMBER BEAMS UNDER CONCENTRATED END LOAD F

1.1 *Procedure of (Baláž & Koleková 2000a, b, 2002a, b) used in prEN 1999-1-1: May 2004 and EN 1999-1-1: May 2007*

General formula:

$$M_{cr} = \mu_{cr} \frac{\pi \sqrt{EI_z GI_t}}{L} \qquad (1)$$

The procedure offers two possibilities how to obtain the relative non-dimensional critical moment μ_{cr}: a) from Figure 1, b) from formula (13). The backgrounds and more details see in (Baláž, I. & Koleková, Y. 2015).

a) The relative non-dimensional critical moment μ_{cr} taken from Figure 1

where:
non-dimensional torsion parameter is

$$\kappa_{wt} = \frac{\pi}{k_w L} \sqrt{\frac{EI_w}{GI_t}}, \quad \kappa_{wt0} = \frac{\pi}{L} \sqrt{\frac{EI_w}{GI_t}} \qquad (2)$$

relative non-dimensional coordinate of the point of load application related to shear center

$$\zeta_g = \frac{\pi z_g}{k_z L} \sqrt{\frac{EI_z}{GI_t}}, \quad \zeta_{g0} = \frac{\pi z_g}{L} \sqrt{\frac{EI_z}{GI_t}} \qquad (3)$$

relative non-dimensional cross-section mono-symmetry parameter

$$\zeta_j = \frac{\pi z_j}{k_z L} \sqrt{\frac{EI_z}{GI_t}}, \quad \zeta_{j0} = \frac{\pi z_j}{L} \sqrt{\frac{EI_z}{GI_t}} \qquad (4)$$

where
I_t = torsion constant; I_w = warping constant; I_z = second moment of area about the minor axis; L = length of the beam between points that have lateral restraint; k_z = buckling length factor for bending perpendicular to axis z; k_w = buckling length factor for torsion.

$$z_g = z_a - z_s \tag{5}$$

where z_a = coordinate of the point of load application related to centroid; z_s = coordinate of the shear center related to centroid; z_g = coordinate of the point of load application related to shear center.

The sign convention for determining z_g is:

 (i) for gravity loads z_g is positive for loads applied above the shear centre,
 (ii) for general case z_g is positive for loads acting towards the shear centre from their point of application.

The cross-section mono-symmetry parameter

$$z_j = z_s - \frac{0.5}{I_y} \int_A (y^2 + z^2) z \, dA \tag{6}$$

$z_j = 0$ mm, ($y_j = 0$ mm) for cross sections with y-axis (z-axis) being axis of symmetry.
The following approximation for z_j may be used:

$$z_j = 0.45 \psi_f h_s \tag{7}$$

where h_s = distance the distance between the shear centre of the upper flange and shear centre of the bottom flange and

$$\psi_f = \frac{I_{fc} - I_{ft}}{I_{fc} + I_{ft}} \tag{8}$$

where I_{fc} = second moment of area of the compression flange about minor axis of the section; I_{ft} = second moment of area of the tension flange about the minor axis of the section; h_s = distance between the shear centre of the upper flange and shear centre of the bottom flange.
Warping constant for an I-section with unequal flanges

$$I_w = (1 - \psi_f^2) I_z \left(\frac{h_s}{2} \right)^2 \tag{9}$$

Warping constant for rectangular cross-section

$$I_w = I_{wt} + I_{wn} = 0 + \frac{1}{144} \left[1 - 4.884 \left(\frac{b}{h} \right)^2 + 4.97 \left(\frac{b}{h} \right)^3 - 1.067 \left(\frac{b}{h} \right)^5 \right] b^3 h^3 \tag{10}$$

Torsion constant for rectangular cross-section

$$I_t = \frac{1}{3} \left[1 - 0.63 \frac{b}{h} + 0.052 \left(\frac{b}{h} \right)^5 \right] h b^3 \tag{11}$$

The sign convention for determining z and z_j is as follows:

 (i) coordinate z is positive for the compression flange. When determining z_j from formula (6), positive coordinate z goes upwards for beams under gravity loads or for cantilevers under uplift loads, and goes downwards for beams under uplift loads or cantilevers under gravity loads
 (ii) sign of z_j is the same as the sign of cross-section mono-symmetry factor ψ_f in (8).

Loading and support conditions	$\dfrac{\pi}{L}\sqrt{\dfrac{EI_w}{GI_t}}$ $= k_w\kappa_{wt} =$ $= \kappa_{wt0}$	$\dfrac{\pi z_g}{L}\sqrt{\dfrac{EI_z}{GI_t}}$ $= k_z\zeta_g =$ $= \zeta_{g0}$	$\overset{\text{(T)}}{\underset{\text{(C)}}{\updownarrow}}\ \overset{\text{(C)}}{\underset{\text{(T)}}{\updownarrow}}$ $z_j < 0$			$\dfrac{\pi z_j}{L}\sqrt{\dfrac{EI_z}{GI_t}} = k_z\zeta_j = \zeta_{j0}$		$\overset{\text{(C)}}{\underset{\text{(T)}}{\updownarrow}}\ \overset{\text{(T)}}{\underset{\text{(C)}}{\updownarrow}}$ $z_j > 0$	
			-4	-2	-1	$z_j = 0$	1	2	4
		4	0,107	0,156	0,194	**0,245**	0,316	0,416	0,759
		2	0,123	0,211	0,302	**0,463**	0,759	1,312	4,024
	0	0	0,128	0,254	0,478	**1,280**	3,178	5,590	10,730
		-2	0,129	0,258	0,508	**1,619**	3,894	6,500	11,860
		-4	0,129	0,258	0,511	**1,686**	4,055	6,740	12,240
		4	0,151	0,202	0,240	0,293	0,367	0,475	0,899
		2	0,195	0,297	0,393	0,560	0,876	1,528	5,360
	0,5	0	0,261	0,495	0,844	1,815	3,766	6,170	11,295
		-2	0,329	0,674	1,174	2,423	4,642	7,235	12,595
		-4	0,364	0,723	1,235	2,529	4,843	7,540	13,100
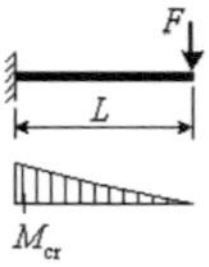		4	0,198	0,257	0,301	0,360	0,445	0,573	1,123
		2	0,268	0,391	0,502	0,691	1,052	1,838	6,345
	1	0	0,401	0,750	1,243	2,431	4,456	6,840	11,920
		-2	0,629	1,326	2,115	3,529	5,635	8,115	13,365
		-4	0,777	1,474	2,264	3,719	5,915	8,505	13,960
		4	0,335	0,428	0,496	0,588	0,719	0,916	1,795
		2	0,461	0,657	0,829	1,111	1,630	2,698	7,815
	2	0	0,725	1,321	2,079	3,611	5,845	8,270	13,285
		-2	1,398	3,003	4,258	5,865	7,845	10,100	15,040
		-4	2,119	3,584	4,760	6,360	8,385	10,715	15,825
		4	0,845	1,069	1,230	1,443	1,739	2,168	3,866
		2	1,159	1,614	1,992	2,569	3,498	5,035	10,345
	4	0	1,801	3,019	4,231	6,100	8,495	11,060	16,165
		-2	3,375	6,225	8,035	9,950	11,975	14,110	18,680
		-4	5,530	8,130	9,660	11,375	13,285	15,365	19,925

Figure 1. Relative non-dimensional critical moment μ_{cr} for cantilever ($k_y = 2$, $k_z = 2$, $k_w = 2$) loaded by concentrated end load F.

For beams with uniform cross-section symmetrical about major axis, centrally symmetric and doubly symmetric cross-section $z_j = 0$ mm.

For rectangular cross-section it may be taken $\kappa_{wt0} = 0$, because $I_w \approx 0\,\text{mm}^6$. For rectangular cross-section it is enough to use numerical values denoted in Figure 1 by bold.

b) The relative non-dimensional critical moment μ_{cr} calculated from formula.

General formula is

$$\mu_{cr} = \frac{C_1}{k_z}\left[\sqrt{1 + \kappa_{wt}^2 + (C_2\zeta_g - C_3\zeta_j)^2} - (C_2\zeta_g - C_3\zeta_j)\right] \tag{12}$$

where C_1, C_2, C_3 are factors depending on the type of loading, end restraint conditions and shape of the cross-section. They are given in tables in (Baláž & Koleková 2000a, b, 2002a, b, 2015) or in EN 1999-1-1: May 2007, Annex I.

For beams with uniform cross-section symmetrical about major axis, centrally symmetric and doubly symmetric cross-section $z_j = 0$ mm. For these beams loaded perpendicular to the mayor axis in the plane going through the shear centre, $z_j = 0$ mm, thus

$$\mu_{cr} = \frac{C_1}{k_z}\left[\sqrt{1 + \kappa_{wt}^2 + (C_2\zeta_g)^2} - C_2\zeta_g\right] \tag{13}$$

For $z_j = 0$ mm; $-4 \leq \zeta_{g0} \leq 4$ $(-2 \leq \zeta_g \leq 2)$ and $\kappa_{wt0} \leq 4$ $(\kappa_{wt} \leq 2)$ the value of μ_{cr} may be calculated from formula (13), where the following approximate values of the factors C_1, C_2 may be used for the cantilever loaded by tip load F:

$$C_1 = 2.56 + 4.675\,\kappa_{wt} - 2.62\,\kappa_{wt}^2 + 0.5\kappa_{wt}^3 \quad \text{if} \quad \kappa_{wt} \leq 2 \tag{14}$$

$$C_1 = 5.55 \qquad\qquad\qquad\qquad\qquad\qquad \text{if} \quad \kappa_{wt} > 2 \tag{15}$$

$$C_2 = 1.255 + 1.566\,\kappa_{wt} - 0.931\,\kappa_{wt}^2 + 0.245\,\kappa_{wt}^3 - 0.024\kappa_{wt}^4 \qquad\qquad \text{if} \quad \zeta_g \geq 0 \tag{16}$$

$$C_2 = 0.192 + 0.585\,\kappa_{wt} - 0.054\,\kappa_{wt}^2 - (0.032 + 0.102\,\kappa_{wt} - 0.013\kappa_{wt}^2)\,\zeta_g \quad \text{if} \quad \zeta_g < 0 \tag{17}$$

For beams with rectangular cross-section

$$C_1 = 2.56\,; \quad C_2 = 1.255 \ \text{if} \ \zeta_g \geq 0 \ \text{or} \ C_2 = 0.192 - 0.032\,\zeta_g \ \text{if} \ \zeta_g < 0 \tag{18}$$

1.2 Procedure used in German standard DIN 1052: 2004 Annex E.3 and in German National Annex to DIN EN 1995-1-1: 2005

The formula valid for doubly symmetric profiles with $I_w = 0$ mm^6 may be written in the form

$$M_{cr} = \frac{\pi\sqrt{EI_z GI_t}}{L_{ef}} = a_1\left(1 - a_2\frac{a_z}{L}\sqrt{\frac{EI_z}{GI_t}}\right)\frac{\pi\sqrt{EI_z GI_t}}{L} \tag{19}$$

where the effective length may be compared with μ_{cr} used in previous procedure as follows $L_{ef} = L/\mu_{cr}$. The numerical values for cantilever loaded by end load F are according to DIN 1052: $a_1 = 1.27$; $a_2 = 1.03$.

The formula taken from the DIN 1052: 2004 is simplifications of the more exact (Baláž, Koleková, 2000b) formula and the following relations are valid $a_1 = C_1/k_z$; $a_2 = \pi C_2/k_z$; $a_z = z_g$. The less exact DIN 1052: 2004 formula gives smaller M_{cr} values comparing with the ones calculted from more exact (Baláž, Koleková, 2000b) and EN 1999-1-1: 2007 formula. The error on the safe side is less than 7% for the values

$$a_2\frac{a_z}{L}\sqrt{\frac{EI_z}{GI_t}} \leq 0.3 \tag{20}$$

1.3 Procedure given in Eurocode EN 1995-1-1: 2004

According to EN 1995-1-1: 2004 effective length $L_{ef} = l_{ef}$, where for cantilever with concentrated load F at the free end $l_{ef}/L = 0.8$. If the load F is applied at the compression edge of the beam, l_{ef} should be increased by $2h$ and may be decreased by $0.5h$ for a load at the tension edge of the beam, where h is height of the rectangular cross-section. This wording is not correct in the case of cantilever (see (31)). The wording should be replaced by the following one: for loads applied above (below) the shear centre l_{ef} should be increased by $2h$ (decreased by $0.5h$).

1.4 Solution of system of differential equations using Bessel functions

System of differential equations for beam in bending with rectangular cross-section was solved using Bessel functions in (Baláž, Koleková, 2015) and in (Vol'mir, 1965). Solution leads for $M_y(x) = -Fx$ to the equation

$$\zeta_g\sqrt{\frac{\mu_{cr}}{\pi}}\Gamma(0.25)\sum_{k=0}^{n}\frac{(-1)^k}{k!\Gamma(k+1.25)}\left(\frac{\pi\mu_{cr}}{4}\right)^{2k+0.25} - \Gamma(0.75)\sum_{k=0}^{n}\frac{(-1)^k}{k!\Gamma(k+0.75)}\left(\frac{\pi\mu_{cr}}{4}\right)^{2k-0.25} = 0 \tag{21}$$

where $\Gamma(z)$ is the gamma function, a shifted generalization of the factorial function to non-integer values. Instead of $n = \infty$ the value $n = 97$ was used in calculations.

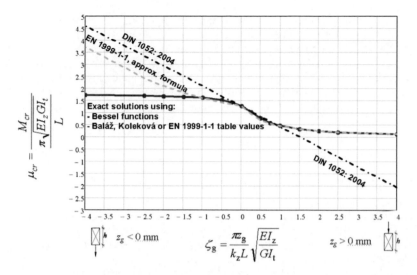

$$\mu_{cr} = \frac{M_{cr}}{\pi \frac{\sqrt{EI_z GI_t}}{L}}$$

Exact solutions using:
- Bessel functions
- Baláž, Koleková or EN 1999-1-1 table values

$z_g < 0$ mm $\zeta_g = \frac{\pi z_g}{k_z L} \sqrt{\frac{EI_z}{GI_t}}$ $z_g > 0$ mm

Figure 2.   Comparison of procedures for calculation of relative non-dimensional critical moment μ_{cr} for cantilever loaded by concentrated load F applied in point of end cross-section defined by relative non-dimensional coordinate of the point of load application related to shear center $-4 \le \zeta_g \le 4$, $(-8 \le \zeta_{g0} \le 8)$.

2   COMPARISON OF DIFFERENT PROCEDURES

The comparable formulae are:

(i) Procedure of (Baláž & Koleková 2000a, b, 2002a, b) used in prEN 1999-1-1: May 2004 and EN 1999-1-1: May 2007. In the following formula (22) the buckling factors for cantilever are $k_z = 2$, $k_w = 2$.

$$M_{cr} = \frac{C_1}{k_z} \left[\sqrt{1 + \kappa_{wt}^2 + (C_2 \zeta_g)^2} - C_2 \zeta_g \right] \frac{\pi \sqrt{EI_z GI_t}}{L} \qquad (22)$$

(ii) Procedure used in German standard DIN 1052: 2004 Annex E.3 and in German National Annex to DIN EN 1995-1-1: 2005

$$M_{cr} = a_1 \left[1 - a_2 \frac{a_z}{L} \sqrt{\frac{EI_z}{GI_t}} \right] \frac{\pi \sqrt{EI_z GI_t}}{L} = \frac{a_1}{k_z} \left[1 - a_2 \frac{k_z}{\pi} \zeta_g \right] \frac{\pi \sqrt{EI_z GI_t}}{L} \qquad (23)$$

(iii) Procedure given in Eurocode EN 1995-1-1: 2004

$$M_{cr} = \frac{1}{0.8 + \frac{2h}{L}} \frac{\pi \sqrt{EI_z GI_t}}{L} \quad \text{if} \quad z_g = -\frac{h}{2}; \quad M_{cr} = \frac{1}{0.8 - \frac{h}{2L}} \frac{\pi \sqrt{EI_z GI_t}}{L} \quad \text{if} \quad z_g = \frac{h}{2} \qquad (24)$$

Formulae (24) give dangerous results, because M_{cr} for $z_g > 0$ is greater than M_{cr} for $z_g > 0$, what is nonsense. See Figure 2 for other procedures. Eurocode formulae are not based on dimensionless parameter ζ_g and therefore their results cannot be compared with others in Figure 2.

3   CONCLUSION

Procedure developed in (Baláž & Koleková 2000a, b, 2002a, b), which was accepted by prEN 1999-1-1: May 2004 and it is used in EN 1999-1-1: May 2007 gives: (i) exact values of relative non-dimensional critical moment μ_{cr} in the Figure 1, (ii) μ_{cr} values which may be calculated using approximate formula. In the Figure 1 it is necessary to use non-linear interpolation (see Figure 2). DIN 1052: 2004 offers formulae which lead to less exact μ_{cr} values. EN 1995-1-1 procedure gives the least exact μ_{cr} values and it is not possible to transform its formula (24) in the comparable form with (22), (23). Figure 2 clearly shows interval in which approximate formulae may be used.

Moreover procedure (Baláž & Koleková 2000a, b, 2002a, b) and EN 1999-1-1: May 2007 enables to calculate critical moment M_{cr} also for cross-section with general shape. This procedure may be used also for timber beams.

ACKNOWLEDGEMENT

Project No. 1/0748/13 was supported by the Slovak Grant Agency VEGA.

REFERENCES

Baláž, I. 2001. *Klopenie drevených nosníkov. Zborník prednášok z konferencie „Výstavba a obnova budov",* *výstava PRO DOMO. Košice 12. marec 2001.* Dom techniky Košice: 27–32.

Baláž, I. 2005. Lateral torsional buckling of timber beams. *Wood Research* 50(1): pp. 51–58.

Baláž, I. & Koleková, Y. 2000a. *Critical Moments of Beams and Girders. Clark-Mrázik formula. Proceedings* *of 19th Czech and Slovak International Conference "Steel Structures and Bridges 2000". Štrbské Pleso,* *High Tatras. September 27–30, 2000*: 87–94.

Baláž, I. & Koleková, Y. 2000b. *Proposals for Improvements of Lateral Torsional Buckling Eurocodes Rules.* *Proceedings of 19th Czech and Slovak International Conference "Steel Structures and Bridges 2000".* *Štrbské Pleso, High Tatras. September 27–30, 2000*: 81–86.

Baláž, I. & Koleková, Y. 2002a. *Critical Moments. Proc. of Internal Colloquium on Stability and Ductility of* *Steel Structures, Budapest*: 31–38.

Baláž, I. & Koleková, Y. 2002b. *Clark-Mrázik Formula for Critical Moments. Proc. of Internal Colloquium* *on Stability and Ductility of Steel Structures, Budapest*: 39–46.

Baláž, I. & Koleková, Y. 2004a. *Factors C1, C2, C3 for computing elastic critical moments Mcr. Zborník VI.* *sympózia Drevo v stavebných konštrukciách so zahraničnou účast'ou. Kočovce 28.–29.10.2004*: 29–34.

Baláž, I. & Koleková, Y. 2004b. *Resistance of timber beams to out-of-plane buckling. Zborník VI. sympózia* *Drevo v stavebných konštrukciách so zahraničnou účast'ou. Kočovce 28.–29.10.2004*: 35–42.

Baláž, I. & Koleková, Y. 2014a. *LTB resistance of beams. Critical moment. Proceedings of full papers on USB.* *Eurosteel 2014: 7th European Conference on Steel and Composite Structures., Napoli, Italy, 10.–12.9.2014.* ECCS: 1-6. ISBN 978-92-9147-121-8.

Baláž, I. & Koleková, Y. 2015. *Lateral Torsional Stability of Timber Beams. Proceedings of 6th International* *Conference on Mechanics and Materials in Design. Recent Advances in Mechanics and Materials in Design.* *Ponta Delgada/Azores, 26-30 July 2015.*

DIN 1052: 2004. Entwurf, Berechnung und Bemessung von Holzbauwerken. Allgemeine Bemessungsregeln und Bemessungsregeln für den Hochbau. Ausgabe: August 2004

DIN EN 1995-1-1: December 2005. Eurocode 5: Bemessung und Konstruktion von Holzbauten; Allgemeines; Allgemeine Regeln und Regeln für den Hochbau.

EN 1995-1-1: November 2004 + A1: June 2008. Design of Timber Structures. Part 1-1 General Rules and Rules for Buildings. CEN Brussels.

EN 1999-1-1: May 2007 + A1: July 2009 + A2: December 2013 Design of Aluminium Structures. Part 1-1 General Rules and Rules for Buildings. CEN Brussels.

Koleková Y. 1999a. *Elastic Critical Moments of the Thin-Walled Beams with Monosymmetric Cross-Section.* (In Slovak). Thesis for Associate Professor (docent) habilitation. Faculty of Civil Engineering. Slovak University of Technology. Bratislava: 1–121.

Koleková Y. 1999b. Kritický moment konzoly s jednoosovo symetrickým I-prierezom. *Zborník prednášok z* *XXV. Celoštátneho aktívu „Stratégia rozvoja ocel'ových konštrukcií." Lipovce 6.–8. október 1999*: 105–110.

Koleková, Y. 1999c. *Klopenie monosymetrickej konzoly (Lateral torsional buckling of monosymmetric* *cantilever). XI. Mezinárodní vědecká konference. VUT Brno, 18.–20. října 1999:* 63–66.

prEN 1999-1-1: May 2004 Design of Aluminium Structures. Part 1-1 General Rules and Rules for Buildings. CEN Brussels.

Vol'mir, A.S. 1967. *Stability of deformable systems.* (In Russian). Izdatel'stvo Nauka (Publisher). Moscow.

Advances and Trends in Engineering Sciences and Technologies – Al Ali & Platko (Eds)
© 2016 Taylor & Francis Group, London, ISBN: 978-1-138-02907-1

Critical moment of timber cantilever under uniform load q

I.J. Baláž & T.J. Živner
Department of Metal and Timber Structures, Faculty of Civil Engineering, STU in Bratislava, Slovakia

Y.P. Koleková
Department of Structural Mechanics, Faculty of Civil Engineering, STU in Bratislava, Slovakia

ABSTRACT: Lateral torsional stability of timber beams with monosymmetric cross-sections. Proposals based on large parametrical studies (Baláž 1999a, Baláž & Koleková 1999a, Koleková 1999a, b) given in (Baláž & Koleková 2000a, b, 2002a, b) for calculation of elastic critical moment M_{cr} of beams under different loadings and various boundary conditions were accepted by prEN 1999-1-1: May 2004 and EN 1999-1-1: May 2007 for design of aluminium structures. They are used also for design of steel structures. This procedure may be used for design of timber structures too. Comparison and evaluation of different procedures for critical moments M_{cr} calculation for timber beams. Examples show advantages of authors procedure which is already used for design of metal structures. Use of Bessel functions for critical moments M_{cr} calculation.

1 PROCEDURES FOR CRITICAL MOMENT CALCULATIONS OF TIMBER BEAMS UNDER UNIFORM LOAD q

1.1 Procedure of (Baláž & Koleková 2000a, b, 2002a, b) used in prEN 1999-1-1: May 2004 and EN 1999-1-1: May 2007

Generally formula:

$$M_{cr} = \mu_{cr} \frac{\pi \sqrt{EI_z GI_t}}{L} \qquad (1)$$

The procedure offers two possibilities how to obtain the relative non-dimensional critical moment μ_{cr}: a) from Figure 1, b) from formula (13). The backgrounds and more details see in (Baláž, I. & Koleková, Y. 2015).

a) The relative non-dimensional critical moment μ_{cr} taken from Figure 1

where:

non-dimensional torsion parameter is

$$\kappa_{wt} = \frac{\pi}{k_w L} \sqrt{\frac{EI_w}{GI_t}}, \quad \kappa_{wt0} = \frac{\pi}{L} \sqrt{\frac{EI_w}{GI_t}} \qquad (2)$$

relative non-dimensional coordinate of the point of load application related to shear center

$$\zeta_g = \frac{\pi z_g}{k_z L} \sqrt{\frac{EI_z}{GI_t}}, \quad \zeta_{g0} = \frac{\pi z_g}{L} \sqrt{\frac{EI_z}{GI_t}} \qquad (3)$$

relative non-dimensional cross-section mono-symmetry parameter

$$\zeta_j = \frac{\pi z_j}{k_z L} \sqrt{\frac{EI_z}{GI_t}}, \quad \zeta_{j0} = \frac{\pi z_j}{L} \sqrt{\frac{EI_z}{GI_t}} \qquad (4)$$

where

I_t = torsion constant; I_w = warping constant; I_z = second moment of area about the minor axis; L = length of the beam between points that have lateral restraint; k_z = buckling length factor for bending perpendicular to axis z; k_w = buckling length factor for torsion.

$$z_g = z_a - z_s \tag{5}$$

where z_a = coordinate of the point of load application related to centroid; z_s = coordinate of the shear center related to centroid; z_g = coordinate of the point of load application related to shear center.

The sign convention for determining z_g is:

(i) for gravity loads z_g is positive for loads applied above the shear centre,

(ii) for general case z_g is positive for loads acting towards the shear centre from their point of application.

The cross-section mono-symmetry parameter

$$z_j = z_s - \frac{0.5}{I_y} \int_A (y^2 + z^2) z \, dA \tag{6}$$

$z_j = 0$ mm, $(y_j = 0$ mm) for cross sections with y−axis (z-axis) being axis of symmetry.
The following approximation for z_j may be used:

$$z_j = 0.45 \psi_f h_s \tag{7}$$

where h_s = distance the distance between the shear centre of the upper flange and shear centre of the bottom flange and

$$\psi_f = \frac{I_{fc} - I_{ft}}{I_{fc} + I_{ft}} \tag{8}$$

where I_{fc} = second moment of area of the compression flange about minor axis of the section; I_{ft} = second moment of area of the tension flange about the minor axis of the section; h_s = distance between the shear centre of the upper flange and shear centre of the bottom flange.
Warping constant for an I-section with unequal flanges

$$I_w = (1 - \psi_f^2) I_z \left(\frac{h_s}{2} \right)^2 \tag{9}$$

Warping constant for rectangular cross-section

$$I_w = I_{wt} + I_{wn} = 0 + \frac{1}{144} \left[1 - 4.884 \left(\frac{b}{h} \right)^2 + 4.97 \left(\frac{b}{h} \right)^3 - 1.067 \left(\frac{b}{h} \right)^5 \right] b^3 h^3 \tag{10}$$

Torsion constant for rectangular cross-section

$$I_t = \frac{1}{3} \left[1 - 0.63 \frac{b}{h} + 0.052 \left(\frac{b}{h} \right)^5 \right] h b^3 \tag{11}$$

The sign convention for determining z and z_j is as follows:

(i) coordinate z is positive for the compression flange. When determining z_j from formula (6), positive coordinate z goes upwards for beams under gravity loads or for cantilevers under uplift loads, and goes downwards for beams under uplift loads or cantilevers under gravity loads

(ii) sign of z_j is the same as the sign of cross-section mono-symmetry factor ψ_f in (8).

For beams with uniform cross-section symmetrical about major axis, centrally symmetric and doubly symmetric cross-section $z_j = 0$ mm.

For rectangular cross-section it may be taken $\kappa_{wt0} = 0$, because $I_w \approx 0$ mm^6. For rectangular cross-section it is enough to use numerical values denoted in Figure 1 by bold.

b) The relative non-dimensional critical moment μ_{cr} calculated from formula.

General formula is

$$\mu_{cr} = \frac{C_1}{k_z} \left[\sqrt{1 + \kappa_{wt}^2 + (C_2 \zeta_g - C_3 \zeta_j)^2} - (C_2 \zeta_g - C_3 \zeta_j) \right] \tag{12}$$

Loading and support conditions	$\dfrac{\pi}{L}\sqrt{\dfrac{EI_w}{GI_t}}$ $= k_w \kappa_{wt} =$ $= \kappa_{wt0}$	$\dfrac{\pi z_g}{L}\sqrt{\dfrac{EI_z}{GI_t}}$ $= k_z \zeta_g =$ $= \zeta_{g0}$	$\begin{array}{c}\downarrow(T)\ \downarrow(C)\\\uparrow(C)\ \uparrow(T)\end{array}$ $z_j < 0$			$\dfrac{\pi z_j}{L}\sqrt{\dfrac{EI_z}{GI_t}} = k_z\zeta_j = \zeta_{j0}$ $z_j = 0$	$\begin{array}{c}\uparrow(C)\ \downarrow(T)\\\uparrow(T)\ \downarrow(C)\end{array}$ $z_j > 0$		
			-4	-2	-1	$z_j = 0$	1	2	4
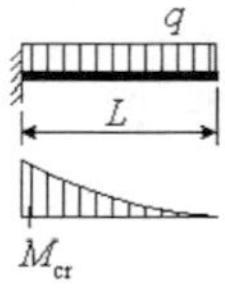	0	4	0,113	0,173	0,225	**0,304**	0,431	0,643	1,718
		2	0,126	0,225	0,340	**0,583**	1,165	2,718	13,270
		0	0,132	0,263	0,516	**2,054**	6,945	12,925	25,320
		-2	0,134	0,268	0,537	**3,463**	10,490	17,260	30,365
		-4	0,134	0,270	0,541	**4,273**	12,715	20,135	34,005
	0,5	4	0,213	0,290	0,352	0,443	0,586	0,823	2,046
		2	0,273	0,421	0,570	0,854	1,505	3,229	14,365
		0	0,371	0,718	1,287	3,332	8,210	14,125	26,440
		-2	0,518	1,217	2,418	6,010	12,165	18,685	31,610
		-4	0,654	1,494	2,950	7,460	14,570	21,675	35,320
	1	4	0,336	0,441	0,522	0,636	0,806	1,080	2,483
		2	0,449	0,663	0,865	1,224	1,977	3,873	15,575
		0	0,664	1,263	2,172	4,762	9,715	15,530	27,735
		-2	1,109	2,731	4,810	8,695	14,250	20,425	33,075
		-4	1,623	3,558	6,025	10,635	16,880	23,555	36,875
	2	4	0,646	0,829	0,965	1,152	1,421	1,839	3,865
		2	0,885	1,268	1,611	2,185	3,282	5,700	18,040
		0	1,383	2,550	4,103	7,505	12,770	18,570	30,570
		-2	2,724	6,460	9,620	13,735	18,755	24,365	36,365
		-4	4,678	8,635	11,960	16,445	21,880	27,850	40,400
	4	4	1,710	2,168	2,500	2,944	3,565	4,478	8,260
		2	2,344	3,279	4,066	5,285	7,295	10,745	23,150
		0	3,651	6,210	8,845	13,070	18,630	24,625	36,645
		-2	7,010	13,555	17,850	22,460	27,375	32,575	43,690
		-4	12,270	18,705	22,590	26,980	31,840	37,090	48,390

Figure 1. Relative non-dimensional critical moment μ_{cr} for cantilever ($k_y = 2$, $k_z = 2$, $k_w = 2$) loaded by uniform load q.

where C_1, C_2, C_3 are factors depending on the type of loading, end restraint conditions and shape of the cross-section. They are given in tables in (Baláž & Koleková 2000a, b, 2002a, b, 2015) or in EN 1999-1-1: May 2007, Annex I.

For beams with uniform cross-section symmetrical about major axis, centrally symmetric and doubly symmetric cross-section $z_j = 0$ mm. For these beams loaded perpendicular to the mayor axis in the plane going through the shear centre, $z_j = 0$ mm, thus

$$\mu_{cr} = \frac{C_1}{k_z}\left[\sqrt{1 + \kappa_{wt}^2 + (C_2\zeta_g)^2} - C_2\zeta_g\right] \tag{13}$$

For $z_j = 0$ mm; $-4 \leq \zeta_{g0} \leq 4$ ($-2 \leq \zeta_g \leq 2$) and $\kappa_{wt0} \leq 4$ ($\kappa_{wt} \leq 2$) the value of μ_{cr} may be calculated from formula (13), where the following approximate values of the factors C_1, C_2 may be used for the cantilever loaded by uniform load q:

$$C_1 = 4.11 + 11.2\kappa_{wt} - 5.65\kappa_{wt}^2 + 0.975\kappa_{wt}^3 \qquad \text{if } \kappa_{wt} \leq 2 \tag{14}$$

$$C_1 = 12 \qquad\qquad\qquad\qquad\qquad \text{if } \kappa_{wt} > 2 \tag{15}$$

$$C_2 = 1.661 + 1.068\kappa_{wt} - 0.609\kappa_{wt}^2 + 0.153\kappa_{wt}^3 - 0.014\kappa_{wt}^4 \qquad\qquad \text{if } \zeta_g \geq 0 \tag{16}$$

$$C_2 = 0.535 + 0.426\kappa_{wt} - 0.029\kappa_{wt}^2 - (0.061 + 0.074\kappa_{wt} - 0.0085\kappa_{wt}^2)\,\zeta_g \quad \text{if } \zeta_g < 0 \tag{17}$$

For beams with rectangular cross-section

$$C_1 = 4.11; \quad C_2 = 1.661 \text{ if } \zeta_g \geq 0 \text{ or } C_2 = 0.535 - 0.061 \zeta_g \text{ if } \zeta_g < 0 \tag{18}$$

1.2 Procedure used in German standard DIN 1052: 2004 Annex E.3 and in German National Annex to DIN EN 1995-1-1: 2005.

The formula valid for doubly symmetric profiles with $I_w = 0 \text{ mm}^6$ may be written in the form

$$M_{cr} = \frac{\pi\sqrt{EI_z GI_t}}{L_{ef}} = a_1\left(1 - a_2\frac{a_z}{L}\sqrt{\frac{EI_z}{GI_t}}\right)\frac{\pi\sqrt{EI_z GI_t}}{L} \tag{19}$$

where the effective length may be compared with μ_{cr} used in previous procedure as follows $L_{ef} = L/\mu_{cr}$. The numerical values for cantilever loaded by end load F are according to DIN 1052: $a_1 = 2.05$; $a_2 = 1.50$.

The formula taken from the DIN 1052: 2004 is simplifications of the more exact (Baláž, Koleková, 2000b) formula and the following relations are valid $a_1 = C_1/k_z$; $a_2 = \pi C_2/k_z$; $a_z = z_g$. The less exact DIN 1052: 2004 formula gives smaller M_{cr} values comparing with the ones calculted from more exact (Baláž, Koleková, 2000b) and EN 1999-1-1: 2007 formula. The error on the safe side is less than 7% for the values

$$a_2\frac{a_z}{L}\sqrt{\frac{EI_z}{GI_t}} \leq 0.3 \tag{20}$$

1.3 Procedure given in Eurocode EN 1995-1-1: 2004

According to EN 1995-1-1: 2004 effective length $L_{ef} = l_{ef}$, where for cantilever under uniform load q is $l_{ef}/L = 0.5$. If the load q is applied at the compression edge of the beam, l_{ef} should be increased by $2h$ and may be decreased by $0.5h$ for a load at the tension edge of the beam, where h is height of the rectangular cross-section. This wording is not correct in the case of cantilever (see (31)). The wording should be replaced by the following one: for loads applied above (below) the shear centre l_{ef} should be increased by $2h$ (decreased by $0.5h$).

1.4 Solution of system of differential equations using Bessel functions

System of differential equations for beam in bending with rectangular cross-section was solved using Bessel functions in (Baláž, Koleková, 2015) and in (Vol'mir, 1965). Solution leads for $M_y(x) = -qx^2/2$ and for $z_g = 0 \text{ mm}$ to the equation

$$\sum_{k=0}^{n} \frac{(-1)^k}{k!\Gamma(k+1-\nu)}\left(\frac{\pi\mu_{cr}}{6}\right)^{2k-\nu} = 0 \tag{21}$$

Graphical interpretation of equation (21) for $n = 97$ instead of ∞ is in Figure 2.

2 COMPARISON OF DIFFERENT PROCEDURES

The comparable formulae are

(i) Procedure of (Baláž & Koleková 2000a, b, 2002a, b) used in prEN 1999-1-1: May 2004 and EN 1999-1-1: May 2007. In the following formula (22) the buckling factors for cantilever are $k_z = 2$, $k_w = 2$.

$$M_{cr} = \frac{C_1}{k_z}\left[\sqrt{1 + \kappa_{wt}^2 + (C_2\zeta_g)^2} - C_2\zeta_g\right]\frac{\pi\sqrt{EI_z GI_t}}{L} \tag{22}$$

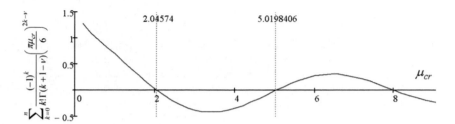

Figure 2. Graphical interpretation of equation (21). The least root is the solution: $\mu_{cr} = 2.04574$.

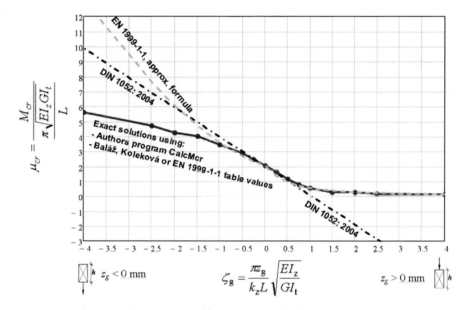

Figure 3. Comparison of procedures for calculation of relative non-dimensional critical moment μ_{cr} for cantilever loaded by uniform load q applied in point of end cross-section defined by relative non-dimensional coordinate of the point of load application related to shear center $-4 \leq \zeta_g \leq 4$, $(-8 \leq \zeta_{g0} \leq 8)$.

(ii) Procedure used in German standard DIN 1052: 2004 Annex E.3 and in German National Annex to DIN EN 1995-1-1: 2005

$$M_{cr} = a_1 \left[1 - a_2 \frac{a_z}{L} \sqrt{\frac{EI_z}{GI_t}} \right] \frac{\pi\sqrt{EI_z GI_t}}{L} \approx \frac{C_1}{k_z} \left[1 - C_2 \zeta_g \right] \frac{\pi\sqrt{EI_z GI_t}}{L} \tag{23}$$

(iii) Procedure given in Eurocode EN 1995-1-1: 2004

$$M_{cr} = \frac{1}{0.5 + \dfrac{2h}{L}} \frac{\pi\sqrt{EI_z GI_t}}{L} \quad \text{if} \quad z_g = \frac{h}{2} ; \quad M_{cr} = \frac{1}{0.5 - \dfrac{h}{2L}} \frac{\pi\sqrt{EI_z GI_t}}{L} \quad \text{if} \quad z_g = -\frac{h}{2} \tag{24}$$

Formulae (24) give dangerous results, because M_{cr} for $z_g > 0$ is greater than M_{cr} for $z_g > 0$, what is nonsense. See Figure 3 for other procedures. Eurocode formulae are not based on dimensionless parameter ζ_g and therefore their results cannot be compared with others in Figure 3.

3   CONCLUSION

Procedure developed in (Baláž & Koleková 2000a, b, 2002a, b), which was accepted by prEN 1999-1-1: May 2004 and it is used in EN 1999-1-1: May 2007 gives: (i) exact values of relative non-dimensional critical moment μ_{cr} in the Figure 1, (ii) μ_{cr} values which may be calculated using approximate formula. In the Figure 1 it is necessary to use non-linear interpolation. DIN 1052: 2004 offers formulae which lead to less exact μ_{cr} values. EN 1995-1-1 procedure gives the least exact μ_{cr} values and it is not possible to transform its formula (24) in the comparable form with (22), (23).

Moreover procedure (Baláž & Koleková 2000a, b, 2002a, b) and EN 1999-1-1: May 2007 enables to calculate critical moment M_{cr} also for cross-section with general shape. This procedure may be used also for timber beams.

ACKNOWLEDGEMENT

Project No. 1/0748/13 was supported by the Slovak Grant Agency VEGA.

REFERENCES

Baláž, I. 2005. Lateral torsional buckling of timber beams. *Wood Research* 50(1): p. 51–58.
Baláž, I. & Koleková, Y. 2000a. *Critical Moments of Beams and Girders. Clark-Mrázik formula. Proceedings of 19th Czech and Slovak International Conference "Steel Structures and Bridges 2000". Štrbské Pleso, High Tatras. September 27–30, 2000*: 87–94.
Baláž, I. & Koleková, Y. 2000b. *Proposals for Improvements of Lateral Torsional Buckling Eurocodes Rules. Proceedings of 19th Czech and Slovak International Conference "Steel Structures and Bridges 2000". Štrbské Pleso, High Tatras. September 27–30, 2000*: 81–86.
Baláž, I. & Koleková, Y. 2002a. *Critical Moments. Proc. of Internal Colloquium on Stability and Ductility of Steel Structures, Budapest*: 31–38.
Baláž, I. & Koleková, Y. 2002b. *Clark-Mrázik Formula for Critical Moments. Proc. of Internal Colloquium on Stability and Ductility of Steel Structures, Budapest*: 39–46.
Baláž, I. & Koleková, Y. 2004a. *Factors C1, C2, C3 for computing elastic critical moments Mcr. Zborník VI. sympózia Drevo v stavebných konštrukciách so zahraničnou účasžou. Kočovce 28.-29.10.2004*: 29–34.
Baláž, I. & Koleková, Y. 2004b. *Resistance of timber beams to out-of-plane buckling. Zborník VI. sympózia Drevo v stavebných konštrukciách so zahraničnou účasžou. Kočovce 28.-29.10.2004*: 35–42.
Baláž, I. & Koleková, Y. 2014a. *LTB resistance of beams. Critical moment. Proceedings of full papers on USB. Eurosteel 2014: 7th European Conference on Steel and Composite Structures., Napoli, Italy, 10.-12.9.2014.* ECCS: 1–6. ISBN 978-92-9147-121-8.
Baláž, I. & Koleková, Y. 2015. *Lateral Torsional Stability of Timber Beams. Proceedings of 6th International Conference on Mechanics and Materials in Design. Recent Advances in Mechanics and Materials in Design. Ponta Delgada/Azores, 26–30 July 2015.*
DIN 1052: 2004. Entwurf, Berechnung und Bemessung von Holzbauwerken. Allgemeine Bemessungsregeln und Bemessungsregeln für den Hochbau. Ausgabe: August 2004
EN 1995-1-1: November 2004 + A1: June 2008. Design of Timber Structures. Part 1-1 General Rules and Rules for Buildings. CEN Brussels.
EN 1999-1-1: May 2007 + A1: July 2009 + A2: December 2013 Design of Aluminium Structures. Part 1-1 General Rules and Rules for Buildings. CEN Brussels.
Koleková Y. 1999a. *Elastic Critical Moments of the Thin-Walled Beams with Monosymmetric Cross-Section.* (In Slovak). Thesis for Associate Professor (docent) habilitation. Faculty of Civil Engineering. Slovak University of Technology. Bratislava: 1–121.
Koleková Y. 1999b. Kritický moment konzoly s jednoosovo symetrickým I-prierezom. *Zborník prednášok z XXV. Celoštátneho aktívu "Stratégia rozvoja ocel'ových konštrukcií." Lipovce 6.-8. október 1999*: 105–110.
Koleková, Y. 1999c. *Klopenie monosymetrickej konzoly (Lateral torsional buckling of monosymmetric cantilever). XI. Mezinárodní vědecká konference. VUT Brno, 18.-20. října 1999:* 63–66.
Vol'mir, A.S. 1967. *Stability of deformable systems.* (In Russian). Izdatel'stvo Nauka (Publisher). Moscow.

Advances and Trends in Engineering Sciences and Technologies – Al Ali & Platko (Eds)
© 2016 Taylor & Francis Group, London, ISBN: 978-1-138-02907-1

Simulation of road tunnel operation

P. Danišovič
University of Žilina, Žilina, Slovakia

ABSTRACT: Safety of road tunnels depends to many factors, e.g. tunnel design elements, safety and technological equipment, etc. Road tunnel operation is also very important part of overall safety. In the Centre of Transport Research at University of Žilina is a device which allows managing technological equipment of virtual two-tube highway tunnel. This device is called Tunnel Traffic & Operation Simulator. Changes of the traffic-operation states and equipment conditions are reflecting in the simulated traffic, as well as simulations of various emergency events in traffic initiate the changes in tunnel on detecting and measuring devices. It is thus possible to simulate emergency states that can be affected by various faults of technology as well as by climatic conditions. The results are in irreplaceable experiences of Slovak road tunnel operators. Possibilities to changing traffic-operation states, visualizations of operator technological display screens, technological devices labeling are acquisition to increase operational safety of road tunnels.

1   INTRODUCTION

Applicant of the project Centre of Transport Research is Transport Research Institute from Žilina and its main partner is University of Žilina. Specific object of University was determined as an applied research of new technologies to increase operational safety of road tunnels. Centre of Transport Research at the University of Žilina has a unique facility in the form of Tunnel Traffic & Operation Simulator. Objectives of workplace "Simulator" are mainly:

– prepare the data and the data collection for the creation of prediction models
– regularly check and verify the operators responses on simulated incidents
– create a space to test different scenarios without direct links to the tunnel technology and with the possibility to connect to existing tunnels of administrative authority.

2   SIMULATOR OF ROAD TUNNEL OPERATION

Basic composition of the Simulator consists of follow parts:

– central control system (CCS) – part of automatic tunnel equipment control
– manual control module (MCM) – part of manual control (separated module)
– software EMUT – evidence of tunnel incidents.

Simulator is developed, in terms of functionality, based on real control algorithms used for management of existing tunnels. At simulations of incidents, which are rare in the real tunnel traffic, is possible to verify correctness and philosophy of management. Visualization surroundings of tunnel management are same as on the real operator workplace of twin-tube tunnel. In contrast to real traffic, simulation of video surveillance shows a virtual traffic in tunnel tubes. As most of operator workstations also operator workstation of simulator is implemented to operate the tunnel by two operators – operator for management technology of tunnel and operator for traffic management. Of course one operator can control entire tunnel. Always be carried out during the recording of faults, incidents, responses and interventions of CCS and operator.

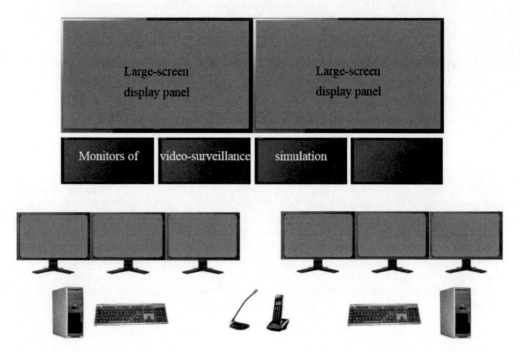

Figure 1. Sheme of simulator control centre.

3   CENTRAL CONTROL SYSTEM (CCS)

CCS simulates the operation of the twin-tube highway tunnel with a length of about 1000 m with connections to contiguous crossroads. Technological equipment of virtual tunnel is in accordance with the Regulation of the Slovak Government No. 344/2006 on minimum safety requirements for tunnels in the road network, TP 11/2011 Fire safety in road tunnels and TP 12/2011 Road Tunnel Ventilation and so on. Central control system has following main parts (Fig. 1) (Schlosser & Danišovič 2014):

- operator workstation
- workstation of coordinator
- simulation of tunnel CCS
- simulation of video surveillance
- simulation of voice communication.

Two large LED panels (Fig. 2) image longitudinal section of tunnel tubes with visualization of technological equipment of tunnel:

- measuring and detection equipment
- evacuation finding equipment
- fire safety equipment
- tunnel reflexes
- vehicle counters
- traffic lights in the tunnel and before the tunnel tubes
- sources of power
- alternative sources of power
- ventilation system of tunnel tubes and cross connections
- and others.

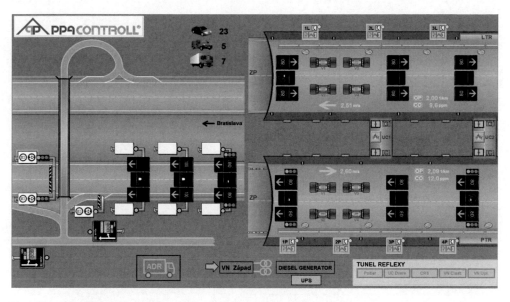

Figure 2.   Detail of LED panel from the wall of simulator control centre.

Figure 3.   Simulated video surveillance on three LCD monitors.

Under the LED panels there are three LCD monitors of simulated video surveillance (Fig. 3). For all monitors can be selected with a view of any camera video surveillance, while the third is divided into four parts, two of which show a view of the cameras sequentially for each tunnel tube separately. First monitor also serves like an alarm monitor, which means that in case of any alarm detection, the corresponding view of the camera is imaged on this monitor.

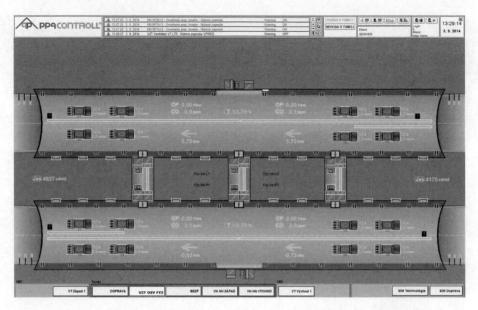

Figure 4.   Operator's technological screen of ventilation, lighting and measurement of physical quantities.

3.1   *Operator workstation*

Everything has to be watched by two operators from two workstations. The first operator is responsible for traffic and the second operator for technology. Each operator has three screens on which he can choose the display up to seven screens:

- traffic management in the tunnel
- adjacent section of traffic management before the west portal
- adjacent section of traffic management before the eastern portal
- ventilation, lighting and measurement of physical quantities (Fig. 4)
- safety
- power supply – west
- power supply – east.

Visualization and operating interfaces of CCS are complemented by typical elements of equipment of real operator tunnel control centre:

- microphone for simulation of announcements to evacuation broadcasting
- telephone of internal net for simulation of connection with (linked to coordinator's telephone):
    - integrated rescue services
    - emergency stations
    - tunnel personnel
    - safety officer
    - other responsible people (Rázga 2013).

3.2   *Workstation of coordinator*

Coordinator can simulate a lot of main emergency events and their combinations with faults of technological devices, for example:

- pedestrian or animal in tunnel
- breakdown of vehicle
- lost cargo
- vehicle in bad direction

Figure 5.   Accident of vehicles from coordinator point of view.

– accident of vehicles (Fig. 5)
– fire after the accident
– fire of the lost cargo
– stopped bus
– demonstration
– terrorist attack
– etc.

These events can be placed randomly or in one of the three parts of the tunnel (in the first period, in the middle and the third period of the tunnel). Events can take place in various conditions such as day and night, fog or smog. System saves all of the alarms, warnings, automatic changes of CCS, activities of operator and coordinator too. It is also possible to create a videorecording from continuance of emergency event. Videorecordings are requisite component of emergency event assessment in road tunnels (Rázga 2014).

4   MANUAL CONTROL MODULE (MCM)

Manual control module (MCM) is autonomous functional device that is used by tunnel specialists as an example of hardware solution of automatic process managing road tunnel technological entities. Configuration of MCM system management comes out from real structure of automatic machine which is used for managing of real road tunnels and highways under the administration of National Motorway Company. Configuration consists of simplified versions of familiarly used communication levels and architectures. Their automatic managing of technology by CCS is divided to three main levels:

– directive – net of data collection from input/output interfaces which complete managing PLC automatic machines (Programmable Logic Controller) and ensure functional technology managing
– procedural – net of communication on the level of managing automatic machines, it ensures transformation of technological device conditions into the electrical form
– visualization – visualization net that determined to ensure connection of human operator with managing technology.

Manual control module has follow parts:

– variable traffic sign
– lamella traffic sign
– traffic lights
– free-standing distribution box
– software for managing tunnel technological devices.

5 CONCLUSIONS

Safety of road tunnel users doesn't depend only from design elements of road communication in tunnel, safety and technological level of tunnel equipment but also from tunnel maintenance and last but not least from the skills, knowledge and experiences of operators. Aforementioned circumstances may have great influence on the progress of emergency event, especially on the result from the event participants' point of view (Danišovič et al. 2015). Skills and experiences from simulated incidents and possibilities of their solutions are irreplaceable for operators from real tunnel control centres of Slovak road tunnels under administration of National Motorway Company.

ACKNOWLEDGEMENTS

This contribution is the result of the project implementation: "Centre of Transport Research" (ITMS 26220220135) supported by the Research & development Operational Programme funded by the ERDF.

"We support research activities in Slovakia/project co-founded from the resources of the EU"
This work was supported by the Slovak Research and Development Agency under the contract NO.SUSPP-0005-07.

REFERENCES

Danišovič, P., Schlosser, F., Šrámek, J. & Rázga, M. 2015. Simulator of road tunnel. Civil and Environmental Engineering CEE, DE GRUYTER Vol. 11, Issue 1/2015, p. 42–47. In process of reviewing.
Rázga, M. 2013. Technological equipment of highways tunnels and increasing of safety operation. Diploma thesis, University of Žilina, Faculty of Civil Engineering, Žilina, Slovakia.
Rázga, M. 2014. Safety in road tunnels. International Scientific Conference Construction Technology and Management CTM 2014, September 09–11, 2014, Bratislava, Slovakia.
Schlosser, F. & Danišovič, P. 2014. Tunnel traffic & operation simulator. International scientific conference Construction Technology and Management CTM 2014, September 09–11, 2014, Bratislava, Slovakia.

Advances and Trends in Engineering Sciences and Technologies – Al Ali & Platko (Eds)
© 2016 Taylor & Francis Group, London, ISBN: 978-1-138-02907-1

Application of selected experimental methods for controlling the homogeneity of reinforced concrete columns

I. Demjan & M. Tomko
Technical University of Košice, Faculty of Civil Engineering, Košice, Slovakia

ABSTRACT: This paper is devoted to the application of selected methods used in the diagnosis of an experimental evaluation of homogeneity in reinforced concrete.

1 INTRODUCTION

Each construction is defined by its own specific requirements. The structural calculation not only defines the effects of load and static analysis, but is an important part of the design process, realization and operation of the structure. For financially and technologically challenging reinforced concrete structures, it is necessary to pay attention to the geometric shape and tolerances, including chemical and thermal changes of concrete, etc. (Sedlakova et al. 2011, Toth ct al. 2014, Toth et al. 2014). Thorough quality control of concrete preparation and the implementation of vibrated concrete demand the proper selection of diagnostic procedures and experimental methods (Guiming et al. 2013, Pintado et al. 2009).

During the casting of the reinforced concrete foundation turbo-machine it has been discovered that the method of casting, in particular the method of concrete vibration, is not in compliance with normative procedures. It is suspected that reinforced concrete columns may exhibit defects, mainly air gaps and an inhomogeneous distribution of concrete in the volume of the column. The paper is focused on the application of relevant methods for the diagnosis and experimental verification of the homogeneity of new reinforced concrete columns turbo-mill, (Figures 1–4).

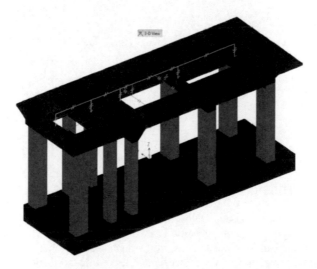

Figure 1.   Reinforced concrete foundation turbine generator.

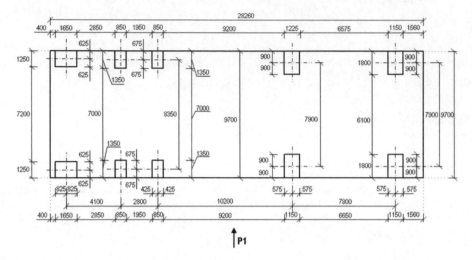

Figure 2. Floorplan of reinforced concrete foundation.

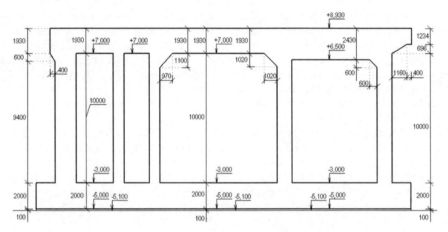

Figure 3. View of reinforced concrete foundation.

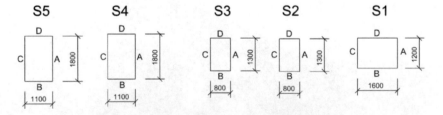

Figure 4. Diagram of analysed columns with individually labelled sides.

2 DETERMINING THE HOMEGENITY OF CONCRETE COLUMNS

The survey was carried using non-destructive methods, wherein reinforced concrete columns were evaluated for their integrity and homogeneity. The extent of the research activities were designed to maximize the identification of weaknesses in the analysed elements (disorders, anomalies of concrete), (Controlling the homogeneity of steel concrete column foundation TBG 5 2011, Garbacz et al. 2009, Stainbruch 2009, Stainbruch et al. 2011).

2.1 Georadar research

Geophysical measurement Georadar (GPR) facilitates high density, high resolution monitoring of changes in the examined environment using a non-destructive manner. The transmitting antenna generates high-frequency electromagnetic pulses that propagate through the environment. The receiver collects a complex impulse response waveform that is formed from previously reflected, surface waves, and the like. The registered quantity is the intensity of the reflected wave at different time points. High density measurement (0.01 m) enables a virtually continuous image exploration of the environment. By knowing the speed of propagation of electromagnetic waves and their arrival time, depth can be calculated from the surface of the reflecting interface. This survey used a propagation speed averaging 0.105 m/ns.

2.2 Impact Echo Measuring method

The Impact Echo (IE) method is typically used for non-destructive testing of concrete structures. IE works by tracking its own resonant frequency vibration of the structure caused by low-energy mechanical impulses. IE is most commonly used to determine the thickness of the investigated structures and for the detection of internal defects (voids, honeycombs) and for determining the depth of open cracks. A CTG measuring machine from Olson Instruments was used in the extended version, which also enables spectral analysis of surface waves (saswati) and sets the speed of propagation of these waves for exploring the test environment.

2.3 Ultrasonic measurement

Ultrasonic measurement (UM), determines the mechanical properties of structures by measuring the ultrasonic wave velocity propagated in the environment. In addition to tracking changes in the strength of the concrete UM can also reveal inhomogeneities such as voids and cracks. The ultrasonic transmitter and receiver are located on opposite sides of the walls of the structure being examined. The measurement was taken using a TICO device manufactured by Proceq.

3 COMPLETED MEASUREMENTS

3.1 Georadar research

The measurement apparatus used was a RAMAC GPR from Swedish producer Mala GeoScience, consisting of a measuring unit X3M Corder and shielded 800 MHz antenna system. The measuring device was set so that: nominal frequency of the antenna was (800 MHz); sampling frequency was (8 310 MHz); number of samples were (350); recording time was (42 ns); with (8) repetitions for each point and a measuring step of (0.02 m).

Column measurements were carried out on selected walls using a pair of parallel profiles L (left) and R (right). Table 1 summarizes the measured GPR profiles.

The measured date was processed using ReflexW version 5.6 KJ Sandmeier, with the aid of various mathematical one and two-dimensional filters. The graphical interpretation of GPR sections was realized in Corel Draw, to highlight individual anomalies. Some radar sections of the interpretation are shown in Figure 5.

3.2 Measurment using the Impact Echo method

The Impact Echo method was conducted by measuring two parallel longitudinal profiles at a distance of 0.4 m from the edge of the column in 0.2 m increments on all of the test columns. Table 2 gives an overview of the measured profiles using the IE method.

3.3 Ultrasonic measurement

The ultrasonic measurement method was carried out on columns S1 to S4. The number of profiles and step measurements were selected individually in Table 3.

Table 1. Measuring range of Georadars.

Column	Side	Profile	Position
S1	A	L	0.4–6.36
		R	0.4–6.36
	C	L	0.4–6.36
		R	0.4–6.36
S2	A	L	0.4–6.99
		R	0.4–6.99
	C	L	−2.6–6.99
		R	−2.6–6.99
S3	A	L	−2.6–5.83
		R	−2.6–5.83
	C	L	−2.6–5.83
S4	A	L	−2.6–5.82
	B	L	−2.6–5.82
		R	−2.6–5.82
	D	L	−2.6–5.82
		R	−2.6–5.82
S5	D	L	0.4–5.95
		R	0.4–5.95

Column S1

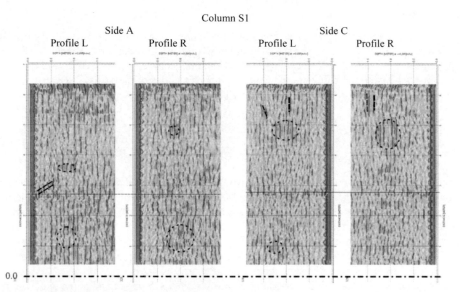

Figure 5.   List of GPR measurements. ⟨⟩ ⋮ Property changes in the environment, ◯ ❘ ❘ Indication of local inhomogeneities.

Changes in ultrasonic wave velocity in the analyzed reinforced concrete columns are shown in Figure 6. In column S1, the main profiles were supplemented by another four sub-sections and the position of 0.2 to 2.0.

4 CONCLUSION

Diagnostic evaluation of results:

Column S1: A change in the environment has been registered inside the column at several heights. This is probably a manifestation of increased humidity inside the un-matured concrete

Table 2.   Measuring range of Impact Echo method.

Column	Side	Profile	Position
S1	C	L–IE 1	0.2–6.0
		R–IE 2	1.0–6.0
S2	C	L–IE 3	−2.6–6.6
		R–IE 4	−2.6–6.6
S3	A	L–IE 5	−2.6–5.6
		R–IE 6	−2.6–5.6
S4	A	R–IE 7	−2.6–5.4
	D	L–IE 8	−2.4–5.4
		R–IE 9	−2.6–5.4
S5	D	L–IE 10	0.2–4.8
		R–IE 11	0.2–4.8

Table 3.   Measuring range of the ultrasonic measurement method.

Column	Side	Profile	Measurement interval	Position
S1	B–D	L	0.2	−2.7–6.2
		C	0.2	2.7–6.2
		R	0.2	0.2–6.2
S2	A–C	L	0.3	−2.7–6.6
		C	0.3	−2.7–6.6
S3	A–C	L	0.3	−2.7–4 85
		C	0.3	−2.7–4.85
S4	A–C	L	0.3	−2.7–4.5

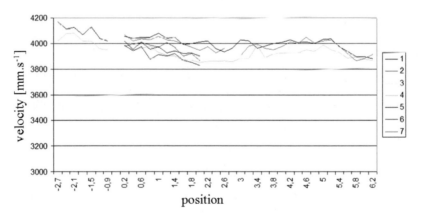

Figure 6.   Velocity of ultrasonic wave propagation in columns S1.

column. The concrete surface is even. An indication of local inhomogeneity was identified on side A – left profile between 2.6 and 3.1 m. This could be a manifestation of a structural element or local failure (cavern, honeycombs, etc.). Other manifestations of local inhomogeneities were found at the column capping (capital) at a height of 5.2 meters. It is probably the appearance of the internal design elements, respectively. un-matured concrete.

Column S2: A change in the environment has been registered inside the column at several height increments. This is probably a manifestation of increased humidity inside the un-matured concrete column. The concrete surface is even. At 6.55 meters internal construction joints were identified. No signs of changes in the properties of concrete or the incidence of local inhomogeneities around the joints were observed indicating that no potential weakening or defect of the structure was detected.

Column S3: A change in the environment was revealed inside the column at several heights. This is probably a manifestation of increased humidity inside the un-matured concrete column. The concrete surface is even.

Column S4: Based on Georadar measurements the environment appears to be homogeneous with only two local anomalies detected (e.g. Increased humidity) identified on side A to profile L 1 and 1.2 m and on side B to profile L 1 at height −0.8 m. All the profiles revealed two rebars starting at the base and extending to a height of about 1.6 meters above the ground plane.

Column S5: Based on Georadar measurements the environment appears to be homogenous with only local anomalies in the environment characteristics (e.g. increased humidity) identified at a height between 3.9 and 4.2 m.

4.1 Evaluation

Based on the measurements, columns S3–S5 can be described as homogeneous with no evidence of local defects. All the observed changes in the nature of the environment (usually within columns) correspond to a likely change in moisture of un-matured concrete.

In column S1, a local inhomogeneity was identified near the corner between walls A and B of the contour line at a height between 2.6 and 3.1 m. The features of the anomalies cannot be clearly determined. Theoretically it could be a manifestation of a structural element or of a local failure (cavern, honeycombs, etc.). Column S2 revealed a structural gap at 6.55 meters without any changes in the properties of concrete in the area.

Despite being an inappropriate method of casting columns, i.e. breaching the correct form of vibrating concrete, it can be concluded that the experimental diagnostics confirmed the expected homogeneity of reinforced concrete columns, ergo, no significant local defects were found.

ACKNOWLEDGEMENTS

The paper is carried out within the project No. 1/0321/12, partially founded by the Science Grant Agency of the Ministry of Education of Slovak Republic and the Slovak Academy of Sciences. The paper is the result of the Project implementation: University Science Park TECHNICOM for Innovation Applications Supported by Knowledge Technology, ITMS: 26220220182, supported by the Research & Development Operational Programme funded by the ERDF.

REFERENCES

Controlling the homogeneity and analysis of chemical composition steel concrete column foundation TBG 5. 2011, Faculty of Civil engineering, Technical university of Kosice, Slovakia

Garbacz, A. & Stainbruch, J. & Hlaváč, Y. & Hobst, L. & Anton, O. 2009: Defect Detection in Concrete Structures with NDT Methods: Impact/Echo Versus Radar. *NDE for Safety*, p. 21–28.

Guiming, W. & Yun, K. & Tao, S. & Zhonghe, S. 2013: Effect of water-binder ratio and fly ash on the homogenity of concrete. *Construction and Building Materials*, Vol. 38, p. 1129–1134.

Pintado, X. & Barragan, B. E. 2009: Homogeneity of self-compacting concretes used in tunnel strengthening – A case study. *Tunnelling and Underground Space Technology*, Vol. 24, p. 647–653.

Sedlakova, A. & Al Ali, M. 2011: Composite materials based on lightweight concrete. *Chemical papers*. Vol. 105, no. special issue 16, p. 445–447.

Stainbruch, J. 2009: GPR Scanner as a Next Step in Detailed 3D Diagnostics. *NDE for Safety*, p. 211–218.

Stainbruch, J. & Anton, O. & Kordina, T. 2011: Development of using georadar for diagnostic analysis of steel concrete constructions. *Concrete*, Vol. 3, p. 66–70.

Toth, S. & Vojtus, J. 2014: Analysis of causes of mold growth on residential building envelopes in central city zone of Košice. *Advanced Materials Research*. Vol. 969, p. 28–32.

Toth, S. & Vojtus, J. 2014: Monitoring and analysis of fungal organisms in building structures. *Advanced Materials Research*. Vol. 969, p. 265–270.

Advances and Trends in Engineering Sciences and Technologies – Al Ali & Platko (Eds)
© 2016 Taylor & Francis Group, London, ISBN: 978-1-138-02907-1

Impact of degree of compaction on void content of warm mix asphalt

M. Dubravský & J. Mandula
Technical University of Košice, Civil Engineering Faculty, Institute of Structural Engineering, Košice, Slovakia

ABSTRACT: Warm mix asphalt is becoming more and more used in asphalt industry. Lowering of the production temperature has many positive environmental and economic impacts. The paper deals with optimization of production temperature and optimization of the number of impacts of an impact compactor on warm mix asphalt with addition of natural zeolite as low temperature additive. Using this additive asphalt mixture can be produced at production temperature lower by 20–30%. Measurements are focused on basic physical – mechanical properties, as density, void content, void filled with bitumen and production temperature. The aim of this paper is to demonstrate advantages resulting from the addition of natural zeolite into warm mix asphalt and in this way to contribute to the research on the influence of the amount of water used to foam WMA on the rheological properties and performance related characteristics of the foamed asphalt. All asphalt mixtures were compared with reference asphalt mixture.

1 INTRODUCTION

This paper focuses on the influence of compaction degree on the final void content of warm mix asphalt (WMA). WMA is becoming more and more used in asphalt industry. It has a wide range of economic, environmental and production benefits. There are several methods for WMA production. The research at our institute is focused on the production of WMA with the addition of natural zeolite. In the paper properties of bituminous specimens produced using hot (145°C) and warm (110–125°C) procedures are compared. Specimens prepared at different rate of compaction using impact compactor -2×35, 2×50 and 2×75 blows of Marshall Compactor were also studied. Compaction temperature was in the range 110–145°C. The differences in void content of WMA (Tomko, M., Demjan, I. 2013) were expected. The results are graphically processed and evaluated in the last part of this paper.

2 WMA PRODUCTION PROCEDURE

WMA production procedure at the temperatures above 100°C results in a very low remaining quantity of water in the mixture. The effective viscosity of the binder is reduced using various technologies which allow complete aggregate coating and subsequent compaction at lower temperatures. Most used technologies are based on:

- organic ingredients,
- chemical additives,
- foam processes.

The development in WMA production in the world in recent years is shown in Figure 1.

2.1 *Technology based on foaming processes*

This technology is based on the addition of a small amount of water which is injected either into the hot binder, or directly into the mixing chamber. If the water is mixed with the hot asphalt, then the

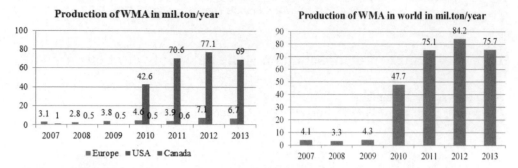

Figure 1. WMA production in the world in recent years (EAPA 2013).

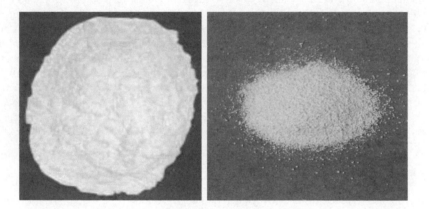

Figure 2. Synthetic zeolite fine powder and granulate.

heat causes that the water is evaporated. This creates a large amount of foam, which temporarily increases the volume and decreases the viscosity of the binder mixture. These effects remarkably improve process ability of the mixture, but its duration is limited. That is, the mixture has to be transported, compacted as soon as possible after production. The amount of water should be just enough to induce foam effect and not to cause problems to aggregates coating. Although the basic process is the same for most WMA products and technologies, the methods for water addition to the production cycle can vary. That is, the foam process can be either water-based technology (direct method) or water containing technology (indirect method).The most used technologies in the world are Aspha-Min® and Advera® – both technologies work in a similar way. They use finely powdered synthetic zeolite that has been hydrothermally synthesized. It contains approximately of 20% water of crystallization that is released when the temperature is increased above 85°C. When the additive is added to the mixture at the same time as the binder, the water is released as water vapor and thus causes foaming of asphalt. The viscosity of the binder is reduced and production and laying temperatures are the lowered. These materials were reported not affect the characteristics of the binder (Drüschner, 2009).

3 NATURAL ZEOLITE

Zeolite is crystalline hydrated aluminosilicate of alkali metals and alkaline earth metals. The unique-ness of the zeolite lies in the fact that the spatial arrangement of atoms creates channels and cavities of constant dimensions (see Figure 3) which can contain solid, liquid and gaseous substances.

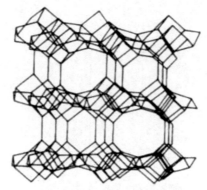

Clinoptilolite Structure

Figure 3.   Structure of natural zeolite.

Table 1.   Chemical composition of natural zeolite.

Oxide	SiO_2	Al_2O_3	TiO_2	Fe_2O_3	MnO	CaO	MgO	K_2O	Na_2O
Quantity (%)	66.97	10.61	0.24	1.72	0.03	2.90	0.73	2.96	0.68

The structure of clinoptilolite consists of three-dimensional lattice which consists of $(SiO_4)_4$ tetrahedrons interconnected via oxygen atoms; a part of the silicon atoms can be replaced by aluminum creating $(AlO_4)_{5-}$ tetrahedrons which bring negative charge to the lattice. Clinoptilolite has then characteristic spatial structure with significant amount of cavities, interconnected by channels, in which metal cations compensating negative charge of the lattice, or water molecules can be stored. The total volume of these cavities is of 24 to 32%.

The extensive use of zeolites is mainly due to their specific physical-mechanical properties:

- high selectivity and ion exchange,
- reversible hydration and dehydration,
- high gas sorption capacity,
- high thermal stability,
- resistance to aggressive media (Dubravský, M., Mandula, J. 2011).

Natural zeolite (product Zeocem 200) from site Nižný Hrabovec was used in our study. Its chemical composition is listed in Table 1, the water content was 6.3% (Dubravský, M., Mandula, J. 2013).

4   MECHANICAL PROPERTIES

4.1   *Void content of asphalt mixtures*

The porosity of the asphalt test specimens (EN 12697-8 [4]) was calculated using the measured values of the maximum and bulk sample densities. The porosity is calculated with an accuracy of 0.1% (by volume) as follows:

$$V_m = (\rho_m - \rho_b)/\rho_m \, x100 \, \% \tag{1}$$

where V_m is void content with an accuracy of 0.1% (volume); ρ_m – maximum bulk density of the test sample in kg/m^3; ρ_b – bulk density of the test sample in kg/m^3.

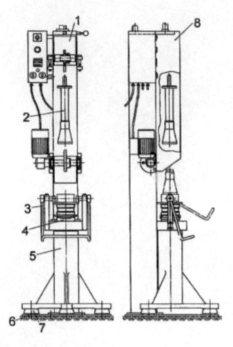

Figure 4.   Impact compactor with wooden block.

4.2   *Compaction of asphalt mixtures with impact compactor (Marshall hammer)*

Warm asphalt mixture (freshly-mixed mixture according to EN 12697-35, or taken to the station or asphalt mixing plant according to EN 12697-27) was allowed to warm before the preparation of test specimens to and poured into a steel mold for compacting. The test specimens are allowed to cool to ambient temperature (Figure 4).

Legend: 1 – basic lifting equipment with motor drive for Compaction proof and surge counter (shown without guard) 2 – Compaction equipment cylindrical guide rod, 3 – eccentric crank clamping device, 4 – backing plate molds for compaction to which the form attached 5 – compaction pedestal 6 – a supporting steel plate, 7 – foot, 8 – protective cover.

Impact Compactor with wooden block (Figure 4) is a motorized device and consists of the following components:

5   MEASUREMENTS

This is the asphalt concrete pavements for subgrade layer with a fraction of up to 16 mm. In the production, we focused on the number of strokes in the compaction impact compactor. The recipe mixture is composed of three types of aggregates, stone meal, warm mix asphalt additives and asphalt. The percentage composition of each type of aggregate is shown in Table 2.

The addition of low-temperature additive (zeolite ZeoCem 200) is given at the expense of filler (VJM Hosžovce). Grading curves of the two ingredients are the same, so the effect of reducing the expense of filler ingredients is irrelevant (Rubio, M. C. 2012).

Several series of samples with the addition of natural zeolite were produced in laboratory conditions. The production temperature was between 105–140°C. Based on the laboratory measurements porosity was evaluated and consequently production temperature. From the experiments, it can be inferred, that porosity decrease as a result of the production temperature increase. The sample was then compacted with different numbers of impact compactor blows. The samples were compacted

Table 2. The composition of bituminous mixtures.

Mixture/signature	Aggregate (fr. – fraction)				Filler	Additive	Asphalt
Type	Hradová	Hradová	Hradová	Hradová	Hošžovce	ZeoCem 200	MOL
Label	fr. 11/16	fr. 8/11	fr. 4/8	fr. 0/4	VJM	Zeolit	50/70
Ref. mix AC 16 P	19.05	19.05	12.38	38.10	6.67	0	4.76
WMA AC 16 P	19.05	19.05	12.38	38.10	5.77–6.47	0.2–0.9	4.76

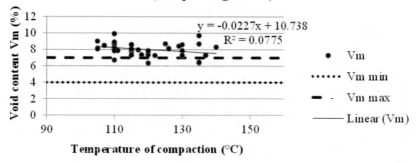

Figure 5. Influence of temperature on void content (2 × 35).

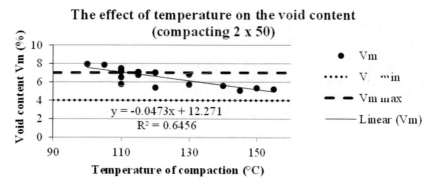

Figure 6. Influence of temperature on void content (2 × 50).

2 × 35, 2 × 50 and 2 × 75 times on each side. For each sample the graph of porosity versus production temperature was made. Figure 5 shows the graph of porosity versus compaction temperature for the samples compacted 2 × 35 times. The compacted samples were then tested for sensitivity to water (EN 12697-12).

The best results were obtained for the sample compacted 2 × 50 times on each side (Figure 6).

These samples were then tested for bulk density (EN 12697-6). Already at 112°C the void content decreased to below 7%, what is according to KLAZ 1/2010 of the maximum possible void content. In Figure 7, the samples compacted 2 × 75 times on each side.

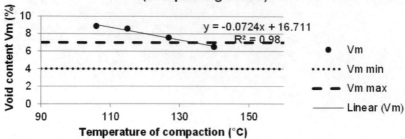

Figure 7.   The influence of temperature on void content (2 × 75).

6   CONCLUSION

Warm mix asphalts are adequate substitutes for conventional hot mix asphalts. They meet all requirements for physical and mechanical properties, as stated in EU standards. The results of a laboratory measurements point to the fact, that the production temperature of the WMA with the addition of natural zeolite can be as low as 115°C. The samples compacted at lower temperatures do not meet the standard requirements for void content. The limit void content in asphalt mixtures is 7%. Warm mix asphalts have a whole range of benefits and they should be used more frequently in asphalt industry. Further research should focus on optimizing the addition of natural zeolite in warm mix asphalts.

ACKNOWLEDGMENT

The contribution was incurred within Centre of the cooperation, supported by the Agency for research and development under contract no. SUSPP-0013-09 by businesses subjects Inžinierske stavby and EUROVIA SK.

The research has been carried out within the project NFP 26220220051 Development of progressive technologies for utilization of selected waste materials in road construction engineering, supported by the European Union Structural funds.

REFERENCES

Drüschner, L. 2009: Experience with Warm Mix Asphalt in Germany. Sonderborg: NVF-rapporter, guest report in conference.
Dubravský, M., Mandula, J. 2011: Nízkoteplotné asfaltové zmesi na báze prírodného zeolitu – Pozemné komunikácie a dráhy. roč. 7, č. 2 (2011), s. 21–30. ISSN 1336–7501.
Dubravský, M., Mandula, J. 2013: Technology of warm mix asphalt based on foaming processes. In: SGEM 2013: 13th International Multidisciplinary Scientific Geoconference: Ecology, Economics, Education and Legislation: conference proceedings: Volume 1, Albena, Bulgaria. – Sofia: STEF92 Technology Ltd., 2013 P. 913–919. ISBN 978-619-7105-04-9 – ISSN 1314–2704.
EAPA 2013 – European asphalt pavement association – http://www.eapa.org/userfiles/2/asphalt%20in%20figures/aif_2013_final.pdf.
Rubio, M. C., Martínez, G., Baena, L., Moreno, F. 2012: Warm mix asphalt: an overview. Journal of Clean Production 24 (2012), s. 76–84.
Tomko, M., Demjan, I. 2013: The dynamic response of structures to random excitation effect. 2013. In: Interdisciplinary in theory and practice. No. 2 (2013), p. 24–27. – ISSN 2344–2409.
Zeocem 2015, http://www.zeocem.sk.

Advances and Trends in Engineering Sciences and Technologies – Al Ali & Platko (Eds)
© 2016 Taylor & Francis Group, London, ISBN: 978-1-138-02907-1

Pedestrian bridge monitoring using terrestrial laser scanning

J. Erdélyi, A. Kopáčik, I. Lipták & P. Kyrinovič
Slovak University of Technology in Bratislava, Department of Surveying, Bratislava, Slovakia

ABSTRACT: This paper presents the deformation monitoring of a pedestrian bridge over the river Malý Dunaj in Bratislava – Vrakuňa. Precise levelling is a conventional surveying method of vertical determinations. Because of the fact that the monitored object is a suspension bridge construction, the levelling is not feasible (the loading of the survey equipments may influence the results). The measurement of the structure was performed by the technology of terrestrial laser scanning (TLS). The advantage of TLS is that this method allows contactless determination of the spatial coordinates of points situated on the surface of the measured object. An application based on software MATLAB® – Displacement_TLS was developed for automated data processing. The designed method represents a new approach of deformation monitoring. The procedure of laser scanning, the data processing (calculation of displacements using orthogonal regression) and the results of deformation monitoring are described in this paper.

1 INTRODUCTION

The weather conditions and the loading during operation cause changes in the spatial position and in the shape of engineering constructions that affect static and dynamic function and reliability of these structures. Because of these facts, geodetic measurements are integral parts of engineering structures' diagnosis.

This paper presents the possibility of using the technology of terrestrial laser scanning (TLS) in the field of deformation monitoring of bridge structures. The deformation monitoring of a steel pedestrian bridge structure is described in this paper.

The advantage of TLS compared with conventional surveying methods is the efficiency of the spatial data acquisition. It allows contactless determination of the coordinates of measured points situated on the surface of the monitored structure. The method of precise levelling is the conventional surveying method of vertical deformation measurement of bridges. But levelling is not feasible in the case of light suspended steel structures because the loading of the levelling equipments may influence the results of the measurement. The accuracy of discrete point coordinate determination by TLS (several millimetres) can be increased by approximation of chosen parts of the monitored object (Kopáčik et al. 2013). In this case the position of the measured point is calculated from tens or hundreds of scanned points (Vosselman et al. 2011) (Soni et al. 2015).

2 CHARACTERISTICS OF THE MONITORED STRUCTURE

The monitored construction was a pedestrian bridge over the river Malý Dunaj in Bratislava – Vrakuňa (Slovakia). The bridge connects the residential zone with the recreational area of Vrakuňa Forest Park. The substructure consists of two abutments from reinforced concrete in which the supporting pylons are anchored (Figure 1). The main section (54 m length) of the bridge is divided into 10 sections by suspensions.

The deck is composed of metal plate (thickness of 10 mm), of steel cross-girders (IPE 270) located in transverse direction in axial distance of 2 m and longitudinal girders (IPE 360 on the

Figure 1. Pedestrian bridge over the river Malý Dunaj in Bratislava, Slovakia.

Figure 2. Detail of the bridge deck (on the left) and the pylons (on the right).

sides and IPE 270). The clearance of the bridge deck is 2360 mm. The superstructure is suspended on four pylons (Figure 2).

3   DEFORMATION MONITORING

Monitoring was performed in two measurement epoch, in December 2014 and in March 2015. TLS Leica ScanStation2 was used. The instrument is able to scan up to the range of 300 m with a scan rate up to 50,000 points / second. The accuracy of single point measurement is 6 mm at 50 m range (one sigma) that is defined by the producer (Leica Geosystems). The spatial position of the measured point is obtained by using polar method where the horizontal and vertical angles are defined by the instrument. The slope distance is measured with help of an in-built electronic distance measurement device of the scanner. The result of TLS is a point cloud in non-regular raster created by points situated on the surface of the scanned object.

The bottom part of the suspension bridge deck was scanned from a single position of the scanner (under the deck approximately in the longitudinal axis). The scanner was located in each epoch approximately in the same position to ensure same conditions of the measurement (distance from the instrument, angle of incidence of measuring signal). The reference network consists of four control points stabilized on the abutments of the bridge. All of them were signalized by Leica HDS

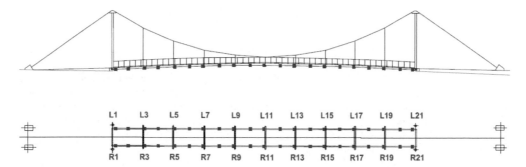

Figure 3. Scheme of the bridge with the position of measured points.

Figure 4. Definition of the measured point position (on the left), fencing box in the point cloud (on the right).

targets. The points of the reference network define a local coordinate system for the deformation monitoring of bridge. The Z axis of the coordinate system is vertical (defined by the dual axis compensator of the instrument) and the X axis is parallel with the longitudinal axis of the bridge. The minimal point density was 3 mm × 3 mm on the surface of the scanned part of the structure. The weather conditions (air temperature) and temperature of the structure were also measured during the process of the scanning.

The data obtained by the TLS were transformed to the local coordinate system of the bridge. The accuracy of the transformation is given by the differences (ΔX, ΔY, ΔZ) between the common identical reference points after the transformation. The main task of the data processing was to determine the vertical displacements of the bridge. It was performed by modelling of the movement in Z direction of the chosen parts (represented by the measured points) of the structure in each measurement epoch. These points are located on the bottom part of the bridge desk on the side girders IPE 360 at the beginning, in the centre and at the end of each bridge section (Figure 3). The total number of observed points is 42.

The Z coordinates of the measured point were modelled using small planar surfaces of 0.1 m × 0.1 m. The position of the measured points in the XY plane is defined by the coordinates of the intersection of the diagonals of these planes (Figure 4). The advantage of the procedure is that the position of the measured points does not change with the thermal expansion of the structure. The Z coordinates of the points were calculated by projecting the points onto regression planes. The standard deviation of the results were calculated using the uncertainty propagation law, from the standard deviation of the vertical component of the transformation error and the standard deviation of the regression planes which were calculated from the orthogonal distance of the points of point cloud from these planes.

Orthogonal regression is calculated from the general equation of a plane by applying Singular Value Decomposition:

$$\mathbf{A} = \mathbf{U\Sigma V}^{\mathrm{T}}$$ (1)

where $\mathbf{A}$ is the design matrix with dimensions $n \times 3$, and n is the number of points used for the calculation. The column vectors of $\mathbf{U}^{n \times n}$ are normalized eigenvectors of the matrix $\mathbf{AA}^{\mathrm{T}}$. The column vectors of $\mathbf{V}^{3 \times 3}$ are normalized eigenvectors of $\mathbf{A}^{\mathrm{T}}\mathbf{A}$. The matrix $\mathbf{\Sigma}^{n \times 3}$ contains eigenvalues on the diagonals. Then the normal vector of the regression plane is the column vector of $\mathbf{V}$ corresponding to the smallest eigenvalue from $\mathbf{\Sigma}$ (Lacko 2008).

The design matrix is following:

$$\mathbf{A} = \begin{pmatrix} (X_1 - X_0) & (Y_1 - Y_0) & (Z_1 - Z_0) \\ (X_2 - X_0) & (Y_2 - Y_0) & (Z_2 - Z_0) \\ \vdots & \vdots & \vdots \\ (X_n - X_0) & (Y_n - Y_0) & (Z_n - Z_0) \end{pmatrix}$$ (2)

where $(X_i - X_0)$, $(Y_i - Y_0)$ and $(Z_i - Z_0)$ are the coordinates of the point cloud reduced to the centroid.

The vertical displacements were calculated as the difference in the height (Z coordinates) of the measured points between the initial and current measurement epoch. The statistical significance of the displacements was determined on the basis of the statistical analysis using interval estimates. An application based on software MATLAB® – Displacement_TLS (Figure 5) was developed for automated data processing. The above mentioned computational procedure is performed and controlled with help of the graphical user interface of the application. The application was created as a standalone app; however the Matlab Runtime is necessary to be installed. The work with the app is as follows: In the first step the user can choose a work directory in which the resulting files will be saved. The second step is the point cloud file loading in *.txt or *.xyz file format which contains the coordinates of scanned points. The measured points can be arranged in *.xls or *.xlsx file defining the coordinates of the monitored points. The vertical displacements are calculated in the points defined by the mentioned file and are transformed to perpendicular displacements using the normal vectors of planar surfaces. The user can load the resulting file of the previous measurement epoch in the Initial / Previous Measurement box (for comparison of point heights). Without this file the result will be an *xlsx file containing the heights of the measured points. The fencing boxes, selecting part of the point cloud around the measured points, are defined by its dimensions along axis X, Y and Z. The standard deviation of the registration (transformation of point cloud) is necessary for the calculation of the standard deviation of the results using the uncertainty propagation law. The results are shown in the table on the right side of the app's dialog window and are saved into an *.xlsx file in the work directory. A figure which shows the point cloud and the displacement vectors in a relative scale is created for the better imagination of the results.

The Figure 6 shows the results of the monitoring. The Z axis represents the vertical displacements of the measured points. The second axis represents the temperature difference between the initial measurement and its change during the scanning (1.5 h). The positive values mean lower temperature in the control measurement epoch. The grey area represents the standard deviation of the displacement in each point (one sigma). If the displacement is in one sigma area, the measurement did not show any displacement. Otherwise the displacement is real with the risk of decision at $5\% - 30\%$ (from 1σ to 2σ) or less than 5% (over 2σ). The measurements show the displacement of the points in the 3rd–8th sections. The displacements towards the centre of the bridge increase and have positive values. In the middle of the bridge they reach values of 7 mm. This is caused by the lower temperature of the structure in the control epoch of the measurement and it corresponds with the theoretical values of deformation. For example: Due to the temperature changes about $3°C$, the theoretical value of a displacement in the middle of the bridge deck at the position of anchorage of suspension cables (point No. R11 and L11) is 4 mm. The displacement determined

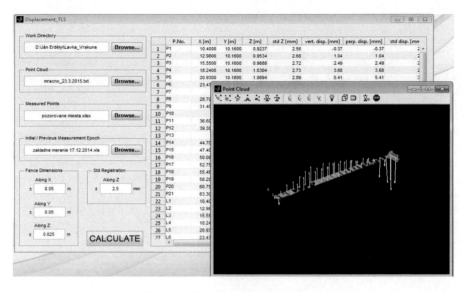

Figure 5. Print screen of the Displacement_TLS dialog window.

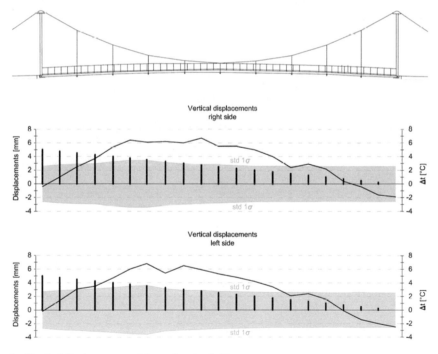

Figure 6. Vertical displacements of the pedestrian bridge.

by the measurement is 5 mm on the both sides of the deck (Figure 6). The difference between the theoretical values and the measured displacements is approximately 20% which is caused by the inaccuracy of static model of the structure. The values of the displacement increase with the temperature difference, too.

4 CONCLUSION

This paper describes the deformation monitoring of a pedestrian bridge over the river Malý Dunaj in Bratislava – Vrakuňa. The measurement was realized in two epochs, in December 2014 and in March 2015. The results of the experimental measurements show that terrestrial laser scanning can be used for determination of the displacements of bridges. To increase the accuracy of the results, selected parts of the monitored structure have to be approximated by single geometric entities using regression. The application of TLS is possible for the deformation monitoring of bridge structures constructed by any technology and from any materials enabling passive reflection of the measuring signal of the in-built electronic distance measurement device of the scanner. It is possible to ensure standard deviation 3 times higher than the accuracy of the single measured point with the given device by modelling the Z coordinates of measured points with help of planar surfaces using orthogonal regression. The standard deviation of the planar surfaces described in this paper was around 1 mm. The accuracy of the transformation of point cloud of the control measurement epoch to the coordinate system of the bridge was 2.5 mm. It leads to standard deviation of the vertical displacement of 3 mm when the uncertainty propagation law is applied. The accuracy of the results is comparable with the accuracy of the conventional surveying methods.

ACKNOWLEDGEMENT

"This work was supported by the Slovak Research and Development Agency under the contract No. APVV-0236-12".

REFERENCES

Kopáčik, A., Erdélyi, J., Lipták, I. & Kyrinovič, P. 2013. Deformation Monitoring of Bridge Structures Using TLS. *2nd Joint International Symposium on Deformation Monitoring [USB]*. Nottingham: University of Nottingham.

Lacko, V. 2008. Singular Value Decomposition and Difficulties of Software Implementation of Golub Algorithm and its Determination. *Student science conference 2008*. Bratislava: Comenius University in Bratislava.

Soni, A., Robson, S. & Gleeson, B. 2015. Structural Monitoring for the Rail Industry Using Conventional Survey, Laser scanning and Photogrammetry. *Applied Geomatics* 7: 123–138.

Vosselman, G. & Maas, H. G. 2010. *Airborn and Terrestrial Laser Scanning*. Dunbeath: Whittles Publishing.

Advances and Trends in Engineering Sciences and Technologies – Al Ali & Platko (Eds)
© 2016 Taylor & Francis Group, London, ISBN: 978-1-138-02907-1

Comparative measurements of acoustic properties of road surfaces

O. Frolova & B. Salaiová
Technical University of Košice, Faculty of Civil Engineering, Košice, Slovakia

ABSTRACT: The results of measurements of sound pressure levels of traffic noise on three road segments with different surfaces are presented in the paper. Based on the comparison of the obtained results the best road pavement from a noise point of view was determined. Measurements were carried out in Slovakia using the SPB (Statistical Pass-By) method which was designed for the traffic noise assessment.

1 INTRODUCTION

The definition of traffic noise as an environmental problem was defined in recent years because of the increasing quantity of vehicles and people living near to the multitude of noise resources. Engineers and scientists around the world are working on the development of new methods for reducing road traffic noise. The reduction of road traffic noise has a good effect on the environment as it limits vibration, which occurs analogously with sound wave propagation (Demjan, Tomko, 2013). One of the possibilities of reducing road traffic noise is by constructing low-noise road paving. Low-noise road surfaces can only be determined after actual sound pressure levels of real road paving are defined. The paper aimed to realize measurements of road traffic noise using the SPB method (Statistical Pass-By). The impact of passenger and heavy truck vehicles on the SPB method is included. The impact of three different categories of vehicles (passenger cars, two-axle heavy vehicles and multi-axle heavy vehicles cars) on road traffic noise can be evaluated by the SPB method. The acoustic properties of road pavements on selected road sections in Kosice and Presov region were measured and the maximal A-weighting sound pressure levels were determined for each vehicle passing the microphone. The SPBI (Statistical Pass-By Index) was calculated for a comparison and assessment of road paving from a noise point of view.

2 SUMMARY OF SELECTED RESULTS IN TRAFFIC ROAD NOISE STUDY

Acoustic scientists are focused on the detection of sound pressure levels and their evaluation in the field of road traffic noise. Firstly, the status of current knowledge and the achievements of scientists in exploring road traffic noise problems should be ascertained for a detailed setting of research objectives.

Authors (Sandberg, Andersen, Bendtsen, Kalman, 2005) presented the results of measurements of sound pressure levels on a few road sections in Denmark and Sweden. Measurements by SPB (Statistical Pass-By) method indicated the reduction of sound pressure levels by the SMA 11 (Stone Mastic Asphalt) surface versus DAC 11 (Dense Asphalt Concrete) with 3 dB for passenger cars and on 1–2 dB for heavy trucks vehicles. Measurements of SMA 8 road paving show a 7 dB reduction of sound pressure levels versus SMA 16 pavement and 3,3 dB versus the original SMA 11 paving. However, other measurements on the same road section after just one year (one winter period with using studded tires) reveal a reduction of traffic road noise of only 0,5 dB.

Acoustic properties of road paving on a test road section in a rural area (highway) were determined by the SPB method and an equivalent sound pressure level was detected (Kudrna, Dašek, 2009). The value of sound pressure levels on asphalt with the application of crumb rubber modified bitumen was lower by 2 dB.

3 IN-SITU MEASUREMENTS OF ACOUSTIC PROPERTIES OF SELECTED ROAD PAVEMENTS

At the Department of Geotechnics and Traffic Engineering, Faculty of Civil Engineering Technical University of Kosice within the research of sound pressure levels practical measurements was realized in the Kosice and Presov region.

3.1 *Characteristics of selected road sections*

The measurements were carried out on selected road sections in the Kosice and Presov regions (Frolova, Salaiova, 2014).

Road section 1: Kosice region, Svinica village, road I/50, width is 7,5 m. Reference speed equals 50 km/h, the average value of the road speed is 52,8 km/h and according to (STN EN 11819-1) the road has "low speed" category. Road surface material – SMA 11 (*Stone Mastic Asphalt*).

Road section 2: Presov region, Haniska near Presove village. Road E50 connects to highway D1. Reference speed is 80 km/h, average value of road speed is 74 km/h and according to (STN EN 11819-1) the road has a "medium speed" category (STN EN 11819-1). Road surface material – SMA 11 (*Stone Mastic Asphalt*).

Road section 3: Kosice region, Mala Ida village, road II/548, width is 7,5 m. Reference speed is 90 km/h, average value of road speed is 71,14 km/h and according to (STN EN 11819-1) road has "medium speed" category (STN EN 11819-1). Road surface material – AC 11 (*Asphalt Concrete*) with 1,2% crush rubber (CR) from used tires (Mandula, 2013).

During the measurements the following experimental equipment was used:

Traffic Detection Device Sierzega SR4 to measure the instantaneous velocity, time of passage and the length of the vehicle. The device was located 1 m from the road surface at an angle of 30° from the direction parallel with the axis of the road. Measurement uncertainty for speed is ±3%, for the length of vehicle ±20%, for time is ±0.2 s. The speed should be measured at the time instant when the vehicle passes the microphone (STN EN 11819-1).

Sound level meter – type 2250 with the module for frequency analysis BZ 7223 was used. It is a sound analyzer with class 1 precision 1 according to EN 61672 with a 120 dB dynamic range without switching. Microphone type 4189 prepolarized 1/2" for free-field was used.

4 RESULTS OF MEASUREMENTS

The measurements of acoustic properties of road surfaces were carried out in accordance with the standard (STN EN 11819-1) using the SPB method. The maximum sound pressure level was recorded when the vehicle passed closest to the microphone, the speed of the vehicle and its length were measured as well. The next step of data processing was data normalization. The values of maximum sound level of the vehicle and the decimal logarithm of the vehicle speed were obtained. The linear regression analysis was then used for data processing. The least-squares method was used for the parameters of regression line calculation for each category of vehicle (STN EN 11819-1). The regression lines are shown in Figure 1, 2, 3. The coordinate of the sound pressure level on the regression line for each vehicle category at the equivalent reference speed is the sound pressure level of the vehicle L_{veh}. For this purpose three values L_{veh} are obtained at the specific location: the first for passenger cars, the second for two-axle heavy vehicles and the third for multi-axle heavy vehicles (STN EN 11819-1). These values were used to calculate the SPBI.

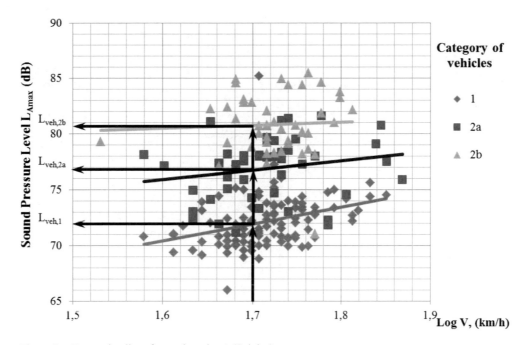

Figure 1.   Regression lines for road section 1 (Svinica).

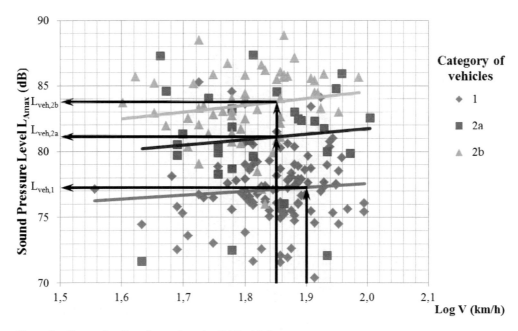

Figure 2.   Regression lines for road section 2 (Haniska).

5   DESCRIPTION OF RESULTS

Figure 4 shows a comparison of the average sound pressure level, which was found on three road sections. The average sound pressure levels (L_{Amax}) were determined for each road section and

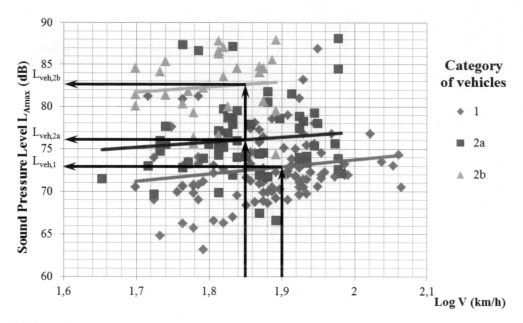

Figure 3.   Regression lines for road section 3 (Mala Ida).

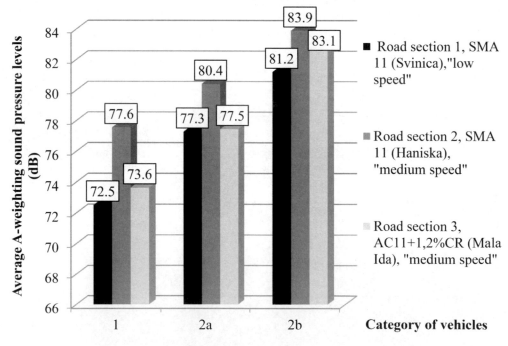

Figure 4.   Comparison of average sound pressure levels.

vehicle category. In the category "medium speed" the best acoustic properties for each category of
vehicles was obtained for the road surface on road section 3 (Mala Ida). The difference of values of
average sound pressure levels is 4 dB for category 1 vehicles 1 (passenger cars); 2,9 dB for category
2a (two-axle heavy vehicles); 0,8 dB for category 2b (multi-axle heavy vehicles).

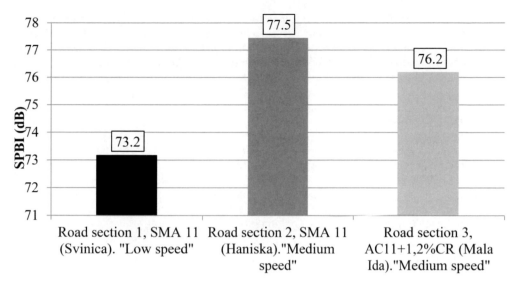

Figure 5.   Comparison Statistical Pass-By Index (SPBI).

The comparison of SPBI (Statistical Pass-By Index) for the tested road sections is shown in figure 5. The SPBI is not an equivalent sound pressure level, but it can be used to describe the relative impact of the road surface of these levels (STN EN 11819-1). For road section 1, Svinica (road pavement SMA 11), SPBI was 73,2 dB; for the road section 2, Haniska (road pavement SMA 11) SPBI was 77,5 dB; for the road section 3, Mala Ida (road pavement AC 11+1,2%CR), SPBI was 76,2 dB.

6   CONCLUSIONS

The detection of the acoustic situation is an important factor in environmental quality assessment. The authors paid attention to acquiring sound pressure levels for the following evaluation of noise influence because of the increasing number of passenger cars and heavy truck vehicles. One of the possibilities for reduced noise levels is to design low-noise road paving which will reduce noise at the source. For example, the SMA cover vs. DMA cover has showed better acoustic properties and has a sound pressure level 3 dB less for passenger cars and 1–2 dB for heavy truck vehicles (research of authors (Sandberg, Andersen, Bendtsen, Kalman, 2005)). In (Kudrna, Dašek, 2009) research of the road cover with an additive of crumb rubber modified bitumen in asphalt mixture has shown better acoustic properties and equivalent sound pressure level was 2 dB lower.

At the Department of Geotechnics and Traffic Engineering the measurements of acoustic properties of road surfaces were carried out using the SPB method on three road sections in Kosice and Presov regions satisfying conditions stated in (STN EN 11819-1). The maximum sound pressure level for each vehicle passing the microphone and vehicle speed were measured. The results obtained were processed and compared. The comparison of sound pressure levels was made for the average sound pressure level and SPBI. Better acoustic properties for all three categories of vehicles (passenger cars, two-axle heavy vehicles and multi-axle heavy vehicles) were determined for the SMA11 surface (road section 1, Kosice region, Svinica) and it can be deduced from the comparison of the average sound pressure levels (fig. 4). The SPBI takes into account the impact of the vehicle category, reference speed of the road and weighting factors. The SPBI values also show that in the "medium speed" category AC 11+1,2%CR cover has better acoustic properties (SPBI was 76,2 dB) versus SMA 11 cover (SPBI was 77,5 dB). The differences of values of road noise on road sections 1 (Svinica, road pavement SMA 11) and on the road section 2 (Haniska, road pavement SMA 11) was 4,3 dB and on the basis of measuring we can conclude that the reduction of speed involved a reduction of traffic road noise (on road section 1 – Svinica, reference speed was 50 km/h, on road section 2 – Haniska, reference speed was 80 km/h, on road section 3 – Mala Ida village reference speed was 90 km/h).

ACKNOWLEDGEMENTS

The contribution was incurred within Centre of the cooperation, supported by the Agency for research and development under contract no. SUSPP-0013-09 by businesses subjects Inžinierske stavby and EUROVIA SK.
The research has been carried out within the project NFP 26220220051 Development of progressive technologies for utilization of selected waste materials in road construction engineering, supported by the European Union Structural funds.

REFERENCES

Demjan, I. & Tomko M. Frequency of the analysis response of reinforced concrete structure subjected to road and rail transport effects. In: *Advanced Materials Research*: *SPACE 013: 2nd International Conference on Structural and Physical Aspects of Civil Engineering*: High Tatras, Slovakia, 27-29 November 2013. Vol. 969 (2014), p. 182–187. – ISBN 978-303835147-4 – ISSN 1022-6680.
Frolova, O. & Salaiová, B. Acoustic properties of road surface containing crumb rubber – 2014. In: SGEM 2014: *14th international multidiscilinary scientific geoconference* – ISBN 978-619-7105-16-2
Kudrna, J. & Dašek, O.: Report about antiskid and antinoises properties of road pavements with crumb rubber modified bitumen (Zpráva o měřeni protismykových a protihlukových vlastností obrusných vrstev s asfaltem modifikovaným pryžovým granulátem). In: *Brno: Vysoké učení technické v Brně, Fakulta stavební*, 2009.
Mandula, J., Active zone impact on deformation state of flexible pavements. In: SGEM 2013: *13th International Multidisciplinary Scientific Geoconference: science and technologies in geology, exploration and mining*, 2013, Albena, Bulgaria. ISBN 978-954-91818-8-3
Sandberg, U. & Andersen, B. & Bendtsen, H. & Kalman, B.: Low-noise surfaces for urban roads and streets. European commission of research, 2005.
STN EN ISO 11819-1:1997. Acoustics. Measurement of the influence of road surfaces on traffic noise. Part 1: Statistical Pass-By method.

Advances and Trends in Engineering Sciences and Technologies – Al Ali & Platko (Eds)
© 2016 Taylor & Francis Group, London, ISBN: 978-1-138-02907-1

Buckling analysis of cold-formed C-shaped columns with new type of perforation

M. Garifullin, M. Bronzova, A. Sinelnikov & N. Vatin
St. Petersburg Polytechnical University, St. Petersburg, Russia

ABSTRACT: This article deals with the buckling analysis of simply-supported cold-formed C-sections subjected to compression. A new type of thin-walled thermoprofile with reticular-stretched perforations actively used in modern construction is considered. First the stability problem of a planar plate simply supported along the entire perimeter is solved to assess the impact of perforations on section web buckling. FE analysis was conducted to compare buckling resistance of a new profile with traditional thermoprofile with longitudinal perforations.

1 INTRODUCTION

Thin-walled cold-formed steel sections are widely used in various fields of industrial and civil engineering, bridges, storage racks, car bodies, railway coaches, transmission towers and poles, various types of equipment (Veljkovic & Johansson 2008; Vatin & Sinelnikov 2012; Heinisuo et al. 2014).

Over the last decade cold-formed constructions have been actively used in the construction of cladding panels in residential housing (Zhmarin 2012). In order to increase thermal effectiveness of these panels cold-formed steel profiles are perforated on web and thus are called thermo-profiles (Fig. 1). Traditionally, thermoprofiles with longitudinal perforations are mostly used in construction.

In this article a new type of thermoprofiles with reticular-stretched perforations (hereinafter – RST-profiles) is considered. The key features of this profile are completely different geometry of perforations and the presence of the intermediate longitudinal stiffener.

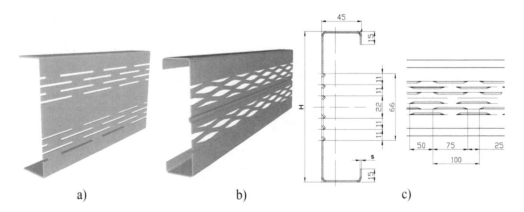

a)   b)   c)

Figure 1.   a) traditional thermoprofile, b) RST-profile, c) geometry of traditional thermoprofile.

Due to the fact that RST-profiles have been designed recently, nowadays there are no building codes which enable the calculation of this type of profiles. Moreover, very little research has been devoted to investigating their mechanical properties.

Sinelnikov (Vatin, Havula, et al. 2014; Vatin et al. 2015) and Nazmeeva (Vatin, Nazmeeva, et al. 2014) investigated the buckling behavior of compressed cold-formed columns made of RST-profiles. Trubina (Vatin, Sinelnikov, et al. 2014; Trubina et al. 2014a; Garifullin et al. 2015; Trubina et al. 2014b) analyzed the problem of local and global buckling of RST-profiles in bending. Rybakov (Lalin et al. 2014; Rybakov & Sergey 2015) presented four types of finite elements to analyze cold-formed members with various boundary conditions on the ends. Belyy (Belyy & Serov 2013) introduced a new method for the approximate estimation of steel structures service life in buildings. Al Ali (Al Ali 2014; Al Ali et al. 2015; Al Ali et al. 2012) presented investigated thin-walled cold-formed steel members with closed cross-sections. Tusnin (Tusnin 2009; Tusnin 2010; Tusnin 2014) introduced thin-walled finite elements to analyze spatial cold-formed structures with open cross sections. Tusnina (Tusnina 2014a; Tusnina 2014b; Danilov & Tusnina 2014) presented a finite element analysis of cold-formed Z-purlins supported by sandwich panels. Prokic (Tusnin & Prokic 2014a; Tusnin & Prokic 2014b) analyzed the behavior of thin-walled open section I-beams under torsion and bending. Björk (Björk & Saastamoinen 2012; Heinilä et al. 2009; Nykänen et al. 2014) studied the influence of residual stresses on the fatigue strength of cold-formed rectangular hollow sections.

Despite the growing interest to RST-profiles, the influence of perforations on the resistance of profiles remains an open issue. Buckling behavior of these sections is also investigated very poorly.

This article deals with buckling performance of cold-formed C-sections made of RST-profile. First buckling of a planar plate simply supported along the entire perimeter is considered to assess the impact of perforations on the web buckling. The article also provides the FE analysis of cold-formed C-sectioned columns and compares the results with those for traditional thermoprofiles and cold-formed profiles without any perforations.

2 BUCKLING PERFORMANCE OF A SIMPLY SUPPORTED ALONG THE ENTIRE PERIMETER PLANAR PLATE

As mentioned above, the main difference of RST-profile is geometry of perforated web and the presence of intermediate stiffeners. From that point of view it seems reasonable to carry out a study of a thin-walled compressed steel planar plate supported along the entire perimeter which simulates a web of RST-profile profile and then compare the data obtained for a RST-profiled plate without intermediate stiffener, a solid plate, and a plate with the traditional longitudinal perforations (Salminen & Heinisuo 2011; Salminen & Heinisuo 2014).

The plate with longitudinal perforations and the solid plate were chosen for comparison because of their prevalence in construction. Analysis of the perforated plate without intermediate stiffeners will enable to making assessments of the influence of reticular-stretched perforation on the stress-strain state of the section web. The geometrical dimensions of profiles for which the plates were created are shown in Figures 1, 2.

Solution of the problem was based on the finite element method. For each plate a finite element shell model in Abaqus was created with 3 mm mesh size. The geometrical dimensions of the plates were set considering the condition of the plates' equality, each up to 200 mm. The thickness of the plates was adopted in accordance with the metal thickness of the investigated profiles: 2.0 and 1.5 mm.

Solution of the problem was based on the finite element method. For each plate a finite element shell model in Abaqus was created with 3 mm mesh size. The geometrical dimensions of the plates were set considering the condition of the plates' equality, each up to 200 mm. The thickness of the plates was adopted in accordance with the metal thickness of the investigated profiles: 2.0 and 1.5 mm.

Finite-element models of studied plates are shown in Figure 3.

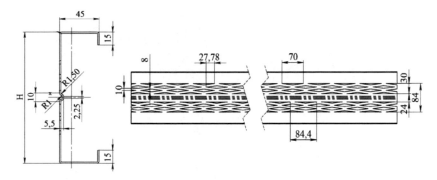

Figure 2.   Geometrical dimensions of RST-profile.

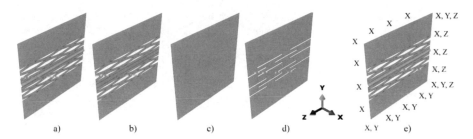

Figure 3.   FE models of plates: à) RST-profile, b) RST-profile no stiffener, c) solid, d) traditional thermoprofile, e) boundary conditions.

Table 1.   Eigenvalues for the tested plates.

| | First eigenvalue, kN | |
Type of plate	2.0 mm	1.5 mm
RST-profile	39.05	20.82
RST-profile, no stiffener	21.46	9.15
Solid	32.25	13.63
Traditional	21.86	9.30

The problem was solved in the general spatial pattern. Boundary conditions were chosen so that the compressed plates were simply supported around the perimeter. The boundary conditions of the plates are shown in the Figure 3e.

The load was set as uniformly distributed along the line on one side of the plates. The direction of the load has been determined so that the plates were in a compressed state.

For each plate linear perturbation analysis in Abaqus was conducted and the first eigenvalue was determined. Every plate experienced buckling with anticlastic surface. The results are shown in the Table 1.

According to the data from the Table 1 the web of traditional thermoprofile has about 30% less load-bearing capacity under compressive than the profile with the solid web.

Reticular-stretched perforations also reduce carrying capacity of the plates by about the same 30%. However, the presence of longitudinal intermediate stiffener increases its load-bearing capacity, making it twice as higher as for a profile without it and even 50% higher than for the solid web profile. This fact might probably mean that this type of perforations geometry might increase eigenvalues in case of local buckling for compressed sections.

Table 2. The results of FE analysis.

Profile	Length, mm	RST-profile Eigenvalue, kN	Eigen-mode	Traditional thermoprofile Eigenvalue, kN	Eigen-mode	Solid profile Eigenvalue, kN	Eigen-mode
TCc 150-45-1.5*	1000	39.59	distortional	24.56	local	46.53	local
	2000	35.82	distortional	24.33	local	47.51	local
	3000	20.94	global	22.99	global	24.00	global
TCc 150-45-2.0	1000	71.51	distortional	56.98	local	109.32	local
	2000	54.64	global	56.38	local	67.16	global
	3000	27.08	global	29.96	global	30.31	global
TCc 175-45-1.5	1000	34.59	distortional	23.75	local	44.76	local
	2000	31.21	distortional	23.48	local	44.24	local
	3000	24.40	global	23.48	local	28.32	global
TCc 175-45-2.0	1000	64.24	distortional	54.89	local	105.19	local
	2000	58.89	distortional	54.88	local	79.04	global
	3000	32.18	global	34.24	global	35.92	global
TCc 200-45-1.5	1000	31.06	distortional	22.20	local	39.10	local
	2000	27.52	distortional	21.89	local	38.57	local
	3000	25.15	global	21.90	local	29.24	global
TCc 200-45-2.0	1000	57.55	distortional	49.79	distortional	91.70	local
	2000	52.69	distortional	49.07	distortional	80.99	global
	3000	33.53	global	35.09	global	37.15	global
TCc 250-45-1.5	1000	21.72	distortional	18.09	distortional	27.66	local
	2000	19.71	distortional	17.55	distortional	27.26	local
	3000	19.10	distortional	17.11	distortional	26.51	local
TCc 250-45-2.0	1000	41.10	distortional	37.75	distortional	61.33	local
	2000	38.08	distortional	37.03	distortional	61.49	local
	3000	33.96	global	35.70	global	38.89	global

*Decoding TCc 150-45-1.5: 150 is web height, 45 is section width, 1.5 is steel thickness.

3   FINITE ELEMENT ANALYSIS OF COMPRESSED COLUMNS

In order to compare the buckling behavior of the new type of profile with the traditional thermo-profiles with longitudinal perforations and the solid web profiles FE buckling analysis of these profiles was conducted.

The analysis of the cold-formed members was performed for models of different lengths (1000, 2000 and 3000 mm), web height (150, 175, 200 and 250 mm) and steel thickness (1.5 and 2.0 mm). Geometrical properties of the tested profiles are presented in Figures 1, 2.

The mesh of the models was created using S4R shell finite elements with 4 nodes and reduced numerical integration. Boundary conditions are shown in Figure 4a. All the models were loaded with concentrated compressive forces acting in Z direction through the rigid bodies at the ends of each model.

For each column a linear perturbation analysis in Abaqus was performed in order to obtain their first eigenmodes and eigenvalues. The results of the FE analysis are presented in Table 2.

For three types of profiles completely different patterns of buckling were observed (Fig. 4). Solid web profiles had clear buckling behavior experiencing local buckling for short (1.0 and 2.0 m long) members and global buckling for the long ones. Traditional thermoprofiles had the similar buckling pattern for members with low web and the prevalence of distortional buckling for columns with large webs (no less than 200 mm).

Oppositely, instead of local buckling, for RST-profiles distortional buckling was the most probable buckling mode, particularly for 1 and 2 m long elements. Moreover, eigenmodes for RST-profiles were significantly higher (by 25–50% for local buckling and 2–20% for distortional

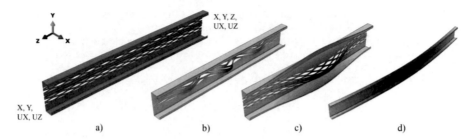

Figure 4.   a) boundary conditions of the tested members, b) local buckling (TCc 150-45-1.5), c) distortional buckling (TCc 175-45-1.5), d) global buckling (TCc 150-45-2.0).

buckling). This is probably due to the fact that the intermediate stiffener of RST-profile prevents the section web from local buckling increasing the eigenvalues. In addition, the probability of distortional buckling mode increases as web height becomes larger (RST-profile with 250 mm web height and 1.5 mm steel thickness experienced only distortional buckling). These results are in a good agreement with the results obtained from the Paragraph 2.

At the same time, 3.0 m long RST-profiles experienced mostly global buckling like their counterparts with longitudinal perforations and the profiles with solid webs. The results for RST-profiles were similar to those for traditional thermoprofiles and 10–14% lower than those for the profiles without any perforations.

4   OVERALL CONCLUSIONS

In this article FE analysis of compressed cold-formed steel C-sections was conducted. Three types of profiles actively used in modern construction were investigated: profile with solid web, thermoprofile with longitudinal perforations on web and relatively new thermoprofile with reticular-stretched perforations on web (RST-profile).

In the first step buckling behavior of a planar plate simply supported along the entire perimeter was considered in order to investigate web buckling of these types of profile. It was shown in Paragraph 2 and proved in Paragraph 3 that due to the intermediate longitudinal stiffener RST-profile has better local buckling performance comparing to traditional thermoprofile. Overall, the described approach might be quite effective for analyzing local buckling processes of various types of other cold-formed sections.

In case of large lengths (more than 3.0 m) RST-profile has 10–14% lower buckling forces than solid web profile and demonstrates almost the same buckling performance as and traditional thermoprofile.

At the same time, in case of short lengths (1.0–2.0 m) RST-profile experiences distortional buckling of the whole section instead of local buckling of web. According to the results of FE modeling RST-profile has significantly higher eigenvalues compared to thermoprofile with longitudinal perforations, although noticeably lower than profile without any perforations. This fact, in view of its lower production cost, makes the possibility of active application of RST-profile in construction feasible, instead of traditional thermoprofile.

REFERENCES

Al Ali, M. 2014. Compressed Thin-Walled Cold-Formed Steel Members with Closed Cross-Sections. *Advanced Materials Research* 969: 93–96.

Al Ali, M. et al. 2012. Thin-Walled Cold-Formed Compressed Steel Members and the Problem of Initial Imperfections. *Procedia Engineering* 40: 8–13.

Al Ali, M., Tomko, M. & Demjan, I. 2015. Experimental Investigation and Theoretical Analysis of Polystyrene Panels with Load Bearing Thin-Walled Cold-Formed Elements. *Applied Mechanics and Materials* In Print.

Belyy, G. & Serov, E. 2013. Particular Features And Approximate Estimation Of Steel Structures Service Life In Buildings And Facilities. *World Applied Sciences Journal* 23(13): 160–164.

Björk, T. & Saastamoinen, H. 2012. Capacity of CFRHS X-joints made of double-grade S420 steel. *Tubular Structures XIV – Proceedings of the 14th International Symposium on Tubular Structures, ISTS 2012*: 167–176.

Danilov, A. & Tusnina, O. 2014. The joints of cold-formed purlins. *Journal of Applied Engineering Science* 12(2): 153–158.

Garifullin, M., Trubina, D. & Vatin, N. 2015. Local buckling of cold-formed steel members with edge stiffened holes. *Applied Mechanics and Materials* 725–726: 697–702.

Heinilä, S., Björk, T. & Marquis, G. 2009. The influence of residual stresses on the fatigue strength of cold-formed structural tubes. *ASTM Special Technical Publication* 1508 STP: 200–215.

Heinisuo, M., Lahdenmaa, J. & Jokinen, T. 2014. Experimental research on modular thin-walled steel structures. *Proceedings of the 12th International Conference on Steel, Space and Composite Structures*: 85–97.

Lalin, V., Rybakov, V. & Sergey, A. 2014. The Finite Elements for Design of Frame of Thin-Walled Beams. *Applied Mechanics and Materials* 578–579: 858–863.

Nykänen, T. et al. 2014. Residual strength at −40°C of a precracked cold-formed rectangular hollow section made of ultra-high-strength steel – an engineering approach. *Fatigue & Fracture of Engineering Materials & Structures* 37(3): 325–334.

Rybakov, V. & Sergey, A. 2015. Mathematical Analogy Between Non-Uniform Torsion and Transverse Bending of Thin-Walled Open Section Beams. *Applied Mechanics and Materials* 725–726: 746–751.

Salminen, M. & Heinisuo, M. 2014. Numerical analysis of thin steel plates loaded in shear at non-uniform elevated temperatures. *Journal of Constructional Steel Research* 97: 105–113.

Salminen, M. & Heinisuo, M. 2011. Shear buckling and resistance of thin-walled steel plate at non-uniform elevated temperatures. *10th International Conference on Steel Space and Composite Structures*: 267–276.

Trubina, D. et al. 2014a. Effect of Constructional Measures on the Total and Local Loss Stability of the Thin-Walled Profile under Transverse Bending. *Applied Mechanics and Materials* 633–634: 982–990.

Trubina, D. et al. 2014b. Geometric Nonlinearity of the Thin-Walled Profile under Transverse Bending. *Applied Mechanics and Materials* 633–634: 1133–1139.

Tusnin, A. 2014. The Influence of Cross-Section Shape Changing on Work of Cold Formed Beam. *Advanced Materials Research* 1025–1026: 361–365.

Tusnin, A.R. 2010. Features of numerical calculation of designs from thin-walled bars of an open profile. *Industrial and Civil Engineering* 11: 60–63.

Tusnin, A.R. 2009. Finite element for numeric computation of structures of thin-walled open profile bars. *Metal constructions* 15(1): 73–78.

Tusnin, A.R. & Prokic, M. 2014a. Bearing capacity of steel I-sections under combined bending and torsion actions taking into account plastic deformations. *Journal of Applied Engineering Science* 12(3): 179–186.

Tusnin, A.R. & Prokic, M. 2014b. Behavior of symmetric steel I-sections under combined bending and torsion actions allowing for plastic deformations. *Magazine of Civil Engineering* 5(49): 44–53.

Tusnina, O. 2014a. A Finite Element Analysis of Cold-Formed Z-Purlins Supported by Sandwich Panels. *Applied Mechanics and Materials* 467: 398–403.

Tusnina, O. 2014b. An Influence of the Mesh Size on the Results of Finite Element Analysis of Z-Purlins Supported by Sandwich Panels. *Applied Mechanics and Materials* 475–476: 1483–1486.

Vatin, N. et al. 2015. Reticular-Stretched Thermo-Profile: Buckling of the Perforated Web as a Single Plate. *Applied Mechanics and Materials* 725–726: 722–727.

Vatin, N., Sinelnikov, A., et al. 2014. Simulation of Cold-Formed Steel Beams in Global and Distortional Buckling. *Applied Mechanics and Materials* 633–634: 1037–1041.

Vatin, N., Havula, J., et al. 2014. Thin-Walled Cross-Sections and their Joints: Tests and FEM-Modelling. *Advanced Materials Research* 945–949: 1211–1215.

Vatin, N., Nazmeeva, T. & Guslinscky, R. 2014. Problems of Cold-Bent Notched C-Shaped Profile Members. *Advanced Materials Research* 941–944: 1871–1875.

Vatin, N.I. & Sinelnikov, A.S. 2012. Long span footway bridges: coldformed steel cross-section. *Construction of Unique Buildings and Structures* 1: 47–53.

Veljkovic, M. & Johansson, B. 2008. Thin-walled steel columns with partially closed cross-section: Tests and computer simulations. *Journal of Constructional Steel Research* 64(7–8): 816–821.

Zhmarin, E.N. 2012. International association of light-gauge steel construction. *Construction of Unique Buildings and Structures* 2: 27–30.

Advances and Trends in Engineering Sciences and Technologies – Al Ali & Platko (Eds)
© 2016 Taylor & Francis Group, London, ISBN: 978-1-138-02907-1

The influence of inaccuracies of soil characteristics on the internal forces in the retaining wall

E. Hrubesova, J. Marsalek & J. Holis
Faculty of Civil Engineering, VSB-Technical University Ostrava, Czech Republic

ABSTRACT: The paper presents the sensitivity analysis of influence of inaccuracies related to strength and deformational soil characteristics on the magnitude of shear forces and bending moments in the retaining wall. The sensitivity analysis is based on the following steps: parametric study (based on the method of dependent earth pressure) to the determination of response surfaces corresponding to shear forces and bending moments in the structure, application of gradient method to the evaluation of global and local sensitivity coefficients, formulation of conclusions related to the influence of input parameters inaccuracies.

1 INTRODUCTION

Safe and economical design of sheeting structure is one of the basic tasks to the stabilization of construction pit. Design of retaining structure is dependent on many factors, to the most significant, in addition to the structure itself, belong geological and hydrogeological conditions at the site and strength and deformation characteristics of the subsoil (Terzaghi et al. 1967). This characteristics of the soil, together with the material and geometric characteristics of designed retaining structure, then determine both stress and deformation behavior of the whole cooperating system. Earth pressure, resulted from the interaction between the soil and structures, is generally dependent on the structure's deformation, which is determined primarily by soil characteristics, type of retaining structures, its stiffness and way of installation in the subsoil. To the lowest reliable characteristics, determining the behavior of retaining structure, belong soil characteristics. The uncertainties of these characteristics follow natural variability and complexity of geological environment, but also our limited level of knowledge about processes in rock mass and other factors. The problem of an uncertainty in determining of the individual characteristics of the rock mass was, among other things, implemented in the partial factors in European standards EC7. The degree of reliability of the input data resulted into the reliability of the calculation results (response of the model), but the response sensitivity of the model, with respect to the variation of certain input parameters, can be generally different.

To the evaluation of internal forces in the retaining wall various calculation methods can be used. Many calculations are based on the classical earth pressure theory (Clough et al. 1991, Yap et al. 2012, Masopust 2013, Vaneckova et al. 20011), other authors introduce to the analysis of soil-structure interaction the application of numerical methods (for example finite element methods (Clough et al. 1971, Potts et al. 1999 etc.).

2 CHARACTERISTICS OF ANALYZED RETAINING WALL

The content of this paper is focused on the sensitivity analysis of the magnitude of the internal forces in the retaining structure with respect to the deformation characteristics of the surrounding rock mass (oedometer module, Poisson's ratio), its strength characteristics (cohesion, friction angle)

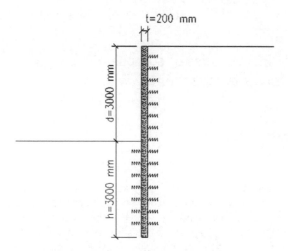

Figure 1.   The basic computation scheme (method of dependent pressure).

and unit weight. Embedded concrete retaining wall with its length of 6 m and thickness 0.2 m is assumed, the wall stabilizes the 3 m deep pit (the depth of embedment is 3 m). Retaining concrete wall is made from the concrete C20/25, zero friction angle δ between the structure and the rock environment is considered in presented calculation.

3   METHODOLOGY OF SENSITIVITY ANALYSIS

To perform a sensitivity analysis of the response of retaining structure, the gradient method is used. This method determines the sensitivity coefficients as partial derivatives of the response function with respect to the input parameters (Saltelli et al. 2000, Saltelli et al. 2004).

Determination of internal forces (both shear forces and bending moments) in retaining structure was performed applying the method of dependent pressures (Fig. 1). This method is, among other things, also implemented in the software system GEO5 produced by the Czech software company FINE (Fine 2011). This method assumes the dependence of the earth pressure on the deformation arising inside the retaining wall, this structure is considered as an elasto-plastic beam element acting on the subsoil (ideal elasto-plastic Winkler material). Earth pressures are not constant, but they can take values within the interval designated by two extreme limits – active resp. passive pressures. In the elastic region deformation of ideal elasto-plastic Winkler materials (and corresponding earth pressure) is characterized by the modulus of subgrade reaction. If the range of the elastic deformation is exceeded, the material behaves as an ideal plastic matter. Corresponding values of earth pressures are then constant and are equal to active resp. passive earth pressure.

Modulus of subgrade reaction is not set out clearly in the ordinary geotechnical survey. It is a function of the subsoil characteristics as well as the analyzed retaining structure and to the determination of this input parameter the various computational methods (according to Schmitt, according to Menard, according Chadeisson etc.) can be used.

For the purposes of sensitivity analysis, which results are presented in this paper, the Schmitt method (Schmitt, 1995) is used. This method evaluates the module of subgrade reaction k_h based on the bending stiffness (EI) and oedometer soil modulus E_{oed} according to the relation:

$$k_h = 2.1 \left(\frac{E_{oed}^{\frac{4}{3}}}{EI^{\frac{1}{3}}} \right) \tag{1}$$

To evaluate the sensitivity of the behavior of the structure with respect to the variation of the input characteristics, the gradient method was used. The sensitivity coefficients k_c are based on the evaluation of the gradient of corresponding response function Y with respect to each input characteristic of the soil – oedometer modulus E_{oed}, unit weight γ, shear strength characteristics (friction angle φ and cohesion c), Poisson's ratio μ.

The sensitivity coefficients are therefore given by the following partial derivatives of the response function $Y(E_{oed}, \gamma, \varphi, c, \mu)$:

$$kc_{E_{oed}} = \frac{\partial Y}{\partial E_{oed}}, kc_\gamma = \frac{\partial Y}{\partial \gamma}, kc_\varphi = \frac{\partial Y}{\partial \varphi}, kc_c = \frac{\partial Y}{\partial c}, kc_\mu = \frac{\partial Y}{\partial \mu} \qquad (2)$$

The shape of the sensitivity coefficients depends on the shape of the response function-assuming the linear shape of the response function, the sensitivity coefficients are constant and indicate the global (absolute) sensitivity of the response with respect to the variation of the input parameters. But the determined absolute sensitivity coefficient may not reflect the actual, local sensitivity response, which varies because of variation of the certain reference input material characteristics $(E_{oed}^{(0)}, \gamma^{(0)}, \phi^{(0)}, c^{(0)}, \mu^{(0)})$.

For example, if the values of the real input characteristics have different order, the established global sensitivity coefficients could also lead to entirely erroneous conclusions. For this reason, it is also considered the relative variation of the input parameter, the variation of response ΔY due to the variation of the relative values of the input soil characteristics is given by the modified equation:

$$\Delta Y = kc_{E_{oed}}^* \frac{\Delta E_{oed}}{E_{oed}^{(0)}} + kc_\gamma^* \frac{\Delta \gamma}{\gamma^{(0)}} + kc_\varphi^* \frac{\Delta \varphi}{\varphi^{(0)}} + kc_c^* \frac{\Delta c}{c^{(0)}} + kc_\mu^* \frac{\Delta \mu}{\mu^{(0)}} \qquad (3)$$

where the local response sensitivity coefficients (marked with an asterisk) are given by relations:

$$
\begin{aligned}
kc_{E_{oed}}^* &= kc_{E_{oed}} * E_{oed}^{(0)} \\
kc_\gamma^* &= kc_\gamma * \gamma^{(0)} \\
kc_\varphi^* &= kc_\varphi * \varphi^{(0)} \\
kc_c^* &= kc_c * c^{(0)} \\
kc_\mu^* &= kc_\mu * \mu^{(0)}
\end{aligned}
\qquad (4)
$$

Using these specified local sensitivity coefficients (LSC) the influence of uncertainties of the input characteristics on the response Y is taken into account.

To determine the response functions, corresponding to the shear forces and bending moments in the retaining structure, the software GEO5 was used. There were performed total 65 variants of parametric calculations with variable soil parameters – oedometric modulus, Poisson's ratio, friction angle, cohesion and unit weight. Oedometric modulus $Eoed$ varied from 10 to 70 MPa, unit weight γ was in the range of 17 to 22 kN/m³, angle of internal fiction ϕ was in the range of 17–37°, cohesion c in the range of 5–19.53 kPa, Poisson's ratio μ in the range of 0.21 to 0.4. Response function for maximum shear force (i = 1) and the maximum bending moments (i = 2) in the structure has been assumed in the shape of a linear polynomial:

$$\alpha_1^i + \alpha_2^i * E_{oed} + \alpha_3^i * \gamma + \alpha_4^i * \varphi + \alpha_5^i * c + \alpha_6^i \mu \ (i=1,2) \qquad (5)$$

Six unknown coefficients of the polynomial (in the case of linear-shaped response surface there are equal to the global sensitivity coefficients) were determined on the basis of the parametric calculations and, subsequently, on the solution of the over determined system of linear equations

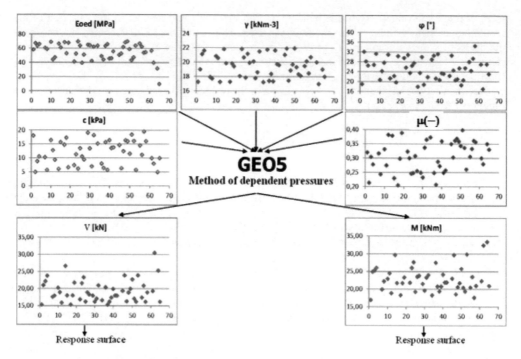

Figure 2. Schematic procedure for the determination of the response surface for maximum shear forces and bending moments inside the retaining structure.

using the least squares method (Fig. 2). Based on the above mentioned methodology the following global sensitivity coefficients for the maximum shear forces V and the maximum bending moment M were evaluated:

$$\nabla V = \left(\frac{\delta V}{\delta E_{oed}}, \frac{\delta V}{\delta \gamma}, \frac{\delta V}{\delta \varphi}, \frac{\delta V}{\delta c}, \frac{\delta V}{\delta \mu} \right) = \left(\alpha_2^1, \alpha_3^1, \alpha_4^1, \alpha_5^1, \alpha_6^1 \right) =$$

$$= (-0.00002947, 1.1883, 0.1378, -0.6432, 6.4544)$$

(6)

$$\nabla M = \left(\frac{\delta M}{\delta E_{oed}}, \frac{\delta M}{\delta \gamma}, \frac{\delta M}{\delta \varphi}, \frac{\delta M}{\delta c}, \frac{\delta M}{\delta \mu} \right) = \left(\alpha_2^2, \alpha_3^2, \alpha_4^2, \alpha_5^2, \alpha_6^2 \right) =$$

$$= (-0.00009973; 1.4477; 0.3546; -0.6166; -4.76604)$$

(7)

Sensitivity analysis is also performed with respect to changes of the reference values for selected input characteristics – $E_{oed}^{(0)} = 40\,\mathrm{MPa}$, $\gamma^{(0)} = 19.5\,\mathrm{kN/m^3}$, $\varphi^{(0)} = 27°$, $c^{(0)} = 12.265\,\mathrm{kPa}$, $\mu^{(0)} = 0.305$. If we evaluate the relative response sensitivity coefficients for the maximum shear force and the maximum bending moments with respect to the selected reference input characteristics of the geological environment, we get, following the equation (3), the local sensitivity coefficients (LSC) corresponding to the maximum shear force (V) and the maximum bending moment (M) in the structure (Table 1).

In the Figure 3 and 4 are shown the relative changes in the sensitivity of the maximum shear forces $\Delta V / V^{(0)}$ and of the maximum bending moments $\Delta M / M^{(0)}$ ($V^{(0)}$ resp. $M^{(0)}$ are the values of the maximum shear forces resp. bending moments corresponding to the reference input characteristics – in this task $V^{(0)} = 20.85\,\mathrm{kN}$, $M^{(0)} = 25.855\,\mathrm{kNm}$) due to the each input parameter. It is assumed, that only one input parameters is varied under assumption of constant values of the rest of them.

Table 1. Table of the local sensitivity coefficients (LSC) for the selected reference input characteristics.

	$k^*_{c\,Eoed}$	$k^*_{c\,\gamma}$	$k^*_{c\,\varphi}$	$k^*_{c\,c}$	$k^*_{c\,\mu}$
LSC for V	1.172	23.1721	3.721	7.88	1.969
Sensitivity order for V*	5	1	3	2	4
LSC for M	3.989	28.231	9.57	7.56	1.453
Sensitivity order for M*	4	1	2	3	5

*The higher the number of sensitivity order, the lower the sensitivity of the response to a given parameter.

Graph of the relative changes in the maximum shear force due to the relative changes in input reference material characteristics

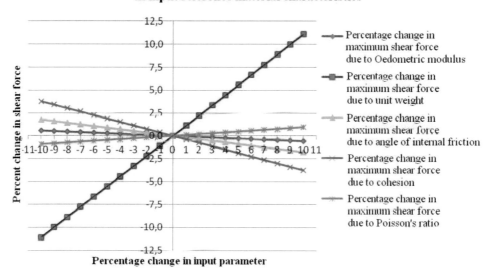

Figure 3. Graph of the relative changes in the maximum shear force due to the relative changes in input reference material characteristics.

4 CONCLUSION

The presented results of sensitivity analysis allow to quantify the response sensitivity of the model response of the retaining structure for each input characteristics of the geological environment and to specify the requirements for the reliability of these input characteristics.

The performed sensitivity analysis showed:

1. the highest sensitivity of shear forces and bending moments has a unit weight, it is followed by the strength characteristics and the low sensitivity indicate the deformational characteristics,
2. in case of shear forces the cohesion is more indicative parameter of sensitivity in comparison with the angle of internal friction,
3. in case of bending moments the friction angle is more indicative parameter of sensitivity in comparison with the cohesion,
4. performed sensitivity analysis showed, among other things, the need to increase the reliability of determination of unit weight of soil also (the most significant local sensitivity coefficient).

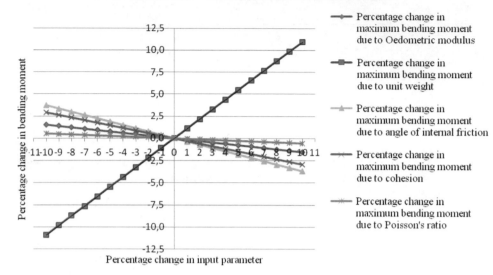

Graph of the relative changes in the maximum bending moment due to the relative changes in input reference material characteristics

Figure 4.   Graph of the relative changes in the maximum bending moment due to the relative changes in input reference material characteristics.

ACKNOWLEDGMENT

The paper was prepared with the support of the Competence Centre of the Technology Agency of the Czech Republic (TAČR) within the project Centre for Effective and Sustainable Transport Infrastructure (CESTI), project number TE01020168.

REFERENCES

Clough, G.W. & Duncan, J.M. 1971. *Finite element analysis of retaining wall behavior*. Journal of Soil Mechanics and Foundation Engineering. ASCE 97-SM12. pp. 657–673.

Clough, G.W. & Duncan, J.M. 1991. Earth pressures. In: H.-Y. Fanf (eds) *Foundation Engineering Handbook*. Chapman Hall, New York. pp. 223–235.

Fine, Ltd. 2011. *GEO5 Sheeting Check – User guide*. http://www.finesoftware.eu/ Prague, Czech Republic.

Masopust, J. 2013. *Navrhovanizakladovych a pazicichkonstrukci*. (in Czech only). CKAIT.

Potts, D.M. & Zdravkovic, L. 1999. *Finite element analysis in geotechnical engineering: theory*. Thomas Telford. London.

Saltelli, A., Chan, K. & Scott, M. 2000. *Sensitivity Analysis. Wiley Series in Probability and Statistics*. New York: John Wiley and Sons.

Saltelli, A., Tarantola, S., Campolongo, F. & Ratto, M. 2004. *Sensitivity Analysis in Practice: A Guide to Assessing Scientific Models*. New York: John Wiley and Sons.

Schmitt. P. 1995. *Estimating the coefficient of subgrade reaction for diaphragm wall and sheet pile wall design*. Revue Française de Géotechnique, N. 71, 2° trimestre 1995. pp. 3–10.

Terzaghi, K. & Peck, R.B. 1967. *Soil Mechanics in Engineering Practice*. New York: John Wiley & Sons.

Vaneckova, V., Laurin, J. & Pruska, J. 2011. *Sheeting Wall Analysis by the Method of Dependent Pressures*. Geotec Hanoi.

Yap, S.P., Salman, F.A. & Shirazi, S.M. 2012. Comparative study of different theories on active earth pressure. *Journal of Central South University*. 19(10). pp. 2933–2939.

Advances and Trends in Engineering Sciences and Technologies – Al Ali & Platko (Eds)
© 2016 Taylor & Francis Group, London, ISBN: 978-1-138-02907-1

Fatigue loading tests of new bolt assembly joints for perspective temporary steel railway bridges

M. Karmazínová, M. Štrba & M. Pilgr
Faculty of Civil Engineering, Brno University of Technology, Brno, Czech Republic

ABSTRACT: The paper deals with the problems of the actual behavior, failure mechanism and load-carrying capacity of the special bolt connection developed and intended for the assembly connections of truss main girders of perspective temporary steel railway bridges. Within the framework of the problem solution, several types of structural details of assembly joints were considered as the conceptual structural design. Following the preliminary evaluation of their advantages or disadvantages, two basic structural configurations – so-called "tooth" and "splice-plate" connections, have been chosen for the subsequent detailed investigation, which is mainly based on the experimental verification of the actual behaviour, failure mechanism and corresponding load-carrying capacity of the connection. This paper is only focused on the cyclic loading (fatigue) test results of "splice-plate" connections and their evaluation already finished. Simultaneously with the fatigue tests, the static loading tests have also been realized, however these ones are not the subject of this paper.

1 INTRODUCTION

Recently, the grant project titled "Research of necessary operating parameters of perspective railway steel temporary bridges" has been solved on the authors' workplace. Within the framework of this research project, one of the important and significant problems is to develop and verify the new type of the bolt assembly connections. The connection has to comply with the set of requirements, that means not only the simple and fast assembly, but also the wear hardness or the possibility of the replacement of the parts worn after repeated use, as well as the reliability in the ultimate limit states and serviceability limit states, including structural reliability in case of fatigue effects. For describing and verifying the actual behaviour of a new bolt assembly connection in the loading process of real structure, the static loading tests (Karmazínová & Štrba 2014, Štrba 2013), as well as the fatigue loading tests, including subsequent elaboration and evaluation of monitored and measured strength quantities, have been realized.

2 TEST SPECIMENS

2.1 *Principle of investigated assembly connections*

The mentioned type of the assembly connections, which has been investigated, consists of two end steel plates forming a single unit together with connected member. Both connecting plates have a pair of narrow opposite flanges and three holes for bolts. The connection also has two splice plates, likewise with narrow flanges and corresponding number of the bolt holes, and six fastening bolts. The behaviour is based on the principle, that at first the tensile force is transferred from one member to the splice plates by mutually lock of their flanges, and then, by splice plates through the gap between end plates to the member on the opposite side of the connection. Fastening bolts prevent the opening or separating of splice plates. The principal scheme of the investigated so-called splice plate connection is shown in Fig. 1. Geometrical parameters of splice plate connection are as follows: the thickness of splice plates is 12 mm and the thickness of end plates is 10 mm.

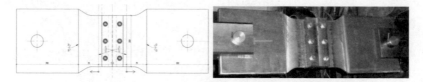

Figure 1. Splice plate connection: scheme of test specimen; real test specimen fixed in testing equipment.

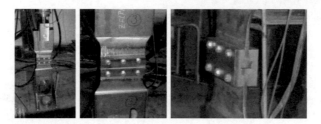

Figure 2. Realization of cyclic loading tests.

2.2 *Configuration of test specimens*

The configuration and dimensions of created connections has been taken as a base for the shape and geometry of the test specimens, which have been experimentally verified when subjected to the cyclic loading effects. The resulting length of the test specimen was about 700 mm and the width was 200 mm in its middle part, while the width on the ends has been decreased about to a half, i.e. 100 mm, because of fixing into the testing equipment. The material of considered assembly connections is steel of S 235 grade. The scheme of the whole test specimen and the real shape and configuration of the test specimen, as it has been produced, are shown in Fig. 1.

3 CYCLIC LOADING TESTS

3.1 *Realization and progress of cyclic tests*

The tests of the "splice-plates" connections subjected to the fatigue effects, caused by the cyclic repeated loading, have been realized in the testing room of the Institute of Metal and Timber Structures (Karmazínová 2013, Karmazínová & Simon 2013, Štrba 2013). In total, 12 test specimens have been verified. For the cyclic loading, the electro-hydraulic testing equipment (hydraulic cylinder) with the maximal capacity of 500 kN, and 400 kN respectively, has been utilized. The illustrations of cyclic loading tests are demonstrated by Fig. 2, where the testing equipment and structural details of the connection are presented.

3.2 *Results of cyclic tests*

During the process of fatigue tests the specimens have been subjected to the repeated cyclic loading effects given by the tensile force, which has been being changed in the range given by the value of the minimum force N_{min}, from which the force was cyclically changed up to the value of the maximum force N_{max}. Then, the loading force amplitude ΔN is given by the difference $\Delta N = N_{max} - N_{min}$. The load frequency has been determined as $f = 5$ Hz. The cyclic loading has been realized up to the failure of test specimens, which corresponds to the maximum number of cycles n, reached. Load parameters, i.e. the minimal level of the loading force N_{min}, the maximal level of the loading force N_{max} and the loading force amplitude ΔN are listed in Table 1, from which the failure mechanisms occurred in the case of particular test specimens are evident.

The first test has been intended for the calibration of lower and upper levels of the loading force and for the verification of the testing equipment and load regime. In this case a small cycle number

Table 1. Overview of all cyclic tests: loading parameters, reached number of cycles, failure mechanisms.

Test	Lower level of loading force N_{min} [kN]	Upper level of loading force N_{max} [kN]	Number of cycles n [–]	Failure mechanism
1	10	232	–	test sample
2a	10	491	53 200	plate fracture
2b	10	490	69 120	in place of
3	10	489	88 600	bolt holes
4a	13	236	1 990 000	plate fracture in
4b	13	236	2 425 000	transition to head
5a	13	243	2 835 000	bolt fracture
5b	13	243	5 321 000	plate fracture in place of bolt holes
6a	12	343	468 000	bolt fracture
6b	12	343	1 014 000	plate fracture
7a	12	344	774 000	in place of
7b	11	342	609 550	bolt holes
8	14	346	503 000	cylinder damage
9	11	239	780 000	weld
10	11	239	630 000	fracture
11	12	237	4 822 000	plate fracture in
12	12	237	6 280 000	place of bolt holes

Figure 3. Typical failure mechanism: plate fracture in place of bolt holes.

has been realized, so the failure has not been monitored and investigated. In the next tests, the most frequent failure mechanism was the plate fracture in the place of bolt holes (see Fig. 3), which occurred in 9 cases. In further cases other failure modes occurred (see Fig. 4): fracture of the plate in transition to the head, bolt fracture, weld fracture.

Unfortunately, in some tests (not included in Table 1) the fracture of testing preparations or even the damage of testing equipment occurred (see Fig. 4). In two cases the fracture of connecting plates in the place of connecting pin occurred. In one special untypical case, the hydraulic cylinder has been damaged. These problems have been caused by the continuous repetition of high-cyclic loading during fatigue tests of particular test specimens. Therefore it was necessary to repair test preparations and mainly testing equipment, so the total tests time was much longer than assumed originally.

Table 2 lists the overview of parameters needed for the test results evaluation that means the minimal and maximal values of stresses in the cross-section failed – σ_{min} and σ_{max}, the corresponding stress amplitude – $\Delta\sigma$, and the plate cross-section area – A_{net} in the place of the fracture. The plate cross-section area A_{net} has been taken as the area in the place of bolt holes.

4 EVALUATION OF CYCLIC TESTS RESULTS

The evaluation of cyclic loading test results is oriented to the study of the fatigue behavior and the determination of the fatigue strength of the splice plate connections, on the base of the test

Figure 4. Other typical failure mechanisms: plate fracture in transition to head; weld fracture. Untypical failure mechanisms: fracture of connecting plate in place of connecting pin; damage of hydraulic cylinder.

Table 2. Cyclic tests results: amplitude of loading force and amplitude of stress.

Test	Amplitude of loading force ΔN [kN]	Logarithm of cycle number log n [–]	Cross-section area A_{net} [mm^2]	Amplitude of stress $\Delta\sigma$ [MPa]
2a	481	4.723		293.3
2b	481	4.840		292.7
3	479	4.947		292.1
5b	229	6.726		139.6
6b	331	6.006	1 640	201.8
7a	332	5.889		202.4
7b	331	5.785		201.8
11	227	6.683		138.4
12	227	6.798		138.4

results mentioned in Table 2. For these reasons the following methods have been applied: the regression analysis for the study of the fatigue behaviour, and probabilistic, respectively reliability analysis for the determination of the fatigue strength (Karmazínová & Melcher 2013, Melcher 1997, Karmazínová 2012, Eurocode 1, 2011).

4.1 Fatigue behaviour

Basic principles of fatigue behavior of the structural detail subjected to many-times repeated cyclic loading are given by the curve of fatigue strength, expressing the dependence of the stress amplitude on the number of loading cycles. To express this relationship, it may be used the logarithmic equation describing Wöhler's curve, which can be transformed into the straight line, if the logarithmic scale of n-axis is used. Plotting the fatigue test results to the graph, the discrete points can be obtained, and then substituted by the straight line based on the linear regression using the least squares method. In the first phase of the evaluation, it therefore has been done tentative determination of the stress amplitude, which corresponds to the reference number of 2 million loading cycles. This value can be taken as the first approximation of the permissible stress amplitude, which can be considered as a base for the determination of detail category. This evaluation arising from the mean value determined by the least squares method does not include more detailed statistical, nor probabilistic evaluation, which should be the base for the determination of the permissible stress amplitude with the required reliability, given by the failure probability. The probabilistic evaluation is mentioned below.

The dependence of the stress amplitude on the number of cycles derived using regression analysis is drawn in the graphs in Fig. 5 (plotted alternatively on linear and logarithmic scales). From the equation of the dependence of stress amplitude on the number of loading cycles

$$\Delta\sigma = 80.463\,(\log n) + 679.65\,,\tag{1}$$

the stress amplitude corresponding to the reference number of 2 million loading cycles (log n = 6.301) has been derived as $\Delta\sigma = 172.7$ MPa.

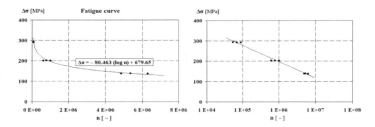

Figure 5.   Dependence of stress amplitude on number of cycles: linear scale; logarithmic scale.

4.2   *Fatigue strength*

The determination of fatigue strength has been done according to the methodology elaborated in the informative annex L of the Czech Standard ČSN 73 1401. The base for the evaluation is logarithmic dependence of the number of cycles realized up to the detail failure, on the stress amplitude, which is determined from the loading force, respectively the difference between extremes of loading force. The essence of the evaluation is the substitution of discrete points obtained plotting the results using the straight line based on linear regression, and subsequent statistical evaluation, aimed to the determination of the difference of extremes of the loading force of one cycle ΔN_C on the fatigue limit, when constant amplitude, defined for 2 million cycles.

The procedure for the determination of the differences of extremes of the force of one load cycle contains the gradual determination of the following quantities and parameters: parameters α, β of the regression line for the failure probability of 50%; amplitude of the force ΔN_P for $n_C = 2 \cdot 10^6$; left-prediction limit n_P (respectively log n_P), the difference of loading force extremes of one cycle ΔN on the fatigue limit for 2 million cycles.

The regression line, for which the independent variable is the logarithm of the difference of loading force extremes of one cycle ($x_i = \log \Delta N_i$), and the dependent variable is the logarithm of the number of cycles ($y_i = \log n_i$), is expressed by the equation

$$y = \alpha + \beta x \,, \qquad (2)$$

$$\text{where} \qquad \beta = S_{xy} / S_{xx}, \quad \alpha = (\Sigma y_i - \beta \Sigma x_i) / p \,, \qquad (3)$$

$$S_{xx} = \Sigma(x_i^2) - (\Sigma x_i^2)/p \,, \; S_{yy} = \Sigma(y_i^2) - (\Sigma y_i^2)/p \,, \; S_{xy} = \Sigma(x_i y_i) - [(\Sigma x_i)(\Sigma y_i)]/p \qquad (4)$$

and p is the number of evaluated tests. For the number of cycles $n_C = 2 \cdot 10^6$, the difference of loading extremes on the regression line and the left-prediction limit for the difference ΔF_P are

$$\Delta N_P = (2 \cdot 10^6 / 10^\alpha)^{1/\beta} \,, \; \log N_P = \log(2 \cdot 10^6) - t \cdot s_R \cdot \sqrt{f} \,, \qquad (5)$$

where t is γ – critical value of Student distribution $t(v, \gamma)$ for the degree of freedom $v = p - 2$ and the probability of $\gamma = 0.05$; s_R is the standard deviation according to the formula

$$s_R = [1/(p-2) \cdot (S_{yy} - \beta \cdot S_{zy})]^{1/2}, \; f = 1 + 1/p + (\log \Delta n_P - \Sigma x_i / p)^2 / S_{xx}. \qquad (6)$$

The difference of loading force extremes of one cycle on fatigue limit for 2 million cycles is

$$\Delta N_C = \Delta N_P (2 \cdot 10^6 / n_P)^{1/\beta} \,. \qquad (7)$$

The values of experimental results, which have been obtained from 9 fatigue tests, are summarily listed in Table 2. From given values, the parameters above have been calculated. On the regression line, for the number of cycles $n_C = 2 \cdot 10^6$ the difference of load force extremes of one cycle is $\Delta N_P = 274.1$ kN. The difference of load force extremes of one cycle for the determination of the fatigue limit, when constant amplitude, for 2 million load cycles is then $\Delta N_C = 251.8$ kN. The stress amplitude on the fatigue limit determined for the effective cross-section area is $\Delta \sigma_C = 153.5$ MPa. For the investigated detail it is possible to assign the detail category 140, in accordance with the spectrum of fatigue categories given by the European Standard EN 1993-1-9 (Eurocode 3, 2012).

5 CONCLUSIONS

Some partial results are related to the specific structural detail, and therefore significantly influenced by this, have already been mentioned in the text above. From particular test results it is seen that the number of load cycles reached is in the range from tens of thousands up to several millions. The aim of the evaluation of test results was to investigate the actual behaviour of the tested structural details, that means to monitor the types of failure mechanisms occurred at the moment of the fracture of test specimens, i.e. at the reaching final number of the loading cycles, and finally, to determine the fatigue category of the developed structural detail. As expected, the most frequent failure mechanism is the fracture of plates in the place of bolt holes, because the connection has been made so intentionally. The fatigue category of the investigated structural detail has been determined using the procedures mentioned above and the results are as follows.

The permissible stress amplitude determined from Wöhler's curve for the reference number of 2 million load cycles, without consideration of the uncertainties caused by small test number, has been calculated as $\Delta\sigma = 172.7$ MPa. Using the methodology described in the paragraph IV. B, taking into account test numbers and considering the reliability with the failure probability of 50%, permissible stress amplitude for 2 millions load cycles has been calculated $\Delta\sigma_C = 153.5$ MPa, which is about by 13% lower than the value $\Delta\sigma$. From there, the fatigue category of the detail can be derived as 140, which correspond with the next lower value given usually in the appropriate European Standards (Eurocode 3-1-1 2008, Eurocode 3-1-9 2012). In conclusion, the performed fatigue tests have shown that the proposed type of assembly connections has the sufficient fatigue resistance, even significantly higher than many other commonly used details.

ACKNOWLEDGEMENT

The paper was elaborated within the research project No. LO1408 "AdMaS UP – Advanced Materials, Structures and Technologies", within the "National Sustainability Programme I".

REFERENCES

ČSN 73 1401 *Design of Steel Structures*, Czech Technical Institute for Normalization, Prague, 1998.
Eurocode 1: EN 1990 *Basis of Structural Design: Annex D – Design assisted by testing* (2nd Edition), CEN Brussels, 2011.
Eurocode 3: EN 1993-1-1, *Design of Steel Structures*, Part 1-1: *General Rules and Rules for Buildings*, CEN Brussels, 2008.
Eurocode 3: EN 1993-1-9 *Design of Steel Structures*, Part 1-9: *Fatigue*, CEN Brussels, 2012.
Karmazínová, M. & Štrba, M. 2014. Static loading tests of new type of bolt assembly connection developed for perspective temporary steel railway bridges, *Applied Mechanics and Materials*, Trans Tech Publications: Zurich, Vol. 590, pp. 331–335. ISSN 1660-9336.
Karmazínová, M. 2013. Fatigue Tests of Assembly Joints of Truss Main Girders of Temporary Footbridge for Pedestrians and Cyclists, *Applied Mechanics and Materials*, Trans Tech Publications: Zurich, Vol. 405–408, pp. 1598–1601. ISSN 1660-9336.
Karmazínová, M. & Simon, P. 2013. Fatigue tests of assembly joints of truss main girders of newly developed temporary footbridges, *Intl. Jl. of Mechanics*, Vol. 7, Issue 4, pp. 475–483. ISSN 1998-4448.
Karmazínová, M. & Melcher, J. 2013. Material testing and evaluation of steel mechanical properties for classification of steel grade of existing civil engineering structure, *Advanced Materials Research*, Vol. 651, pp. 274–279. ISSN 1022-6680.
Karmazínová, M. 2012. Design assisted by testing – a powerful tool for the evaluation of material properties and design resistances from test results, *International Journal of Mathematical Models and Methods in Applied Sciences*, Vol. 6, No. 2, pp. 376–385. ISSN 1998-0140.
Melcher, J. 1997. Full-Scale Testing of Steel and Timber Structures: Examples and Experience, In *Structural Assessment – The Role of Large and Full Scale Testing*, E & FN SPON, London, pp. 301–308.
Štrba, M. 2013. On the problems of testing methodology used in case of the temporary steel through truss footbridge development, *Intl. Jl. of Mechanics*, Issue 2, Vol. 7, pp. 73–80. ISSN 1998-4448.

Advances and Trends in Engineering Sciences and Technologies – Al Ali & Platko (Eds)
© 2016 Taylor & Francis Group, London, ISBN: 978-1-138-02907-1

Analysis of the effectiveness of using composite materials in bridge construction

K.A. Kokoreva & N.D. Belyaev
St. Petersburg State Polytechnical University, St. Petersburg, Russia

ABSTRACT: The developed infrastructure of any country is impossible without construction of reliable and durable bridges. As the world practice shows that the real durability of the bridges, which were built from the common used materials in bridge construction (wood, metal and reinforced concrete) significantly decreases in modern conditions. As a result, when the service life of the bridge can be calculated to one hundred years, in fact they can be operated normally not more than 50 years, and their need for major repairs or strengthening occurs after 30 years. Such issues are relevant to the entire world bridge engineering, force to search for effective method to increase the longevity of bridges. The solution is wider application of modern composite materials instead of wood, concrete and metal. The article presents a comparative analysis of using traditional and composite materials in bridge design.

1   INTRODUCTION

Currently about 70% of bridges are reinforced concrete structures. The undoubted benefits of such structures are their strength, the acceptable production cost, availability of well-established technologies and quality standards. The disadvantages include the need for regular maintenance and expensive repair work. Due to the influence of hostile environment, extreme temperatures, exposure to chemicals and high-intensity operation actual service life of reinforced concrete bridges is not more than 40–50 years (Design Guide 2002). It is obvious that to solve the problem of premature wear of bridge structures, it is necessary to increase corrosive, chemical and temperature resistance of the used materials (Ponomarev & Mospan 2011, Kokoreva et al. 2015, Kolgushkin & Belyaev 2007 a, b). Polymer composites can meet these requirements (Kokoreva et al. 2014, Kokoreva &Yaleshev 2014). Composite materials are widely used in construction (Kearns & Shambaugh 2002, Manchado et al. 2005, Thwe & Liao 2000, Blaznov et al. 2006, Loktev et al. 2010).

The main tasks for the production in Russia in the current economic conditions are the following: to reduce production cycles; to improve product quality; to cut down manufacturing costs (Ptuhina et al. 2014). The paper substantiates the possibility of using ultra modern composite materials to meet these challenges in bridge construction.

2   BRIDGES FROM TRADITIONAL MATERIALS

From the beginning to the middle of the twentieth century massive and buried abutments were used in bridge construction. Cumbersome, resource-demanding and labor-consuming designs required the sophisticated production technology works: the device of pits and their fastenings, drainage and insulation concreting in the cold season. Conical designs, in which the bearing elements of foundations (a pile, a rack, a frame, etc.) were in cone soil, became the alternative constructional decision.

At the same time, with the advent of a cone, additional bridge spans for laying the overlapping length of the cone were needed. For small, medium and large bridges the cost of the cone buried abutment device ranges from 20 to 40% of the estimated cost of the bridge. Together with additional spans two additional supports and mounting cones are required.

Thus, negation of old, bulky and time-consuming designs with a new tapered buried abutment brought in a bridge engineering a number of essential drawbacks: elongation of the bridge, no unloading abutment bearing elements, the device of a cone and its fastening (El Sarraf et al. 2015).

3 NEWEST COMPOSITE MATERIALS IN BRIDGE CONSTRUCTION

Bridges are exposed not only to the different loads (static, dynamic, cyclic) and temperatures, but also aggressive environments. Besides, in case of the long use of constructions, the aging processes leading to the considerable change of mechanical properties of steel are activated. Frequently these factors may act together and in various combinations, significantly reducing the load carrying capacity and reducing the durability and safety of constructions.

The solution is wider application instead of traditional concrete and metal, new composite materials.

The advantages of the composite reinforcement are:

– high tensile strength (compared to steel reinforcement above three times);
– inertness of magnet and radio transparency;
– low heat conductivity;
– resistance to any mechanical influences;
– increased humidity;
– no susceptibility to corrosion;
– small weight, approximately four times lighter, than steel fittings;
– resistance to low temperatures;
– it is possible to deliver reinforcement practically of any length;
– high firmness to chlorine-containing antifreeze additives, of road chemistry and different deicing means;
– not a susceptibility to influence of acids, alkalis and various salts (including sea salt).

It is necessary to carry to shortcomings of composite fittings:

1. The value of the modulus of elasticity that is four times lower than the steel reinforcement. It causes the necessity to perform additional calculations for flexural elements.
2. Low fire resistance, because when the heat reaches over 200 degrees, sharp decrease in physical and mechanical indicators occurs. But such heating is possible only in case of direct exposure to open fire, which is practically impossible in reinforced concrete structures.

Comparison of metallic and composite reinforcement on basic physical and mechanical characteristics is given in table 1 (STO 2013).

Nowadays production of bridge spans from composite materials is carried out with application of two technologies (STO 2013).

The first is the pultrusion, which allows making profiles (http://www.pultrusion.ru/).

The second is infusion, it provides production of large-size structures. Vacuum infusion technology provides the absence of deformation occurring from exposure to high temperatures, corrosion resistance, etc. Service life of such bridge will really make more than 100 years without the need for capital repairs throughout this time. Disadvantage of pultrusion is long installation work, since the final design is formed by joining.

The essence of the infusion is to form the material by impregnating it with a low viscosity resin to the vacuum. The main advantages of this technology are the creation of large-scale structures for one approach, and cost reduction during assembly.

Table 1. Physical and mechanical properties of metallic and composite reinforcement.

Features	Reinforcing steel class A-III (AC)	Non-metallic composite reinforcement	
		NCA-C	NCA-B
Material	Hot-rolled steel	Glass fibers connected by polymer	Basalt fibers connected by polymer
The ultimate strength in tension (tensile strength), MPa	590	600–1750	800–1850
The modulus of elasticity tensile, MPa	200000	45000–70000	50000–80000
Elongation, %	14	1.5–3.0	
Nature of behavior under loading (dependence "tension deformation")	Curve with the fluidity platform under loading	Straight line with elastic and linear dependence under loading before destruction	
Corrosion resistance to hostile environment	Corrodding with the release of rust products	Stainless material of the first chemical resistance group, including to the alkaline environment of concrete	
Heat conductivity	Thermally conductive	Not heat conductive	
Electrical conductivity	Electroconductive	Dielectric	
The outer diameters of the produced profiles, mm	6–40	4–80	
Rod length, m	6–12	Any length on request	
Environmental friendliness	Environmentally friendly	No harmful and toxic substances during storage and operation	
Durability	Depending on service conditions and anticorrosive protection	Not less than 50 years, even in sea water	Not less than 80 years, even in sea water

4 RECOMMENDATIONS FOR THE APPLICATION OF GEOCOMPOSITES

Reinforced soil as a composite material can arrange steeps or vertical slopes, eliminating the need for the device of the cone and its mounting, overlapping of a cone isn't required by additional flights, and there is no need in the abutments inside the cone. Extreme support of the bridge is completely discharged from soil pressure.

The fastening of the geogrid's sheet to the blocks is carried out by a simple method using two rods. General view of the structure ground reinforced walls with blocks is presented in Figure 1.

For strengthening bridges' cones two types of modern materials are used: the Reno mattresses and their analogues, as well as gabion structures. To prevent corrosion processes, various sheetings for metal are used. Besides, the device of cones with retaining walls (Fig. 2) gradually gains distribution. At the same time polymeric geosynthetic materials, which durability is more than 100 kNn/m, are used.

Water and wind erosion of slopes, embankments, cones bridges and overpasses can cause serious damage. Natural vegetation is the best natural anti-erosion protection. However, very often vegetable layer under the influence of external factors, such as flooding and high water, waves, the steepness of slopes, insufficient time for formation of strong root system, can't cope with the task (Krisdani et al. 2008, Kuhn 2005). In these cases, the most efficient, reliable and cost-effective event, providing protection of embankments' slopes, cones bridges and overpasses from erosion, is the use of upstream materials.

Geocomposite can consist of two or three constructive layers, which provide the above properties (Giroud et al. 2004). Nonwoven geotextile acts as a filter and protect the drainage core from damage and siltation (Report 2012). The drainage kernel serves for ensuring water conductivity in its plane, to exert pressure from the overlying material (Fig. 3).

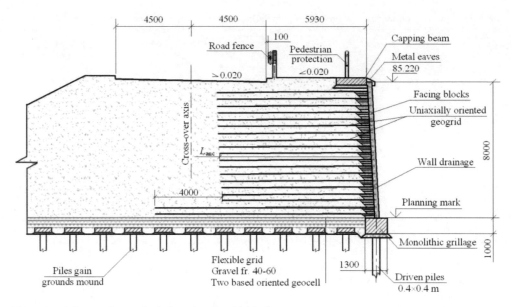

Figure 1. Scheme of ground reinforced walls with blocks.

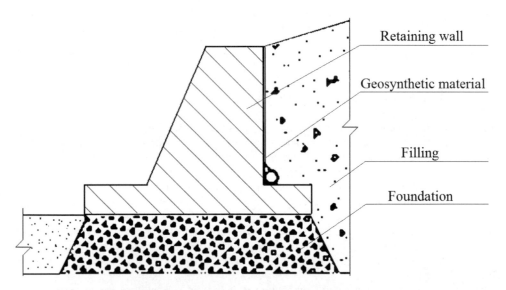

Figure 2. The cone device with retaining wall.

5 CONCLUSION

The use of polymer composite materials in bridge construction is profitable solution (Report 2013). This article shows that the adaptation of composites in bridge construction is favorable decision as they demand smaller expenses than steel or concrete bridges, have the smaller weight, high durability, aren't subject to corrosion, are inert to solutions of acids, alkalis and de-icing solutions

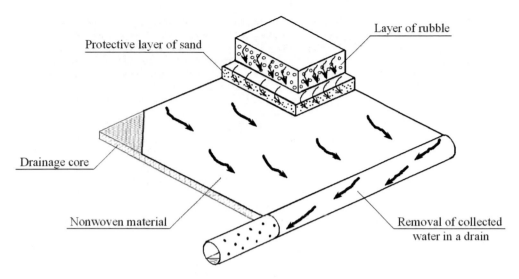

Figure 3.   Operating principle of the drainage geocomposite.

(Iluhin 2012). Therefore, durability of the bridge makes more than 100 years Application of polymeric composites due to the minimum maintenance costs leads to lower total cost of ownership and reduce construction time (Seredina 2013). The overall economic effect is 50–65%.

REFERENCES

Blaznov, A.N., Volkov, Yu, P., Lugovoy, A.N., Savin, V.Г. 2006. Ispytaniya na dlitelnuyu prochnost sterzhney iz kompozitsionnykh materialov [Tests for long-term strength of rods made of composite materials] Zavodskaya laboratoriya. *Diagnostika materialov*. Issue 2. pp. 44–52.
Design Guide. 2002. 7210-PR/113 Composite Bridge Design for Small and Medium Spans 141 p.
Giroud, J.P., Zhao, A., Tomlinson, H.M. and Zornberg, J.G. 2004. Liquid flow equations for drainage systems composed of two layers including a geocomposite. *Geosynthetics International*, No 11, pp. 43–58.
GOST 5781-82 Stal' gorjachekatanaja dlja armirovanija zhelezobetonnyh konstrukcij. Tehnicheskie uslovija [GOST 5781-82 Hot-rolled steel for reinforcement of ferroconcrete structures. Specifications].
El Sarraf, R., Iles, D., Momtahan, A., Easey, D., Hicks, S. 2013. Steel-concrete composite bridge design guide, NZ Transport Agency research report 525, 252 p. http://www.pultrusion.ru/ – The official web site of JSC "Pultruded technology".
Iluhin, D.A. 2012.Svai iz ultrakompozitnogo materiala – novaya era v stroitelstve [Piles of ultra composite material – a new era in the construction]. *Gidrotekhnika*, 2 (27), pp. 66–67.
Kearns, J.C., Shambaugh, R.L. 2002. Polypropylene Fibers Reinforced with Carbon Nanotubes. *Journal of Applied Polymer Science*. Issue 2. 53 p.
Kokoreva, K., Belyaev, N., Yaleshev, A. 2015. Sheet Pilings from Ultra Composite in Hydraulic Engineering.*Construction of Unique Buildings and Structures*, No 4(31) (in print).
Kokoreva, K.A., Kolgushkin, A.V., Yaleshev, A.I. 2014. Analiz vozmozhnosti primeneniya shpunta iz ul'trakompozitnyh materialov v gidrotekhnicheskom stroitel'stve [Analysis of the possibility of using sheet pilings from the ultra composite materials in hydraulic engineering] *Materialy Vserossiyskoy mezhvuzovskoy konferentsii, Sbornik dokladov molodezhnoj nauchno-prakticheskoj konferencii v ramkah Nedelinauki XLIII*, SPbPU, NOC "Vozobnovlyaemye vidy ehnergii i ustanovki na ih osnove", pp. 109–111.
Kokoreva, K.A., Yaleshev, A.I. 2014. Raschetnoe obosnovanie primeneniya shpunta iz ul'trakompozitnyh materialov v gidrotekhnicheskom stroitel'stve [Settlement justification of the using sheet pilings from the ultra composite materials in hydraulic engineering] *Materialy Vserossiyskoy mezhvuzovskoy konferentsii, Sbornik dokladov molodezhnoj nauchno-prakticheskoj konferencii v ramkah Nedelinauki XLIII*, SPbPU, NOC "Vozobnovlyaemye vidy ehnergii i ustanovki na ih osnove", pp. 112–114.

Kolgushkin, A.V., Belyaev, N.D. 2007a. Inzhenernyie meropriyatiya po uvelicheniyu dolgovechnosti skvoznyih gidrotehnicheskih sooruzheniy [Engineering measures to increase longevity through waterworks] *Nauchno-tehnicheskievedomostiSPbGPU*, 49 (1), pp. 185–193.

Kolgushkin, A.V., Belyaev, N.D. 2007b. Ob uchete korrozii pri proektirovanii skvoznyih metallicheskih GTS [Taking into account corrosion through the design of of metal HES]*Sbornik nauchnyih trudov "Predotvraschenie avariy zdaniy i sooruzheniy"*, Magnitogorsk, pp. 159–168.

Krisdani, H., Rahardjo, H., and Leong, E.C. 2008. Measurement of geotextile-water characteristic curve using capillary rise principle.*Geosynthetics International*, No. 15 (2), pp. 86–94.

Kuhn, J.A., McCartney, J.S., and Zornberg, J.G. 2005. Impinging flow over drainage layers including a geocomposite. In: *Proceedings of the Sessions of the Geo-Frontiers Congress*, Austin, Texas, 7 p.

Loktev, M.J., Abanin, V.A., Savin, V.F., Ermolaev, D.A., Suranov, A.Y. 2010. Automated information measurement system for determination of mechanical characteristics for polymer composite materials by longitudinal bending. *IEEE 2nd Russia School and Seminar on Fundamental Problems of Micro/Nanosystems Technologies*, MNST'2010 Novosibirsk, pp. 26–28.

Manchado, M.A.L., Valentini, L., Biagiotti, J., Kenny, J.M. 2005. Thermal and mechanical properties of single-walled carbon nanotubes-polypropylene composites prepared by melt processing. Carbon. Vol. 43. Issue 7. pp. 1499–1505.

Ponomarev, A.N., Mospan, Ye, A. 2011. Analiz napravleniy ispolzovaniya nanokompozitnoy armatury "astrofleks" v promyshlennom i transportnom stroitelstve [Analysis of the uses of the nanocomposite reinforcement "astrofleks" in the industrial and transport construction] *Stroitelnayamekhanikainzhenernykhkonstruktsiy i sooruzheniy*. Issue 3. pp. 69–74.

Ptuhina, I.S., Dalabayev, A.S., Turkebayev, A.B., Tleukhanov, D.S., Bizhanov, N., Zh, Dalabayeva, A.E. 2014. Effektivnost' ispol'zovanija innovacionnyh kompozitnyh materialov v stroitel'stve [Efficiency of innovative composite materials in construction] *Construction of Unique Buildings and Structures*, No. 9(24), pp. 84–95.

Report 2012. Geocomposite moisture barriers in roadway applications Report FHWA-CFL/TD-12-003, 124 p.

Report 2013. Composite Bridge Decking Final Project Report FHWA-HIF-13-029 March 2013 98 p.

Seredina, O.S. 2013. Stekloplastikovaya armatura v sovremennom stroitelstve [Fiberglass rebar in modern construction] *Molodezh i nauchno-tekhnicheskiy progress v dorozhnoy otrasli yuga Rossii materialy mezhdunarodnoy nauchno-tekhnicheskoy konferentsii studentov, aspirantov i molodykh uchenykh FGBOU VPO "Volgogradskiy gosudarstvennyy arkhitekturno-stroitelnyy universitet"*, pp. 63–70.

SP 46.13330.2012. Mosty i truby [Code of Practice 46.13330.2012. Bridges and pipes].

STO NOSTROI 2.6.90-2013. Primenenie v stroitel'nyh betonnyh i geotehnicheskih konstrukcijah nemetallichesko jkompozitnoj armatury [Standards of organization, National Association of Builders 2.6.90-2013. Application of non-metallic composite reinforcement in building concrete and geotechnical structures].

Thwe, M.M., Liao, K. 2000. Characterization of bamboo-glass fiber reinforced polymer matrix hybrid composite. *Journal of Materials Science Letters*. Vol. 19. Issue 20. pp. 1873–1876.

Advances and Trends in Engineering Sciences and Technologies – Al Ali & Platko (Eds)
© 2016 Taylor & Francis Group, London, ISBN: 978-1-138-02907-1

Seismic analysis of cylindrical liquid storage tanks considering of fluid-structure-soil interaction

K. Kotrasová, S. Harabinová, E. Panulinová & E. Kormaníková
Technical University of Košice, Faculty of Civil Engineering, Košice, Slovakia

ABSTRACT: Ground-supported tanks are used to store a variety of liquids. The fluid develops hydrodynamic impulsive and convective pressures on walls and bottom of tank during an earthquake. This paper provides theoretical background for specification of hydrodynamic actions of fluid in liquid storage cylindrical container by using analytical methods. Numerical model on seismic response of fluid-structure-soil interaction of cylindrical tank was obtained by using of Finite Element Method (FEM).

1   INTRODUCTION

Storage tanks containing liquids can be found in many industries, including:

- petroleum production and refining,
- petrochemical and chemical manufacturing,
- bulk storage and transfer operations, and
- other industries consuming liquids.

Tanks are used to store a variety of liquids, e.g. water for drinking and fire fighting, petroleum, chemicals, and liquefied natural gas. Several cases of damage to tanks have been observed in the past as a result of earthquakes. Ground-supported circular tanks are critical and strategic structures, and damage to them during earthquakes may endanger drinking water supply, cause failure in preventing large fires and contribute to substantial economic loss.

The seismic analysis and design of liquid storage tanks is, due to the high complexity of the problem, in fact, really complicated task. Number of particular problems should be taken into account, for example: dynamic interaction between contained fluid and tank-fluid/structure interactions (FSI), sloshing motion of the contained fluid; and dynamic interaction between tank and sub-soil-soil/structure interaction (SSI). The knowledge of fluid effects acting onto walls and the bottom of containers during an earthquake plays essential role in reliable and durable design of earthquake resistance structure/facility – tanks (Mihalikova 2010).

2   SEISMIC ANALYSIS OF LIQUID FILLED TANKS

Seismic analysis of liquid – containing tanks differs from typical civil engineering structures. Liquid inside tank exerts during seismic excitation hydrodynamic force on the tank walls and base due to liquid-tank interaction (Kotrasova et al. 2014, Jendzelovsky et al. 2014).

The motion of fluid in the tank is possible to define using the simple quasistatic model, in which the inertial forces are defined by hydrostatic and hydrodynamic pressure on the tank wall. The tank-liquid system may be modeled by spring-mass models as shown in Figure 1. The fluid is replaced by an impulsive mass m_i (denotes the mass of the contained fluid which moves together with the walls) that is rigidly attached to the tank container wall and by the convective masses m_{cn} that are connected

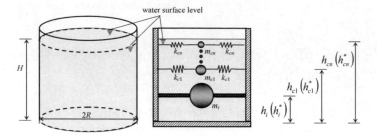

Figure 1.   Spring-mass analogy for ground supported circular tanks.

to the walls through the springs having stiffness (k_{cn}). According to the literature, although only the first convective mass may be considered additional higher-mode convective masses may also be included for the ground-supported tanks. The dynamic characteristics of this model are estimated by using the expressions given by Eq. (1)–(8).

$$m_i = m 2\gamma \sum_{n=0}^{\infty} \frac{I_1(v_n/\gamma)}{v_n^3 I_1'(v_n/\gamma)} \tag{1}$$

$$h_i = H \frac{\sum_{n=0}^{\infty} \frac{(-1)^n I_1(v_n/\gamma)}{v_n^4 I_1'(v_n/\gamma)} \left(v_n(-1)^n - 1\right)}{\sum_{n=0}^{\infty} \frac{I_1(v_n/\gamma)}{v_n^3 I_1'(v_n/\gamma)}} \tag{2}$$

$$h_i^* = H \frac{\frac{1}{2} + 2\gamma \sum_{n=0}^{\infty} \frac{v_n + 2(-1)^{n+1} I_1(v_n/\gamma)}{v_n^4 I_1'(v_n/\gamma)}}{2\gamma \sum_{n=0}^{\infty} \frac{I_1(v_n/\gamma)}{v_n^3 I_1'(v_n/\gamma)}} \tag{3}$$

where $v_n = \frac{2n+1}{2}\pi$; $\gamma = H/R$, $I_1(\cdot)$ and $I_1'(\cdot)$ denote the modified Bessel function of order 1 and its derivate. The derivate can be expressed in term of the modified Bessel functions of order 0 and 1 as $I_1'(\cdot) = \frac{dI_1(x)}{dx} = I_0(x) - \frac{I_1(x)}{x}$.

The convective component of the response may be obtained from that of oscillators having masses m_{cn}, attached to the ridig tank through springs having stiffnesses k_{cn}.

$$m_{cn} = m \frac{2 \tanh(\lambda_n H/R)}{(\lambda_n H/R)(\lambda_n^2 - 1)} \tag{4}$$

$$h_{cn} = H \left(1 + \frac{1 - \cosh(\lambda_n \cdot H/R)}{(\lambda_n \cdot H/R) \cdot \sinh(\lambda_n \cdot H/R)} \right) \tag{5}$$

$$h_{cn}^* = H \left(1 + \frac{2 - \cosh(\lambda_n H/R)}{(\lambda_n H/R) \cdot \sinh(\lambda_n H/R)} \right) \tag{6}$$

$$\omega_{cn}^2 = \lambda_n \tanh(\lambda_n H/R)(g/R) \tag{7}$$

$$k_{cn} = m_{cn} \omega_{cn}^2 \tag{8}$$

λ_n are the roots of the first-order Bessel function of the first kind ($\lambda_1 = 1.8412$; $\lambda_2 = 5.3314$; $\lambda_3 = 8.5363$, $\lambda_4 = 11.71$ and $\lambda_{5+i} = \lambda_5 + 5i (i = 1, 2, \dots)$). H is the depth of fluid filling. h_{cn} and h_{cn}^* are the level where the oscillators of convective masses, h_i and h_i^* are the heights of impulsive mass. R is the inner radius of container. m is the total mass of the fluid. Only the first oscillating, or sloshing, mode and frequency of the oscillating liquid ($n = 1$) needs to be considered for design purposes.

Total base shear V of ground supported tank at the bottom of the wall can be also obtained by base shear in impulsive mode and base shear in convective mode, Eq. (9). Total base shear V^* of ground supported tank at the bottom of base slab is given also by base shear in impulsive mode and base shear in convective mode too, Eq. (10).

The overturning moment M of ground supported tank immediately above the base plate is given also by, Eq. (11) and the overturning moment M^* of ground supported tank immediately below the base plate is given also by, Eq. (12).

$$V = (m_i + m_w + m_r)S_e(T_i) + (m_c)S_e(T_c) \tag{9}$$

$$V^* = (m_i + m_w + m_b + m_r)S_e(T_i) + (m_c)S_e(T_c) \tag{10}$$

$$M = (m_i h_i + m_w h_w + m_r h_r)S_e(T_i) + (m_c h_{c1})S_e(T_c) \tag{11}$$

$$M^* = (m_i h_i^* + m_w h_w + m_b h_b + m_r h_r)S_e(T_i) + (m_c h_{c1}^*)S_e(T_c) \tag{12}$$

where m_i is the impulsive mass of fluid, m_c the convective mass of fluid, m_w mass of the tank wall; m_b mass of the tank base plate; m_r mass of the tank roof; h_w the height of center of gravity of wall mass; h_b the height of center of gravity of base plate mass; h_r the height of center of gravity of roof mass. $S_e(T_i)$ is impulsive spectral acceleration, is obtained from a 2% damped elastic response spectrum for steel and prestressed concrete tanks, or a 5% damped elastic response spectrum for concrete and masonry tanks; $S_e(T_c)$ convective spectral acceleration, is obtained from a 0.5% damped elastic response spectrum.

3 FEM – FLUID STRUCTURE INTERACTION

For the fluid-structure interaction analysis, there are possible three different finite element approaches to represent fluid motion, Eulerian, Lagrangian and mixed methods. In the Eulerian approach, velocity potential (or pressure) is used to describe the behavior of the fluid, whereas the displacement field is used in the Lagrangian approach. In the mixed approaches, both the pressure and displacement fields are included in the element formulation.

In fluid-structure interaction analyses, fluid forces are applied into the solid and the solid deformation changes of the fluid domain. For most interaction problems, the computational domain is divided into the fluid domain and solid domain, where a fluid model and a solid model are defined respectively, through their material data, boundary conditions, etc. The interaction occurs along the interface of the two domains. Having the two models coupled, we can perform simulations and predictions of many physical phenomena.

In many fluid flow calculations, the computational domain remains unchanged in time. Such the problems involve rigid boundaries and are suitable handled in Eulerian formulation of equilibrium equations (Zmindak 1997). In the case where the shape of the fluid domain is expected to change significantly, modified formulation called Arbitrary Lagrangian-Eulerian (ALE) formulation was adopted to simulate the physical behavior of the domain of interest properly. The ALE description is designed to follow the boundary motions rather than the fluid particles. Thus, the fluid particles flow through a moving FE-mesh. Basically there are two different algorithms available for generation of possible moving mesh.

4 NUMERICAL EXPERIMENT OF FLUID/STRUCTURE INTERACTIONS (FSI)

In this study, a ground supported cylindrical storage tank, without a roof, with inner radius $R = 6$ m, and height $H_w = 6.25$ m. The walls have the uniform thickness 0.25 m and the base slab of the tanks is 0.4 m. The water filled tank is grounded on gravel sub-soil. The following material characteristics of tank are: Young's modulus E = 35 GPa, Poisson ratio $v = 0.2$, density $\rho = 2550$ kg/m^3. The height of water filling is 6 m. H$_2$O is given by bulk modulus $B = 2,1 \cdot ,10^9$ N/m^2, density $\rho_w = 1\,000$ kg/m^3.

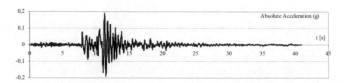

Figure 2.   Accelerogram Loma Prieta, California.

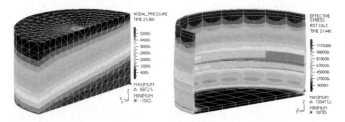

Figure 3.   Pressure of fluid domain in time 21.36 s at the left side and Von Mises stress of solid domain (tank) in time $t = 21.44$ s on the right side.

Table 1.   Standard characteristics of soils.

| Characteristics of soils | | Soil group | | | | |
		G1	G2	G3	G4	G5
Unit weight	γ (kN.m^{-3})	21.0	20.0	19.0	19.0	19.5
Angle of internal friction	ϕ_{ef} ($^\circ$)	38.5	35.5	32.5	32.0	30.0
Cohesion of soil	$c_{ef(kPa)}$	0.0	0.0	0.0	4.0	6.0
Deformation modulus	$E_{def(MPa)}$	320	145	85	70	50
Poisson's ratio	ν	0.20	0.20	0.25	0.30	0.30

As the excitation input we consider horizontal earthquake load given by the accelerogram (Figure 2) of the earthquake in Loma Prieta, California (18.10.1989). Seismic excitation acts along y – direction.

Dynamic time-history response of concrete open top cylindrical liquid storage tank was performed by application of Finite Element Method (FEM) utilizing software ADINA. Arbitrary-Lagrangian-Eulerian (ALE) formulation was used for the problems. Two way Fluid-Structure Interaction (FSI) techniques were used for simulation of the interaction between the structure and the fluid at the common boundary.

The maximum value $p_{max} = 68721$ Pa measured at node "RD" (right down edge of fluid region) in time 21.36 s as in the Figure 3. Peak Von Mises stress measured at the inside circumference of the solid wall. The peak 1.35472 MPa measured at the time point 21.44 s. Time variation of Von Mises stress at area of concern (plane of symmetry) is shown in Figure 3.

5   NUMERICAL EXPERIMENT OF SOIL/STRUCTURE INTERACTION (SSI)

The cylindrical tank is founded at depth of 0.5 m below the surface on the circular base with diameter of 6.5 m with a thickness of 0.5 m. The concrete tanks – water reservoirs – resting on gravel subsoil have been analyzed. The subsoil has been modeled using 5 various types of gravel subsoil – soil group G1 – G5 and three basic models – load conditions:

1. empty tank – static analysis,
2. water filled tank – static analysis,
3. water filled tank – seismic analysis.

Table 2. Results of II. the serviceability limit state.

Settlement and rotation of foundation		Load condition		
		1	2	3
soil group G1				
Foundation settlement	(mm)	0.4	1.5	1.5
Depth of influence zone	(m)	6.79	10.87	10.87
Max. rotation of foundation	–	0.0	0.0	0.019
Max. compress. foundation edge settlement	(mm)	0.2	0.7	0.9
Min. compress. foundation edge settlement	(mm)	0.2	0.7	0.6
soil group G2				
Foundation settlement	(mm)	0.9	3.3	3.3
Depth of influence zone	(m)	7.04	11.10	11.10
Max. rotation of foundation	–	0.0	0.0	0.040
Max. compress. foundation edge settlement	(mm)	0.4	1.7	1.9
Min. compress. foundation edge settlement	(mm)	0.4	1.7	1.4
soil group G3				
Foundation settlement	(mm)	1.6	5.4	5.4
Depth of influence zone	(m)	7.31	11.33	11.33
Max. rotation of foundation	-	0.0	0.0	0.060
Max. compress. foundation cdge settlement	(mm)	0.6	2.7	3.1
Min. compress. foundation edge settlement	(mm)	0.6	2.7	2.3
soil group G4				
Foundation settlement	(mm)	1.7	5.8	5.8
Depth of influence zone	(in)	7.31	11.33	11.33
Max. rotation of foundation	–	0.0	0.0	0.064
Max. compress. foundation edge settlement	(mm)	0.7	2.9	3.4
Min. compress. foundation edge settlement	(mm)	0.7	2.9	2.5
soil group G5				
Foundation settlement	(mm)	2.3	8.0	8.0
Depth of influence zone	(m)	7.17	11.21	11.21
Max. rotation of foundation	–	0.0	0.0	0.093
Max. compress. foundation edge settlement	(mm)	0.9	4.1	4.6
Min. compress. foundation edge settlement	(mm)	0.9	4.1	3.4

The tank has been verificated according to theories of Limit States I. – the ultimate limit state (ULS) and II. – the serviceability limit state (SLS) under EC 7. It was computed vertical and horizontal bearing capacity, settlement and rotation of a footing (Kuklik 2011).

The resulting deformation depends on the deformation properties of the subsoil (Table 1) and on the size of the tensions in the foundation soil – stress in soils (σ_{or}) and stress from the external load (σ_z). Stresses in thesoilwere determinedaccording to the theoryof elasticity (Boussinesqtheory). When computing settlement below the footing bottom,the first is calculated the stress in the footing bottom and then is determined the overall settlement and rotation of foundation (Kralik at al. 2013, Melcer at al. 2014).

The general approach draws on subdividing the subsoil into layers of a different thickness based on the depth below the footing bottom or ground surface. Vertical deformation of each layer is then

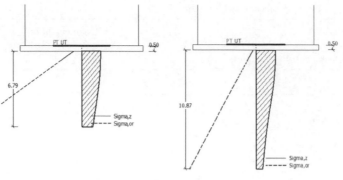

a) Depth of influence zone for 1.load condition   b) Depth of influence zone for 3. load condition

Figure 4.   Settlement of foundationin Soil group G1.

computed – the overall settlement is then defined as a sum of partial settlements of individual layers within the influence zone (deformations below the influence zone are either zero or neglected). Equation to compute compression of ith soil layer below foundation having a thickness h arises from the definition of oedometric modulus E_{oed}.

The results of numerical solutions have been presented and compared on Figure 4 and in Table 2. Evaluation of the results:

– The settlement of the circular plate calculated for 2. load condition was 5 times higher than for 1. load condition – for all types of gravel.
– By comparing 2. and 3. load condition, it can be seen that the torque effect of seismic loading may cause "lifting" of the tank edge.
– Maximum rotation of foundation is growing with the reduction of the stiffness of the subsoil.

ACKNOWLEDGEMENTS

Preparation of the paper was supported by the ScientificGrantAgency of the Ministry of Education of SlovakRepublic and the Slovak Academy of Sciences under Project 1/0477/15.

REFERENCES

Jendzelovsky, N. & Balaz, L. 2014.Modeling of a gravel base under the cylindrical tank. *Advanced Material Research* Volume 969: 249–252.
Kralik, K. & Kralik, J. 2013. Probability assessment of analysis of high-rise buildings seismic resistance. *Advanced Materials Research* Volume 712–715: 929–936.
Krejsa, M. & Janas, P. & Krejsa, V. 2014. Software application of the DOProC method. *International Journal of Mathematics and Computers in Simulation* Volume 8, No. 1: 121–126.
Kotrasova, K. & Grajciar, I. & Kormanikova, E. 2014. Dynamic Time-History Response of cylindrical tank considering fluid – structure interaction due to earthquake. *Applied Mechanics and Materials.* No. 617: 66–69.
Kuklik, P. 2011. Preconsolidation, Structural Strength of soil, and its effect on subsoil upper structure interaction. *Engineering Structures* No. 33: 1195–1204.
Melcer, J. & Kuchárová, D. 2014. Frequency response functions of a lorry. *Advanced Material Research* Volume 969: 188–191.
Mihalikova, M. 2010. Research of strain distribution and strain rate change in the fracture surroundings by the video extensometric methode. *Metalurgija* Volume 49, No. 3: 161–164.
Zmindak, M. & Grajciar, I. 1997. Simulation of the aquaplane problem. *Computers and Structures.* Volume 64, Issue 5–6: 1155–1164.

Advances and Trends in Engineering Sciences and Technologies – Al Ali & Platko (Eds)
© 2016 Taylor & Francis Group, London, ISBN: 978-1-138-02907-1

Behavior of welded T joints

V. Kvočák & P. Beke
Technical University of Košice, Civil Engineering Faculty, Košice, Slovakia

ABSTRACT: The contribution deals with experimental tests focused to behaviour of joints composed of various types of sections. Special attention is paid to T-joints that consist of the chord member of a single rectangular section and brace members of different kinds of cross section. The evaluation procedure monitors the resistances and deformations of such joints.

1 INTRODUCTION

Hollow sections are more and more commonly used and not only have they found their place in the machine industry, but also in the field of construction. Thanks to their properties they are now used in fields where classical open rolled-steel sections formerly dominated.

From the constructional point of view, hollow sections are particularly suitable in elements subjected to uniaxial compression and torsion. Aesthetically, this kind of sections is frequently used to form so-called direct-finish or visual elements, as the appearance of the whole structure is not impaired.

However, the use of such sections brings along a problem of their connections, which is one of the limiting factors in the design of structures. A joint determines the behaviour of the structure as a whole and thus its importance in the design of a structure is unquestionable.

The resistance of joints in hollow sections depends to a large degree on the ratio β (the ratio between the widths of a vertical brace member and a straight chord member), the wall thickness of the connecting elements, and the quality of the material used. In the evaluation of such joints in compliance with the Slovak standard currently in force (Kvočák, Beke 2008a) the essential part in the design of the joint is the ultimate breaking strength which the wall of a tension or compression chord can withstand when loaded locally by brace members. In that case the evaluation of joints in hollow sections is relatively simple, but nevertheless, in the majority of cases it results in the non-economical design (Packer et al 1992), (Freitas et al 2008).

In view of the European standard (Kvočák, Beke 2008b) the design of joints is based on a principle of prevailing failure that leads to the collapse of the whole system. In the evaluation of the joint, this standard, unlike STN, takes into consideration as many as six different modes of failure:

- Plastic failure of the chord wall or the chord cross-section at the joint with brace members;
- Punching shear failure of the horizontal chord wall at the joint with brace members (crack initiation leading to rupture of the brace members from the chord member);
- Cracking in the tension brace member or its welds;
- Local buckling failure of the compression brace member at the joint location;
- Shear failure of the chord;
- Local buckling or crippling of the chord wall behind the foot of the tension brace member.

Despite the relatively good approximation of actions in joints of rectangular and circular hollow sections and the specified methods of making calculations regarding such joints in (Kvočák, Beke 2008b) not all combinations of sections in joints can be unambiguously evaluated. Therefore, an experimental programme has been established to focus on the determination of resistance of selected

Table 1. Tensile test results.

Specimen	R_{eH} [MPa]	$R_{p0,2}$ [MPa]	R_{el} [MPa]	R_m [MPa]	A_5 [%]
28-1	324	318	317	388	31.5
28-2	321	317	316	389	31.5
28-3	323	320	319	391	32
29-1	324	318	317	388	31.5
29-2	321	317	316	389	31.5
29-3	323	320	319	391	32
30-1		305		384	33
30-2		312		388	32.5
30-3		310		388	32.5

joint types, the resistance of which according to (Kvočák, Beke 2008a) and (Kvočák, Beke 2008b), is incompatible, or where there are no evaluation methods available for specific combinations of members in the joint.

2 EXPERIMENTAL PROGRAMME

Relatively extensive research into the above-mentioned issues is currently being carried out at the Technical University, Civil Engineering Faculty, in Košice. Besides the experimental research, the programme is focused on the theoretical modelling and determination of joint resistances as well as on the comparison of achieved theoretical and experimental results with the design resistances of the joints according to standard procedures.

The first part of the experiment concentrates on joints in rectangular hollow sections, where three T-shaped joint types are tested. The chords in the specimens used are composed of a rectangular hollow section RHS 140 × 140 × 4. Perpendicularly, a rectangular brace member with a variable section is fillet-welded to the chord. The dimensions of the brace member were, in the first case, identical to the chord dimensions, i.e. the rectangular hollow section RHS 140 × 140 × 4 was used. In the second case, the brace member section was RHS 100 × 100 × 3 and, in the third case, RHS 60 × 60 × 3. The geometry of the joints was selected in the effort to cover, as far as possible, the whole spectrum of β-parameters. In order to maintain the correctness of the obtained results, three specimen subsets were made for each joint type, i.e. nine experimental specimens were used.

Three specimens were taken from the material used in all types of sections and tensile tests were carried out to determine the real mechanical properties of the materials used. Mainly, the yield strength and ductility were measured during the experiments. The final values of the yield strength for individual materials are presented in Table 1.

It can be seen from the measured yield strength values in the tensile test that these considerably exceeded the values given in standards for the material S235 of which the specimens were made.

Due to the real behaviour of the joint in a lattice structure, it was necessary not only to take into account the vertical loading, but also the horizontal loading of the joint. The vertical load was simulated by means of a hydraulic press, while the action was gradually increased until the specimen failure.

The horizontal load was also imposed on the specimen using a hydraulic press. Unlike the vertical loading, the magnitude of the horizontal load remained constant during the whole experiment. By means of the horizontal loading an axial force in the lattice structure chord was actuated.

Owing to the geometry of the specimen tested and the equipment used, it was necessary to form a testing frame to activate the horizontal load. The frame was made in such a manner that it could fit, along with the horizontal press and the specimen, the hydraulic press used for activating the vertical loading. The whole assembly consisting of the specimen and the frame is depicted in Fig. 1.

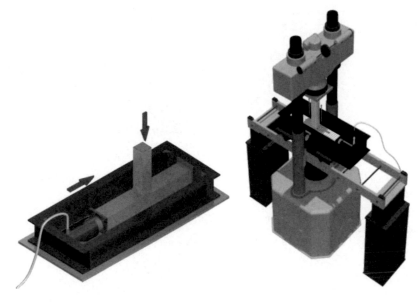

Figure 1.   A test assembly.

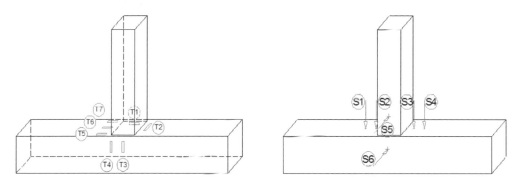

Figure 2.   Arrangement of strain gauges T1 – T7 and inductive sensors S1 – S6.

Before putting the specimen into the hydraulic press, a network of points was marked on each test specimen. This network served the topographical measurement of strain (deformation per unit of length) in the individual measurement points. The measurements were performed before and after the test by means of which precise values of permanent sets after the specimen failure were detected.

In the experiment, relative deformations/strains were measured by means of strain gauges fixed on selected points of member sections. The arrangement of the strain gauges was determined by the calculations (FEM) and can be seen in Fig. 2. By means of inductive sensors and mechanical indicators the overall horizontal and vertical deformations were measured.

3   OBTAINED RESULTS

The types of joints selected were observed for both stress and deformation. In the following section of the article, attention will be directed to the deformation of the joints. The following sections

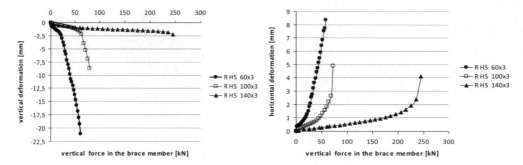

Figure 3.   Comparison of the experimentally measured vertical and horizontal deformations in the joints composed of the brace members with rectangular hollow sections.

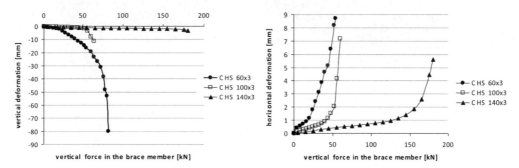

Figure 4.   Comparison of the experimentally measured vertical and horizontal deformations in the joints composed of the brace members with circular hollow sections.

provide a very sharp picture of the real behaviour of the individual types of joint. The figures presented compare both vertical and horizontal experimentally measured deformations of the joints.

3.1   *Comparison of deformations in the joints consisting of the identical type of the brace member*

The joints with the identical type of the brace member are compared. First, when there is linear deformation, the joint is in the elastic range. Later, as the load is increased, the joint deviate from this linear proportionality and the deformation goes through the elasto-plastic and eventually into its plastic range of action.

Of all the types of brace members, the most resistant certainly seem to be those with $b_0 = b_1$ ($\beta = 1$). From the deformation point of view, the joints with a width of 60 mm are the least suitable. These exhibited excessive deformations even under minimal load.

3.2   *Comparison of deformations in the joints with the same widths of chord and brace members*

The stiffness of the joint does not depend only on its dimensions but also on the type of the brace member used. The comparisons are presented in the form of the following figures 6 to 8.

In the first type of joint ($\beta = 1$), the stability of the chord web (wall) was crucial for the overall resistance of the joint. Figure 6 presents the distribution of values of vertical and horizontal deformation. Significant deformation of the vertical chord web (wall) occurred even under a relatively light vertical load. The joint collapsed completely with the gradual increase in load due to the buckling of the vertical web (wall) of the horizontal chord member. When the resistance of rectangular, circular and open sections was compared, the rectangular sections proved to be the stiffest.

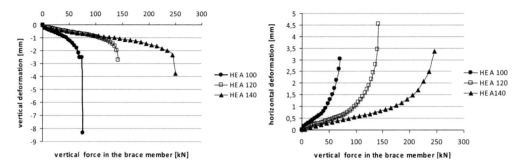

Figure 5.   Comparison of the experimentally measured vertical and horizontal deformations in the joints composed of the brace members with open HEA-sections.

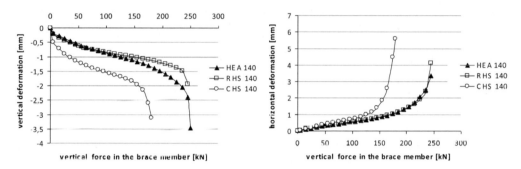

Figure 6.   Vertical and horizontal deformation of the T-joint with $\beta = 1.00$ depending on the type of brace member used.

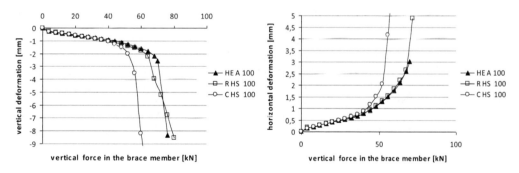

Figure 7.   Vertical and horizontal deformation of the T-joint with $\beta = 0.714$ depending on the type of brace member used.

As can be seen from the figure 7, the buckling effect of the chord web (wall) on the overall resistance of the joint can be observed also in the joints with $\beta = 0.714$. The overall resistance of the joint was influenced by the loss of stability of the vertical web (wall) although the difference between the vertical and horizontal deformation was less significant than in the first case. When comparing the types of brace member used, the rectangular section appeared to be the most resistant of all. However, the difference between the rectangular and open HEA-section was minimal regarding vertical and horizontal deformation.

In the third type of joint (figure 8) with the most slender brace members, the overall resistance of the joint was to a great extent affected by the stiffness of the horizontal chord web (wall). The difference between the horizontal and vertical deformation in this type of joint is the biggest and

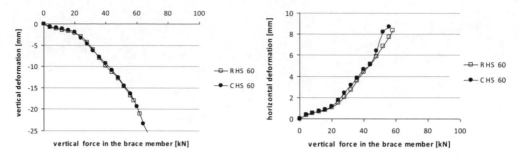

Figure 8.   Vertical and horizontal deformation of the T-joint with $\beta = 0.428$ depending on the type of brace member used.

the vertical deformation the greatest. The influence of the type of the brace member used on the overall resistance of such joints is virtually negligible. Due to the limited possibilities of the HEA rolling programme, only circular and rectangular hollow sections were compared. For $\beta = 0.714$ the behaviour of open sections was identical to that of rectangular sections and, obviously, it does not change for lower.

4   CONCLUSION

The scientific research results and evaluations presented characterize the correlations regarding the global resistance of joints in lattice structures in the light of the latest scientific knowledge that should be responsibly taken into consideration in their reliable and cost-effective design. The article presented points to the significance and topicality of the issue of investigating the joints in lattice structures. The priority of the research was to acquire knowledge of the real behaviour of the joints and supplement it with more effective methods for the design of joints composed of rectangular and circular hollow sections, and open HEA-sections.

From the results obtained some patterns of behaviour of T-joints may be identified. With respect to the geometry and type of section, it can be concluded that the resistance of a joint with $\beta = 1.00$ is greatly influenced by the type of brace member. This influence sharply diminishes with the decreasing value of a β-parameter. With very low β-parameters, the influence of the type of brace member becomes virtually negligible and unimportant.

The conclusions presented in this article represent only part of a number of results obtained in the experiments. The authors would like to continue in the analysis of such joints, while the main emphasis should be placed on the verification of the obtained results using an appropriate finite model for the joints in question.

REFERENCES

Kvočák, V., Beke, P. 2008 a. "Experimental analysis of "T"-joints created from various types of sections", In Selected Scientific Papers: Journal of Civil Engineering. 3(2).51–60.ISSN 1336-9024.
Kvočák, V., Beke, P. 2008 b. "Experimental verification of welded hollow section joints", In Zeszytynaukowe-PolitechnikiRzeszowskiej: BudownictwoiinzynieriaSrodowiska. vol. 256, no. 50 (2008).185–192.ISSN 0209-2646.
Packer, J. A., Wardenier, J., Kurobabne, Y., Dutta, D., Yeomans, N. "Design Guide for Rectangular Hollow Section Joints under Predominantly Static Loading", Verlag TUV Rheiland, 1992.
Freitas, A. M. S., Mendes, F. C., Freitas, M. S. R. 2008. "Finite Elements Analyses of Welded T-Joints", In Proceedings from 5th European Conference on Steel and Composite Structures, Eurosteel, 2008 Graz, 2008. 555–560.

Advances and Trends in Engineering Sciences and Technologies – Al Ali & Platko (Eds)
© 2016 Taylor & Francis Group, London, ISBN: 978-1-138-02907-1

Evaluation of results from push-out tests

V. Kvočák & D. Dubecký

Technical University of Košice, Faculty of Civil Engineering, Institute of Structural Engineering, Košice, Slovakia

ABSTRACT: Creative innovation in the construction industry along with the improvements of material properties and the quality of technologies make it possible to construct buildings of greater height and bridges of greater span. Steel-concrete composite structures have been among the most economical structural systems used in tall buildings and in small- and mid-spanned bridges. In the latter type of construction, transfer of shear forces between a steel beam and concrete slab is achieved by means of mechanical devices called shear connectors. To date, a variety of types of composite action have been investigated at the Civil Engineering Faculty of the Technical University of Košice; therefore, the article presented shows some of the test results from push-out tests performed on various specimens.

1 INTRODUCTION

First mentions of the employment of shear connectors in building construction appeared in 1894. The increasing level of knowledge later enabled development of composite structures in buildings (Viest et al. 1992) and subsequently in bridges.

Various types of shear connectors include, among others, studs and strips. Composite action provided by stud connectors, which are considered to be standard, automatically manufactured technological devices, has several drawbacks. Their main disadvantage is loose or non-continuous force transfer between composite structural members that leads to the concentration of forces. The locations where studs are welded are frequently sources of initiation of cracks that propagate into the flanges of the main beams. From the point of view of fatigue failure, such structural detail is the most inappropriate detail in a composite beam, especially in bridge structures. Another drawback of stud shear connectors is their production cost and laboriousness, which prolong the construction period. In most cases, the ultimate load capacity of composite member is given and directly determined by the load capacity of the shear stud; therefore, it is impossible to raise the load-bearing capacity of a composite beam using higher strength concrete. Unlike studs, the majority new type of shear connectors are not symmetrical regarding the axis perpendicular to the direction of shear flow. The orientation of such shear connectors must be adjusted to the orientation of the prevailing shear flow. Strip shear connectors welded to the flange of a steel beam produce composite action in a longitudinal direction only; therefore, they must be supplemented with elements preventing the uplift of a concrete slab. One of the first lying strips used in combination with studs was designed and applied in practice by Leonhardt in 1938 (Leonhardt, 1938).

A standing perforated strip, first developed in Germany in 1985, is frequently used to ensure composite action both in vertical and horizontal directions (Andra, 1985). Research into shear connection strips has gradually progressed in other countries. The first results were achieved in Canada in 1988 (Antunes, 1988), and their complete analysis complemented with other results was published in 1992 (Veldana, Hosain, 1992). Some other results were published in Australia in 1992, Japan in 1994 (Li, Cederwall, 1996), and in the Czech Republic in 1996 (Studnička, 1996). In Slovakia, the first investigations of shear strip connectors were reported in 1996 (Rovòák et al. 2006).

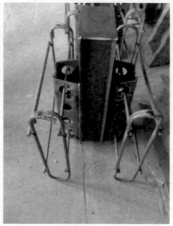

Figure 1. Preparation of experiments.

This article directs attention to the composite action provided by strip type connectors, since they seem to be more suitable than traditional headed shear studs with respect to their resistance and flexure. Moreover, they exhibit better fatigue strength and longer life and shear force transfer to the flange of a beam is continuous. Once again, the labour intensity in the production of strip connectors is lower and the method of assembly of steel construction is simpler than that with stud connectors. All these properties make this type of composite action more suitable for mid-span composite bridges, which are more economical and structurally effective than pre-stressed concrete bridges (Kvočák et al. 2011).

2 EXPERIMENTAL PROGRAM

Intensive research into strip shear connectors in the Slovak Republic started in 1995 at the Civil Engineering faculty in Košice. It was supported by *Inžinierske stavby*, a local public limited company, first within the framework of its own research activity and later within the framework of a widely conceived scientific engineering project. Experiments were focused on various differences in composite action by means of several types of concrete mixtures.

This article pertains to the composite action of a closed perforated strip. All preparations and the conduct of the experiment took place at the Laboratory of Excelent Research. The experiment itself was preceded by formwork preparation. In order to simplify and shorten the preparation process, the formwork was made from large formatted foiled plywood panels connected with screws. Steel sections were supplied and reinforcement bars were bent by specialised companies. The ready-made steel reinforcement cages (Fig. 3b) were then inserted into the framework. The bottom edge of the steel beam was protected with a polystyrene plate to prevent direct contact of the front steel part with the concrete block during the experiment.

The specimens were then encased in fresh concrete. In order to maintain the proportioning in concrete mix design, the concrete for the specimens was delivered from a central mixing plant, and to ensure the good quality of the specimens, all of them were properly compacted and their surface levelled and smoothed.

2.1 *Actual values of material characteristics*

Concrete cubes, cylinders and blocks were made to determine the real parameters of the delivered concrete during the experiment. The actual values of material characteristics of the steel were

Table 1. Tensile test results of structural steel.

Test No.	f_y [MPa]	$f_{y,Avg.}$ [MPa]	f_u [MPa]	$f_{u,Avg.}$ [MPa]	A [%]
1	317	315,3	397	396,0	41,7
3	313		397		41,7
4	316		394		40,0

Table 2. Compressive strength of concrete.

Cube strength				Cylindrical strength			
Test No.	Weight [kg]	Max. Load [kN]	$f_{ck,max}$ [MPa]	Test No.	Weight [kg]	Max. Load [kN]	$f_{ck,max}$ [MPa]
1	7,679	840	37,38		10,70	590	33,26
3	7,244	870	38,88		10,65	595	33,97
4	7,605	880	38,87		10,85	590	33,66
		Average	38,56			Average	33,63

determined from the stress-strain diagrams obtained by tensile tests. Three equivalent tensile tests were performed for a given type of steel beam (Kmeť et al. 2010). Subsequently there were determined their average values, which are considered as being nominal to calculate theoretical ultimate loads and evaluation of test results.

Tensile tests of three samples determined the average yield strength of structural steel fy = 315,3 MPa.

Concrete was brought from the central mixing plant. Three pieces and samples for strength testing of concrete were produced. It was necessary to verify the actual compressive strength of concrete. That strength test was a part of the certificate from the supplier. Simultaneously, there were conducted cylinder and cube strength tests separately for each batch of concrete. Tests of concrete compressive strength were realized in the timeframe t = 28 days and then on the date of the test. The selected measured values are shown in Table 2.

At the time of testing concrete compressive strength fc = 33,63 MPa was found out by the test of concrete compressive strength for cylindrical samples. Yield strength of reinforcement was considered by tabulated values fsk = 490 MPa.

2.2 Test specimens of closed steel sections

All test specimens were manufactured in compliance with STN EN 1994-1-1 Annex B. Composite action was ensured by means of a closed steel section made by welding a U-shaped steel plate 6 mm thick to the bottom flange with protruding ends. Holes 50 mm in diameter were cut by flame in the webs at an axial distance of 100 mm. Transverse reinforcement 12 mm in diameter was threaded through every third hole in the beam. Identically, holes 50 mm in diameter were cut by flame at an axial distance of 100 mm in the upper flange. The holes were arranged alternately either in the webs or flange of the beam in each cross-section. The length of the composite section was 500 mm. The steel part of the specimen was made of Steel S 235. The two connecting sections ensuring composite action were fillet-welded to each other with the help of steel plates (Fig. 2). The box section created was able to transfer load equally into both strip connectors. Steel sections were not coated with any protective paint so that the results of the experiment were not distorted.

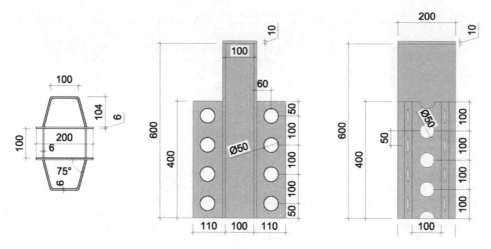

Figure 2.   Steel section of a composite member.

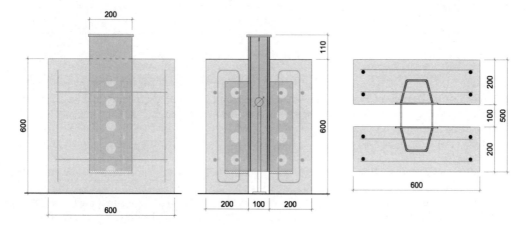

Figure 3.   Push-out test setup – Scheme.

2.3   *Push – out tests*

The framework was removed from the specimens after the concrete had hardened and the required strength reached. Next, composite action was experimentally verified. Static load was applied in the experiment. DrMB300 hydraulic press was used producing a force up to 3000 kN (Fig. 4). Displacement between concrete blocks and steel members was measured by inductive and mechanical sensors with a precision of 0.01 mm. Contact between the steel section of the specimen and strip connectors was provided by means of a steel plate.

Specimens of standard shape and dimensions according to STN EN 1994-1-1 Annex B were used in push-out tests. The dimensions of the concrete block were 600 × 600 mm and the length of the steel (composite) part embedded in concrete was 500 mm. The force applied was monitored and controlled during the experiment. The load was gradually raised with an increment of 10% of the estimated load-bearing capacity of the specimen to 40% of its previously calculated overall load-bearing capacity. At this stage the specimen was repeatedly unloaded and loaded within the range between 10 and 40% of its estimated load capacity. After 25 loading and unloading cycles, the experiment proceeded by incremental advances until the ultimate strength of the specimen was reached. Apart from the amount of slip, a rate of its increase was recorded at each loading

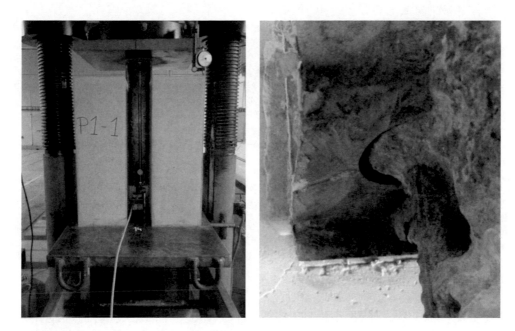

Figure 4.   Hydraulic press and Specimen after the termination of the experiment.

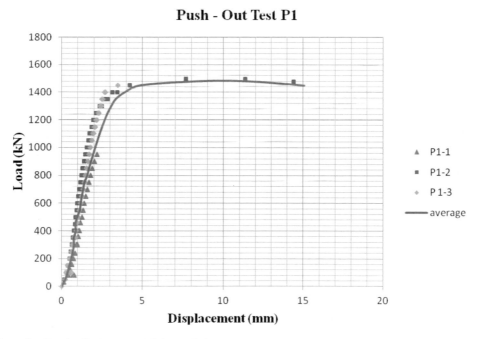

Figure 5.   Load – displacement (slip) correlations.

stage. It was detected after the first minute following the required loading level and then every 5 minutes (if necessary) until the slip consolidated. Slip was regarded as steady when the rate of its increase dropped below 0.01mm/min. The experiment was terminated when the loading force became uncontrollable and it was impossible to increase it any longer.

The mode of deformation was similar in all test specimens; however, failure of the box section occurred in the first specimen. As a result, all other specimens were then reinforced by adding steel plates on both sides of the section so that deformation of the steel part not acting compositely was prevented. Sound emissions were registered in the course of the experiment caused by cracking of concrete. That was temporarily accompanied by a decrease in force and a sudden increase in slip by several thousand of a millimetre. After a particular time the force stabilized and returned to its original magnitude. The results of push-out tests are presented in Fig. 5.

3 CONCLUSION

The article presents some results from push-out tests of steel-concrete composite members. It is obvious from the experiment that it is concrete that fails first in the loaded specimen, whereas a steel strip shear connector deforms only slightly. Critical elements in this type of composite action are concrete studs providing material cohesion. When these are deformed, total failure of a composite member occurs.

ACKNOWLEDGEMENTS

This article was written with the support of the operational research and development programme for the project: University Science Park TECHNICOM for innovative applications supported by knowledge technologies, Code ITMS: 26220220182, co-funded from the funds of the European Fund of Regional Development.

REFERENCES

Viest, I. M., Easterling W. S. and Morris, W. M. K., (1992), "Composite Construction: Recent Past, Pre sent and Near Future, in Composite Construction in Steel and Concrete, 2nd., ASCE, New York, 1–16.

M. Rovňák, A. Ďuricová, K. Kundrát, Ľ. Naď, (2006), Spriahnuté oceľ'o-betónové mosty, 1. vyd. – Košice: Elfa, – 2006. – 270 s. – ISBN 80-8073-485-2.

Leonhardt, F. (1938) Die Autobahnbrucke uber den Rhein bei Koln-Rodenkirchen, Bautechnik, ISSN 0932-8351, Vol. 28, No. 1, 1951, 289–291.

Andra, H.P. (1985) Neuartige Verbundmittel fur den Anschluss von Ortbetonplatten an Stahltrager, Beton-und Stahlbetonbeu, ISSN 0005-9900, Vol. 80, No. 12, 1985, 325–328.

Antunes, P.J. (1988) Behavior of Perfobond Rib Shear Connector in Composite Beams, Thesis (B.Sc.) Univer sity of Saskatchewan, 1988, Saskatoon, Canada.

Veldana, M.R., Hosain, M.U. (1992) Behavior of Perfobond Rib Shear Connectors in Composite Beams, Push out tests, Canadian Journal of Civil Engenieering, ISSN 0315-1468, Vol. 19,

Studnička, J. Macháček, J., Peleška, K. (1996) Nové prvky spražení pro ocelobetónové konstrukce, Stavební ob zor, ISSN 1210-4027, roč. 5, č. 2, 1996, 42–45.

Kvočák, V. – Kožlejová, V. – Dubecký, D. (2012) Analysis of Encased Steel Beams with Hollow Cross-Sections. 23rd Czech and Slovak International Conference "Steel structures and bridges 2012". Elsevier, e-Journal Procedia Engineering, vol. 40/2012, p. 223–228.

Li, A., and Cederwall, K., (1996), Push-Out Tests on Studs in High Strength and Normal Strength Con crete, Journal of Constructional Steel Research, vol. 36 No. 1, 15–29.

Kvočák, V., and Kožlejová, V. (2011) Research into Filler-Beam Deck Bridges With Encased Beams of Var ious Sections. 2011. In: Technical Gazette. Vol. 18, no. 3, p. 385–392. – ISSN 1330-3651.

Kmeť, S., Tomko, M., Demjan, I. (2010), Modelling and analysis of steel structures damaged by ef fects of fire, In: Selected Scientific Papers : Journal of Civil Engineering. Roč. 5, č. 1 (2010), s. 17–30. – ISSN 1336-9024.

Kmeť, S., Tomko, M., Demjan, I. (2012), Assessment of Fire-Damaged Steel Struc tures, In: Transac tions of the Universities of Košice. Č. 4 (2012), s. 25–36. – ISSN 1335-2334.

Advances and Trends in Engineering Sciences and Technologies – Al Ali & Platko (Eds)
© 2016 Taylor & Francis Group, London, ISBN: 978-1-138-02907-1

Electrical resistance spot brazing of two different alloys

V. Lazić, D. Arsić & M. Djordjević
Faculty of Engineering, University of Kragujevac, Kragujevac, Serbia

R.R. Nikolić & B. Hadzima
Research Center, University of Žilina, Žilina, Slovakia

ABSTRACT: The general problems related to electrical resistance brazing of various metals were considered in this paper. In brazing different electrodes and technological parameters are used. For experimental purposes, electrodes were made of highly alloyed tungsten steel and of copper with graphite and tungsten inserts, so the symmetrical temperature field was obtained. Success of brazing was determined by testing the mechanical properties, micro hardness and microstructure of the brazed joint. Hardness measurement results suggested the strain hardening of the steel thin sheets, what imposed necessity of recrystallization annealing. It was necessary to apply both the flux and the silver solder for joining, since copper and iron are poorly mutually soluble. When the metallurgy problems were solved and the optimal brazing parameters selected, the optimal mechanical properties were achieved, which were experimentally confirmed.

1 INTRODUCTION

Due to low solubility of copper-based alloys' and steel's basic components, i.e., copper and iron, brass and steel are difficult to join by the welding procedures. This problem can be solved by inserting the inter-layer made of the third metal, which is soluble both in copper and in iron. In this case, that is the silver solder in the form of a foil with addition of the flux in the form of the thin coating. To enable successful joining of those alloys it was necessary to perform numerous tests, which made possible selection of the soldering parameters, which gave the needed mechanical properties of the brazing joint. Besides that, it was also necessary to check the micro hardness and microstructure of the joint. Hardness measurements results shown hardening of the steel thin sheet, due to action of the compressive force during the brazing, what imposed the necessity for recrystallization annealing of the joint for improving the structure, namely the mechanical properties and the corrosion resistance.

Electric resistance brazing is the process of joining metals where the Joule's effect – RI^2 is used as the heat source. There are three methods of electrical resistance brazing: *spot wise*, used for joining steel and brass thin sheets, for joining parts in electronics, etc.; *seam wise*, used for making the box-like parts where the hermetic joints are required and *butt brazing*, used for brazing the sintered cutting platelets onto the steel holder of the cutting tools, Radomski and Ciszewski (1985), AMS-Metals Handbook (1979), JMM Brazing materials and applications (1967).

Heat amount (Q), released within the closed power circuit of the total electric resistance (R), through which flows the electric current (I), during the time (t), amounts to $Q = RI^2t$. The released heat amount is not the same in all the parts of the circuit, but it is proportional to the electric resistance within particular part. This is why one can obtain, by construction measures and selection of the electrode material, the highest resistance at the thin sheets overlap. Since it is not possible to prevent completely the heating of the rest of the circuit, it has to be reduced by water-cooling of the electrodes. Electrodes material's characteristics are given in Table 1.

Table 1. Characteristics of the electrodes' materials.

Electrode type	Electrode material	Own electric resistance Ωmm	Thermal conductivity W/mK	Hardness HB	Softening temperature °C
Metal	Hardened Cu	0.000019	394.0	95	150
	Hardened Cu alloys	0.000021	–	110–150	250–450
	Sintered Cu-W alloys	0.000056	–	200–280	1000
	Tungsten	0.000055	201.0	450–500	>1000
	Molybdenum	0.000057	146.5	150–190	>1000
Graphite	Soft graphite	0.010	150.6	–	–
	Medium hard graphite	0.018	50.2	–	–
	Hard graphite	0.061	33.5	–	–

2 ELECTRIC RESISTANCE SPOT BRAZING-ELECTRODES AND DEVICE

Electric resistance spot brazing was done on the devices for the spot welding. Energy for the resistance brazing is the alternate current, with voltage less than 20 V. The power of the apparatus ranges between 5 and 200 kVA.

Electrodes for brazing are different from the welding electrodes. The most applied are the electrodes made of the strain hardened copper alloys which posses very high electrical conductivity, Agapiou and Perry (2013), as well as of sintered alloys Cu-W and W-Mo. Sintered alloys Cu-W are especially convenient for brazing of copper and its alloys, since they possess relatively higher ohm resistance and lower thermal conductivity, what creates very favorable conditions for heating. Additional advantage of those electrodes is their significantly higher thermal stability with respect to copper electrodes. Besides the metal electrodes, the graphite electrodes are also used for electric resistance brazing. They are characterized by relatively higher own electric resistance (0.01 to 0.06 Ωmm, depending on the graphite type) and by the lower thermal conductivity, what allows for brazing with the lesser intensity current. Sometimes, the brazing electrodes are made of carbon or stainless steels (Cr-Ni steel – has higher electric resistance), Kianersi et al. (2014). For joining of materials with different thermal conductivity, electrodes made of different materials are applied; the metal with higher thermal conductivity comes to contact with electrode with the higher electric resistance.

To make the soldering process successful, one must apply fluxes. The flux is being deposited in the form of the water or alcohol solution, immediately prior to soldering, to prevent the flux evaporation. The amount of flux must not be too large, since it soils the working surfaces and thus slows the heating process. Recommendation related to soldering of the copper and steel parts, i.e., the copper parts need to be silver coated, while the steel parts should be zinc coated, either by the electroplating or by immersion. During the soldering, copper is alloyed by zinc due to heating, thus, by composition the joint point corresponds to brass. This type of soldering is widely applied in electronics, Radomski and Ciszewski (1985), JMM Brazing materials and applications (1967).

3 SOLDERING PARAMETERS AND TECHNOLOGY

3.1 Soldering equipment and base metals properties

Experimental investigations were conducted with the aim to determine soldering parameters and technology (current intensity, time, force, clearance, type of solder and flux, etc.), so the joints would correspond to required pressure characteristics. Tests were done on the stable apparatus for the spot welding TA-60. In Figure 1 is presented the general schematics of soldering used in experiments.

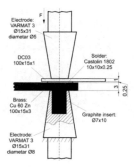

Figure 1.　Soldering schematics.

Table 2.　Characteristics of the electrodes' materials.

Electrode type	Electrode material	Own electric resistance Ωcm	Thermal conductivity W/mK	Hardness HB	Softening temperature °C
Metal	Hardened Cu	0.0000019	394.0	95	150
	Hardened Cu alloys	0.0000021	–	110–150	250–450
	Sintered Cu-W alloys	0.0000056	–	200–280	1000
	Tungsten	0.0000055	201.0	450–500	>1000
	Molybdenum	0.0000057	146.5	150–190	>1000
Graphite	Soft graphite	0.0010	150.6	–	–
	Medium hard graphite	0.0018	50.2	–	–
	Hard graphite	0.0061	33.5	–	–

Tests were conducted by soldering the brass thin sheet CuZn37, of dimensions $100 \times 20 \times 3$ mm and the steel DC 01 thin sheet of dimensions $100 \times 20 \times 1$ mm, with application of the silver solder Castolin 1802 of 0.25 mm thickness. Prior to soldering the contact surfaces of samples were degreased and ground with the sand paper 3M of granulation P180, to remove dirt.

The graphite insert at the bottom side plays the role of an electrode, which is placed in the hollow copper holder (Fig. 1). Electrode with the graphite insert has the increased electric resistance, what is convenient, from the aspect of thermal balance, since the brass is far better heat conductor than steel (Table 2). The electrode with the graphite insert should always be placed at the bottom side, to prevent graphite crumbling and reducing the contact surface between the electrode and the working piece. The compressive force should not be too large, so that the molten solder would not be squeezed out and the thin-walled copper holder.

3.2 Electrodes for resistance brazing

Special electrodes were produced for the needs of this experiment, made of copper alloys with graphite and tungsten inserts, as well as electrodes of highly alloyed high-speed steel HS 18-0-1 (Table 2). The purpose of built-in inserts is to increase the specific resistance of electrodes, i.e., when the current flows through the electrode the larger amount of heat is to be released, with respect to the case when there are no inserts on electrodes. Considering that the solder in the electric circuit – the brazing zone has the lowest melting point, only the solder would melt, while the thin sheets are being heated, though remaining in the solid state. This is why the lower intensity current is needed for soldering than for welding. It is possible to control energetically the process of electric resistance brazing by optimal combination of current intensity and time, Spitza et al. (2015).

Shapes and materials of electrodes used in this experiment are shown in Figure 2. At position 1 is presented the VARMAT 3 (Cu-Cr-Zn) alloys, which has the highest specific resistance of $17.2 \, \mu\Omega$mm. The total electric resistance of an electrode is increased by built-in insert made of

Figure 2. Various forms of used electrodes: 1 – copper electrodes with graphite insert, 2 – electrode of steel HS 18-0-1, 3 and 4 copper electrodes with tungsten inserts, 5 – copper holder of the lower electrode.

Table 3. Basic brazing parameters of steel and brass.

Brazing parameters	Electrode tip diameter d_e mm		Current intensity I_z A	Brazing time b_t		Compressive force F_z daN
	Copper	With graphite insert		per	s	
Regime 1	5	8	5700	12	0.24	280
Regime 2	5	8	5070	12	0.24	280

graphite EG01, with specific electric resistance of $10000\,\mu\Omega$mm. Electrodes at positions 3 and 4 in Figure 2 with the tungsten inserts are constructed according to the same principle.

Tungsten has specific resistivity of $55\,\mu\Omega$mm, which is significantly higher than that of the VARMAT 3 alloy, but still for another order of magnitude lower with respect to graphite, so the tungsten electrodes are scarcely used in these experiments.

To avoid problems related to manufacturing the electrodes with inserts, electrodes made of pure graphite or the highly alloyed tungsten steels are used sometimes. In this case, the electrode made of the steel HS 18-0-1 (0.75% C, 1.1% V, 18% W and the rest Fe) was used, Figure 2, position 2.

3.3 Energetic parameters for resistance brazing

Base metals taken for practical tests were the steel and brass thin sheets (melting temperatures $T_{tst} = 1525°C$ and $T_{tbr} = 920°C$, respectively), while the solder was CASTOLIN 1802 (with $T_{tsol} = 550°C$). Since the solder's melting temperature is significantly lower than the melting temperatures of both base metals, there is no danger of zinc evaporation from the brass. The brazing energetic parameters are being selected in such a way that the current intensity (driving energy) would be lower than for welding, while the brazing time should be as long as possible. The compressive force, i.e., the clamping force of the working pieces, must also be lower than for welding. Taking all the aforementioned into account and since there are no empirical expressions for selecting the brazing regime, the brazing parameters were determined based on numerous practical trials. The two regimes (Table 3), with different current intensity were selected that provided for the brazed joints of satisfactory carrying capacity.

3.4 Results of experimentally executed joints

Checking of the brazed joint, and indirectly the validity of the selected brazing regimes, was done for all the joints, first by visual inspection of the samples' and electrodes' surfaces, then by tearing tests and finally, by macro inspection of the fractured joints and metallographic investigations. The trace of the electrodes indentation was checked by visual inspection, as well as the depth of the imprint, deformation of the thin walled copper electrode holder, etc. Results of those investigations enabled successive corrections of the brazing parameters during the brazing experiments. Mechanical investigations were done by tearing of the brazed samples, with comparison of obtained results to mechanical characteristics of the base metals (Table 4).

Table 4.   Stresses and the tearing forces of the brass and steel samples.

Material	Cross-section dimensions mm	Cross-section area mm^2	Tensile strength R_m MPa	Yield stress R_{eH} MPa	Force at the yield point F_{eH} kN
CuZn37	20×3	60	300–380	max 200	max 12
DC01	20×1	20	280–420	250	≈ 5

Table 5.   Values of the tearing forces of the brazed joints.

Sample number		1	2	3	4	5	6	7	8	9	10	Average value
Force kN	Regime 1	5.45	5.82	5.46	5.60	5.71	5.32	5.48	5.38	5.65	5.60	5.55
	Regime 2	5.42	4.58	5.34	4.79	5.18	5.02	5.10	5.35	5.25	5.31	5.13

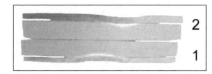

Figure 3.   Slits for metallographic investigations: 1 – Brazed joint realized according to regime 1, 2 – Brazed joint realized according to regime 2.

By reviewing the results presented in Table 4, one can notice that the force of 5 kN corresponds to the yield stress of steel, i.e., the beginning of the plastic deformation. That force can serve as an indicator for quality estimates of the executed brazed joints. In the ideal case of rational design, it is assumed that the carrying capacity of the brazed joint should correspond to that of the weaker of the two materials. However, usual request is that the brazed joint strength should somewhat supersede the strength of the weaker material.

Results of the shear tests, obtained by testing ten samples are presented in Table 5.

From Table 5 one can see that even the average values of the shearing force of the brazed joints were exceeding the limiting force that corresponds to the yield stress of the steel thin sheet. In the other words, the permanent deformation of the steel thin sheet would occur before the shearing of the brazed "spot", i.e. the nugget. Macro inspections of the nuggets showed that melting of the total volume of the solder did not occur in any of the cases, but only in the middle portion around the electrode. It was not possible to determine precisely the nugget's diameter since it varied between 5 and 7 mm regime 1 and between 4 and 6 mm for regime 2.

3.5   Metallographic analysis of brazed joints

For the purpose of the more precise analysis of the brazed joints, metallographic investigation of one joint executed by each of the regimes 1 and 2 was performed. The slits were ground from each joint in such a way that samples were cut longitudinally through the center of the brazed joint and casted into the same plastic, Figure 3. The light zones beneath the imprints (dents) on the steel thin sheet represent the brazing points. The electrode imprint on the steel thin sheet, in the case of regime 2, has uneven profile due to worn tip. The compressive force caused deformation, not only the steel thin sheet, but the brass thin sheet as well. The thickness of the solder was variable, from the initial thickness of 0.25 mm, to just 0.074 mm, at the end.

On certain samples on steel sheet was performed micro hardness measuring (Fig. 4). Change of hardness in the zone of the pressure action, with respect to the unchanged base metal is very prominent, especially in the middle portion of the sheet. This increase of hardness is a consequence of the cold plastic deformation of the steel thin sheet. Homogeneous hardness distribution and

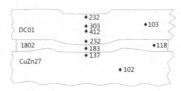

Figure 4.   Hardness HV0.2 distribution within the brazed joint.

uniform size of metal grains are obtained by the recrystallization annealing at temperature that does not jeopardize the brazed joint (T < 550°C). Increased hardness, for over 50%, was noticed in the solidified solder that was completely molten during the brazing process.

4   DISCUSSION OF RESULTS AND CONCLUDING REMARKS

Possibilities for electric resistance spot brazing of metals and alloys that are insoluble in the solid state, i.e., which cannot be welded by melting, were investigated in this paper, theoretically and experimentally. Within the experimental part, the possibility for brazing of brass and steel was investigated. It was shown that joints executed in laboratory conditions have complied with all the requirements related to mechanical properties of the brazed joints. Results of these voluminous investigations could serve as a basis for manufacturing application of resistance brazing of different metals.

Advantages of the resistance brazing, with respect to other brazing procedures, mainly consist of: fast heating of the joining zone, high productivity of the process, high quality of joints, possibility for easy regulation of temperature and heating time, possibility for monitoring the process flow, process simplicity, lower price, etc. The basic disadvantage of the electric resistance brazing is that it could be done only on parts of smaller dimensions, of a simpler shape and with application of costly silver solder.

Further research directions, for the purpose of improving this joining method, would be in proper selection of the brazing parameters, use of graphite inserts of the higher electric resistance and application of a solder in the form of a foil with thickness less than 0.1 mm. Additionally one must consider application of special auxiliary tools for controlling the optimal clearance between parts that are being brazed.

ACKNOWLEDGEMENT

This research was partially financially supported by European regional development fund and Slovak state budget by the project "Research Center of the University of Žilina" – ITMS 26220220183 and by the Ministry of Education, Science and Technological Development of Republic of Serbia through grants: ON174004, TR32036, TR35024 and TR33015.

REFERENCES

Agapiou J.S. & Perry, T.A. 2013. Resistance mash welding for joining of copper conductors for electric motors. *Journal of Manufacturing Processes* (15): 549–557.
ASM-Metals HandBook Vol 6-Welding-Brazing-Soldering. 1979. Metals Park Ohio, USA: ASM.
JMM Brazing materials and applications: Resistance brazing 1967. London: Johnson Matthey Metals Ltd.
Kianersi D., Mostafaei A., Amadeh A. A. 2014. Resistance spot welding joints of AISI 316L austenitic stainless steel sheets: Phase transformations, mechanical properties and microstructure characterizations, *Materials and Design* (61): 251–263.
Radomski, T. and Ciszewski, A. 1985. *Lutowanie*, Warzsawa: WNT.
Spitza M., Fleischanderlb M., Sierlingerb R., Reischauerb M., Perndorferb F., Fafilek G. 2015. Surface lubrication influence on electrode degradation during resistance spot welding of hot dip galvanized steel sheets. *Journal of Materials Processing Technology* (216): 339–347.

Advances and Trends in Engineering Sciences and Technologies – Al Ali & Platko (Eds)
© 2016 Taylor & Francis Group, London, ISBN: 978-1-138-02907-1

Experimental application of timber 3D elements on a slope affected by rill erosion

M. Lidmila, V. Lojda & H. Krejčiříkova
Czech Technical University in Prague, Faculty of Civil Engineering, Prague, The Czech Republic

ABSTRACT: In the framework of the Ministry of Industry and Trade of The Czech Republic project no. FR-TI4/330 new 3D timber post elements named "slope rosettes" were designed, manufactured and experimentally installed on a slope adjacent to a railway track. The slope rosettes were designed to ensure retention of soil and rock materials carried by water runoff on slopes affected by water erosion. In the article, different types of slope rosettes are described as well as the way of their installation on the test site. Furthermore, evaluation of the effectiveness of the observed slope rosettes 2.5 years after their installation is shown.

1   INTRODUCTION

In cases where the railway track is routed in a valley or on a slope (Figure 1), there is a need to deal not only with the stability of the slope of the railway substructure but also with the following valley slope. The condition and the structure of a valley slope may directly threaten the safety of the railway traffic. As an example, can be mentioned the incident which became at the railway track Všetaty – Děčín in km 416.600 416.750. In 2007 due to the cloudburst the track no. 2 was flooded with the mixture of soil, rocks and organic material (in volume of about 200 m^3) from the adjacent slope (Lidmila & Pýcha 2012). In 2008 there were realized extensive technical arrangements (rockfall barriers and fences) to minimize the threat caused by soilstone flow. The goal of the implemented arrangements was to catch soilstone flow in the lower part of the valley slope. It was just partially focused on the surface stabilization in the whole area of the valley slope. The research team took the advantage of its knowledge of this area and designed and also brought into the practice the innovative 3D stabilization features called the "slope rosettes". It was done within the project FR-TI4/330 supported by The Czech Ministry of Industry and Trade of The Czech Republic titled: "Research and development of innovative 3D features to protect the stability of slopes threatened by erosion". The railway track and the adjacent slope form together the area called "The experimental locality Libochovany".

2   DESCRIPTION OF THE EXPERIMENTAL LOCALITY LIBOCHOVANY

The experimental locality for installation of the slope rosettes is located on the right side of the river Elbe between the villages Libochovany and Velké Žernoseky and lies in the charming region called Porta Bohemica. The railway line no. 072 (according to the traffic order of The Czech Republic) is double railway track. In the observed section is cut into the slope. The observed area lies in the steep slope above the railway track, where the soilstone flows arise and where these flows may threaten the railway transport as it has already happened before.

The area is classified as the slope instability of nature origin in the database of The Czech Geological Survey (2015). The panoramic view of the experimental locality Libochovany is shown in Figure 2.

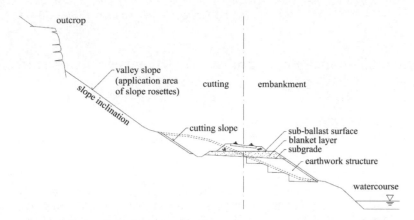

Figure 1.   The cross section of a track bed cut into a slope.

Figure 2.   The panoramic view of the experimental locality Libochovany – the slope with four marked erosion rills, where the slope rosettes were installed.

The valley slope is a part of both the crystalline complex Oparenske Valley (Opárenské údolí in Czech language) and The Czech Gate (Česká brána). The pre-quaternary soil consists of mica gneiss to magmatites of Early Paleozoic – Proterozoic ages. Quarter is formed of irregular position of the diluvia soilrock deposits with a thickness of 1.5 meters. There are many rock outcrops in the slope (Lidmila 2007).

The slope inclines to the west at angle of 30° to 40°, 33° on average. From the north as well as from the south the area is surrounded by the old quarries that came from the past times (1870–1874) when the railway line was constructed. The slope is terminated by the slope step in the east and there is also the ancient settlement Hrádek (264 above sea level). The bench had been built into the slope already during the construction of the railway line. The bench lies in the height of 3 to 7 meters above the track No. 2 which is from 7 to 15 meters from the centre of the track. The bench edges were provided with walls and posts. There is no watercourse or spring in the slope.

The valley slope is a part of Landscape Park (protected area) called Bohemian Central Uplands (České středohoří). The slope is covered with oak forest with disrupted tree canopy, i.e. there are gaps and space between the treetops.

The main causes of development of the soilrock flows in the mentioned locality are (Lidmila 2007):

• inclination of the slope higher than 25°;
• downpours;
• oak forest with disrupted tree canopy;
• grassland grazed by overpopulated deer – mouflons;
• existence of areas with uncovered rock coating that becomes eroded.

Figure 3.   Formation of the unaffected slope step in the experimental locality Libochovany.

Furthermore it is necessary to mention the negative impact of preferential trajectories of flows in the valley slope area. Those preferential trajectories of flows are called within the research project the erosion rills.

In 2008 the company STRIX Chomutov implemented three main technical steps to reduce the risk of further threat to transportation. For the highest part of the slope the erosion control measures were designed. They consisted of spatial degradable geomat anchored to the slope using steel mesh with mesh size 60 × 80 mm. In the middle part of the slope high strength intercepting fences were designed and implemented. In the lowest part of the slope – at the edge of the old slope bench – one meter high barrier was designed and implemented. It was made of discarded concrete sleepers SB8. Technical details of the individual measures states Pýcha & Lidmila (2012).

The steps implemented in 2008 do their protective function. However in the first place they were focused on catching the rocks, boulders and other rock objects. The measures did not sufficiently solve the situation in individual erosion rills, where the erosion went on influenced by downpours. The research team of the project no. FR-TI4/330 used their detailed knowledge of this area and proposed experimental verifying of the retention effectiveness of the slope rosettes that had been installed straight into the erosion rill.

3   SLOPE ROSETTE DESIGN

The main idea of designing the slope rosettes was to create such an element that would replace supporting and retention effect of trees and shrubs in the erosion rill (Figs. 3–4). Trunk and root system of a tree growing in the axis of the slope naturally help to create an unaffected slope step, which is formed out of soil, branches and stones carried by water running over and thus way the longitudinal inclination of the rill is locally reduced (Fig. 3).

For the structure of slope rosettes were determined following general characteristics:

- high stability of an element in the slope with angle α of 30° to 40°;
- high resistance to overturning caused by animals;
- simple structure enabling installation in difficult conditions;
- high retention efficiency;
- nature friendly materials.

Both the "Czech hedgehog" – an anti-tank obstacle used since the 30s of the 20th century and the obstacle called "Spanish rider" or "Spaniard" used against cavalry in the Middle Ages has inspired the research team to find out the solution – the slope rosette. Both of these structures are highly resistant to overturning. Even if they turn over, they still fulfill its function.

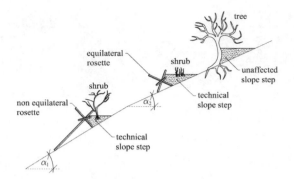

Figure 4.   Formation of slope steps in the erosion rill.

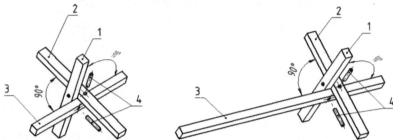

Figure 5.   The illustration of the slope rosette: (left) equilateral slope rosette; (right) non equilateral slope rosette; 1 & 2 – rod, 3 – strut; 4 – pin.

A bar is the basic structural element of the slope rosette. Three equally long bars in the space can form a geometric axis of the body which is called an octahedron. In terms of structural mechanics it is a bar structure with one joint where the three equally long bars are jointed at the mutual angle 90°. From the perspective of the possibility of catching the rocks and soil the individual rods have no effect. Therefore the slope rosette structure was supplemented by a technical retention membrane, which is mechanically fastened to the structure of rosette. The slope rosette is therefore composed of a supporting rod structure and a technical retention membrane. The structure is determined by rods and strut. The description of the slope rosette elements is shown in Figure 5. Timber spruce prism bars with cross-sectional dimensions from 30×30 mm to 90×90 mm were chosen as the basic material for prototyping of the slope rosettes. There were no other requirements on wood (such as moisture content, impregnation, strength or modulus of elasticity). The joint was designed with three beech pins with dimension 20 mm in diameter (Kulhánek 2013).

At the beginning there were made six types of slope rosettes, different shapes and dimensions. These types were identified by Roman numerals. The specifics of each type are summarized in the Table 1.

The slope rosette labeled as "Non equilateral" means that one rod directed along the fall line of the slope was extended (Fig. 5). This rod has been marked as a strut. Production of supporting structures slope rosettes took place in the workshops of the company STRIX Chomutov, which produced a total of 300 pieces.

4   FIELD INSTALLATION

4.1   *Installation*

Installation in the experimental locality Libochovany was held during November and December 2012. Four significant erosion rills were selected and identified by letters A–D in the area (Fig. 2).

Table 1. Types, shapes and dimensions of the slope rosette structure.

| | | Dimensions | |
| | | Rods | Strut |
Type of the structure	Shape	mm	mm
I	Equilateral	300	300
II	Equilateral	600	600
III	Equilateral	900	900
IV	Equilateral	1200	1200
V	Non-equilateral	1200	900
VI	Non-equilateral	2400	1200

Table 2. Material types of the retention membrane.

Type	Technical specification
Fine nylon meshes	PE-HD, knitted mesh
Antierosion biodegradable geomat	Biomac C
Coir (coconut net)	Mesh 50 × 50 mm
Steel mesh + anti-erosion geomat	plastic coating, mesh 60 × 60 mm + MACMAT, Maccafferri
Geotextile	non-woven, MACTEX N

Figure 6. Slope rosettes installed in the Experimental locality Libochovany: (left) rosettes without the technical membrane; (right) the rosettes installed in the assembly with the ensure membrane made of coconut net.

Different kinds of rosettes systems were tested in each erosion rill. Moreover five different types of retention membrane were tested (Tab. 2).

To ensure the variety of results the rosettes were arranged on the slope individually. All factors were changed, i.e. the type of supporting structures and their placing individually or in assemblies as well as using different types of technical retention membrane. Technological installation procedure was divided into three main parts. In the first part the rosette structure was assembled afterwards it was fitted in the rill. In the second technological part the technical retention membrane was fastened on the rosette structure. In the last part of the installation the geodetically surveying was performed for each unique marked rosette.

4.2 The event in 2013

The slope rosettes were checked for its further progress in 2013 (after 1 year). It was found, that the smallest rosettes (type I) have been displaced in relation to the effects of erosion or local action of animals. None of the rosettes were significantly damaged or destroyed.

4.3 The event in 2015

The experimental locality Libochovany with the slope rosettes was checked in February 2015 (almost 2.5 years after installation) and detailed passportization of each rosette was conducted. The control was done first because of evaluating their retention functions, then checking their positions on the slope using the surveying and at last because of describing the technical condition of rosettes and the retention membrane. The slope rosettes were also evaluated visually. Regarding retention functions there were defined three possible conclusions: no effect, satisfactory effect or excellent effect of the retention ability.

5 CONCLUSION

The initial results of the long-term observation show that the slope rosettes are able to ensure retention of soilrock flow caused by erosion. The retention effect was not the same to all types of the rosettes. The excellent effect was achieved by the rosettes installed in the assembly across the entire width of the erosion groove. The most effective installation was formed with slope rosettes of structure type V and VI, combined with retention membrane made from steel mesh with anti-erosion geomat. That assembly could capture even a small rock block of size $0.6 \times 0.3 \times 0.2$ m. The slope rosettes type I, II without installed retention membrane, proved to be unstable and inefficient from the perspective of the retention or catching effect. Furthermore, the authors negatively evaluated the capture impact of non-woven geotextile membrane and anti-erosion biodegradable geomat. Since the retention membrane has a small stability to UV, it caused the degradation and thus the membrane was unable to fulfill its function.

ACKNOWLEDGEMENTS

This research was supported by the program TIP of Ministry of Industry and Trade of The Czech Republic in the framework of the project "Research and development of innovative 3D features to protect the stability of slopes threatened by erosion", the project no. FR-TI4/330.

REFERENCES

Czech Geological Survey, 2015. *Ground instabilities.* http://mapy.geology.cz/svahove_nestability/index_EN.html?config=config_EN.xml (accessed Mar. 20, 2015)

Kulhánek, A. 2013. The drawing documentation of prototypes hedgehogs. Partial research project report FR – TI4/330 – Production and testing of prototypes "hedgehogs" and "clayonnage". *STRIX Chomutov.*

Lidmila, M. & Pýcha, M. 2012. Efficiency of high strength intercepting fences in an alternative use. *Silnice-železnice* 7:113–116.

Lidmila, M. 2007. The ensuring of the slope stability km 416.600 to 416.750 Vsetaty – Decin. Preparatory construction documentation. *Stavební geologie – Geotechnika.*

Pýcha, M. 2013. Partial research project report FR – TI4/330 – Production and testing of prototypes "hedgehogs" and "clayonnage". *STRIX Chomutov.*

Structural health monitoring of bridges using terrestrial radar interferometry

I. Lipták, A. Kopáčik, J. Erdélyi & P. Kyrinovič
Slovak University of Technology in Bratislava, Faculty of Civil Engineering, Department of Surveying, Bratislava, Slovakia

ABSTRACT: Bridge dynamics is actually very often discussed topic in structural health monitoring and bridge engineering. Innovative approaches in measurement provide important information about behavior of monitored structures. The paper describes the terrestrial radar interferometry, which is innovative technology for monitoring of bridge dynamics in real time. Paper presents basic principles of radar interferometry, possibilities of usage in bridge monitoring and methodology of analysis of the measured deformation. Radar interferometry is able to determine the structural deformation of whole monitored structure in relatively high geometric and time resolution. The paper describes some experiences from realized case study at real bridge structure.

1 INTRODUCTION

Bridge structures are usually exposed to the greatest extent by external influences such as weather conditions and loading by some objects. These factors have a significant influence on the behavior of the structure which results in deformation of the whole structure or its parts. Changes in deformations of the structure have typically a cyclical behavior which reflects the influences of the surroundings. Therefore, the rate of the structure's stress and the magnitude of the impact of individual factors on the structure can be determined.

Nowadays, knowledge of the dynamic characteristics of behavior of a bridge structure is increasingly important. They are mainly caused by wind and moving objects on the structure (pedestrians, cyclists, vehicles). They affect the resonant behavior of the structure which results in the dynamic deformation of the structure. They are usually described by the modal characteristics of the structure's deformation (vibration modes). For the safe operation of the structure it is necessary to design these deformations by computational modelling and to monitor them during loading tests. Also long-term monitoring of the structure using suitable methodologies is essential. These data significantly contribute to the stable and safe operation of the bridge and can be used for the calibration of a structural numerical model (Braun et al. 2014).

The technology of terrestrial radar interferometry is innovative method for deformation monitoring of engineering structures. The technology supports long-term monitoring of static and dynamic deformations of structures at the high relative accuracy level of 0.01 mm. Principle is based on the satellite radar interferometry which is known as InSAR (Interferometric Satellite Radar).

Applications of terrestrial radar interferometry in structural health monitoring are very wide. It can be used for a landslide monitoring, monitoring of water dams and mining structures. Dynamic monitoring is applicable for a monitoring of tall buildings (skyscrapers, towers etc.) and bridge structures (Owerko et al. 2012).

The paper is focused on general possibilities of this innovative technology. Paper describes the methodology of measurement and data analysis based on the real experience of research at real bridge structure. Research experiments were realized on the Cycle Bridge over Morava River in Devínska Nová Ves. This structure is very flexible cable-stayed steel structure with considerable

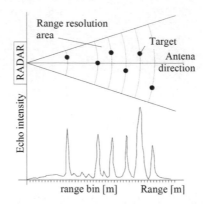

Figure 1.   Radar range resolution principle.

deformations. Devínska Nová Ves – Schlosshof cycling bridge is constructed on a cycling route between an urban area of Bratislava – Devínska Nová Ves and the Austrian village of Schlosshof. The bridge spans the Morava River at the 4.31 km mark. The river forms the border of Slovakia and Austria. The bridge is a steel truss structure consisting of three parts with a full length of 525.0 m. The main steel structure spans the Morava River and its length is 180.0 m. The main steel structure which is the subject of the dynamic deformation monitoring is a symmetrical cable-stayed steel structure with three sections. The bridge girder's height is from 2.00 m to 2.80 m. The girder has a ring shape with a radius of 376.35 m in the middle section. The bridge deck is orthotropic and cable-stayed by loading rods which are anchored at the top of all four steel pylons with heights of 17.85 m.

2   TECHNOLOGY OF RADAR INTERFEROMETRY

Ground-based radar is an innovative measurement approach for the dynamic deformation monitoring of the large structures such as bridges (Gentile 2009). This chapter focuses on basic principles of the technology.

2.1   *Basic principle of radar measurements*

Radar measurements use the Stepped Frequency Continuous Wave (SF-CW) technique in microwave frequency band. This approach enables the detection of target displacements in radar's line of sight. The basic principle of the technique is the transmission of a set of sweeps which consists of many electromagnetic waves at the different frequencies. Pulse radar generates short-term duration pulses to obtain a range resolution which is related to the pulse durations according to

$$\Delta r = \frac{c\tau}{2},$$

(1)

where c is the speed of light in a free space and τ is time of the pulse's flight. At each time interval of the measurements the components of the received signals represent a frequency response measured at the number of discrete frequencies. The frequency response is transformed to a time domain by the application of an Inverse Fourier Transformation (IFT). The system then builds one-dimensional image – a range bin profile where the reflectors are resolved with a range resolution according to their distance from the radar (Fig. 1).

When the range profile is generated, the displacements of the targets are detected by the Differential Interferometry technique. This approach compares the phase delay of the emitted and

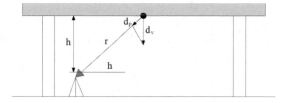

Figure 2.   Radial displacement and projected vertical displacement.

Figure 3.   Interferometric radar IDS IBIS-S and antenna.

reflected microwaves. Radial displacement is therefore linked with the phase delay between two acquisition phases by the following

$$d_p = \frac{\lambda}{4\pi}(\varphi_2 - \varphi_1),$$

(2)

where λ is the wavelength of the signal. When the displacement at the point is zero, there is no difference of phase delays.

The displacement can be measured in selected radial resolution area. Displacement is defined as an average displacement of all points in a selected radial resolution area. Maximum displacement between two acquisitions is restricted by the phase ambiguity at the level of 4.38 mm.

2.2   Principle of radar measurements in bridge monitoring

Displacements in radial resolution areas are measured in a radial direction. Generally, radar measurements are oriented in a different direction in comparison with the required direction of bridge deformation which is vertical direction most often. This problem can be solved by transformation of radial displacements to vertical displacements based on the known geometric relation between radar and bridge structure. This approach is illustrated in Fig. 2.

For higher accuracy of displacement transformation it is necessary to measure these parameters by land surveying approaches, especially by total station.

2.3   Interferometric radar IDS IBIS-S

Ground-based radar IDS IBIS-S measures dynamic displacements by comparing the phase shifts of reflected radar waves collected at the same time intervals. Displacement is measured in a radial direction (line of sight). The minimal radial range resolution of the radar is 0.5 m. The accuracy of the measured displacements is at a level of 0.01 mm but it depends on the range and quality of the reflected signal (Bernardini et al. 2007).

Frequency range of measurements is at the level up to 200 Hz. Maximum range of measurement is up to 1000 m. System works at microwave frequency band from 17.1 GHz to 17.3 GHz. The

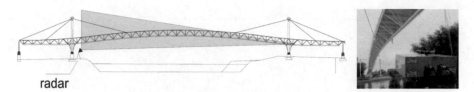

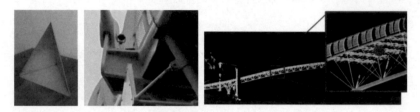

Figure 4.   Possibilities scan area performed by interferometric radar.

Figure 5.   Steel beams as natural reflector (left) and corner reflectors (center and right).

measurements and data registration are managed by the IBIS-S operational software installed in a notebook (Bernardini et al. 2007).

3   MEASUREMENT REALISATION

Radar interferometry is the very strong tool for the deformation monitoring of bridges. In practical applications we have to accept some risk and specific behavior of this technology which are caused by physical principles of this technology.

Strong advantage of this technology is the possibility of synchronous measurement of the whole structure as illustrates Fig. 4. Deformation can be measured in very high resolution. On the other hand in this case it is important to pay particular attention on bridge geometry in connection with the radar position.

Lower radar tilt angles ($<5°$) affect lower accuracy of transformation radial displacements into vertical direction. Steel structures affect risk of multipath effect. This effect decreases the accuracy of the measured deformation; however Signal to Noise Ratio (SNR) of signal is high. These factors affect planning of radar position.

Increasing of SNR of signal can be performed by corner reflectors (Fig. 5 right). Corner reflectors at steel structures lost the importance. The reason is bad identification in radial profile. Good possibility is to use natural reflectors such as steel beams, etc. (Fig. 5 left and center).

Selection of measurement parameters (frequency, range and resolution) has to be designed in suitable ratio. Increasing of measurement range and radial resolution affects decreasing of frequency range of measurements. The optimal design of parameters has to be performed according to the application.

4   PROCESSING AND ANALYSIS THE DATA

Data processing is realized by several steps which are focused on measuring the amplitude and frequency of dynamic deformation in selected parts of a structure. Data analysis is generally divided into following steps:

- transformation of radial displacements in to vertical or horizontal direction,
- measuring points identification,
- modal analysis.

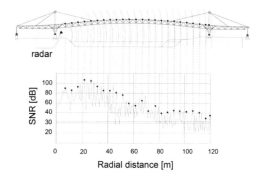

Figure 6.   Range bin profile.

4.1   *Displacements transformations*

Raw displacement data are measured in a radial direction (direction in the line of sight). For the most purposes it is required to obtain vertical deformations. The first step in the data processing of the radar measurements is therefore to set the geometry of the structure and to define the position of the radar and measuring points (Fig. 2).

4.2   *Measuring points identification*

We can define measured points manually using corner reflectors or by finding of such parts on the structure which have acceptable reflection parameters. A good reflection of the signal from the structure defines the range bin profile (Fig. 8).

The selected peaks correspond with the good signal reflection. Fig. 6 visualizes the estimated signal to noise ratio (SNR) of the signal's reflection depending on the structure's range which we obtain by geodetic measurements of structure's geometry. The SNR can be presented in linear or logarithmic scale. A signal with a higher SNR above the threshold has a better quality and there is an assumption of a higher degree of accuracy in determining the displacements. Points included to modal analysis are presented by peaks in the range bin profile.

4.3   *Modal analysis*

In the case of bridge modal analysis, spectral analysis methods are used. The most often used is the Fourier transformation. This approach describes a time-dependent signal by harmonic functions which can be used to transition of the signal from the time to the frequency domain. The signal can be expressed continuously or in a discrete form.

In practical applications a finite number of the data is analyzed by the numerical method of the Fourier transformation known as the Discrete Fourier Transformation (DFT). Calculation of the DFT can be realized by several algorithms. In the case of the dynamic deformation of bridges, the Fast Fourier Transformation (FFT) is the most often used method (Cooley et al. 1965). An alternative is the application of the Welch method which uses the FFT algorithm. In this case, the spectral density of the time series is computed from overlapped segments. These segments are analyzed by the FFT method. The results give a smooth periodogram and higher accuracy of the determined frequencies. However, the resolution of the magnitude spectrum is unfortunately lower (Welch 1967).

The spectral analysis of the longer time series is based on a determination of the common spectral density of the time synchronized measurements. The distribution of the spectral density in separate time series is often strongly differentiated. The resulting average normalized spectral density is calculated as the arithmetical average of all the normalized periodograms. The Average Normalized Power Spectral Density (ANPSD) describes the distribution of the spectral density

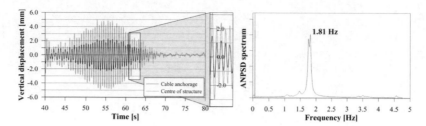

Figure 7.   Displacements at two different parts of structure measured synchronously (left) and average normalized spectrum (ANPSD) of whole structure's vibration obtained by interferometric radar (right).

of each time series included in the analysis. This fact enables to get a complex view of the static and dynamic properties of the monitored structure. The radar synchronous measurement of the whole structure gives possibility to calculate the average periodogram of a vibration of the whole structure.

Fig. 7 (left) illustrates the measured vertical displacements at two different parts of the bridge structure. These deformations are measured synchronously. Surveyor can select the part of structure which will be analyzed. Fig. 7 (right) gives information about structural deformation of the whole structure in a frequency domain using ANPSD spectrum.

Maximum measurable frequency is at the level of 50 Hz which is sufficient for the significant structural dynamic deformation identification. Vibration frequencies can be determined at the high accuracy level by using an average spectrum.

5   CONCLUSION

The paper describes the radar interferometry based on real research experience from Cycle Bridge in Devínska Nová Ves. This presented technology is shown as a suitable application for dynamic structural health monitoring. It is a perspective non-contact technology which is able to monitor entire deformation of structures synchronously with the high frequency. The accuracy of measured deformation is depended on a system configuration and quality of reflected signal. This can be improved by corner reflectors, especially when the surface of the structure is low reflective.

Measured deformations can significantly contribute to the prediction of possible failures of structures which can be reflected by anticipated temporal changes in the modal frequencies of structures. These failures can also be investigated by other surveying methods such as terrestrial laser scanning.

REFERENCES

Braun, J. & Štroner, M. 2014. Geodetic Measurement of Longitudinal Displacements of the Railway Bridge. *Geoinformatics FCE CTU 2014* 12(2014): 16–21. ISSN 1802-2669.

Cooley, J.W. & Tukey, J.W. 1965. An algorithm for the machine calculation of complex Fourier series. In *Mathematic Computation.* 19(90): 297–301.

Gentile, C. 2009. Radar-based measurement of deflections on bridges and large structures: advantages, limitations and possible applications. In *IV. ECCOMAS Thematic Conference on Smart Structures and Materials (SMART'09)*: 1–20.

Owerko, T., Ortyl, Ł., Kocierz, R., Kuras, P., & Salamak, M. 2012. Investigation of displacements of road bridges under test loads using radar interferometry–case study. In *Proceedings of the Sixth International IABMAS Conference.* 181–188. Stresa: Italy.

Welch, P. D. 1967. The Use of Fast Fourier Transform for the Estimation of Power Spectra: A Method Based on Time Averaging Over Short, Modified Periodograms. In *IEEE Transactions on Audio Electroacoustics* 2(1967): 70–73.

Advances and Trends in Engineering Sciences and Technologies – Al Ali & Platko (Eds)
© 2016 Taylor & Francis Group, London, ISBN: 978-1-138-02907-1

The impact of the severe corrosion on the structural behaviour of steel bridge members

M. Macho & P. Ryjáček
Faculty of Civil Engineering, CTU in Prague, Czech Republic

ABSTRACT: The process of the degradation affects the behaviour and the load capacity of many bridge structures. However, the behaviour of corroded and degraded steel members is still not fully investigated. The purpose of the presented research is to deepen the knowledge about the effects of degradation on the behaviour of steel structures. The paper is based especially on own measurements, that were done on real structures and their elements.

1 INTRODUCTION

The industrial development in the beginning of the last century would not be successful without the transport infrastructure. At those days, many steel bridges were built in the Czech Republic and also all over the world. Those bridges are usually at the end of their service life, however, the lack of the financial sources forces the railway owners for operating them much longer. Unfortunately, the bridges are often damaged by heavy corrosion, and also they can be affected by fatigue. Nevertheless, the exact process how the corrosion weakening affects deformational properties and load-bearing capacity of steel structures is not investigated yet. Although some research works were done earlier, see [2, 3, 4, 5, 6], there are still opened questions. If we know how such structures behave, we will be able to reconstruct them effectively and also provide the effective maintenance. In order to analyse this behaviour, it was decided to find suitable steel samples with heavy corrosion and thoroughly analyse their behaviour. The search of the samples was focused especially on steel railway bridges, that are common in the Czech republic.

2 DESCRIPTION OF THE BRIDGE

Finally, the most suitable bridge was chosen thanks to the cooperation with SŽDC, s.o., the owner of the railway infrastructure. The final choice between bridges was done with following objectives:

- to find samples with heavy corrosion,
- to find a relatively small structure, where the samples would be in the reasonable scale,
- the bridge should be under reconstruction.

The samples were taken from the bridge, located in the east Bohemia, close to the town Opočno. It is a typical steel girder railway bridge with upper deck with sleepers, that is a part of the railway track Týnište nad Orlicí – Meziměstí. The bridge was built in 1874. The main load-bearing structure is composed of two separate steel structures for each track. Both are the same and symmetrical about the axis. The steel structure consists of two main riveted girders of height 595 mm together with riveted stringers and crossbeams. The height of stringers is 210 mm and height of crossbeams is 380 mm. The full cross-section is shown in Figure 2. The superstructure is equipped with rails of shape S49. The rails are placed on timber sleepers (14 pieces on each structure) with dimensions $240 \times 240 \times 2500$ mm. The span of the bridge is 10.3 m (the view on the bridge can be seen in Fig. 1). The construction height is 0.61 m. Total width of the bridge is 11.6 m.

Figure 1.   View on the bridge.

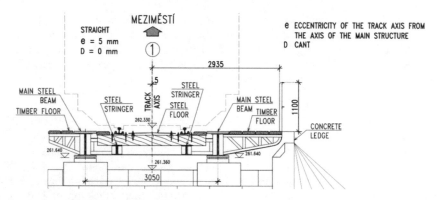

Figure 2.   Cross-section of the bridge.

Figure 3.   Steel stringers during the load test.   Figure 4.   Laser scan of steel stringer.

3   SAMPLES

The bridge was thoroughly inspected and the weak places were identified. The heavies corrosion was found on stringers, so that four of them were taken to the laboratory. The choice was done in order to have great variety of corrosion weakening. All stringers are riveted steel beams with span 1260 mm (see Fig. 3). The upper flange of the beams is composed of steel plate and two flanges of squares. The plate is 220 mm width and 10 mm thick. The squares have dimensions L100 × 80 × 8 mm. The total thickness of the upper flange is 18 mm (see Fig. 5). Steel plate with dimensions 12 × 200 mm forms the web of stringer. The bottom flange consists of two flanges of L100 × 80 × 8. And the total height of each stringer is 210 mm. All samples are affected by small to severe corrosion, especially the bottom flanges (see Fig. 5).

Totally 16 small samples from the webs and flanges were taken from all beams to find out the material properties in the tensile test. Then all beams were sandblasted and complete laser scanning was performed for the specification of the exact geometry (see Fig. 4).

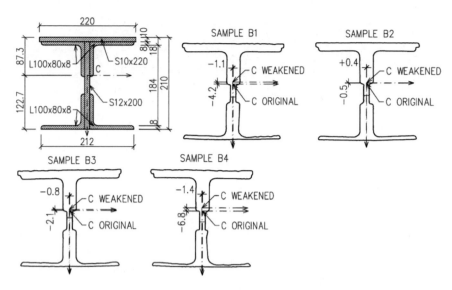

Figure 5.   Cross-section of original and weakened steel samples.

Table 1.   Comparison of cross-section characteristics of weakened and original beams.

Sample		Min. tbf mm	Corrosion weakening %	Min. tuf mm	Corrosion weakening %	A mm²	I_y 1×10^6 mm⁴	$W_{el,b}$ 1×10^3 mm³	$W_{el,u}$ 1×10^3 mm³	W_{pl} 1×10^3 mm³
Weakened	1	4.0	50.0	16.0	11.1	9750	58.440	474.35	703.84	665.69
cross-	2	6.7	16.3	17.3	3.9	10018	64.250	519.07	740.21	714.41
section	3	5.5	31.3	17.0	5.6	9973	61.761	501.67	725.07	696.26
	4	3.6	55.0	15.7	12.8	9229	54.461	432.61	676.62	615.79
Full cross- section		8	–	18.0	–	10156	65.344	532.33	748.93	727.27

On the Figure 5 the cross-sections of original and corrosion weakened stringers are shown. It can be noticed, that the centre of gravity moves because of corrosion loss. It is caused by non-symmetric corrosion loss especially on the bottom flanges. The corrosion weakening of the upper and the bottom flange in % can be seen in Table 1. Cross-section characteristics of the original and the corroded samples were computed: sectional area A, moment of inertia Iy and elastic section modulus Wel.

4   EXPERIMENTAL TESTS AND MEASUREMENTS

The standard static load test was applied on all samples. Each sample was equipped with several strain gages to see the distribution of relative deformation during loading. The arrangement of strain gauges is shown in Figures 6 and 7. Groups of strain gauges were added on places where corrosion pits were situated – usually on the bottom flange (see Fig. 6). Load-bearing capacity of each sample was founded out by test in bending. The arrangement of load tests can be seen in Figure 6. During the load tests relative deformations were measured in places where strain gauges were applied. In the middle of span sags were measured on the left and the right side of the bottom flange. Theoretical span of such supported beam was 1250 mm. The load was carried into the samples by couple of steel cylinders and the distance between them was 150 mm.

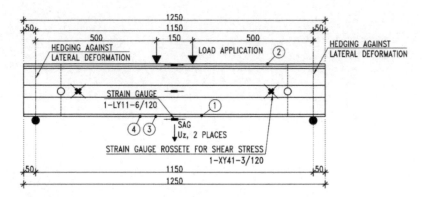

Figure 6.   Arrangement of load test.

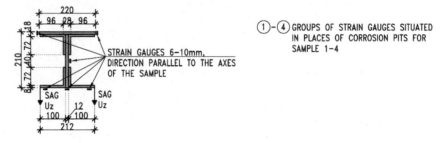

(1)-(4) GROUPS OF STRAIN GAUGES SITUATED
IN PLACES OF CORROSION PITS FOR
SAMPLE 1-4

Figure 7.   Arrangement of strain gauges.

5   EVALUATION OF THE EXPERIMENTS AND RESULTS

From each stringer four samples of material were taken – two samples from web and two samples from L100 × 80 × 8, so that the stress-strain curve and the yield points of steel was determined. From these results the design value of yield point was computed according to the standard ČSN EN ISO [1]. The value was computed as:

$$f_{yd} = m_x \cdot (1 - k_n \cdot V_x) / \gamma_R \tag{1}$$

where m_x = average value; k_n = coefficient for 5% quintile of characteristic value (here 1.69 for 16 members); V_x = variation coefficient; γ_R = partial coefficient of resistance. Partial coefficient of resistance can be expresses as:

$$\gamma_R = \exp(-1.645 V_x) / \exp(-\alpha_R \cdot \beta \cdot V_x) \tag{2}$$

where $\alpha_R = 0.8$; β = required index of reliability (here $\beta = 2.996$ for the estimated residual life 10 years and age of the samples 90 years). The design value of yield point is then $f_{yd} = 261 \cdot (1 - 1.69 \cdot 0.14)/1.14 = 180.2$ MPa.

Another design value of yield point was taken according to [1] for mild steel as: $f_{yd} = f_{yk}/\gamma_R$, where f_{yk} = characteristic value of yield point ($f_{yk} = 210.0$ MPa for mild steel); γ_R = coefficient of resistance ($\gamma_R = 1.1$ for mild steel according to [7]). The design value of yield point is: $f_{yd} = 210/1.1 = 190.9$ MPa. Then the design values of bending moment capacity were computed as $M_{el,Rd} = W_{el}.f_{yd}$, where W_{el} is the elastic cross-section modulus (see Table 1); and f_{yd} is the design value of yield point and as $M_{pl,Rd} = W_{pl}.f_{yd}$, where W_{pl} is the plastic cross-section modulus (see Table 1). The value $M_{el,Rd}$ and $M_{pl,Rd}$ was determined for both values of f_{yd}. All computed values can be seen in Table 2.

126

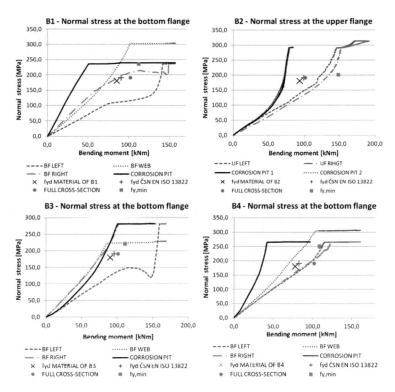

Figure 8.   Relation between normal stress on bending moment.

Table 2.   Yield strength and bending moment capacity (values with marks are shown in Figure 8).

Element		Design value of yield point (real)			ČSN EN ISO 13822 (mild steel)			Minimum value of yield point (real)			Test capacity
		f_{yd} MPa	$M_{el,Rd}$ kNm	$M_{pl,Rd}$ kNm	f_{yd} MPa	$M_{el,Rd}$ kNm	$M_{pl,Rd}$ kNm	$f_{y,min}$ MPa	M_{el} kNm	M_{pl} kNm	M_{pl} kNm
Weakened	1	180.2	85.5×	119.9	190.9	90.6+	127.1	236.0	111.9■	157.1	133.7
cross-	2*	180.2	133.4×	128.7	190.9	141.3+	136.4	201.0	148.8■	143.6	163.4
section	3	180.2	90.4×	125.4	190.9	95.8+	132.9	221.0	110.9■	153.9	143.0
	4	180.2	77.9×	110.9	190.9	82.6+	117.6	251.0	108.6■	154.6	137.9
Full cross-s.		–	–	–	190.9	101.6●	138.8	–	–	–	–

*Values of $M_{el,Rd}$, $M_{pl,Rd}$, M_{el} and M_{pl} for sample 2 are computed for the upper flange.

The relation between normal stress σ on relative deformation ε (measured during the test) was founded out from the working diagram of tested materials. The curve was appropriately interspersed with lines. The values of relative deformations ε up to the yield point were determined with the application of the Hooke's law, with modulus of elasticity E=200,000 MPa. This procedure was used for all samples, separately for material of the webs and material of the flanges. This derived the relation between σ and ε that was applied to the load tests results.

Figure 8 shows the relation between normal stresses on bending moment of samples 1–4. The values of bending moments were computed from the measured values of load forces and from dimensions used for load tests (see Figure 6). The green points on Figure 8 show load-bearing capacity of non-corroded cross section. The value of yield strength was taken as

$f_{yd} = 210/1.1 = 190.9$ MPa. The blue points characterize load-bearing capacity of weakened samples with the minimum values of yield strength which were founded from material tests for each sample.

6 CONCLUSION

Several significant conclusions can be found in the experimental study. Some of them are uncommon and differs from the typical understanding of the steel bridges and structural behaviour.

- The value of normal stress is much higher in corrosion pits that in other parts of steel elements. The normal stress in these places is growing faster and material starts to plasticize by smaller values of applied bending moment. This effect can be found in the parts in tension and compression. The relation between normal stress and bending moment is linear, but the material stops behave elastically much earlier. This effect can be significant especially for the fatigue behaviour, if the pit is located in the fatigue risk area.
- The design value of bending moment capacity ($M_{el,Rd}$ and $M_{pl,Rd}$) computed for the value of yield point is mostly smaller or equals to real load-bearing capacity of the tested samples. It means that there exists some safety reserve.
- The real load capacity of corroded stringers (M_{el} and M_{pl}) is higher, than the design load capacity of full cross section without corrosion.
- There is a difference in the stress in the web and the bottom flange angles. Although general assumption is, that the riveted cross section acts fully as a homogenous cross section, the reality differs significantly. The main cause can be found in the slip in the riveted connection and also small lateral bending due to the corrosion non-symmetry.

The above mentioned conclusions show, that many questions arises in the field of assessment of old structures and should be answered. Next research work will be focused on the numerical simulation of the samples behaviour and establishment of their function. Future research should be focused on the fatigue properties of the corroded members and fatigue reduction because of the surface deterioration and principles of the load transfer between different parts of the cross section.

ACKNOWLEDGEMENTS

Research reported in this paper was supported by Competence Centres program of Technology Agency of the Czech Republic (TA CR), project Centre for Effective and Sustainable Transport Infrastructure (No. TE01020168). Authors also greatly acknowledge the help and cooperation with the SŽDC, s.o.

REFERENCES

[1] ČSN EN ISO 13822, Bases for design of structures – Assessment of existing structures.
[2] Appuhamy, J., Ohga, M., Kaita, T., Dissanayake, R. Reduction of Ultimate Strength due to Corrosion – A Finite Element Computational Method. *International Journal of Engineering*. 2011, vol. 5, no. 2.
[3] Kayser, Jack R. and Andrzej S. Nowak. Reliability of corroded steel girder bridges. *Structural safety*. 1989, vol. 6, issue 1, pp. 53–63.
[4] Sharifi, Yasser and Reza Rahgozar. Remaining moment capacity of corroded steel beams. *International Journal of Steel Structures*. 2010, vol. 10, issue 2, pp. 165–176.
[5] François, Raoul, Inamullah Khan and Vu Hiep Dang. Impact of corrosion on mechanical properties of steel embedded in 27-year-old corroded reinforced concrete beams. *Materials and Structures*. 2012, vol. 46, issue 6, pp. 899–910.
[6] Beaulieu, L.-V., F. Legeron and S. Langlois. Compression strength of corroded steel angle members. *Journal of Constructional Steel Research*. 2010, vol. 66, issue 11, pp. 1366–1373.
[7] Koteš, P. and Vičan, J. Reliability levels for existing bridges evaluation according to Eurocodes. *Procedia Engineering*. 2012, vol. 40, pp. 211–216.

Advances and Trends in Engineering Sciences and Technologies – Al Ali & Platko (Eds)
© 2016 Taylor & Francis Group, London, ISBN: 978-1-138-02907-1

Influence of aggregate gradation curve on properties of asphalt mixture

J. Mandula & T. Olexa
Technical University of Košice, Civil Engineering Faculty, Institute of Structural Engineering,
Košice, Slovakia

ABSTRACT: Pavement construction materials as asphalt mixtures still need to be improved because of increasing traffic volumes. Present research is focused mostly on different kinds of chemical or natural additives which can improve their basic properties. However, there are also options how to make better asphalt mixtures without additives. The most important for asphalt mixtures is mainly the quality of components like aggregates, asphalt binders and fillers. The appropriate ratio of these components can result in an asphalt mixture of high quality without any chemical additives. As far as aggregates are concerned there is also option to use aggregates with different shapes of gradation curve which has also effect on the quality of asphalt mixture. This paper is focused on two different shapes of gradation curve for two different asphalt mixtures and their influence on the total quality of asphalt mixture. As comparison criteria basic asphalt mixture properties were chosen.

1 INTRODUCTION

The importance of correct design of asphalt mixture is very often emphasized in the research concerning pavement construction materials. The knowledge of design of pavement and asphalt mixtures is collected in manuals and directives of each country but basic principles are the same (Laboratoire central des ponts et chaussées 1997). There are also many publications which are focused on correct design of asphalt mixtures and suitability of different types of asphalt mixtures in pavement layers (Queensland Department of Main Roads Pavements & Materials Branch 2009). Each layer is characterized by thickness and properties of construction material. Requirements for each type of asphalt mixture are stated in European standards STN EN 13108 (SUTN 2013) and also in national appendix or national catalogue of asphalt mixtures such as KLAZ 1/2010 (MDPT SR 2010). These standards are based on long term research of asphalt materials and they state the best properties for lifespan extension of pavement. The endurance of pavement depends on material resistance to different negative influences of traffic or weather (Demjan I. & Tomko M. 2013).

One of requirements for asphalt mixture is percentage of passing aggregate through the sieve. The percentage of passing aggregates determines the gradation curve of asphalt mixture. This requirement is defined by maximum and minimum limit for each fraction of aggregate. The middle curve which represents ideal gradation curve for specific asphalt mixture is described by Fuller parable. See Equation 1 below:

$$Y = \left(\frac{d}{D}\right)^{0.5} * 100 \ \%$$

(1)

where Y = percentage of aggregate fraction passing (%); d = aggregate fraction size (mm); and D = maximal fraction size in mixture (mm).

The design of asphalt mixture should include gradation curve between minimum and maximum limits. Even though the position of gradation curve is limited there is still enough space to study the influence of gradation curve on basic properties of asphalt mixture (Airey, G.D. et al. 2008), (Elliott, R.P. et al. 1991). For example in case of aggregate size of 2 mm passing limits are from 25% to 50%. Gradation curve can be also changed by skipping one aggregate fraction which is not allowed by limits. The fraction skipped causes different arrangement of gaps and aggregate in the asphalt mixture. This option is mostly used in porous asphalt mixtures and asphalt concrete mixtures.

The other option could be also customization of binder volume in asphalt mixture. However, in many studies the best results were obtained using a standard minimum percentage of asphalt binder or percentage of binder calculated according to equation (2) for the calculation of optimum binder content. Equation 2 mainly uses surface of each fraction of aggregate.

$$p = n * \sqrt[5]{\varepsilon} * \frac{2.65}{\rho_a} \tag{2}$$

where p = optimal percentage of binder in asphalt mixture (%); n = coefficient of saturation; ε = specific surface of aggregate (m²/kg); ρ_a = bulk density of aggregate (mg/m³).

Low content of binder could cause bad aggregate coating which leads to insufficient compaction of mixture. However, a high content of asphalt binder could cause low temperature and deformation resistance of asphalt mixture.

2 ASPHALT MIXTURES STUDIED

This paper is focused on two types of asphalt mixtures for surface course of pavement. The studied asphalt mixtures are in the second quality class and experiments were aimed at improvement of their properties so they are comparable with asphalt mixtures of the first quality class. The first tested mixture was asphalt concrete with maximum grain size of 8 mm (AC 8) and the second tested mixture was asphalt concrete with maximum grain size of 11 mm (AC 11). In both these mixtures road bitumen with penetration range 50/70 (CA 50/70) is used as a binder. Mixtures do not contain any other additives.

These types of asphalt mixture are used in construction of pavements and mainly in surface layer of pavement. They are used in construction of local roads, low – loaded roads, sidewalks and car parks. Two types of asphalt mixture were tested in order to verify the different maximum grain sizes and different aggregate structures in the mixture. The source of used aggregate was the stone quarry Hradová near Košice city in Slovakia. The components used meet all requirements for using in asphalt mixtures.

2.1 Asphalt concrete – AC 8

Two variants of asphalt concrete AC 8 were studied. The first one has aggregate gradation curve which is located near the Fuller's curve. Gradation curve could not be exactly the same as the Fuller's curve because aggregate composition is never ideal and percentage of aggregate fractions could only reduce deviation. There are expectations that location of gradation curve close to the Fuller's curve can cause low air void content because of higher volume of fine aggregate. The air void content has a straight influence on other properties of mixtures such as water resistance, stiffness and indirect tensile strength. In the case of the first variant there were expectations of low indirect tensile strength and low stiffness because of higher volume of fine aggregate.

Gradation curve of the second variant of mixture is located close to the minimum limit curve. This limit curve set up minimal passing of each fraction. This kind of aggregate is mixture contains higher volume of coarse aggregate and the volume of middle and fine grains reduced. This variation of mixture should have higher air void content and also it should have better deformation resistance. These properties are expected because coarse aggregate in the mixture form a bearing skeleton of the mixture (Schlosser et al. 2011).

Table 1. Percentage of aggregate and asphalt binder (AC 8).

Mixture	Fraction (mm)	Proportion in aggregate (%)	Proportion in the mixture (%)
AC 8 1.variant	4–8	25.0	23.6
	0–4	68.0	64.2
	0–0.063	7.0	6.6
	Binder	–	5.6
	Total:	100.0	100.0
AC 8 2.variant	4–8	35	33.0
	0–4	58	54.8
	0–0.063	7	6.6
	Binder	–	5.6
	Total:	100.0	100.0

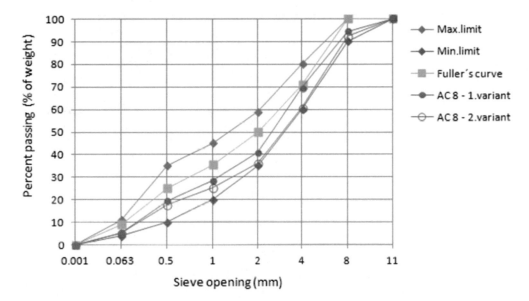

Figure 1. Gradation curve of both variants of asphalt concrete AC 8.

The content of bitumen binder was based on minimum recommended value but for both variants theoretical optimal content of binder was calculated. The calculated optimum content was 5.4% and 5.3% of mixture weight but minimum recommended content was 5.6% which was chosen for all mixtures studied.

The most important aspect for the quality of mixture design is a correct ratio of mixture components. The percentages of each aggregate fraction of mixtures and other components of asphalt mixtures are listed in Table 1.

Based on the percentages of aggregate fractions gradation curve of asphalt mixture was created. For each mixture there are specified minimum and maximum limit values of percentages passing through the characteristic sieve opening. Gradation curves for both variants of asphalt concrete AC 8 and also maximum and minimum limit values can be seen in Figure 1. Fuller's curve is located between the limit curves. This figure also illustrates that it is rather complicated to create gradation curve with same shape as limit or Fuller's curve.

2.2 Asphalt concrete – AC 11

This kind of asphalt mixture was also prepared in two variants. The difference between these two variants consisted in different shapes of their gradation curves. The first variant had gradation curve

Table 2. Percentage of aggregate and asphalt binder (AC 11).

Mixture	Fraction (mm)	Proportion in aggregate (%)	Proportion in the mixture (%)
AC 11 1.variant	8–11	15	14.2
	4–8	20	18.9
	0–4	58	54.8
	0–0.063	7	6.6
	Binder		5.6
	Total:	100	100.0
AC 11 2.variant	8–11	22	20.8
	4–8	22	20.8
	0–4	49	46.3
	0–0.063	7	6.6
	Binder		5.6
	Total:	100	100.0

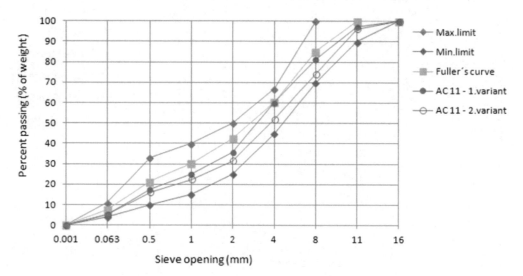

Figure 2. Gradation curve of both variants of asphalt concrete AC 11.

close to the Fuller's curve which means that proportion of coarse aggregate is low and proportion of fine aggregate is slightly higher. The total proportion of coarse aggregate is about 33% of the total mixture weight. Gradation curve of the second mixture variant was located near the minimum limit curve and proportion of coarse aggregate increased to 42% of the total mixture weight.

The optimum binder content calculated for these mixture variants was 5.3% of the total weight. Minimum standard content of bitumen should be 5.4% but in order to make it possible to compare all tested mixtures the used bitumen content was 5.6% of the total mixture weight. In comparison with AC 8 mixture AC 11 mixture has a higher content of coarse aggregate because it contains one more fraction of aggregate (8–11 mm). The influence of percentage of coarse aggregate could be seen not only on two variants of mixture but also between two types of asphalt concrete AC 8 and AC 11. In general higher percentage of coarse grains should cause better deformation resistance but there is also possibility that air void content in the mixture will be too high. From economical point of view mixtures with higher content of coarse aggregate are better because they need less binder and binder is the most expensive component of asphalt mixture. Proportions of AC 11 mixture components are listed in Table 2.

In Figures 1–2 it could be seen that gradation curve of mixture with higher proportion of fine aggregate can copy the shape of limit curve better than mixtures with high proportion of coarse

Table 3. Results of measurements.

		AC 8 (1.v)	AC 8 (2.v)	AC 11 (1.v)	AC 11 (2.v)	Limit
Binder content	(%)	5.6	5.6	5.6	5.6	5.6
Air void content	(%)	7.9	3.5	6.9	4.0	2.5–4.5
Voids filled by bitumen	(%)	61.7	78.6	64.7	76.4	75–83
Maximal density	(kg/m³)	2,462	2,351	2,425	2,417	–
Indirect tensile strength	(MPa)	0.9	1.1	0.8	1.1	–
Indirect tensile strength ratio	(%)	61.8	76.5	61	87.6	70
Stiffness 20°C (IT-CY)	(MPa)	5,317	6,730	4,950	5,424	–

Figure 3. Sample surface of first and second mixture variant (AC 8).

aggregate. On the other hand the mixture with high proportion of coarse aggregate can better copy the shape of Fuller's curve.

3 EVALUATION OF LABORATORY TESTS

Each sample of both asphalt mixtures was tested for basic asphalt properties and the results obtained were compared with standard limit values according to current European standards. The testing conditions such as temperature and others were strictly kept. The samples were prepared by Marshall hammer and the number of blows was different depending on the laboratory test for which the sample was prepared. For example the test of indirect tensile strength requires the sample compaction of 70 blows. Production and compaction temperature was 150°C for all samples. The results for each mixture variation are listed in Table 3.

The samples of mixture AC 8 can be seen in Figure 3. There could be seen that the sample of the first mixture variant has closed surface mostly because of high content of fine aggregate and the sample seems to have low air void content but test results showed high content of air voids. On the other hand the sample of the second mixture variant has open surface and it seems to have high content of air voids but the measured value is slightly under maximal limit value.

There is also possibility that in case of the mixture with higher content of fine aggregate, binder covers mostly fine grains and causes clumps. This effect could cause bad coating of coarse grains and in general this could cause bad compaction of this sample. In case of AC 11 mixture there was no visual difference between two variants of the mixture. The samples of AC 11 mixture have mostly slightly open surface because of high percentage of coarse grains.

4 CONCLUSION

The research in this paper was focused on the influence of gradation curve location between minimum and maximum limit curves on asphalt mixture properties. Two asphalt mixtures were chosen for the study because there is still need for more laboratory tests on different types of asphalt concrete.

In general most of the expectations mentioned in the beginning of this study were not fulfilled. Lower values of air void content were measured in mixtures with gradation curve located near the minimum limit curve. This means that the amount of coarse grains in mixture affects the value of air void content in the samples and it can also reduce maximal mixture density. The expectations of good deformation resistance of the second variant for both asphalt mixtures were confirmed. The indirect tensile strength increased by about 0.25 MPa for both mixtures. The indirect tensile strength ratio which describes asphalt mixture water resistance was better for the second variant of mixtures. There was difference of about 20% in comparison with the samples of the first variants of mixtures.

As far as stiffness of asphalt mixtures is concerned there are also unexpected results. The stiffness was measured by indirect tensile method with half sine loading shape and at temperature of about 20°C. There was an increase of mixture stiffness observed when the amount of coarse grains increased. The increase of stiffness was about 1,000 MPa. Another unexpected fact was that the stiffness of asphalt mixture AC 8 was higher than the stiffness of AC 11 mixture which could be the consequence of grain and air void allocation in the mixture structure.

Based on this research it can be concluded that location of gradation curve has a clear effect on basic properties of asphalt mixtures. Location of gradation curve close to the minimum limit curve has a positive effect on all studied properties.

ACKNOWLEDGEMENTS

The contribution was incurred within Centre of the cooperation, supported by the Agency for research and development under contract no. SUSPP-0013-09 by businesses subjects Inžinierske stavby and EUROVIA SK.

The research was performed within project NFP 26220220051, Development of Progressive Technologies for Utilization of Selected Waste Materials in Road Construction Engineering, supported by the European Union Structural Funds.

REFERENCES

Airey, G.D. & Hunter, A.E. & Collop, A.C. 2008. The effect of asphalt mixture gradation and compaction energy on aggregate degradation. *Construction and Building Materials. Volume 22, Issue 5:* 972–980.
Elliott, R.P. & Ford, Jr.M.C. & Ghanim, M. & Tu Y.F. 1991. Effect of aggregate gradation variation on asphalt concrete mix properties. Washington: USA.
Demjan, I. & Tomko, M. 2013. Frequency of the analysis response of reinforced concrete structure subjected to road and rail transport effects. *In: Advanced Materials Research : SPACE 013: 2nd International Conference on Structural and Physical Aspects of Civil Engineering:* High Tatras, Slovakia, 27–29 November 2013. Vol. 969 (2014), p. 182–187. – ISBN 978-303835147-4 – ISSN 1022-6680.
Laboratoire central des ponts et chaussées. 1997. French design manual for pavement structures – Collection Guide Technique.
MDPT SR. 2010. KLAZ 01/2010 : Cataloque of asphalt mixtures.
Queensland Department of Main Roads Pavements & Materials Branch. 2009. Pavement design manual. Queensland: Australia.
Schlosser, F. & Zgútová, K. & Híreš, V. & Nemec, B. & Križovenská, E. & Šrámek, J. 2011. Deformation Parameters of the Asphalt Mixtures in Pavement Construction Layers. In: XXIVth World Road Congress Roads for a better life Mobility Sustainability and Development. September 26th to 30th 2011 – Centro Banamex – Mexico City. DVD 12 s.
SUTN. 2013. EN 13108 – 1/O1: Bituminous mixtures – Material specifications – Asphalt Concrete.

Advances and Trends in Engineering Sciences and Technologies – Al Ali & Platko (Eds)
© 2016 Taylor & Francis Group, London, ISBN: 978-1-138-02907-1

Systematic approach to pavement management on regional and municipal road networks

J. Mikolaj, L. Remek & L. Pepucha
University of Žilina, Žilina, Slovakia

ABSTRACT: The article deals with pavement management system (PMS) on low class and municipal road networks. It describes road administration, which is characteristic to eastern European countries and contemporary problems of PMS integration from the viewpoint of these administration bodies. The article describes the reasons for the present unfortunate current state of affairs and their consequences. Consecutively, the article presents three distinct approaches for how to deal with the current situation. Implementation of PMS with data collecting capacity enhanced by the private sector, fixed repair plans, and pre-computed repair plans. These three different approaches to substitution of regular PMS and their implementation are explained addressing limited pavement data collection capacities, and/or lack of technical experience of regional road administrations. Overall, the article describes the problematic of pavement management and gives an insight into road network management system (PMS used for freeways and main arterial roads) and implementation of this system or its substitution for low class and municipal road network management.

1 INTRODUCTION

1.1 *Road network administration*

Road administrators differ significantly with available budged, length of roads they are responsible for, demands put on their assets, demands put on acquisition of new assets and many other issues, yet, their task is the same. Their task is to develop and maintain a safe, eco-friendly and efficient transport system. This may be seen as securing a fluent and safe transport on them entrusted roads by providing maintenance, winter service, repair, reconstructions and acquisition of new assets, according to concept of development of road network of Slovakia (Mikolaj, et al, 2012).

The main functions of road network administrators are:

– road management and creating conditions for safe traffic on responsible road network;
– increasing traffic safety and reducing harmful environmental impact of vehicles (Celko, Decky and Kovac 2009);
– organization of traffic and public transport;
– state and owner's supervision over road construction and road maintenance, road usage, the service level of roads and organizing state supervision over compliance with the requirements established by legislation;
– keeping road databank of roads, vehicles and public transport; observing special requirements established by legislation;
– participating in the elaboration of policies, strategies, and development plans of road development.

1.2 *Pavement Management Systems*

Systematic approach to the maintenance of rural road network is a very important issue from the viewpoint of public costs. Most countries developed custom PMS (pavement management systems) based on deterministic or probabilistic approach.

PMS used in Slovakia – "Road Network Management System" (RNMS). – The outputs of RNMS development are the different sophistication levels of operational funding allocation, with the main aim to maintain the road network at the required technical and operational level.

1.3 Peculiarities of low level road network management

The road network of Slovakia consists of 391 km of limited access roads (motorways and express roads) and 174,367 km of 1st, 2nd and 3rd class roads. The main objective of motorway network is to provide transit according to Pan-European transport corridors, namely the IV, V and VI corridor. The purpose of express road network is to collect and transfer the transport generated by Slovak republic's regions and contra wise to distribute transport from foreign countries from motorways to the body of Slovak Republic. The 1st class trunk roads fulfill the service task of transportation between regions of Slovak republic. 2nd and foremost 3rd class roads compose a rural road network. On top of this network, a network of urban communications and minor purpose communication is connected. Different types of roads have different owners and administrators with their executive offices. Their general task is to securing a fluent and safe transport on them entrusted roads by providing maintenance, winter service, repair, reconstructions and acquisition of new assets according to concept of development of road network of Slovakia.

Although road administrators of self-governing regions no doubt have similar operational and organizational needs and face the same general challenges as National Highway Agency or Slovak Road Administration, they lack any sophisticated planning and decision making for pavement management. The RNMS is successfully applied on motorway network, limited access road network and 1st class road network. Municipal road administrators and road administrators of self-governing regions however still rely on reactive repair planning. We know for a fact that this approach is wasting resources as repairs are conducted on roads near the end of their lifespan, repair in this advanced state of deterioration means a high-cost repair or full-blown reconstruction action. Regardless of whether this is due to the lack of adequate resources to establish the initial database and set up the system, or whether there is a general lack of technical expertise to implement the program, local agencies are in need of a methodology for effectively managing the various components of their pavement network.

There are three approaches how to implement a PMS for road administrators of self-governing regions:

– Implementation of the existing RNMS
– Use of a simplified PMS (fixed repair plans)
– Preparation of pre-computed repair plans and their variants using an external software tool

1.4 RNMS implementation for regional road administrators

Currently the internal data regarding pavement are being stored in a data warehouse They are currently collected by Slovak road administration for 1st class roads in a road databank. They also collect the data for limited access roads as contractual agreement with National Highway Agency. The division of Road databank of Slovak road administration is also responsible for keeping the database up to date. The problem lays in 2nd a 3rd class road network which volume exceeds the capabilities of Slovak road administration and their data collecting capacities. The solution to this problem may lie in procurement of Automated Data Collection Vehicle (Fig. 1). The vehicle should be equipped with Road Profiler, Line Scan Camera, Thermo camera, Ground Penetrating Radar and Global Positioning System. This would greatly expand the effectiveness of data collection and expand the capacity of roads which can be surveyed and kept updated (PIARC, 1997).

1.5 Use of simplified PMS (fixed repair plans)

Simplifying PMS to the form of fixed repair plan may be a good subsidiary and temporary solution (Pepucha and Remek, 2014) A maintenance standard is a schedule of repair and maintenance works

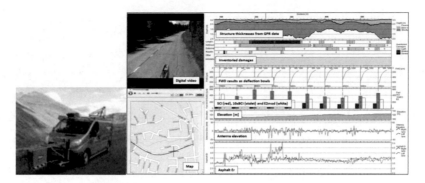

Figure 1.   Automated Data Collection Vehicle and its output (Road Scanners, 2014).

Table 1.   Common pavement repair technologies.

Technology	Period	Description	Effect
Basic surface treatment	1 year	Pothole patching and Crack sealing according to administrators available technologies.	A local defective part of pavement is treated for re-acquiring of lost geometry and structural properties.
Microsurfacing regeneration	5 year	Microsurfacing is a cold mixed polymer modified thin asphalt layer lied with traveling paving truck.	Microsurfacing restores lost surface properties and protects and preserves, extending pavement life.
Cover layer exchange	10 year	Upper part of the road is milled off and replaced with a new bituminous layer. The thickness may vary.	Continuous regaining of geometric, structural and surface properties.

which also represents the allowable limit for road deterioration. A standard is based on road class, characteristics of traffic and general operational practice. Generally, when roughness reaches close to the standard (fixed International Roughness Index, IRI), any treatment is required to restrain road roughness to go beyond the standard. Alternatively Rut depth may also be included as a factor (Gavulová, 1999). Standards have to optimum considering cost and road condition, and should be set at network level. The fixed maintenance standard prescribes the maintenance and repair procedures to certain years.

The maintenance and repair procedures prescribed by fixed maintenance standard don't always correspond with the actual needs of the road conditions nor do they take into account the budget possibilities of the road administrator. Nevertheless, it is an empirically based schedule of pavement treatment works which guarantees a good condition of the road throughout its whole life cycle. The downsides are obvious; the overall idea doesn't correspond with the procedures described in pavement management system theory with all the impacts that fact has on effective road administration.

The suitability of lower-cost maintenance standards for rural road network had to be evaluated to make recommendations for this approach to work. The condition of pavement related to service life for different variants is shown in Fig. 2.

Six Maintenance repair and rehabilitation (MR&R) variants were evaluated, each with different cost and effects on pavement conditions: Variant 1: current maintenance variant—very expensive variant appropriate only for road sections with heavy traffic load.

– Variant 2: microsurfacing based variant—safe to use on all rural roads.
– Variant 3: balanced cover layer exchange based variant—may be appropriate even for 3rd class roads with traffic load under 1000 annual average daily traffic (AADT) especially, if they are not subject to excessive high load vehicles encumbrance.

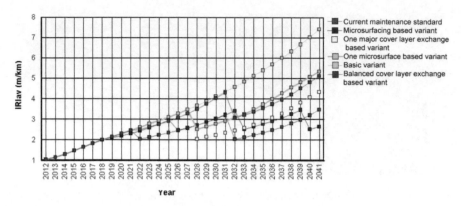

Figure 2. Average roughness by project for various fixed repair plans.

Table 2. Maintenance standard effects, NPV (net present value) and IRR (internal rate of return) (economic effectiveness ranking).

Name	Description	Costs €	NPV €	IRR %
1 Current maintenance standard	5, 15 and 25th year 25 mm microsurfacing with 10 and 20th year surfacing replacement.	2 480 085	1 691 695	13.9
2 Basic variant	Whole lifetime of only basic surface treatment.	2 643	0	0
3 Microsurfacing based variant	Basic surface treatment with 25 mm microsurfacing in 7th 16th and 25th year.	977 588	1 723 229	12.1
4 One major cover layer exchange based variant	Basic surface treatment with 40 mm cover layer exchange in 14th year	502 654	2 045 110	30.9
5 One microsurface based variant	Basic surface treatment with 25 mm microsurfacing in 14th year.	327 279	1 588 528	37.9
6 Balanced cover layer exchange based variant	Basic surface treatment with 20 mm cover layer exchange and 25 mm microsurfacing in 8th 18th and 28th year.	760 652	1 236 838	15.6

– Variant 4: one major cover layer exchange based variant—fairly safe to use on all rural roads
– Variant 5: one microsurface based variant—may be appropriate even for 3rd class roads with traffic load under 1000 AADT especially if they are not subject to excessive high load vehicles encumbrance.
– Variant 6: basic variant—is appropriate only for 3rd class roads where the 1000 AADT limit is not exceed and are not subject to excessive high load vehicles encumbrance.

The variants have impact on the pavement serviceability through the whole life cycle – shown in Fig. 2, represented through International Roughness Index (IRI). The variants are ranked for cost, technical suitability, economic effectiveness (Table 2) and from overall point of view. The ranking of these MR&R variants is shown in Table 3.

1.6 Pre-computed repair plans

This approach deals primarily with the prioritization of a defined long list of candidate road projects into a one-year or multi-year work program under defined budget constraints. It is essential to note that here; we are dealing with a long list of candidate road projects selected as discrete segments of a road network. The selection criteria will normally depend on the maintenance, improvement or development standards that a road administration may have defined (for example from the output produced by the strategy analysis application). Examples of selection criteria that may be used to

Table 3. Rankings of MR&R variants – cost and economic effectiveness according to Table 2, technical suitability based on Figure 2 (final IRI in year 2041), overall average by weight.

Viewpoint	Variant 1	Variant 2	Variant 3	Variant 4	Variant 5	Variant 6
Cost	6th	5th	4th	3rd	2nd	1st
Technical suitability	1st	2nd	5th	3rd	4th	6th
Economic effectiveness	4th	5th	3rd	2nd	1st	6th
Overall	3rd	4th	5th	1st	2nd	6th

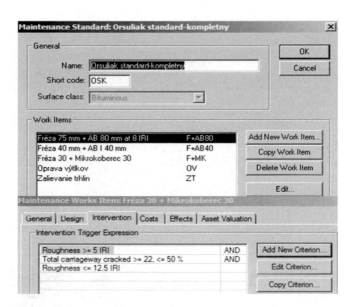

Figure 3. Maintenance thresholds for repair actions.

identify candidate projects such as periodic maintenance thresholds are shown in Fig. 3. HDM-4 software was used to draft the repair plan according to Fig. 3, th maintenance standard is MR&R standard proposed by Slovak Road Administration for trunk roads.

When all candidate projects have been identified, the HDM-4 programe analysis application can then be used to compare the life cycle costs predicted under the existing regimen of pavement management (that is "without project case") against the life cycle costs predicted for the periodic maintenance, road improvement or development alternative (that is "with project case"). This provides the basis for estimating the economic benefits that would be derived by including each candidate project within the budget timeframe.

For this approach, the problem can be posed as one of seeking that combination of treatment alternatives across a number of sections in the network that optimizes an objective function under budget constraint. Each budget constraint defines a Budget Scenario.

The amount of financial resources available to a road agency determines what road investment works would be affordable. The level of budget is not always constant over time due to a variety of factors. This variation of budget levels over time affects the functional standards as well as the size of road network that can be sustainable. Pre-computed repair plans are prepared for several budget scenarios to be specified and optimized simultaneously. This permits comparisons to be made between the effects of different budget scenarios and to produce desired reports. A budget scenario represents the available resources defined for each budget period, excluding the costs of annual routine maintenance and special works (PIARC, 2006).

Table 4. Example – Repair plan with limited budget for 5 year period.

Section	Road Class	Length (km)	AADT	Surface Class	Year	Work Description	NPV/CAP	Financial Costs	Cumulative Costs
583/B	I. + II. trieda	8.2	1788	Bituminous	2014	Oprava výtlkov	0.000	0.00	0.000
59006	III. trieda + MK	4.4	3385	Bituminous	2014	Fréza 40 mm + AB I 4	13.575	0.57	0.574
59017/E	III. trieda + MK	0.7	1020	Bituminous	2014	Fréza 40 mm + AB I 4	4.981	0.06	0.632
59017/D	III. trieda + MK	5.5	1020	Bituminous	2014	Oprava výtlkov	0.000	0.00	0.633
	III. trieda + MK	5.5	1020	Bituminous	2014	Zalievanie trhlín	0.000	0.00	0.633
59022/B	III. trieda + MK	4.3	648	Bituminous	2014	Zalievanie trhlín	0.000	0.00	0.633
7001/A	III. trieda + MK	1.4	741	Bituminous	2014	Fréza 40 mm + AB I 4	3.385	0.12	0.752
7001/B	III. trieda + MK	2.9	741	Bituminous	2014	Oprava výtlkov	0.000	0.00	0.752
	III. trieda + MK	2.9	741	Bituminous	2014	Zalievanie trhlín	0.000	0.00	0.752
7002	III. trieda + MK	1.7	1020	Bituminous	2014	Oprava výtlkov	0.000	0.00	0.752

2 CONCLUSION

This paper describes the road network of SR and its road administrators. It describes the problematic of pavement management and gives an insight into road network management system. As presented, the implementation of pavement management system for road network administrators of self-governing regions is possible. An educated guess would be, that best solution would be to adopt the RNMS, the two other possibilities presented in this article could make for temporary or backup solutions. Road Network Management System was accepted by international financial institutions on providing investment credits for rehabilitation of road network. At present, when the system is in operation for more than 15 years, we can draw on results achieved in practice.

ACKNOWLEDGEMENT

The research is supported by the European Regional Development Fund and the Slovak state budget for the project "Research Centre of University of Žilina", ITMS 26220220183.

"We support research activities in Slovakia/project co-founded from the resources of the EU".

REFERENCES

Čelko, J., Decky, M., and Kovac, M. 2009: An analysis of vehicle – road surface interaction for classification of IRI in frame of Slovak PMS. In: Maintenace and reliability. Polish Maintenance society, Nr 1 (41) 2009, pp. 15–21, ISSN 1507-2711.

Gavulová, A. 1999: Factors influencing bituminous layers rutting. In: 3rd. International Conference TRANSCOM '99, Žilina, jún 1999, str. 97–100.

Komacka, J., and Celko, J., 1996: CANUV – Computer program for analysis of the bearing capacity on the deflection bowl basis. Communications: Scientific Letters of the University of Žilina, 19, 63–70.

Mikolaj, J. 1996: The Road Network Management System in Slovakia. Transport Reviews, 16 (4), 313–321.

Mikolaj, J., Trojanova, M., and Pepucha L. 2012: International Federation of Municipal Engineering Conference Proceedings, Helsinki, Finland.

Pepucha, L., and Remek L. 2014: Sustainable Maintenance of Rural Roads in Slovakia Journal of Civil Engineering and Architecture, ISSN 1934-7359, USA, pp. 486–491.

PIARC 1997: Highway Performance Monitoring Systems (HPMS). World Road Association PIARC, Paris.

PIARC 2006: Highway Development and Management Series. Volumes one – five. World Road Association PIARC, Paris. ISBN: 2-84060-058-7.

Zgutova, K., Decky, M. and Ďurekova, D. 2012: Implementation of static theory of impulse into correlation relations of relevant deformation characteristics of earth construction. In SGEM 2012: 12th international multidisciplinary scientific geoconference, Albena, Bulgaria. ISSN 1314–2704, pp. 107–115.

Advances and Trends in Engineering Sciences and Technologies – Al Ali & Platko (Eds)
© 2016 Taylor & Francis Group, London, ISBN: 978-1-138-02907-1

Application of Huber's estimation method in the Decourt's extrapolation method validation for various foundation piles

Z. Muszyński & J. Rybak
Wrocław University of Technology, Wrocław, Poland

ABSTRACT: Static Load Test can be used for the calibration of other design methods like: static calculations based on the field tests of ground samples or dynamic testing of piles. Precise calibration requires that the pile's ultimate resistance should be known from the prior static test, whereas it may often be impossible to determine that value due to insufficient kentledge or limited capacity of the testing station. The engineer is then forced to calculate the failure load on the basis of the available data by means of the "extrapolation". Previous studies proved that extrapolation methods are entirely dependent on the range of the performed test and on outliers resulting from random errors. Selected robust estimation method was than checked in order to give some recommendations concerning the use of Decourt's pile test extrapolation method in the case of small diameter Controlled Modulus Columns and large diameter bored piles.

1 INTRODUCTION – SHORT DESCRIPTION OF DECOURT'S METHOD

Static load test (SLT) is a testing procedure of primary importance in pile designing. In general, load should increase until the pile reaches the limit state (unhindered increment in settlements with respect to increase of load applied), which would allow to determine the ultimate capacity. In practice, due to limited strength of testing station, insufficient capacity of anchoring piles or too small kentledge, it is often impossible to create the loads which would cause that pile operates in soil within plastic range. Even if the force applied during the test is larger than the design capacity, and the structure safety is confirmed, the range of test provides no information about pile ultimate capacity and would not allow to calibrate the calculation model. When the ultimate capacity is unknown, we do not make use of the margin which would make it possible, for instance, to reduce the number of pile in foundation, or to reduce their length. Due to uncertainty of extrapolation methods, the limitation consisting in assuming the maximum load effected in field testing as the ultimate capacity is often used in designing practice. Such recommendation was given by Fellenius (1975, 1980, 2001). Currently, economic reasons are of utmost importance, hence while making static tests (which are in themselves costly and time-consuming), attempts are made to get as much information about capacity as possible. This paper outlines how the number of measurements (for the range of test load effected) affects the accuracy of determining the ultimate capacity Q_u. The review was based on test load data from Controlled Modulus Columns (CMC) and bored piles. The slenderness of these elements differed significantly as the columns and the bored piles had only 36 cm and 120 cm in diameter, respectively. The main goal of the study was to compare the accuracy of Decourt's method applied to piles or columns of different slenderness. The method was originally published by Decourt's (1999). The original settlement-loading relation ($s - Q$) from the field tests is transformed into the coordinate system: $Q =$ abscissae and $Q/s =$ ordinates (Fig. 1) for the last points of the performed test at which the settlement reached stabilization.

The following linear dependence is found (by way of approximation): $Q/s = A \cdot Q + B$ where: $A =$ slope of regression straight line; $B =$ intersection point between regression straight line and 0Y axis. The capacity can be than derived from formula (1) and the settlement-load relation can be

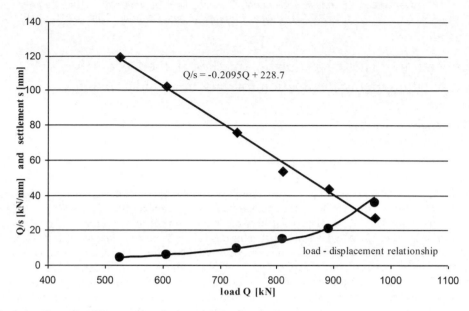

Figure 1.   Example of Decourt's method graph – 6 points for linear regression.

transferred to equation (2). In the case shown here below (Fig. 1), the value of $A = -0.2095$ and $B = 228.7$ respectively. It means that the pile capacity equals $Q_u = 1092\,\text{kN}$.

$$Q_u = -B/A \qquad (1)$$

$$Q = \frac{B \cdot s}{1 - A \cdot s} \qquad (2)$$

2   BASIC METHODS OF APPROXIMATION

The most frequently used methods of approximation are based on the so called least squares method (LS). This method was previously used by the authors (Muszyński & Rybak 2013) for the analysis of Brinch Hansen 80% extrapolation method. Nowadays the majority of spreadsheets is equipped with built-in solvers (using that method) which make it possible to describe the set of points under analysis in terms of a linear dependence (or a dependence of higher order). In more advanced instances of the implementation of the least squares method it is also possible to assign weights to particular observations.

In each case under scrutiny, while calculating the ultimate resistance, we eliminate those points (by way of "zero-weighting") whose distribution falls into the range of the elastic work of a pile in the subsoil. The data gained from the range of the elastic-plastic pile behavior may also be divided into the more and less significant ones. Intuition tells us that the calculation result is influenced most significantly by the last readouts of the load test, and in particular those which belong to the range of $(0.8 \div 1.0)\,Q_{max}$.

A separate issue entails the effort at minimizing the influence of the aforementioned systematic and random errors, as well as at eliminating the gross errors. One may apply with that respect methods of robust estimation, known from mathematical statistics and frequently used in geodesy. The essence of those methods consists in the application of the least squares methods accompanied by the iterative modification (decreasing) weights of the observations considered to be outliers.

The points are thought to be outliers when the deviation of their approximation exceeds the value assumed as permissible. As a result, the outlying observations have a smaller (and in the extreme cases – null) impact on the obtained solution. That makes it possible to fit the assumed theoretical model of approximation into the largest number of congruent points. In this article one method of robust estimation (the Huber's method) is being considered.

3 PRINCIPLES OF THE HUBER METHOD

The Huber's method was proposed by Huber in work (Huber 1964) and repeated in article (Huber 1972). It consists in the combination of the least squares method and the method of the least standard deviation. For the i-th observation, the iterative modification of weights follows the formula (3):

$$\hat{p}_i = \begin{cases} p_i & \text{for} \quad |v_i| \leq f \\ p_i \dfrac{f}{|v_i|} & \text{for} \quad |v_i| > f \end{cases} \tag{3}$$

where $\hat{p}_i$ = modified weight; p_i = weight of the i-th observation from the previous iteration (in the initial step assumed from the least squares method); v = fitting deviation (derived from the approximation); f = controlling parameter, determining the range of permissible values of fitting deviation v.

4 ASSESMENT OF THE MODEL FITTING

The Decourt's extrapolation method is generally based on the concept of the fitting of the theoretical model to the certain range of field test results population. To evaluate the accuracy of this fitting, several statistic parameters were computed on the basis of load test logs:

– standard deviation of the residual component (S_e) given by the formula (4):

$$S_e = \sqrt{\dfrac{\sum\limits_{i=1}^{n}(y_i - \hat{y}_i)^2}{n-k}} \tag{4}$$

where y = the real value of the dependent variable; $\hat{y}$ = the theoretical value of the dependent variable (from the model); n = sample size; k = the amount of the estimated model parameters;
– residual coefficient of variation (V_e) expressed by formula (5):

$$V_e = \dfrac{S_e}{\bar{y}} \cdot 100\% \tag{5}$$

where $\bar{y}$ = the arithmetic mean of the empirical values of dependent variable;
– mean square error (MSE) according to the formula (6):

$$MSE = \dfrac{\sum\limits_{i=1}^{n}(y_i - \hat{y}_i)^2}{n} \tag{6}$$

– mean absolute error (MAE) explained by the formula (7):

$$MAE = \dfrac{\sum\limits_{i=1}^{n}|y_i - \hat{y}_i|}{n} \tag{7}$$

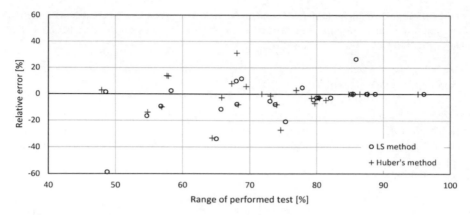

Figure 2.   Estimated inaccuracy of ultimate capacity for the CMC piles.

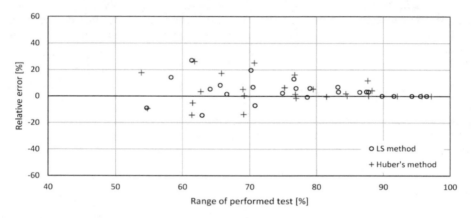

Figure 3.   Estimated inaccuracy of ultimate capacity for bored piles.

5   CALCULATION RESULTS

The calculations of ultimate resistance were carried out on the basis of the results of static tests of 6 CMC piles and 5 bored piles, tested at the construction sites of large commercial centers in central and southern Poland, respectively. Using the Mathcad environment, an algorithm was developed for the purpose of calculating the ultimate resistance by means of the following methods: the least squares method (LS), assuming identical weights for all the points under scrutiny; the Huber's method (H), assuming identical weights in the first iteration.

The calculations were based on the group of 4 observations from the elastic-plastic work of piles. The last four observations were called as "range 0". To verify the result of calculation from unfinished pile test, recent observations were subsequently rejected. Data without the last observation were indicated as "range 1", without the last two observations – as "range 2", etc.

The application of the aforementioned robust estimation methods requires the assumption of certain values of controlling parameters which define the permissible range of fitting deviation (approximation). Those parameters are usually assumed empirically, appropriately for particular implementations of robust estimation methods. In this article, for Huber's method the value of parameter f was set at the level of 50% of the value of the standard deviation of approximation inaccuracies from the previous iteration (at the initial stage of calculations – from the least squares method with identical weights). The test for convergence of the iteration process was set at the

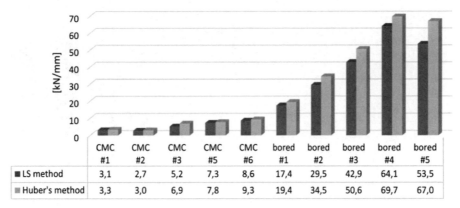

Figure 4. The standard deviation of the residual component (S_e).

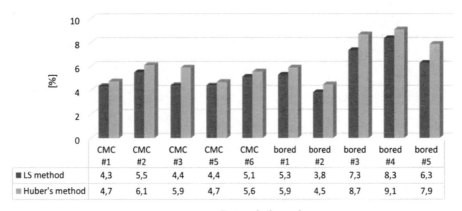

Figure 5. Residual coefficient of variation (V_e).

following level: the alteration of the values of both searched parameters A and B in three subsequent iterations should not exceed $1 \cdot 10^{-7}$ (to provide a proper accuracy of results).

The results shown above (Figs 4–7) clearly show that all the applied measures confirmed higher accuracy of Decourt's method in the case of the CMC columns as compared with large diameter bored piles.

6   BRIEF SUMMARY AND CONCLUSION

Extrapolation of the test results always brings the risk of overestimation or underestimation of pile capacity. Robust estimation enables to reduce the influence of random errors, however further studies are necessary to provide the recommendations based on larger population of pile test results for various piling technologies. The first attempt were done in view of pile slenderness. It can be clearly noticed that the accuracy is much better in the case of the CMC columns which are much more slender. It might suggest that Decourt's method is more suitable for friction piles with a relatively small amount of base capacity which can be achieved at larger displacements. Further studies should be based mainly on the influence of piling technologies.

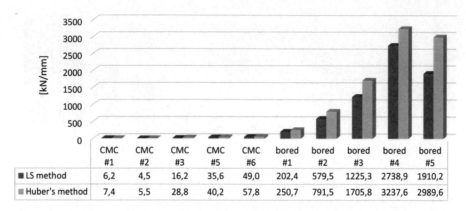

	CMC #1	CMC #2	CMC #3	CMC #5	CMC #6	bored #1	bored #2	bored #3	bored #4	bored #5
■ LS method	6,2	4,5	16,2	35,6	49,0	202,4	579,5	1225,3	2738,9	1910,2
■ Huber's method	7,4	5,5	28,8	40,2	57,8	250,7	791,5	1705,8	3237,6	2989,6

Type and pile number

Figure 6.   Mean square error (*MSE*).

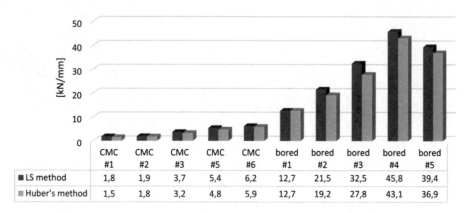

	CMC #1	CMC #2	CMC #3	CMC #5	CMC #6	bored #1	bored #2	bored #3	bored #4	bored #5
■ LS method	1,8	1,9	3,7	5,4	6,2	12,7	21,5	32,5	45,8	39,4
■ Huber's method	1,5	1,8	3,2	4,8	5,9	12,7	19,2	27,8	43,1	36,9

Type and pile number

Figure 7.   Mean absolute error (*MAE*).

REFERENCES

Decourt, L. 1999. Behavior of foundations under working load conditions. *Proceedings of the 11th Pan-American Conference on Soil Mechanics and Geotechnical Engineering, Foz DoIguassu, August 1999, Brazil* 4: pp. 453–488.

Fellenius, B.H. 1975. Test loading of piles. Methods, interpretation, and new proof testing procedure. *ASCE* 101, GT9: pp. 855–869.

Fellenius, B.H. 1980. The analysis of results from routine pile loading tests. *Ground Engineering*, London 13 (6): pp. 19–31.

Fellenius, B.H. 2001. What capacity value to choose from the results a static loading test. We have determined the capacity, then what? *Deep Foundation Institute*, Fulcrum Winter: pp. 19–22 and Fall: pp. 23–26.

Huber, P.J. 1964. Robust Estimation of a Location Parameter. *The Annals of Mathematical Statistics* 35 (1): pp. 73–101.

Huber, P.J. 1972. The 1972 Wald Lecture Robust Statistics: A Review. *The Annals of Mathematical Statistics* 43 (4): pp. 1041–1067.

Muszyński, Z. & Rybak, J. 2013. Application for robust estimation methods for calculation of piles' ultimate resistance. *International Journal for Computational Civil and Structural Engineering* 9 (3): pp. 61–67.

Advances and Trends in Engineering Sciences and Technologies – Al Ali & Platko (Eds)
© 2016 Taylor & Francis Group, London, ISBN: 978-1-138-02907-1

Use of close range photogrammetry for structure parts deformation monitoring

K. Pavelka, J. Šedina & Z. Bílá
Czech Technical University in Prague, Faculty of Civil Engineering, Department of Geomatics, Prague, Czech Republic

ABSTRACT: This paper presents an application of photogrammetrical methods for measuring the deformation of structural elements at sequential loading. Several load conditions were documented using correlation techniques and intersection photogrammetry. Two types of material were tested: timber beams (linear part) and concrete beams. The progress of deformation was measured with a non-contact photogrammetrical method with the use of the image correlation technique (image based modeling – IBM). On the object, cca 50 measured points were monitored for dynamical deformation. Photomodeler and Agisoft Photoscan software's were used in this case. As comparative models, measurements by the precise laser scanner *Surphaser* were done such as the direct measurement of deformations by potentiometers. The outputs shown, which all used non-traditional technology, are good enough for deformation measurement in sub-mm range.

1   INTRODUCTION

1.1   *Deformation measurement*

At the CTU (Czech Technical University) in Prague, Laboratory of Photogrammetry, new methods of 3D objects documentation are tested. There are two main approaches in documentation: precise documentation by several techniques such as special instruments (potentiometers, tenzometers), precise close range photogrammetry or 3D scanning and simple documentation by using non expensive devices and instruments (Robson & Setan, 1996). In this case, we tested photogrammetric methods, laser scanning and direct deformation measuring (using potentiometers) on a timber beam and a concrete beam. Results from the concrete beam were not processed at this time, only the timber beam as a case project).

1.2   *Photogrammetric system*

Several photogrammetrical systems were tested (Hampel & Maas, 2003). The first technology was based on image-based modeling. It is a very simple technology, which needs only a (good)digital camera, a set of overlapped images and software for image correlation. We used Agisoft Photoscan. As a digital camera, the middle format professional Pentax 645D camera with 40MPix resolution (CCD 44 × 33 mm) was used. For the measured beams, about 10 images were taken for one load case. The second system consists of three calibrated digital high resolution (10–15MPix) cameras (Canon 450 with mounted lens Canon 18–55 mm and Canon Mark IID with 400mm lens in an intersection configuration). All cameras were calibrated on calibration field in our Laboratory of photogrammetry (Řezníček, 2014). The cameras were installed about 2 m from the measured object on tripods. All cameras were controlled remotely by computer and the images were saved to the hard disk in raw and jpeg format. The orientation of all cameras was constant throughout (Řezníček &, Pavelka, 2010). The images were taken synchronized in steps, based on load (special devices for testing controlled-deformation load, see Figure 1). The measured object was lit with

Figure 1.   Test area.

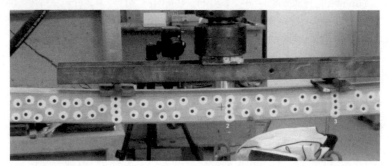

Figure 2.   Marked points on the beam.

one 500W lam. On the tested object (beam) there were approximately 50 measured marked points overall (Fig. 2).

1.3   *Laser scanning*

For creating a 3D model of the object, laser scanning or image-based modeling was used. Both technologies resulted in a point cloud. By use of a professional laser scanner, the acquired 3D model is referenced in metric units (usually in m or mm). By image based modeling, there is a problem with the units; originally the resulting point cloud is in a local coordinate system. For the model scale, it is necessary to measure the control points using geodetical methods. These points are used for 3D transformation. In our case, the laser scanner Surphaser, with precision 0,6 mm on 10m distances, was used. The point clouds that were reached for every load step were processed in Geomagic Studio software. Outputs from the laser scanner Surphaser are shown in Figure 3.

1.4   *Direct measurement*

All experiments are performed in the Experimental centre of the Faculty of Civil Engineering. The Experimental Centre is used by the staff members of all departments and specializations

Figure 3.   Mash created from laser scanner measurement.

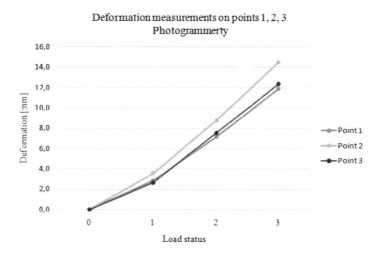

Figure 4.   Outputs from intersection photogrammetry.

focusing on the instructions for building structures and materials (Zatloukal & Konvalinka, 2013, Strejček et al., 2011). The Testing Laboratories are equipped with technological apparatus and machinery necessary for performing mechanical, static, and dynamic tests of building structures and materials. On a measured beam, which was fastened on a load case stool, three direct measuring sensors are placed (Fig.2). They measure vertical deformations with a very high frequency and save it in a data set with a high accuracy. The disadvantage is only low number of sensors are on a measured object (Litoš et al. 2013).

2   TECHNOLOGY TESTING

2.1   *Technology*

Different technologies were tested on measured and deformed beams. As a primary measurement, direct vertical deformation measurements based on special sensors with accuracy better than 0,1 mm was used, but at several points. In our case project, three sensors on most of the exposed parts of the beam were used. On the other hand, laser scanning with a precise phase scanner gave us a point cloud with hundreds of thousands of points, but its accuracy in position was not better

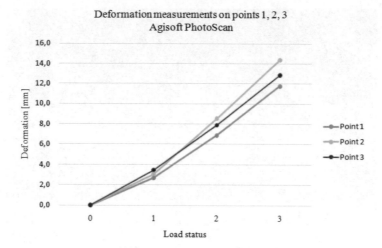

Figure 5.   Outputs from image based modeling (IBM) method.

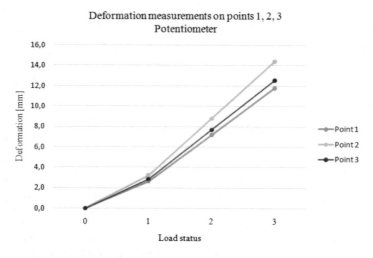

Figure 6.   Outputs from potentiometers.

Figure 7.   Point cloud created by Agisoft Photoscan.

Figure 8.   Joining of two point clouds created by Agisoft Photoscan software; noisy edge is a problem for precise shift detection (without erasing of the noise, measured shift reached 18 mm; after the noise reducing and surfaces approximations the shift was 14,5 mm only, which is very good result comparable with other measurement methods.

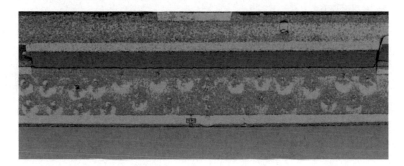

Figure 9.   Joining of two point clouds created from laser scanner Surphaser measurement (Gemagic Studio); the shift after load reaches 18 mm.

than 0,6 mm. For photogrammetrical technology (Hamouz et al., 2014), signalized object points on small adhesive labels were used for deformation measurement. For technical reasons and for comparison, signalized points directly on the location of the sensors (potentiometers) were used for deformation monitoring. Results from the stereo or intersection photogrammetrical methods are good enough and have a better accuracy level than the aforementioned laser scanning. Our case project reaches an intersection photogrammetry accuracy of 0,1 mm. From image based modeling, the results depend on the texture of the measured and documented objects. Timber and concrete parts have a good texture; with other materials (e.g. steel) it is necessary to use synthetic textures (for example, projecting on a measured object with an ordinary projector).

2.2   Experiments

The experiment was extensive and we did not have all the results. For comparison of all the methods for deformation measurement, the standstill and three load conditions of the tested beam were recorded. Final vertical deformations were measured for this reason on three points only. We can conclude that all deformation results are comparative and have approximately the same size in mm (Figs 4–6). There was a big problem to measure vertical deformation distances with the laser scanning method in the obtained point clouds. It was necessary to joint both point clouds and to find the same points in the standstill point cloud and the point cloud after deformation. We reached vertical deformation 18 mm by laser scanner measurement (Fig.9). Similar problem is by point clouds created using image based modeling technology (Agisoft Photoscan). There is problem with the noise on modeled beam edge, we cannot precise measure shift of the beam after load from two point clouds – it is necessary to use surface approximation (Figs 7–8).

3 CONCLUSIONS

The photogrammetrical measuring of the deformation of the beams can be used as an independent technology. Intersection photogrammetry with three or more cameras can be used for the dynamical deformation of objects, but it depends on the speed of deformation. We have to account for the necessary time for the registration of photos (a few seconds per image, depending on the type of camera). Image based modelling technology is simple, but it needs to take a large number of acceptable overlapped photos. This can be done in several minutes with one camera. For this reason, this technology cannot be not used for dynamical deformation with a one camera configuration, but it is good enough for deformation in steps. The same is true for the use of the laser scanner – one scan takes several minutes. For future consideration, we will test single image photogrammetry, which can be used for planar objects only, but with a very high resolution dependent on the type of camera and the distance to the object.

REFERENCES

Hamouz, J., Braun, J., Urban, R., Štroner, M. & Vráblík, L. 2014. Monitoring and Evaluation of Prestressed Concrete Element Using Photogrammetric Methods. 14th International Multidisciplinary Scientific Geoconference SGEM 2014, Conference Proceedings vol. III. Sofia: STEF92 Technology Ltd., 2014,. ISSN 1314-2704. ISBN 978-619-7105-12-4. pp. 231–238

Hampel, U. & Maas, H.G., 2003. Application of Digital Photogrammetry for Measuring Deformation and Cracks During Load Tests in Civil Engineering Material Testing. Optical–3D. pp. 25–38

Litoš, J., Vejmelková, E. & Konvalinka, P. 2013. Monitoring of Deformation of Steel Structure Roof of Football Stadium Slavia Prague. Measurement Technology and its Application.

Uetikon-Zurich: Trans Tech Publications, ISSN 1660-9336. ISBN 978-3-03785-545-4 pp. 622–630

Robson, S. & Setan, H.B., 1996. The Dynamic Digital Photogrammetric Measurement and Visualisation of a 21m Wind Turbine Rotor Blade Undergoing Structural Analysis. International Archives of Photogrammetry and Remote Sensing. Vol. XXXI, Part B5. Vienna, pp. 493–498

Řezníček, J. & Pavelka, K. 2010. Photogrammetrical measuring of the dynamical deformation of the joint and the column web panel at elevated temperature. In Proceedings of 31st ACRS Conference. Hanoi: ACRS, , vol. 1, p. 1859–1863. ISBN 978-1-61782-397-8

Řezníček, J. 2014. Measuring Repeatability of the Focus-variable Lenses. Geoinformatics, vol. 9, no. 12-2014, art. no. 2, pp. 9–18

Strejček, M.,Wald, F., Řezníček, J. & Tan, KH. 2011. Behaviour of Column Web Component of Steel Beam-to-Column Joints at Elevated Temperatures. Journal of Constructional Steel Research., vol. 67, no. 67, ISSN 0143-974X. pp. 1890–1899.

Zatloukal, J. & Kovalinka, P. 2013. Moment capacity of FRP reinforced concrete beam assessment based on centerline geometry. In: 51st Annual of the International Scientific Conference on Experimental Stress Analysis, EAN 2013; Applied Mechanics and Materials book series, Volume 486, 2014, ISSN: 16609336 ISBN: 978-303785977-3. pp. 211–216

Advances and Trends in Engineering Sciences and Technologies – Al Ali & Platko (Eds)
© 2016 Taylor & Francis Group, London, ISBN: 978-1-138-02907-1

Adaptive tensegrity module: Adaptation tests 06 and 07

P. Platko & S. Kmeť
Technical University of Košice, Faculty of Civil Engineering, Institute of Structural Engineering, Košice, Slovakia

ABSTRACT: This paper describes two adaptation experimental tests (nBS 06 and nBS 07) which were carried out on a newly developed adaptive tensegrity module in the form of a double symmetrical pyramid. The module is formed by thirteen members – eight pre-stressed cables, four edge compressed struts and one centrally positioned compressed strut which is designed as an actuator. This system contains sensors that sense forces from the environment and an actuator that adjusts its shape accordingly, depending upon the load applied. The results obtained from the experimental tests are compared with the results obtained theoretically by using geometrically nonlinear and physically linear FEM analysis.

1 INTRODUCTION

In general, tensegrity systems are spatial structures, composed of only tensioned members – cables and compressed members – struts (e.g. Fuller 1975, Motro 2003). The members of tensegrity systems can simultaneously work as load-carrying members, sensors and actuators (Skelton, Sultan 1997). Therefore their implementation is very promising in various projects that require active (adaptive) or deployable structural systems.

Stiffness and stability of a tensegrity system depends upon its geometry and levels of initial pre-stress applied to tensioned members at its self-equilibrium state. The existence of pre-stress in members of the adaptive tensegrity structure promotes the modification of its stiffness when an excessive external load is applied and includes the ability to change its behavior below the predefined safety and mainly serviceability limit.

The level of pre-stress will not only affect the geometric configuration and stiffness of the tensegrity system, but also determine its serviceability and load bearing capacity. The structure could fail if cables slacken or structural members become overstressed at a certain pre-stress level or certain external load level.

Research of tensegrity structures is mainly focused on theoretical studies of non-adaptive regular and non regular systems (research related to their geometry, form-finding, static analysis, etc.). Research or tests on full-scale adaptive tensegrity structures or models, as presented in this paper, are rare. Active or adaptive tensegrity systems equipped with sensors and actuators provide the potential to control their shape and adapt to changing load and environmental conditions (e.g. Shea et al. 2002, Fest et al. 2003, 2004).

This paper describes a recently developed adaptive tensegrity module, with the ability to alter its geometrical form and stress properties in order to adapt its behavior in response to current loading conditions (see also Kmeť, Platko 2014a, 2014b). Results of two adaptation experimental tests (nBS 06 and nBS 07) are presented and compared with results obtained numerically by using geometrically nonlinear and physically linear FEM analysis. The tests was aimed at a verifying the tensegrity system's ability to adapt its geometry and state of stress to changing load cases in order to maintain the reliability of the system.

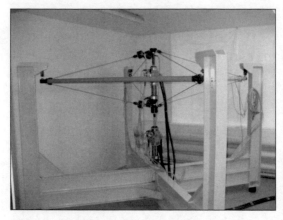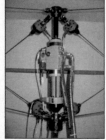

Figure 1. Adaptive tensegrity module: a) the module suspended on a self-supporting frame, b) detailed view of the central strut as the action member, c) hydraulic load cylinder.

Table 1. Members of the adaptive tensegrity module and their properties.

Member	Cross-section	A [mm^2]	Material	E [GPa]
Compressed members	CHS 51/3.2 mm	$A_s = 475.9$	steel S 235	$E_s = 210$
Bottom cables	d = 6 mm	$A_c = 15.14$	steel cable 7 × 7*	$E_c = 120$
Top cables	d = 6 mm	$A_c = 15.14$	steel cable 7 × 7*	$E_c = 120$
Central active member	–	–	–	$E_s = 210$

*7 strands with 7 wires per strand.

2 ADAPTIVE TENSEGRITY MODULE

A full-scale prototype of an adaptive tensegrity module or tensegrity unit, as well as the whole test facility was developed at the Institute of Structural Engineering at the Technical University of Košice's, Faculty of Civil Engineering (Figure 1). Its production was performed in cooperation with INOVA Praha Ltd.

The chosen tensegrity module consists of a strut that is centered in a rectangular base and stiffened by crossed cables. This module is also known as a tensegric unit cell of type I (Saitoh, M. 2001) or like a crystal pyramid (Wang, B. 1999) and it is suitable for the generation of line structures or plate structures with a straight or curved central line.

The theoretical dimensions of the square base of the unit are 2.000 × 2.000 mm and its theoretical height is $L_{AM,0} = 750$ mm. This basic structural bearing system consists of thirteen members (four circumferential compressed members, four bottom and four top cables and one central strut) as is presented in Table 1. The unit is also equipped with six strain gauges SG1 to SG6 and four force transducers FT1 to FT4 (Figure 2). All members are mutually connected in nodes by hinge joints.

3 EXPERIMENTAL ANALYSIS

The duration of both preprogrammed fully automated block tests was set at 360 s. Required values of the initial pre-stress in the cables of the tensegrity module were attained by extending the action member with the corresponding axial force F_{AM} (Figure 4b, 7b). At the time 55 s a load P_{LC} was applied, which gradually increases as is shown in Figure 4a and Figure 7a.

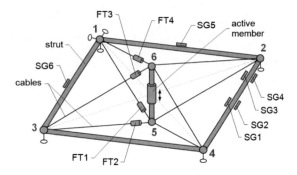

Figure 2.   Isometric diagram of the module and position of force transducers and strain gauges.

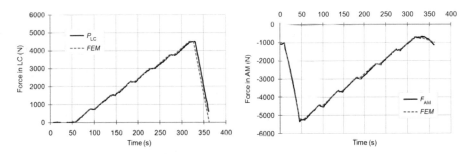

Figure 3.   Finite element model of the adaptive tensegrity unit and assignment of finite-element types for individual members.

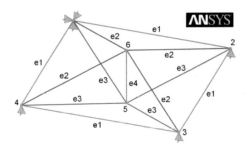

Figure 4.   Test nBS 06 – comparison of the experimentally and theoretically obtained forces in the load cylinder (a – left) and in the action member (b – right) at the investigated times.

3.1   Adaptation test nBS 06

The aim of the first adaptation test was to keep tensile forces in the bottom cables (force transducers FT1 and FT2) of the tensegrity module subjected to a gradual increase in load on the approximately constant at a level of $N_b = 5000$ N (Figure 6).

The value of these forces were maintained and controlled by the contraction of the action member. Therefore tensile forces in the top cables (force transducers FT3 and FT4) decrease at successive load increments as is shown in Figure 5.

3.2   Adaptation test nBS 07

The aim of the second adaptation test was to keep tensile forces in the top cables (force transducers FT3 and FT4) of the tensegrity module subjected to a gradual increase in load at an approximately constant level of $N_t = 2200$ N (Figure 8).

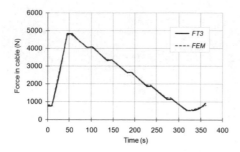
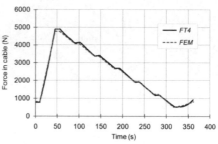

Figure 5.   Test nBS 06 – comparison of the experimentally and theoretically obtained forces in the top cables recorded from force transducers FT3 (a – left) and FT4 (b – right) at the investigated times.

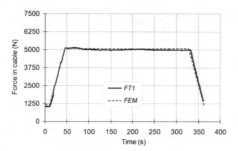
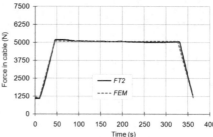

Figure 6.   Test nBS 06 – comparison of the experimentally and theoretically obtained forces in the bottom cables recorded from force transducers FT1 (a – left) and FT2 (b – right) at the investigated times.

The value of these forces were maintained and controlled by the elongation of the action member. Therefore tensile forces in the bottom cables (force transducers FT1 and FT2) increase at successive load increments as is shown in Figure 9.

4   NUMERICAL ANALYSIS

In order to verify results obtained by the experimental tests, a finite element analysis of the adaptive tensegrity model was performed. A geometrically nonlinear and physically linear computational model was created in ANSYS 12 Classic software and the following types of finite elements were used (ANSYS, Inc. 2005):

– LINK10 compression-only spar for the compressed members (e1),
– LINK10 tension-only spar for the top cables (e2) and bottom cables (e3),
– LINK11 linear actuator for the active member (e4).

LINK10 and LINK11 are two-node rectilinear spatial elements with three degrees of freedom at each node (displacements in the nodal x, y and z directions). LINK10 possesses a unique feature which is a bilinear stiffness matrix resulting in a uniaxial tension or compression-only element. LINK11 may be used to model hydraulic cylinders and other applications. Stress stiffening and large deflection capabilities are available for these elements.

The finite element model was supported at nodes 1, 2, 3 and 4 (corner nodes of the square base) as is shown in Figure 3. The real constants and material properties of the members are shown in Table 1. In the FEM analysis the self-weight of the members, as well as the weight of the action member were considered.

The initial self equilibrium state of the adaptive tensegrity module was determined by using the dynamic relaxation method and ΔDRM software developed by Mojdis (2011). In this state

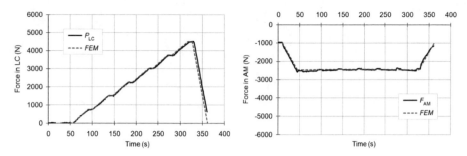

Figure 7.   Test nBS 07 – comparison of the experimentally and theoretically obtained forces in the load cylinder (a – left) and in the action member (b – right) at the investigated times.

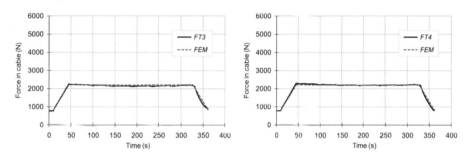

Figure 8.   Test nBS 07 – comparison of the experimentally and theoretically obtained forces in the top cables recorded from force transducers FT3 (a – left) and FT4 (b – right) at the investigated times.

compression force in the action member is $P_{AM,0} = 1000\,N$, tension force in the top cables is $N_{t,0} = 816\,N$ and in the bottom cables is $N_{b,0} = 1154\,N$.

5   COMPARISON OF RESULTS

5.1   *Adaptation test nBS 06*

The experimental results obtained by the first adaptation test of the adaptive tensegrity module are compared with results obtained by the nonlinear finite element analysis. The comparison and validation of the forces in the load cylinder, action member, top and bottom cables at the investigated times, is shown in Figure 4, Figure 5 and Figure 6.

5.2   *Adaptation test nBS 07*

The experimental results obtained by the second adaptation test of the adaptive tensegrity module are compared with results obtained by nonlinear finite element analysis. The comparison and validation of the forces in the load cylinder, action member, top and bottom cables at the investigated times, is shown in Figure 7, Figure 8 and Figure 9.

6   CONCLUSION

A newly developed adaptive tensegrity module in the form of a double symmetrical pyramid was presented in the paper. This unit has the ability to alter its geometrical form and stress properties in order to adapt its behavior in response to current loading conditions.

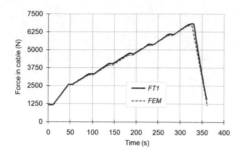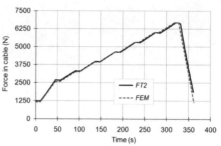

Figure 9.   Test nBS 07 – comparison of the experimentally and theoretically obtained forces in the bottom cables recorded from force transducers FT1 (a – left) and FT2 (b – right) at the investigated times

The results of two adaptation experimental tests (nBS 06 and nBS 07), and their numerical verification are presented. Results of the experimental and theoretical analyses of the tensegrity system confirmed a very good agreement between individual approaches.

ACKNOWLEDGEMENT

This work is part of Research Project No. 1/0321/12, partially founded by the Scientific Grant Agency of the Ministry of Education of the Slovak Republic and the Slovak Academy of Sciences. The present research has been carried out within the project: Centre of excellent integrated research for progressive building structures, materials and technologies, supported by European Union Structural funds.

REFERENCES

ANSYS, Inc. (2005). "Release documentation of Ansys 11.0", Southpointe, 275 Technology Drive, Canonsburg.

Fest, E., Shea, K., Domer, B., and Smith, I.F.C. (2003). "Adjustable tensegrity structures." *Journal of Structural Engineering*, 10.1061/(ASCE)0733-9445(2003)129:4(515), 515–526.

Fest, E., Shea, K., and Smith, I.F.C. (2004). "Active tensegrity structure." *Journal of Structural Engineering*, 10.1061/(ASCE)0733-9445(2003)130:10(1454), 1454–1465.

Fuller, R.B. (1975). *Synergetics explorations in the geometry of thinking*. Collier Macmillan Publishers, London.

Kmet', S., Platko, P. (2014a). "Adaptive tensegrity module. Part I: Closed-form and finite element analyses." *Journal of Structural Engineering*, 10.1061/(ASCE)ST.1943-541X.0000957, 04014055.

Kmet', S., and Platko, P. (2014b). "Adaptive tensegrity module. Part II: Tests and comparison of results." *Journal of Structural Engineering*, 10.1061/(ASCE)ST.1943-541X.0000958, 04014056.

Mojdis, M. (2011): "Analysis of adaptive cable domes." PhD. Thesis, Technical University of Kosice, Slovakia (in Slovak).

Motro, R. (2003). *Tensegrity: Structural systems for the future*. Hermes Science Publishing, Penton Science, U.K.

Saitoh, M. (2001): "Beyond the tensegrity, a new challenge toward the tensegric world," *Theory, design and realization of shell and spatial structures* – H. Kunieda (ed.), International Symposium of IASS, Nagoya, Japan, IASS, TP 141.

Shea, K., Fest, E., and Smith, I.F.C. (2002). "Developing intelligent tensegrity structures with stochastic search." *Advanced Engineering Informatics*, 16(1), 21–40.

Skeleton, R.E., and Sultan, C. (1997). "Controllable tensegrity, a new class of smart structures." *Proc., Conf. on Mathematics and Control in Smart Structures – Smart Structures and Materials 1997*, Vol. 3039, SPIE, Bellingham, 166–177.

Wang, B. (1999): "Simplexes in tensegrity systems", *Journal of the IASS*, 40, 1, 57–64.

Advances and Trends in Engineering Sciences and Technologies – Al Ali & Platko (Eds)
© 2016 Taylor & Francis Group, London, ISBN: 978-1-138-02907-1

Impact of surfactants used in extinguishing agent to corrosiveness of firefighting equipment

K. Radwan, J. Rakowska & Z. Ślosorz
Research Centre for Fire Protection – National Research Institute, Józefów, Poland

ABSTRACT: Surfactants have a wide range of application in many industries as components in many products, i.e. extinguishing agents. They may affect the corrosiveness of fire extinguishing systems and firefighting fitting used during rescue and firefighting actions. Some of surfactants may also be corrosion inhibitors and used to prevent the metals. Corrosion is considered as the destructive disintegration of a metal by electrochemical means in which a metal is destroyed by a chemical reaction. This is a very serious problem which generates huge economic costs every year. The aim of this review is to examine the role of the surfactants for the protection of various metals or alloys commonly used as the material of a firefighting equipment. The corrosion was evaluated by determination the rate of metal mass loss and observation of surface of samples.

1 INTRODUCTION

Firefighting fittings and components of fixed firefighting systems are usually made from steel, brass and aluminium alloys. Extinguishing media as concentrates or solutions, stored in containers, fixed firefighting systems or flowing through the firefighting equipment due to their chemical composition may cause corrosion of metal fittings. Devices for the firefighting should be characterized by a high effectiveness and an operational reliability, therefore studies on a corrosion protection are being conducted (Lemańska, 2013).

Corrosion of metals and alloys is an electrochemical process which is influenced by parameters such as moisture content, chemical environment and electrochemical state of material (Hihara, 2015), (Rakowska, et al., 2011). Stored metal parts of firefighting equipment are subject to corrosion. The various types of corrosion can be classified by different causes, and various types of corrosion have been defined, based generally on morphology or mechanism. Usually some variation of the following categories of corrosion are distinguished: uniform, galvanic, crevice, pitting, localized, intergranular, dealloying, erosion-corrosion, environmentally assisted cracking (EAC) and stress corrosion cracking (SCC) (Revie, 2011; Sharma, 2011).

In spite of progress related to the improvement of corrosion protection occurring losses caused by the corrosion of the materials are still one of main causes of technical failures and economic losses. Corrosion not only causes economic loss but also has an impact on safety or technology or health injury. While corrosion is inevitable it can be controlled and many of these losses could be saved by use methods of corrosion protection (Sharma, 2011).

A corrosion protection method is a technique used to minimize corrosion. There are several ways to eliminate or decrease the amount of corrosion on a material. The most obvious solution is to choose a corrosion insensitive material. However, many other aspects have influence the choice of a material (Bombara, 1980). The usual the application of protective coating, cathodic and anodic protection or corrosion inhibitors have a beneficial effect to make metal resistant to corrosion.

– An electrochemical protection (cathodic protection, anodic protection). An electrochemical protection is a method to control the corrosion of a metal surface by making that surface the cathode

Table 1. Commercial surfactants.

Surfactant	Active substance (purity)	Density $\mathrm{g\,cm^{-3}}$	Producer
CTAB	N-cetyl-N,N,N-trimethylammonium bromide (100%)	0.39	Merck
EOT	Sodium di(2-ethylhexyl) sulphosuccinate (61.9%)	1,000	Huntsman
BET	Coconut oil acid amidopropylbetaine (30%)	1,040	PCC Rokita
TXT-100	Octhylphenol ethoxylate (100%)	1,057	Sigma Aldrich

of an electrochemical cell or a method in which the potential of the protected metal surface is polarized more negative until the metal surface has a uniform potential (Bombara, 1980).

– Inhibitors. Corrosion inhibitors are chemical agents that effectively reduce corrosion of a metal surface by reacting with the metal surface directly or by reacting with or modifying a preexisting corrosion product film or by scavenging constituents from the solution before they react with the metal. Interface inhibition is generally defined as adsorption of the inhibitor directly onto the metal surface (McCafferty, 2010). As successful corrosion inhibitors due to their ability to adsorb onto surfaces, forming an interfacial film and protecting them from corrosive agents surfactant molecules have been studied. Several parameters affect the mechanism of corrosion inhibition, such as the type of surfactant, temperature, and composition of the chemical systems used (Hihara, 2015).

– Protective coating. Corrosion resistant coatings protect metal components against degradation due to moisture, salt spray, oxidation or exposure to a mixture of environmental or industrial chemicals. Anticorrosion coating allows for protection of metal surfaces and act as a barrier to inhibit the contact between chemical compounds and corrosive materials. Thin coatings made up of metals and certain nonmetallic substances perform an effective barrier between the solid substrate to be protected and its working environment (Hihara, 2015).

2 MATERIALS AND TEST METHODS

Corrosion tests provide the appropriate procedures for carrying out corrosion tests on specified metallic materials and alloys. These tests are conducted to examine and evaluate the behaviour in corrosion environment and resistance of materials to corrosion.

2.1 Materials

Corrosion test was carried out on steel S235JRG2 (EN 10025:2004), brass CW612N (EN 1412:1996) and aluminium AW 2017A (EN 573-3:2013) tiles with dimensions $105 \times 50 \times 1$ mm. In all tests 5% aqueous surfactant solutions as corrosive environments were used. In the Table 1 characteristic parameters of the surfactants used in the experiment were shown.

2.2 Test methods

Influence of surfactants during corrosion test was determined on the following samples: steel S235JRG2, brass CW612N, aluminium AW 2017A, which were placed in studied solution for 28 days. Prepared and marked metallic tiles were of the following sizes $105 \times 50 \times 1$ mm. Before conducting the tests, the tiles were cleaned using acetone and their edges were protected with paraffin (ISO 11845:1995). The tiles mentioned above were placed in the solution for 28 days in such a manner as for the entire surface to be submerged. Once the corrosion test was ended, the samples were cleaned with distilled water. Then the samples were immersed in different etching solutions according to the standard (ISO 8407:2009). In order to remove loose corrosion products, samples were washed with tap water and wiped with a soft brush. They were then washed with distilled water, dried and placed in a desiccator with calcium chloride.

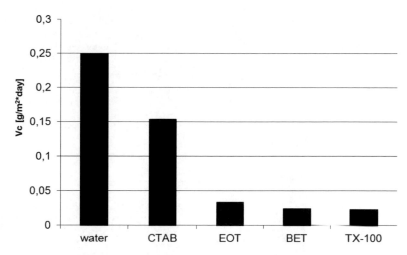

Figure 1.   Corrosion rate of brass in surfactant solutions.

In order to determine the loss of weight of a sample, it was weighed before and after corrosion with an accuracy of 0.0001 g. To eliminate the influence of external factors, also carried out was a blank test, the results of which were included in registering the mass loss of samples. For each case, three tests were carried out; the given results were the average of three measurements.

Corrosion rate V_m, in $g \cdot m^{-2} \cdot day^{-1}$, was calculated according to the following formula:

$$V_m = \frac{\Delta m}{A \cdot t} \tag{1}$$

where: $\Delta m = m_0 - m_1$, m_0 – weight of sample before test, m_1 – weight of sample after test, A – sample surface, t – time of test.

After corrosion examinations all metallic tiles were subjected to microscopic observations, in which morphology changes were being compared on the corroded material surfaces.

Effect of extinguishing agents containing surfactants on corrosion fire extinguishers were tested according to the described method. Extinguishers with water based extinguishing medium were subjected 8 times to the temperature cycle. Each cycle consist of 4 stage of duration 24 hours. Temperatures of stages were given in (EN 3-7:2008). On completion of the temperature cycles, the firefighting agent was drained of and examined for refraction index, and extinguisher body was cut for internal examination.

3   RESULTS

For evaluating the ability of studied specimens to resist corrosion when immersed in 5% an aqueous solution of surfactants continued exposure to the analysed solutions were used. The figure 1 shows corrosion rate of brass in 5% solution and results of the blank test. The results of the blank tests had the biggest corrosion rate. The corrosion rate for brass was bigger in CTAB solution than in other surfactant solutions. For another solutions the corrosion rate were less than $0.03\ g \cdot m^{-2} \cdot day^{-1}$.

In the Figure 2, a comparison of corrosion resistance studied alloys in solutions of two surfactants: CTAB and EOT were shown. The considered environments significantly affect the corrosion of steel, and have a small influence on the corrosion of aluminium and brass.

After corrosion tests, the tiles were analyzed by being placed under an optical microscope Olympus BX51. Figures 3–5 show the results of the surface investigation. In the pictures centre of the corrosion are visible, the biggest corrosion centre are found on low alloy steel tiles.

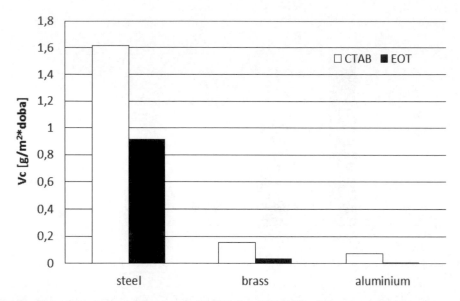

Figure 2.   Corrosion rate versus alloy type in 5% CTAB and 5% EOT solutions.

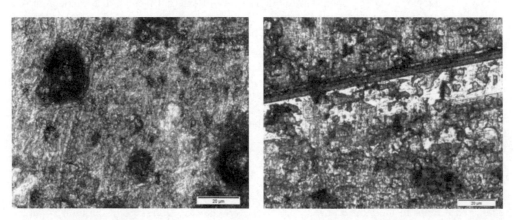

Figure 3.   Corrosion rate of brass in surfactant solutions.

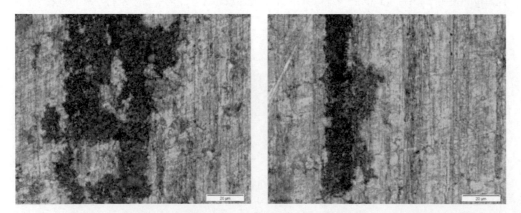

Figure 4.   Brass tiles after corrosion test.

162

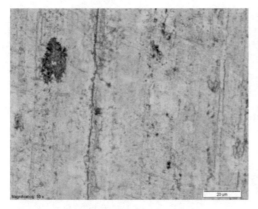 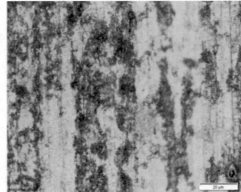

Figure 5.   Aluminium tiles after corrosion test.

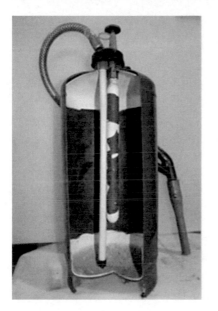 

Figure 6.   Internal corrosion of extinguisher.

The corrosivity of metallic elements and alloys influenced by an extinguishing agent, whose main ingredients were surfactants, determines durability and reliability of firefighting equipment. The corrosive effect of firefighting agents on fire extinguishers were tested according to standard method (EN 3-7:2008). On completion the temperature cycles, the extinguishing medium was drained off. Next the body of tested extinguishers were cut into two sections in a manner which permits internal observation. Examples of internal examinations in Figure 6 were presented.

4   CONCLUSION

Corrosion of the firefighting equipment is a serious problem. The results of corrosion can have a negative effect on the efficiency of extinguishing equipment and can cause a danger to the user. The efficiency to fire extinction may be decreased because of pipe clogging by corroded material petals.

Considered surfactants do not increase the corrosion rate of the tested alloys. On the contrary – they show an inhibitory effect. The increase of corrosion in firefighting equipment may be caused by other components of fire-extinguishing composition or the phenomenon may occur only in the presence of some chemical systems.

In the analyzed 5% surfactant solutions rate of corrosion of tested alloys was lower than corrosion rate in water. The steel was most corroded alloys in the presence of 5% surfactant solutions and in water.

The matter of corrosion of firefighting equipment requires further tests.

ACKNOWLEDGMENTS

The research was financially supported by MNiSW as project 016/BC/CNBOP-PIB/2011-2016.

REFERENCES

Bombara, G. & Bernabai, U., 1980. Use of electrochemical techniques for corrosion prevention and control in the process industries, *Anti-Corrosion Methods and Materials*, 27 (3), 6–10.

EN 10025:2004 *Hot rolled products of structural steels.*

EN 1412:1996 *Copper and copper alloys. European numbering system.*

EN 3-7:2008 *Portable fire extinguishers – Part 7: Characteristics, performance requirements and test methods.*

EN 573-3:2013 *Aluminium and aluminium alloys. Chemical composition and form of wrought products. Chemical composition and form of products.*

Hihara, L., 2015. Electrochemical Aspects of Corrosion-Control Coatings. [In:] Tiwari, A., Hihara, L., Rawlins, J., (eds), *Intelligent Coatings for Corrosion Control*, 1–15 Oxford: Elsevier.

ISO 11845:1995 *Corrosion of metals and alloys. General principles for corrosion testing.*

ISO 8407:2009 *Corrosion of metals and alloys. Removal of corrosion products from corrosion test specimens.*

Lemańska, K. & Główka, S., 2013. Review, application and development trends of firefighting equipment, *Safety and Fire Technique,* 30 (2), 91–99.

McCafferty, E., 2010. *Introduction to Corrosion Science.* New York: Springer.

Rakowska, J. & Ślosorz, Z., 2011. Corrosion of fire extinguishing systems and fire fighting fittings, *Safety and Fire Technique*, 24 (4), 113–120.

Revie, R. & Uhlig, H., 2011. *Uhlig's Corrosion Handbook,* Hoboken, New Jersey: Wiley.

Sharma, S., 2011. *Green Corrosion Chemistry and Engineering: Opportunities and Challenges.* Chichester, John Wiley & Sons.

Advances and Trends in Engineering Sciences and Technologies – Al Ali & Platko (Eds)
© 2016 Taylor & Francis Group, London, ISBN: 978-1-138-02907-1

Effectiveness of degreasing agents in removing petroleum compounds from asphalt pavement

J. Rakowska, K. Radwan & D. Riegert
Research Centre for Fire Protection – National Research Institute, Józefów, Poland

ABSTRACT: The tests were carried out in the Department of Chemical and Fire Research at the Research Centre for Fire Protection – National Research Institute. The aim of the studies was to evaluate the effectiveness of an innovative formulation intended for the removal of contaminants and pollutants from road infrastructure in terms of reducing the negative impact of petroleum compounds on asphalt durability. Oil spilt on a road and after penetration into its layers can cause elution of the asphalt binder, decrease the strength properties of asphalt and consequently lead to premature destruction of the road pavement. The study investigates the physicochemical properties of the analysed formulation and their impact on the structure of road pavement exposed to petroleum substances.

1 INTRODUCTION

Along with the growth of industrial technologies, human impact on the environment is also increasing. Humans' constantly expanding needs and increasing industrial technology capacity have resulted in increases in transport, which results then in the threat of accidents on roads and spills of hazardous substances (Kielin, 2012; Ślosorz, 2013).

The removal of these dangerous products is a difficult and iterative process requiring people with specialized knowledge and experience. The method for removing contaminants and the choice of this method largely depends on the incident that has occurred and the type and quantity of the substance removed (Mizerski, 2013; Twardochleb et al., 2012; Radwan, 2012). For removing oily hydrocarbons, chemical agents containing emulsifying surfactants are used.

The cleaning process can be divided into two stages. The first step is to collect the excess oil from the road surface with a loose sorbent or sorption mat. Then, the contaminated sorbent material should be removed from road. The second step consists of cleaning the surface using washing agents, called degreasers (Rakowska, 2012). After covering the affected area with a degreasing agent, the pavement should be brushed for 5 minutes. The next step is to rinse clean the area with water.

Asphalt pavements are the main paving type of road. Asphalt mixtures have been widely used in road pavements because of such advantages as good skid resistance, riding comfort, low noise, improved comfort convenience of maintenance and recyclability. However, asphalt pavements are subject to distress such as deformation, cracking and rutting under the effects of heavy use (Trzaska, E., 2014) (Fig. 1).

In road building a mixture of mineral aggregates and asphalt components are used. Asphalt performs as a binder. Asphalt is added to the road surface in small proportions. It has two important functions (Trzaska, E., 2014):

a) perform as a binding agent for the aggregate;
b) act as a layer protective the aggregate from external factors (water, de-icing agents, oil).

Figure 1. The destruction of the road surface – ruts in the asphalt.

Table 1. Characteristics of the degreasing agents.

Degreasing agent	Main components	Density g cm^{-3}	Concentration of use %
Degreaser 1	Ethoxylated lauryl alcohol, alkyl polyglucosides	1.007	1.5
Degreaser 2	Ethoxylated lauryl alcohol, alkyl polyglucosides alcohols C12-C14 – alkoxylate	1.004	1.5
Degreaser 3	Etoxylated C6-C12 alkohols, sodium dodycylbenzenesulfonate	1.001	100

2 MATERIALS AND TEST METHODS

Studies was carried out on two types of pavement: concrete and asphalt. As degreasing agents were used composition based on different types of surfactants.

2.1 *Materials*

In the Table 1 the characteristics of the degreasing agents used in the experiment are shown.

2.2 *Test methods*

Contact angle
Measurements of contact angle for aqueous solutions of degreasing agents were carried out using the sessile drop method with the instrument KSV CAM 200. Measurements were performed at $20 \pm 0.5°C$. In the preparation of solutions, distilled water with a surface tension of 72 mN/m and conductivity equal to $10\,\mu S/cm$ was used. Contact angle was investigated on the two surfaces: asphalt and concrete.

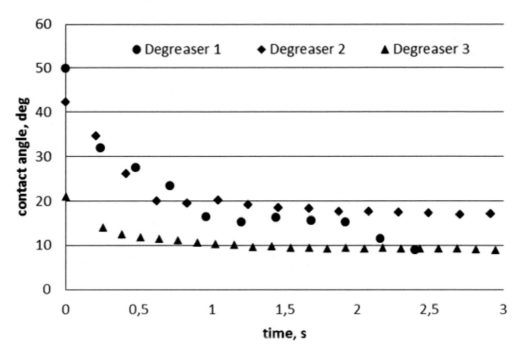

Figure 2. Dynamic contact angle on asphalt.

Wash out the oil from the surface of the asphalt
Asphalt samples were weighed and then submerged in 50 cm^3 of diesel oil. After 24 hours, the asphalt samples were drained of oil and washed with degreasing solutions and using a soft brush. Next, the samples were left to dry. The change in weight after cleaning was determined. Samples were subjected to macroscopic observation.

3 RESULTS

Effective elimination of liquid oil depends on the balance of contact angles. Oil forms a low contact angle with a substrate. To increase the contact angle between the oil and substrate, one has to increase the solid/water interfacial tension. The addition of a surfactant allows for an increase in the contact angle in the soil/solid/water interface.

In Figs. 2 and 3, the contact angle of the studied degreasing agents on asphalt and concrete are presented. All analysed degreasing agents show good wetting properties. The best wetting ability for degreaser 3 was found.

Both asphalt and diesel oil are petroleum products. Their chemical similarity causes an interaction that results in the dissolution of the binder. An oil spill on asphalt pavement can cause the aggregate to wash away from the asphalt, and penetration of the oil into the layers loosens and weakens the strength parameters of the asphalt mixture. As a consequence, this can lead to premature destruction of the road surface (Fig. 4). The majority of these incidents involve diesel oil. The impacts of these hydrocarbon spills are well known, e.g., environmental degradation, expensive cleanup costs, and community impacts through long traffic delays (average of 5 hours). Another important impact that has not been widely reported is the degradation of asphalt pavement (Fig. 5).

As stated, resilient modulus testing was carried out to assess the mechanical strength of the core specimens after saturation with diesel (i.e., a diesel spill). A reduction in modulus values would

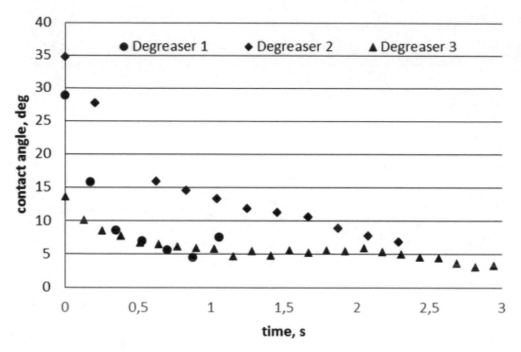

Figure 3. Dynamic contact angle on concrete.

Figure 4. Asphalt surface after 10 minutes, 1 hour and 24 hours exposure to diesel oil.

indicate that pavement strength was being affected by the addition of diesel oil into the core of the pavement (Baldwin, 2005).

The principal degradation mechanism is diesel dissolving the asphalt binder. The *liquefied* binder then flows through the core profile to collect at the sides and bottoms. As a result, binder integrity is adversely affected and the asphalt's performance is severely reduced. With the loss of asphalt binder, the aggregates in the core become loose and are easily displaced. This causes significant cracking and rutting in the core, thereby creating "residual" (or secondary) instability in the asphalt (Fig. 6).

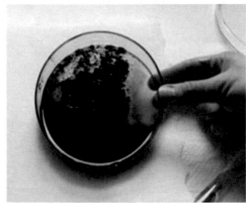

Figure 5.   Asphalt surface after 24 hours exposure to diesel oil.

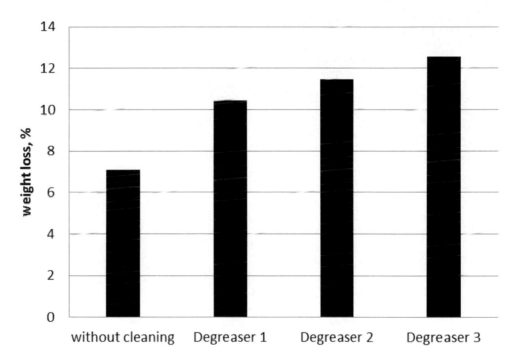

Figure 6.   Change in weight of the asphalt surface after 24 hours of exposure to diesel oil.

4   CONCLUSIONS

An oil stain can be very hard to clean up from the road surface since the asphalt or concrete will absorb some of the oil.

The study confirmed the disadvantageous impact of diesel oil on the asphalt surface. It is important to remove it as soon as possible through the use of sorbents and degreasing agents. These agents create an aqueous emulsion with oil and discharge it from the surface. The influence of the type of agent used for cleaning the surface varies slightly, but in every case, the application of the cleaning procedure reduces the degradation of the surface.

ACKNOWLEDGEMENT

The research was partially supported by project No 0004/R/ID3/2011/01 funded by the NCBR and 016/BC/CNBOP-PIB/2011-2016 financed by the MNiSW.

REFERENCES

Balwin, B., Carmod, O., Collins, T., 2005. Degradation of Asphalt Due to Diesel Spills on Roads, *Road System & Engineering Technology Forum- August 2005* [http://www.academia.edu/6479274/Degradation_of_Asphalt_Due_to_Diesel_Spills_on_Roads] 11.03.2015

Kielin, A. (2012). Chemical and ecological rescue. Analysis of the action in Cracow at the Powstańców str. *Safety & Fire Technique*, 27 (1), 107–112

Mizerski, A. (2013). Foams as carriers of chemicals for neutralizing contamination. *Safety & Fire Technique,* 31 (1). 87–93

Radwan, K., Rakowska J. & Ślosorz Z. (2012). Environmental effects of oil pollutants. *Safety & Fire Technique*, 29 (3), 107–114

Rakowska, J., 2012. Usuwanie substancji ropopochodnych z dróg i gruntów. [in:] *Usuwanie substancji ropopochodnych z dróg i gruntów.* Józefów: CNBOP-PIB, p. 139 [in polish].

Ślosorz, Z., Radwan, K., Rakowska, J., Statystyka zdarzeń niebezpiecznych z udziałem substancji ropopochodnych, [in:] *Problemy usuwania zanieczyszczeń ropopochodnych z infrastruktury drogowej oraz przemysłowej*, Conference Proceeding, CNBOP-PIB, Józefów 2013

Trzaska, E., 2014. Asfalty drogowe – produkcja, klasyfikacja oraz właściwości; *Nafta – Gaz*, LXX (5) [http://archiwum.inig.pl/inst/nafta-gaz/nafta-gaz/Nafta-Gaz-2014-05-07.pdf] 12.03.2015

Twardochleb et al., 2012, – Twardochleb, B., Jaszkiewicz A., Semeniuk I., Radwan K. & Rakowska J. 2012. Effect of anionic surfactants on the properties of the formulations used for removal oil derived substances. *Przem. Chem.* 10 (91), 1918–1921.

Advances and Trends in Engineering Sciences and Technologies – Al Ali & Platko (Eds)
© 2016 Taylor & Francis Group, London, ISBN: 978-1-138-02907-1

Shear strengthening of concrete elements by unconventional reinforcement

P. Sabol & S. Priganc
Technical University of Košice, Košice, Slovakia

ABSTRACT: This article presents the results of experimental research focused on the shear strengthening of reinforced concrete structures with one of the latest and most promising methods: NSM – Near surface mounted reinforcement. In recent years, research in this area focuses on combination of epoxy and FRP (fiber reinforced polymer) materials suitable for strengthening concrete elements. A traditional material -stainless steel- was also used in this study but in non-traditional T-cross section in terms of strengthening of concrete members.

1 INTRODUCTION

1.1 *Strengthening of concrete structures with adhesive bonded reinforcement*

In practice, it is often required to increase or restore the resistance of reinforced concrete structures. There are several quick and effective solutions for strengthening that can be classified into two groups. In the first method, EBR (externally bonded reinforcement), the reinforcement is glued to the surface of strengthened members. The second method, NSM, involves reinforcement embedding into the glue filled grooves cut out within the cover layer of strengthened members. Both methods use modern materials such as epoxy adhesives and the reinforcement from the fiber reinforced polymers. The decisive factor for effective strengthening is full transfer of stresses between the additional reinforcement and a concrete element. Stresses transfer efficiency increases within the surface enlarging involved in this transfer. Based on these facts the idea was developed (Sabol 2014) to take advantage of both methods where proposed cross-section of reinforcement with T shape maximizes contact with surface bonded parts while at the same time embedded web of in the groove allows better anchoring of reinforcement. Another aim of the above mentioned research was to verify use of commercially available materials – stainless steel, the characteristics of which can compete with expensive modern materials mainly in corrosion resistance and stiffness.

2 PARAMETERS OF THE EXPERIMENTAL PROGRAM

2.1 *Concrete beams*

The beams dimensions $150 \times 250 \times 1500$ mm and design of reinforcement determined their failure in shear. Beams were reinforced only with longitudinal reinforcement placed on the bottom of the beam in two layers with a total of 6 pieces with a diameter of 12 mm and concrete cover of 20 mm (Figure 1). Reinforcing bars were made of steel B420 ($f_{yk} = 420$ MPa, $f_t = 453.6$ MPa, Es $= 200$ GPa). Two bars of 8 mm diameter of steel were placed on top of the beam fixed at the support ends by a stirrup, diameter of 6 mm. Strength characteristics of concrete beams on the day of testing were $f_{ck,cyl} = 29.12$ MPa, $f_{ctm} = 2.89$ MPa and $E_c = 33.42$ GPa.

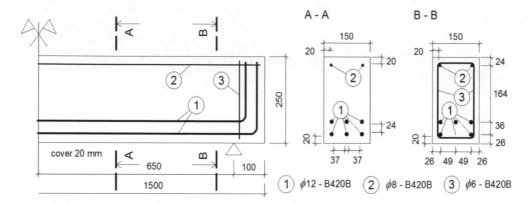

Figure 1. Beam reinforcement arrangement.

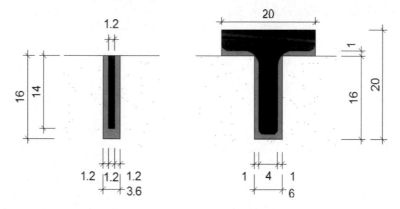

Figure 2. Dimensions of the grooves and shape of additional reinforcement.

2.2 *Additional reinforcement, structural adhesive and grooves*

For additional reinforcement we used two different kinds of materials, moreover reinforcing bars had different cross-section. The first one was stainless steel (grade 1.4301) hereinafter referred to as S.STEEL. This reinforcement had T cross section with dimensions $20 \times 20 \times 4$ mm. The T cross section was connected to concrete element through the web and the underside of the flange (Figure 2). S.STEEL reinforcement had tensile strength $R_m = 649$ MPa, nominal yield strength $R_{p0.2} = 412$ MPa and modulus of elasticity $E_s = 200$ GPa.

The second type of reinforcement was FRP strip with carbon fibers CarboDur S512 from Sika Company, hereinafter referred to as CFRP. This reinforcement had a square cross-section with dimensions 1.2×14 mm obtained by longitudinal cutting of 50 mm wide strip. CFRP reinforcement had tensile strength $f_t = 3100$ MPa and modulus of elasticity $E_f = 165$ GPa, these data were declared by the manufacturer.

Structural adhesive Sikadur 30 from Sika Company was used for bonding of additional reinforcement. Hardened glue, after 24 hours at $23°C$, had tensile strength $f_{tg} = 22.5$ MPa and modulus of elasticity $E_g = 9.6$ GPa. These data were declared by the manufacturer.

Grooves for embedding additional reinforcements were created by cutting, using an angle grinder and diamond cutting disc. Dimensions of the grooves had been derived from the dimensions of the additional reinforcement taking into account the requirements for the adhesive thickness of approximately 1 mm. In both cases the groove depth was 16 mm, whereas the thickness of the concrete cover was 20 mm (Figure 2).

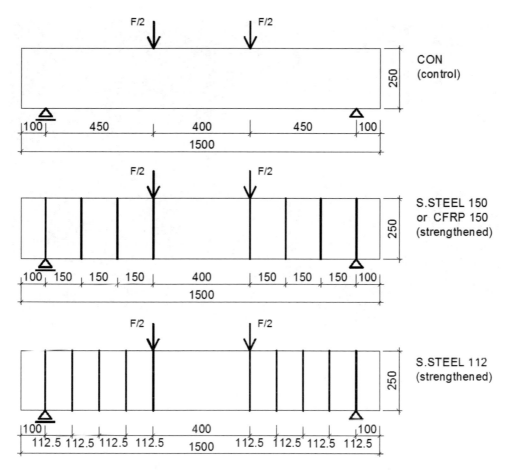

Figure 3.   Arrangement of additional reinforcement.

2.3   *Scheme of placing additional reinforcement*

Concrete beams with a deficit of shear resistance were strengthened in shear regions (Figure 3). Series of beams consisted of four species, differing in strengthening ratio of the shear field with a width 450 mm given by a load and supporting elements of the beam. The beam with an indication CON was not strengthened and served as a reference sample. The beam with an indication CFRP 150 was strengthened by CFRP reinforcement at axial distance of 150 mm with a strengthening ratio $\rho_{add} = 0.0015$, similarly the beam S.STEEL 150 was strengthened by T sections S.STEEL reinforcement with a strengthening ratio $\rho_{add} = 0.0130$ and beam S. STEEL 112 was strengthened by T sections S.STEEL at axial distance 112.5 mm with a strengthening ratio $\rho_{add} = 0.0174$.

3   TESTING OF BEAMS

3.1   *Test setting and equipment*

Testing of beams was carried out under four-point bending using pressing machine DrMB300 (Figure 4). Beam was placed on the steel supports and the load was applied over the steel plates to avoid local damage of the beam. The load was applied in steps 5, respectively 10 kN in cycles of 2 to 3 minutes, during which deflection and cracks formation were recorded. Deflections were

Figure 4. Testing of a beam and equipment.

measured with a dial gauge indicator located on the bottom edge of the beam at its mid span. Crack width was measured by using a standard foil crack width gauge. Illustration of scheme of cracks (Figure 6) was prepared based on detailed photos taken during the test.

3.2 *Results*

Strengthening effect was reflected in the significant increase of strength and stiffness of the tested beams (Figure 5). In all cases failure in shear occurred. Reference beam CON failed at overall load force 126.41 kN and increase in resistance of the strengthened beams was almost twice. The CFRP 150 beam failed at load force 216.48 kN, S.STEEL beam failed 150 at load force 205,41 kN and S.STEEL beam 112 failed at load force 224.13 kN. In addition to increased resistance of the beams, effect of strengthening was also reflected in the width of CSC (critical shear crack) at beams failure. CSC was the first shear crack formed on the beams particularly crucial to the failure of the beam. In all cases CSC was formed between the second and third rod of additional reinforcement (Figure 6).

For beam CFRP 150 width of CSC 1.0 mm was recorded, for the other two beams strengthened with S.STEEL reinforcement width of CSC recorded was only 0.4 mm. The beam failure was manifested in increase of the deflection and evidence in pulling off additional reinforcement. Slip or debonding of reinforcement did not occur in this test (Figure 7).

3.3 *Summary*

Summary of above mentioned data is presented in the following Table 1.

4 CONCLUSIONS

The results of the laboratory tests have shown high efficiency of the NSM strengthening system, manifested by significant increase in strength and stiffness of strengthened beams as well as in

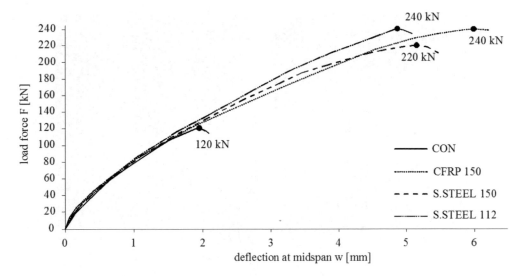

Figure 5.   Load-deflection curves.

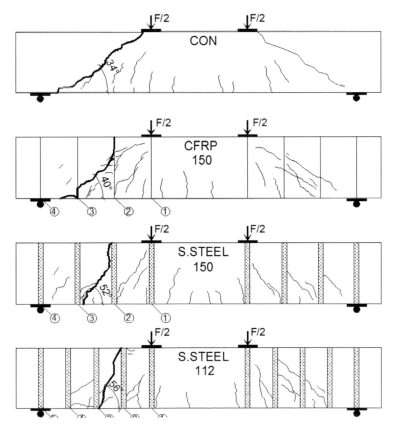

Figure 6.   System of cracks at failure of beams.

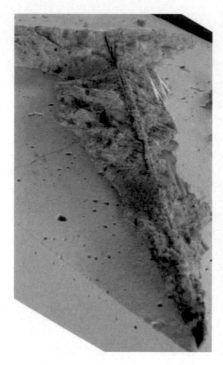 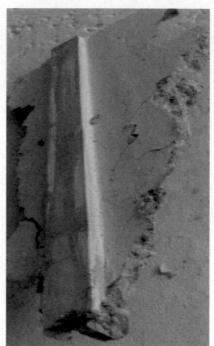

Figure 7.   Pulling of additional reinforcement.

Table 1.   Summary of the results.

Beam	F kN	w mm	ΔF %	ρ_{add}	w_k mm	α °
CON	120.0	1.86	–	–	0.4	34
CFRP 150	240.0	5.86	100.0	0.0015	1.0	40
S.STEEL 150	220.0	5.15	83.30	0.0130	0.4	52
S.STEEL 112	240.0	5.02	100.0	0.0147	0.4	56

F – maximum load by the failure of beam
w – deflection at midspan by the failure of beam
ΔF – increasing of the load capacity
ρ_{add} – strengthening ratio
w_k – maximum crack (CSC) width
α – angle slope of CSC

widths restrictions of CSC at failure of the beams. Generally, it can be said that the use of stainless steel reinforcement with T-cross section is more preferable as the increase of resistance ranged from 83 to 100%, as well as width of CSC was from only 0.4 mm. Slip or debonding of reinforcement did not occur in this test and the tensile strength of concrete seems to be the weakest element of strengthening system.

REFERENCE

Sabol, P. 2014. Stress and strain of strengthened concrete members at elevated temperatures, Dissertation thesis, Kosice: SvF TU Kosice.

Advances and Trends in Engineering Sciences and Technologies – Al Ali & Platko (Eds)
© 2016 Taylor & Francis Group, London, ISBN: 978-1-138-02907-1

Numerical analysis of concrete elements strengthened in shear by NSM method

P. Sabol & S. Priganc
Technical University of Košice, Košice, Slovakia

ABSTRACT: This paper is focused on the numerical analysis of shear resistance strengthened concrete beams. Numerical analysis was carried out by the software package ATENA from Cervenka Consulting Company and obtained results were compared with laboratory tested beams. Beams were strengthened by NSM (Near surface mounted reinforcement) method by combination of epoxy and CFRP (carbon fiber reinforced polymer) strips or T-cross section bars from stainless steel.

1 INTRODUCTION

1.1 *Strengthening of concrete beams with adhesive bonded reinforcement and FEM analysis*

In practice, it is often required to increase or restore the resistance of reinforced concrete structures. There is a quick and effective solution – NSM method, which involves reinforcement embedment into the glue filled grooves cut out within the cover layer of strengthened members. The decisive factor for effective strengthening is full transfer of stresses between the additional reinforcement and the concrete element. Stresses transfer efficiency increases within the surface enlarging involved in this transfer. Based on these facts the idea was developed (Sabol 2014) that the proposed cross-section of reinforcement with T shape maximizes contact with surface bonded parts while at the same time embedded web of in the groove allows better anchoring of reinforcement. Another object of the above mentioned research was to verify use of commercially available materials – stainless steel, the characteristics of which can compete with expensive modern materials mainly in corrosion resistance and stiffness. The numerical simulations of behavior of the tested beams using software Atena were carried out in parallel with laboratory tests. The results confirmed the correctness of proposed numerical models and thus approved usage of mentioned models to be used as an alternative to laboratory tests with sufficient accuracy.

2 STRENGTHENED BEAMS AND MATERIAL CHARACTERISTICS

2.1 *Concrete beams*

The beams dimensions $150 \times 250 \times 1500$ mm and design of reinforcement determined their failure in shear. Beams were reinforced only with longitudinal reinforcement placed on the bottom of the beam in two layers with a total of 6 pieces with a diameter of 12 mm and concrete cover of 20 mm (Figure 1). Reinforcing bars were made of steel B420 ($f_{yk} = 420$ MPa, $f_t = 453.6$ MPa, $E_s = 200$ GPa). Two bars of 8 mm diameter were placed on top of the beam fixed at the support ends by a stirrup, diameter of 6 mm. Strength characteristics of concrete beams on the day of testing were $f_{ck,cyl} = 29.12$ MPa, $f_{ctm} = 2.89$ MPa and $E_c = 33.42$ GPa.

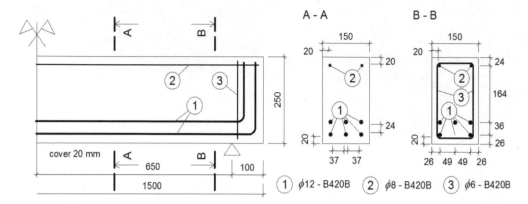

Figure 1. Beam reinforcement arrangement.

2.2 NSM reinforcement, structural adhesive and grooves

For NSM reinforcement were used stainless steel (grade 1.4301) hereinafter referred to as S.STEEL. This bars had T cross section with dimensions 20 × 20 × 4 mm. The T cross section was connected to concrete element through the web and the underside of the flange. S.STEEL reinforcement had tensile strength $R_m = 649$ MPa, nominal yield strength $R_{p0.2} = 412$ MPa and modulus of elasticity $E_s = 200$ GPa. The second type of reinforcement was FRP strip with carbon fibers CarboDur S512, hereinafter referred to as CFRP. This reinforcement had a square cross-section with dimensions 1.2 × 14 mm. CFRP reinforcement had tensile strength $f_t = 3100$ MPa and modulus of elasticity $E_f = 165$ GPa. For bonding of NSM reinforcement was used adhesive Sikadur 30. Hardened glue, after 24 hours at 23°C, had tensile strength $f_{tg} = 22.5$ MPa and modulus of elasticity $E_g = 9.6$ GPa. Grooves for embedding NSM reinforcements were created by cutting. Dimensions of the grooves had been derived from the shape of the NSM reinforcement taking into account the requirements for the adhesive thickness of approximately 1 mm. In both cases the groove depth was 16 mm.

2.3 Scheme placing of additional reinforcement

Concrete beams with a deficit of shear resistance were strengthened in shear region. Series of beams consisted of four species, differing in strengthening ratio of the shear field with a width 450 mm given by a load and supporting elements of the beam. The beam with an indication CON was not strengthened and served as a reference sample. The beam with an indication CFRP 150 was strengthened by CFRP reinforcement at axial distance of 150 mm with a strengthening ratio $\rho_{add} = 0.0015$, similarly the beam S.STEEL 150 was strengthened by T sections S.STEEL reinforcement with a strengthening ratio $\rho_{add} = 0.0130$ and beam S. STEEL 112 was strengthened by T sections S.STEEL at axial distance 112.5 mm with a strengthening ratio $\rho_{add} = 0.0174$.

3 FEM ANALYSIS

3.1 Atena

Atena software is based on the finite element method and allows non-linear analysis of concrete structures. Material models reflect their real behavior such as crushing and cracking of concrete, plastic deformation of reinforcement etc. These specifications allow detection of stress-strain characteristics, the position of the origin and development of cracks. The modeling was carried out in the preprocessing phase, which involved the formation of an input file in the software environment GiD: geometry, material characteristics, mesh, calculation and boundary conditions. Input file was then exported to Atena software where it run the calculation followed by the postprocessing phase.

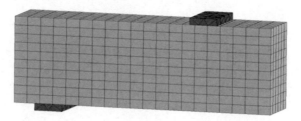

Figure 2. Beam CON.

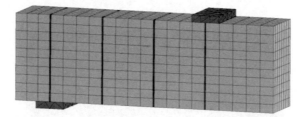

Figure 3. Beam CFRP 150.

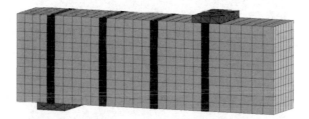

Figure 4. Beam S.STEEL 150.

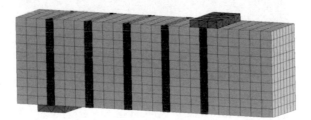

Figure 5. Beam S.STEEL 112.

3.2 *Numerical models*

Mesh topography of the finite elements and its parameters were specified with respect to the available computing performance. As the symmetry was used for numerical simulations of the beams it was sufficient to model only one half of the total length of the beam (Figure 2–5).

3.3 *Parameters of numerical models and simulations*

The mesh of the finite elements (Figure 2–5) was generated by hexahedral elements. For the solution of the nonlinear problems an iterative Newton-Raphson method had been chosen, where

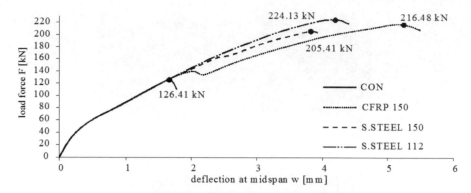

Figure 6.   Load-deflection curves.

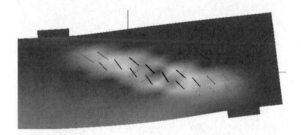

Figure 7.   Beam CON after failure.

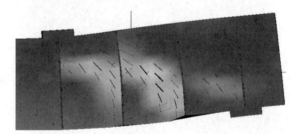

Figure 8.   Beam CFRP 150 after failure.

the load is applied in incremental steps. Loading of beams was carried out in a movement of steel load plates placed on top of the beams. Numerical models of the beams are composed of several types of materials, so as tested beams. Contacts between components of NSM system were modeled as perfectly rigid. For concrete was used material Cementitious 2 which combines the basic models for tensile and compressive concrete behavior. Reinforcement was modeled as a discrete 1D Reinforcement. Here bi-linear diagram with increasing branch was used. NSM reinforcement from stainless steel was modeled using material Steel Von Mises 3D. CFRP reinforcement was represented by material Solid Elastic as well as supporting and loading plates. Epoxy adhesive was modeled using 3D Steel Von Mises material.

4   RESULTS

4.1   *FEM analysis*

Strengthening effect was reflected in the significant increase of strength and stiffness of the tested beams (Figure 6). In all cases failure in shear occurred Reference beam CON failed at overall load

Figure 9.   Beam S.STEEL 150 after failure.

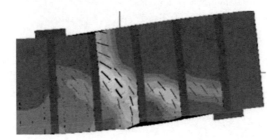

Figure 10.   Beam S.STEEL 112 after failure.

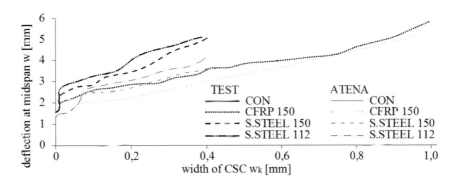

Figure 11.   Development of crack width of CSC in strengthened beams TEST/ATENA.

force 126.41 kN. The CFRP 150 beam failed at load force 216.48 kN, S.STEEL beam 150 at load force 205.41 kN and S.STEEL beam 112 failed at load force 224.13 kN. In addition to increased resistance of the beams, effect of strengthening was also reflected in the width of CSC (critical shear crack) at beams failure. CSC was the first shear crack formed on the beams particularly crucial to the failure of the beam. In all cases CSC was formed between the second and third rod of additional reinforcement (Figure 7–10).

Width of CSC 0.96 mm was recorded for the beam CFRP 150. For the other two beams strengthened with S.STEEL reinforcement recorded width of CSC was only 0.4 and 0.41 mm respectively (Figure 11).

4.2   *Comparison of FEM analyzes and laboratory tests*

Comparison of the results of laboratory tests and numerical simulation show high accuracy in developing width of CSC (Figure 11). Similarity can be also seen in the load-deflection curves of all types of the beams (Figure 12). Table 1 below provides the summary of obtained data.

Table 1. Summary of the results.

Beam	F_T (kN)	F_A (kN)	F_T/F_A	w_T (mm)	w_A (mm)	w_T/w_A	w_{kT} (mm)	w_{kA} (mm)	w_{kT}/w_{kA}
CON	120.0	126.41	0.95	1.86	1.69	1.10	0.4	0.38	1.05
CFRP 150	240.0	216.48	1.11	5.86	5.24	1.12	1.0	0.96	1.04
S.STEEL 150	220.0	205.41	1.07	5.15	3.83	1.34	0.4	0.41	0.97
S.STEEL 112	240.0	224.13	1.07	5.02	4.19	1.20	0.4	0.40	1.00

F_T – maximum load by the failure of beam – Test
F_A – maximum load by the failure of beam – Atena
w_T – deflection at midspan by the failure of beam – Test
w_A – deflection at midspan by the failure of beam – Atena
$w_{k,T}$ – maximum crack (CSC) width – Test
$w_{k,A}$ – maximum crack (CSC) width – Atena

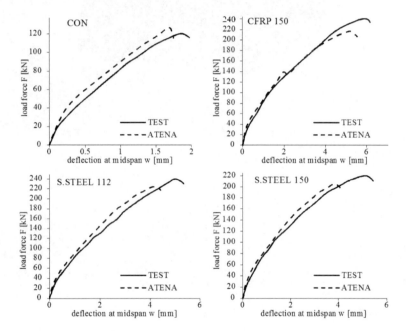

Figure 12. Comparison load-deflection curves TEST/ATENA.

5 CONCLUSIONS

Conduced analyses showed high efficiency of NSM strengthening system, manifested by significant increase in strength and stiffness of strengthened beams as well as in widths restrictions of CSC at failure of the beams. Generally was the increase of resistance to 88%, as well as width of CSC was from only 0.4 and 0.41 mm for S.STEEL reinforcement. The results confirmed the correctness of numerical models and thus proved they can be used as an alternative to laboratory tests with sufficient accuracy. Differences in the results represent 5% for load capacity of beams 21% for deflections and 2% for width of CSC.

REFERENCE

Sabol, P. 2014. Stress and strain of strengthened concrete members at elevated temperatures, Dissertation thesis, Kosice: SvF TU Kosice.

Advances and Trends in Engineering Sciences and Technologies – Al Ali & Platko (Eds)
© 2016 Taylor & Francis Group, London, ISBN: 978-1-138-02907-1

Experience with a spiral roundabout in Slovakia

B. Salaiová
Technical University of Košice, Faculty of Civil Engineering, Košice, Slovakia

ABSTRACT: Experience with a spiral roundabout in Košice (Slovakia) is presented in the paper. The results of the analysis of the use of each entry lane to the roundabout are discussed. The results indicate that the spiral roundabout is accepted by drivers in Kosice and that they can use it properly. This leads to the conclusion that this type of junction is effective in practice and can be recommended as an alternative to light-controlled intersections.

1 INTRODUCTION

The constantly increasing traffic volumes in urban areas force traffic engineers to face a particular problem in managing intersections. Intersections have to have sufficient capacity while traffic flow has to be of the highest possible continuity and safety. This can be achieved using a large roundabout (with 2 lanes on the circle). The drawback which reduces the expected capacity is poor use of the left-hand lane at the entries to the roundabout, which drivers often explain by uncertainty about the possibility of using a particular exit (Novotný 2014). Sufficient capacity and continuous traffic flow on a roundabout will be achieved only if there is a proper geometric design of the intersection, which depends on the traffic volume on the roundabout. A direct consequence of optimal design is reduction in the delay of vehicles negotiating the roundabout. The crucial fact for traffic engineers in the decision-making process concerning the layout of the intersection is its capacity. The theoretical value is evident from the literature (TP 234 2011, TP 10/2010 2010, Highway Capacity Manual 2000), but the real capacity is often different. Knowledge of the characteristics that affect roundabout capacity therefore plays an important role. The modern type of intersection, which is designed to increase the intersection capacity, is a roundabout with a spiral arrangement, rarely found in Slovakia. It is a spiral intersecting (Bartoš 2008) junction type 2-2-1, with two lanes on the entry, 2 lanes on the circle and one lane on each exit. However, its effective use is achieved only if the lanes are used in the appropriate way. This means that the drivers select the proper lane for entering, which depends on the desired exit.

This paper presents the results obtained on a roundabout with a spiral arrangement in Kosice, particularly the analysis of the use of lanes at the entries to the circle obtained from direct measurements in a traffic survey, as a parameter affecting the capacity of the intersection

2 MEASUREMENTS – TRAFFIC SURVEY

The aim of the traffic survey was to obtain information on the actual use of the lanes at the entries to the roundabout at different "supply pressure". For this reason a special survey was carried out which included automobile traffic and trams in a typical working day at the time of the morning peak and the saddle hour (the time of the survey was chosen as part of a long-term monitoring of traffic volumes). The traffic survey methods used made it possible to obtain data on:

- traffic volumes on the circle and all entries to the roundabout,
- traffic volumes on separate lanes for a right turn (bypasses),

Figure 1.   View of the intersection.

- traffic volumes of trams across the roundabout,
- composition of traffic flows, local (Kosice, Kosice-surroundings) and foreign vehicles,
- waiting time of vehicles at the entry to the circle,
- waiting time of trams input / output to / from the roundabout, and
- volumes of pedestrians on crossings.

2.1   *Characteristics of the roundabout*

The intersection of Trieda SNP –and Alejova streets with the Moldavska road including four-way tram lines (Figure 1) consists of a roundabout with external diameter D = 103 m with two lanes on the circle with a spiral design indicated by white road markings. There are 4 – lane local roads of functional group B coming to the roundabout mouth, arranged into two lanes at the entry to the circle, one lane at the exits and a single lane for vehicles branching right and bypassing the roundabout proper. A specific feature of this intersection is the tramway, with two tracks in each direction across the center island. Pedestrian movement is possible over crossings at all entries to the roundabout. The scheme of the intersection geometry can be seen in Figure 2.

2.2   *Methodology of the traffic survey*

The traffic survey was conducted using two methods: a dotted method for the direct right turn, and the method of recording vehicle registration plates. A special transportation survey was used to determine the waiting time. The dotted method assessed vehicle types and the number of vehicles which used a separate right turn lane. The method of recording registration plates was used to find the direction of vehicles crossing the intersection, track traffic flows and traffic volumes at the entries and exits and on the circle. A separate survey assessed the waiting time of vehicles before entering the roundabout. The survey was conducted during a working day, at the time of the morning peak (7:00–8:00) and during the traffic saddle (11:30–12:30). The following roundabout entries (Figure 2) were surveyed:

1. Moldavska road – towards the stadium (direction Stadion).
2. Trieda SNP street.
3. Moldavska road – towards the airport (direction Airport).
4. Alejova street.

The survey of tram traffic was conducted by recording the number and direction of trams, and waiting time by the automatic recording of tram standing time. The survey of pedestrians was conducted by recording the number of pedestrians at the crossings.

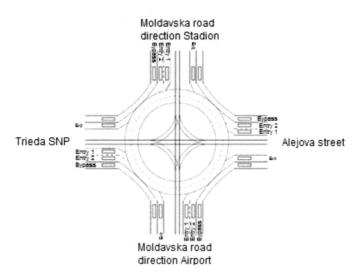

Figure 2.   Scheme of the intersection with surveyed entries.

Table 1.   The traffic volumes of the vehicles in a direct right turn – peak hour 7:00 to 8:00 (veh./h).

| FROM | TO | | | | |
	1.	2.	3.	4.	$\sum 1$
1.	–	96	–	–	96
2.	–	–	275	–	275
3.	–	–	–	276	276
4.	264	–	–	–	264
$\sum 3$	264	96	275	276	911

Table 2.   The traffic volumes of the vehicles in a direct right turn – saddle hour 11:30 to 12:30 (veh./h).

| FROM | TO | | | | |
	1.	2.	3.	4.	$\sum 1$
1.	–	125	–	–	125
2.	–	–	223	–	223
3.	–	–	–	313	313
4.	172	–	–	–	172
$\sum 3$	172	125	223	313	833

2.3   *Analysis of use of particular lanes*

Traffic volumes at all entries are relatively high because this is the main entrance to the city of Kosice from the south; the entry of road I/50, from the direction of Rožňava (Bratislava) connecting with local streets Trieda SNP and Alejová and the continuation of the Moldavska road.

The traffic volumes in particular lanes turning right at the exit are listed in Tables 1 and 3. The volume and direction of vehicles passing through the area at the time of the survey are shown in Tables 2 and 4. The results are shown graphically in the cartograms in Figure 3.

Table 3. Vehicles passing through area – peak hour 7:00 to 8:00 (veh./h).

	Entry	Lane	TO				Σ1	
			1.	2.	3.	4.		
FROM	1.	1	12	0	11	219	242	589
		2	0	1	329	17	347	
	2.	1	229	5	1	73	308	924
		2	12	0	0	604	616	
	3.	1	58	304	1	1	364	826
		2	453	7	1	1	462	
	4.	1	0	63	109	30	202	661
		2	0	449	10	0	459	
	Σ3		764	829	462	945	3000	

Table 4. Vehicles passing through area – saddle hour 11:30 to 12:30 (veh./h).

	Entry	Lane	TO				Σ1	
			1.	2.	3.	4.		
FROM	1.	1	19	0	24	183	226	580
		2	3	0	327	24	354	
	2.	1	100	5	0	42	147	551
		2	16	1	3	384	404	
	3.	1	31	182	4	0	217	616
		2	379	19	1	0	399	
	4.	1	1	27	122	16	166	631
		2	4	426	32	3	465	
	Σ3		553	660	513	652	2378	

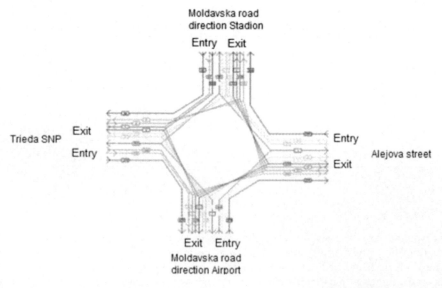

Figure 3. Cartogram of traffic volumes – routing scheme of vehicles passing through area – peak hour 7:00–8:00.

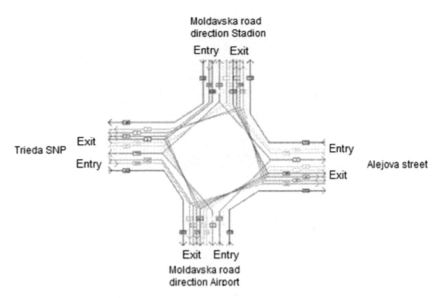

Figure 4. Cartogram of traffic volume – routing scheme of vehicles passing through area – saddle hour 11:30–12:30.

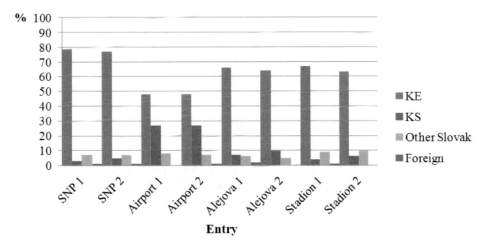

Figure 5. Local and foreign vehicles at entries in % (7:00 to 8:00).

It was found that only about 6% of the drivers at the entries to the roundabout used the wrong lanes, and thus it can be concluded that drivers in Kosice know the proper and effective use of this type of intersection.

2.4 *Determination of the proportion of local and foreign vehicles*

Since the survey included recording vehicle registration plates, it was possible to identify local (Kosice region) and foreign vehicles. The results can be seen in Figures 5 and 6.

Based on the obtained results it can be concluded that local vehicles from the Kosice area accounted for about 70% of the traffic, which suggests that the majority of drivers know the intersection well.

187

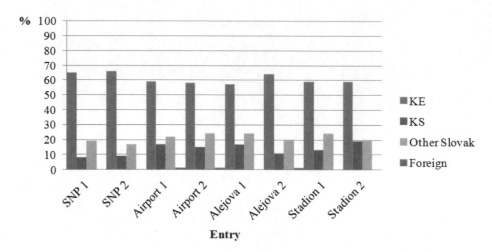

Figure 6. Local and foreign vehicles at entries in % (11:30 to 12:30).

3 CONCLUSION

The results presented provide information on the actual use of the different lanes for entry into the spiral roundabout in Kosice. Since it is a unique junction of this type in Slovakia, it was not possible to compare the results obtained with the results of other measurements. These types of junctions are more common abroad (Brilon 2005, Smelý et al. 2011, Engelsman et al. 2007) and spiral roundabouts are considered to be generally well accepted by drivers. Due to their advantages in terms of capacity and safety, they should be used more frequently in practice. The geometric design parameters should take into account the movement of pedestrians as well as public transport (tram traffic).

ACKNOWLEDGMENT

The paper presents results of the research activities of the Centre for Progressive Constructions and Technologies in Transportation Engineering. The Centre has been supported by the Slovak Research and Development Agency under contract No. SUSPP-0013-09 and the companies Inžinierske stavby and EUROVIA SK. The research has been carried out within the project NFP 26220220051 Development of progressive technologies for utilization of selected waste materials in road construction engineering, supported by the European Union Structural Funds.

REFERENCES

Barto, L., Rozsypal, V. 2008. Application of spiral roundabouts. Road review, Vol. 69, Prague
Brilon, W. 2005. Roundabouts: A State of the Art in Germany, National Roundabout Conference, Vail, Colorado
Engelsman, J.C., Eken, M. 2007. Turbo roundabouts as an alternative to two lane roundabouts. Proc. of the 26th SATC, Pretoria, South Africa
Highway Capacity Manual 2000. Transportation Research Board. National Research Council. Washington, D.C.
Novotný, P. 2014: Insights and experiences with spiral roundabouts – year 2013. Road review, Vol. 75, Prague
Smelý, M., Radimský, M., Apeltauer, T. 2011. Capacity of roundabouts with multilane entries. Traffic engineering, No. 01/2011
TP 10/2010 2010. Calculation of road capacity, MDPaT SR, Bratislava
TP 234 2011. Assessing the capacity of roundabouts, Edip s.r.o., Liberec

Advances and Trends in Engineering Sciences and Technologies – Al Ali & Platko (Eds)
© 2016 Taylor & Francis Group, London, ISBN: 978-1-138-02907-1

Numerical investigation of aerodynamic characteristics of vertical cylinder with rivulet

R. Soltys & M. Tomko

Technical University of Košice, Faculty of Civil Engineering, Institute of Structural Engineering, Košice, Slovakia

ABSTRACT: Rain-wind induced vibration is a well-known phenomenon affecting structures. It occurs under windy conditions with a moderate rain and causes large amplitude vibrations of slender structures with a low damping ratio. Such vibrations have been observed on horizontal and inclined cables and rarely on vertical hangars. This paper presents a numerical method based on fluid-structure interaction for simulating a rivulet motion on cylinder surface. The numerical simulation has been applied with a uniform inlet flow velocity to identify the aerodynamic forces acting on the vertical cylinder with a rivulet. An incompressible fluid flow with Navier-Stokes equations has been applied. Considering the flow velocities and stresses in a very close area of the cylinder surface, the former position of the water rivulet has been considered leeward. The water rivulet oscillated in a narrow region near the initial position. Aerodynamic forces lightly decreased comparing to the flow past the dry cylinder.

1 INTRODUCTION

A significant part of cable-stayed bridges are inclined cables which are sensitive to dynamic loading due to their low structural damping and flexibility. In special weather conditions cables may vibrate with large amplitudes and low frequency as was observed on several bridges (Meikonishi Bridge, Hikami & Shiraishi, 1988; Fred Hartman Bridge, Zuo et al., 2007). This aeroclastic phenomenon is known as rain-wind induced vibration (RWIV). During the last decades numerous wind-tunnel experiments have been realized focusing on the formation of rivulets and aerodynamic effects on the vibration of cables (Matsumoto et al., 2003; Zhang et al., 2008).

Fluid mechanical interpretation has been deduced from mechanical models (Yamaguchi, 1990; Peil & Nahrath, 2003; Seidel & Dinkler, 2006) and analytical solutions of an airflow-rivulet have also been presented (Seidel & Dinkler, 2006). Using numerical simulations two different approaches have been presented. At first, the rivulet has been considered as a rigid protuberance (Li & Gu, 2006; Soltys & Tomko, 2014). In the second approach, the thin-film model has been used to investigate the formation of rivulets which modify the effective shape of the body (Taylor & Robertson, 2011).

According to the complexity and the three-dimensionality of the RWIV problem, fluid-structure interaction (FSI) approach has been used for the investigation of airflow-induced rivulet movement and frequency of the movement, related influence on the airflow and corresponding changes of cylinder aerodynamic characteristics. This approach could be considered to be a very realistic, which involves dynamics of the rivulet, boundary layer of the cylinder and the rivulet.

2 APPLIED PHYSICAL AND BOUNDARY CONDITIONS FOR THE NUMERICAL MODEL

Conditions in which the rivulets can be formed have been investigated by several studies (Lemaitre et al., 2010; Consentino et al., 2003). Cable vibrations may occur under conditions when

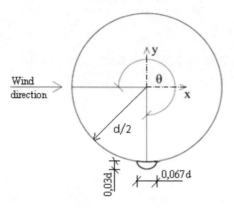

Figure 1. Cylinder and rivulet relative position and dimensions.

wind velocity ranges between 5–17 m/s, cable diameter $d = 0.1-0.25$ m and Reynolds number $0.5 \cdot 10^5 - 1.5 \cdot 10^5$. Considering the cable orientation to the wind direction and the angle of cable inclination, vibrations may also be initiated.

In this study a vertically positioned circular cylinder with a diameter of $d = 0.12$ m using a two-dimensional coupled airflow-rivulet interaction model using fluid-structure interaction (FSI) has been investigated. The FSI model consisted of two interacting models. The first one was the fluid model representing the air flow. The second one represented the cylinder and rivulet with rotational degree of freedom. Air with uniform speed of $u = 13$ m/s has been considered with a corresponding Reynolds number of Re $= 10^5$ and the air has been represented by an incompressible viscous fluid flow the in computational fluid dynamics (CFD) model with applied Navier-Stokes equations in turbulent regime with Spalart-Allmaras Detached Eddy Simulation model (SA-DES). Dimensions of the rivulet have been adopted according to Matsumoto et al. (2003) as is shown in Figure 1. The initial position of the rivulet has been considered leeward of the vertically mounted cylinder with the angle related to inlet flow direction of $\theta = 180°$.

It has been considered, that the initial position of the rivulet has been affected by wind flow forces. The gravitational force direction was perpendicular to the modelled plane. The mesh of the coupled computational models and applied boundary conditions are shown in Figure 2. The structural model for computational structural dynamics (CSD) used the Newmark integration method for transient dynamic response calculations. An attached circularly-shaped water rivulet has been considered with a density of 1000 kg/m^3 and with nearly infinitesimal rigidity, therefore no surface tension forces have been applied. A rivulet mesh has been fixed to cylinder mesh via shared nodes. Rivulet and cylinder meshes consisted of 128 and 2029 elements, the fluid mesh consisted of 94,172 elements (of which 43,380 elements were in a boundary layer near cylinder wall). Between the boundary layer mesh and the rest of the fluid mesh the sliding mesh interface was applied to ensure both the meshes do not deform. Therefore, the accuracy of the computation will remain as was required. The thickness of the first element layer was $d \cdot 1.875 \cdot 10^{-4}$. A nearly infinitesimally small mass and infinitesimal rigidity has been considered for the cylinder model. Since the FSI interaction has been applied and cylinder mesh has been fixed rigidly through its central node, the aerodynamic characteristics could be effectively calculated from resulting forces in this node. Fluid flow model mesh density decreased with increasing distance from cylinder surface. Therefore, high mesh density has been achieved in a boundary layer and a less time consuming model was acquired, observing a sufficient accuracy (according to experimental measurements) as can be seen in Table 1.

Table 1 summarizes the computed aerodynamic characteristics (c_d, where is the drag coefficient, S_t is the Strouhal number and f_{sh} is the vortex shedding frequency), compared with numerically and experimentally obtained results by Breuer (2000), Wieselsberger (1921) and Roshko (1961).

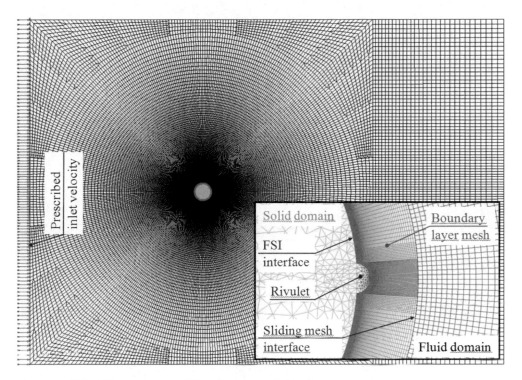

Figure 2.   CFD, CSD and rivulet model meshes and applied boundary conditions.

Table 1.   Aerodynamic characteristics of flow past circular cylinder with and without rivulet.

Method (Re $= 10^5$)	Mesh size	c_d (–)	S_t ( )	f_{sh} (Hz)
Without rivulet, simulation – SA-DES	200 × 400	1.224	0.201	21.74
With rivulet, simulation – SA-DES	200 × 400	1.191	0.196	21.27
Breuer (2000), (Re $= 1,4 \cdot 10^5$), simulation – LES, 3D	165 × 165	1.218	0.217	–
	325 × 325	1.286	0.203	–
Wieselsberger (1921), experiment		$\doteq 1.2$	$\doteq 0.2$	–
Roshko (1961), experiment	–	–	$\doteq 0.19$	–

A boundary layer mesh has been fixed to a movable rivulet mesh. Therefore, to ensure the relative movements of the meshes, the sliding mesh interface between two fluid meshes has been applied.

3   EQUATION OF MOTION FOR THE RIVULET

The motion of the rivulet can be described according to Seider & Dinker (2006) in polar coordinates using equations of motion

$$m\left(\ddot{\theta}d + \ddot{x}\sin\theta + \ddot{y}\cos\theta\right) = T , \quad m\left(-\dot{\theta}^2 d - \ddot{x}\cos\theta + \ddot{y}\sin\theta\right) = N , \qquad (1), (2)$$

where m is the mass of the rivulet; T is the tangential force and N is normal force.

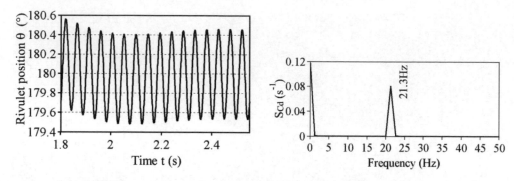

Figure 3. Rivulet position in time domain (left) and corresponding power spectral density (right).

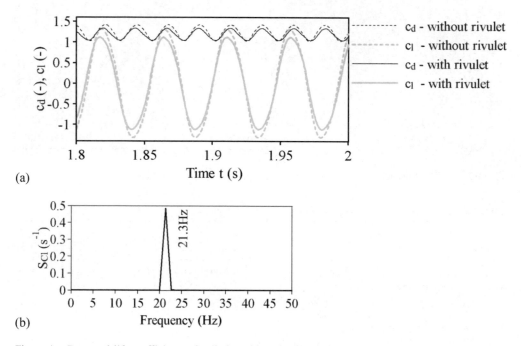

(a)

(b)

Figure 4. Drag and lift coefficients of cylinder with and without rivulet (a), power spectral density of lift coefficient (b).

4 AEROELASTIC RESPONSE

Figure 3 displays the temporal evolution of rivulet position on the cylinder surface and corresponding power spectral density. As can be seen, the rivulet has remained in a downwind location and it moved periodically with frequency 21.3 Hz in the leeward region of the cylinder. It has been assumed, that this periodical movement was particularly coordinated by vortex shedding ($f_{sh} = 21.27$ Hz) and gravitational force, whose direction was perpendicular to the cylinder cross-section.

Time courses of drag and lift coefficients are shown in Figure 4a and a power spectral density of lift coefficient can be seen in Figure 4b. Periodicity with a frequency 21.3 Hz of lift coefficient has been observed which was in coincidence with periodical movement of the rivulet and vortex shedding as well. Consequently, the rivulet occurrence directly influences the aerodynamic behaviour of

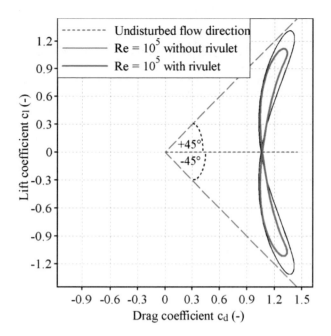

Figure 5. Dependence of drag and lift coefficients of cylinder with and without rivulet.

the cylinder. Therefore, the aerodynamic force direction and intensity was subjected to slight time-dependant changes.

When comparing the values of drag coefficients of the cylinder with the rivulet and without it (Table 1), slightly changed values have been observed. The rivulet occurrence did not change the vortex shedding frequency significantly, but it slightly improved the aerodynamic characteristics (the uniform inlet flow applied). This effect may be different when applying turbulent wind velocities. Figure 5 shows the dependency of the drag and lift coefficients of the cylinder with and without the rivulet. The described slight changes of aerodynamic characteristics are evident.

5 CONCLUSIONS

Numerical simulation based on a coupled model of airflow-rivulet interaction has been created using commercial software Adina 8.7. Simulations with and without water rivulet on vertical circular cylinder surface have been investigated. As the vertically mounted cylinder has been considered, the gravitational force has been considered as a perpendicular force to the cylinder cross-section. It has been assumed, the vortex shedding phenomenon and partially gravitational force have an influence on periodical rivulet movement. Periodicity of the occurred rivulet movement has been driven by vortex shedding frequency.

Concerning the leeward initial rivulet position, the rivulet did not influence aerodynamic forces of cylinder significantly. The rivulet reduced (when comparing to dry cylinder) the drag force slightly and it also reduced amplitudes of drag and lift forces, caused by vortex shedding.

5.1 *Future objectives*

As well as the vertical cylinder, the other cylinder positions will be investigated in the future. Attention will be paid to an initially formed rivulet position. As the galloping phenomenon is well known partially on skew wet cables, numerical models for the solution of this kind of problem are being developed.

ACKNOWLEDGEMENT

The paper is carried out within the project No. 1/0321/12, partially founded by the Science Grant Agency of the Ministry of Education of Slovak Republic and the Slovak Academy of Sciences. Paper is the result of the Project implementation: University Science Park TECHNICOM for Innovation Applications Supported by Knowledge Technology, ITMS: 26220220182, supported by the Research & Development Operational Programme funded by the ERDF.

REFERENCES

ADINA, 2010. Theory and modeling guide volume I: ADINA and volume III: ADINA CFD & FSI, Version 8.7, ADINA R&D, Inc., Watertown, MA, USA.

Breuer, M. 2000. A challenging test case for large eddy simulation: high Reynolds number circular cylinder flow. *International Journal of Heat and Fluid Flow*, 21, pp. 648–654.

Cosentino, N., Flamand, O., Ceccoli, C. 2003. Rain-wind induced vibration of inclined stay cables. Part II: mechanical modeling and parameter characterisation. *Wind and Structures*, pp. 485–498.

Hikami, Y., Shiraishi, N. 1988. Rain-wind induced vibrations of cables in cable stayed bridges. *Journal of Wind Engineering and Industrial Aerodynamics*, 29 (1–3), pp. 409–418.

Lemaitre, C., de Langre, E., Hémon, P. 2010. Rainwater rivulets running on a stay cable subject to wind. *European Journal of Mechanics B/Fluids*, 29, pp. 251–258.

Li, S.Y., Gu, M. 2006. Numerical simulations of flow around stay cables with and without fixed artificial rivulets. *Proc. 4th Int. Symp. Comp. Wind Eng., (CWE2006)*, Yokohama, Japan, pp. 307–310.

Matsumoto, M., Yagi, T., Goto, M., Sakai, S. 2003. Rain-wind-induced vibrations of inclined cables at limited high reduced wind velocity region. *Journal of Wind Engineering and Industrial Aerodynamics*, 92, pp. 1–12.

Peil, U. & Nahrath, N. 2003. Modeling of rain-wind induced vibrations. *Wind and Structures*, 6 (1), pp. 41–52.

Roshko, A. 1961. Experiments on the past a circular cylinder at very high Reynolds number. *Journal of Fluid Mechanics*, 10, pp. 345–356.

Seidel, C. & Dinkler, D. (2006) Rain-wind induced vibrations – phenomenology, mechanical modelling and numerical analysis. *Computer and Structures*, 84, pp. 1584–1595.

Soltys, R. & Tomko, M. 2014. Influence of airflow-rivulet interaction on circular cylinder in uniform flow. Engineering Mechanics 2014: 20th International Conference, Svratka, Czech Republic, pp. 584–587.

Taylor, I.J. & Robertson, A.C. 2011. Numerical simulation of the airflow-rivulet interaction associated with the rain-wind induced vibration phenomenom. *Journal of Wind Engineering and Industrial Aerodynamics*, 99, pp. 584–587.

Wieselsberger, C. 1921. Neuere feststellungen über die gesetze des flüssigkeits und luftwiderstands. *Phys. Z.*, 22, pp. 321–328, (in German).

Yamaguchi, H. 1990. Analytical study on growth mechanism of rain vibration of cables. *Journal of Wind Engineering and Industrial Aerodynamics*, (33), pp. 73–80.

Zhang, S., Xu, Y.L., Zhou, H.J., Shum, K.M. 2008. Experimental study of wind-rain-induced cable vibration using a new model setup scheme., *Journal of Wind Engineering and Industrial Aerodynamics*, 96, pp. 2438–2451.

Zuo, D. Jones, N.P., Main, J.A. 2008. Field observation of vortex and rain-wind-induced stay-cable vibrations in a three-dimensional environment. *Journal of Wind Engineering and Industrial Aerodynamics*, 96 (6–7), pp. 1124–1133.

Advances and Trends in Engineering Sciences and Technologies – Al Ali & Platko (Eds)
© 2016 Taylor & Francis Group, London, ISBN: 978-1-138-02907-1

A shape correction of an active four-strut tensegrity pyramid

M. Spisak & S. Kmeť
Technical University of Košice, Faculty of Civil Engineering, Institute of Structural Engineering, Košice, Slovakia

ABSTRACT: Since the 50s of the 20th century, interest for studying a behaviour of tensegrities has noticed a significantly increasing tendency in almost all spheres of science. Tensegrity systems have a huge potential in creation of progressive bearing systems due to their special structural and mechanical properties and capabilities. In architecture and civil engineering, implementation of a tensegrity idea may induce a large revolution in designing and constructing of civil engineering structures. Tensegrities are perfect candidates for becoming adaptive regarding to surrounding environment because they can be relatively easily handled. Shape correction of an active tensegrity system with respect to defined objectives, constraints and bounds is presented in the article. Results showed an acceptable level in accordance with chosen criteria.

1 INTRODUCTION

Tensegrity systems are progressive and since their inception, they have pervaded into many spheres of research and development. R. B. Fuller, K. D. Snelson (both from the USA) and independently on them D. G. Emmerich (France) are official authors of idea and principle of tensegrity (Motro 2003). In addition, R. B. Fuller may be denoted as a creator of the word tensegrity itself, which comes from coupling of two words tension and integrity (Fuller 1975).

A principle of tensegrity may be defined by an equilibrium between discontinuous internal compression surrounded by continuous external tension. However, a location of tension and compression is always invariable under changing environment (Motro 2003).

Therefore, tensegrity systems are composed of isolated compressed members (struts) placed inside of a reticulum created from tensioned members (cables). Synergy between both kinds of members defines a stable self-stressed state of the tensegrities (Motro 2003).

Tensegrity systems may be of various geometrical shapes comprising one or more cells assembled together. Tensegrity pyramids (T-pyramids – more accurately Sn T-pyramids with n number of struts) pertaining to pure tensegrity systems are one of such cell configurations (Burkhardt 2008, Motro 2003).

Structural and mechanical properties, such as flexibility, kinematics, light weight, pin-jointed one directionally stressed members, predestinate tensegrities to become interesting active or even intelligent (adaptive) systems. A notable progress in exploring of such tensegrity systems has been achieved by means of a researchers' group at Swiss Federal Institute of Technology in Lausanne (École polytechnique fédérale de Lausanne - EPFL), e.g. works of authors Adam & Smith (2008), Domer & Smith (2005).

The article deals with a shape correction of an active four-strut tensegrity pyramid (S4 T-pyramid) with all compressed members chosen to be active (i.e. four active members – linear actuators). A non-linear static analysis coupled with design exploration procedure for searching of optimal length modifications (strokes) for active members with respect to predefined goals was employed.

2 EXPLORATION PROCEDURE FOR ACQUIRING AN OPTIMAL SOLUTION

The most essential question as well as challenge in engineering is to achieve an effective and economic design, product or to find an adequate control decision with saving as much computer time as possible until a global optima is uncovered (Singiresu 2009). Regarding to a size of a search (design or solution) space and difficulty of an optimisation problem, verifying of each possible solution could be awfully time-consuming process.

Therefore, a more convenient and sophisticated approach is to utilise a design optimisation (DO) for beforehand formulated problem with expected convergence until an optimum is reached or a design exploration (DE) for design development based on a belief that system evolves during searching but without any proof about convergence (Jenkins 2014). As it has been mentioned before, the key question is a computer time required for finding an optimal solution. In some cases overall DO or DE procedure may last for days, weeks or even months, but still lesser comparing to exhaustive deterministic 'what-if' analyses which might take centuries.

Before executing a DE one has firstly to define input (design) variables and output (state) variables of the interest. The latter can be also called as performance indicators. DE procedure for discovering of optimal strokes of active members for selected tensegrity system was established based on ANSYS DE procedure by employing a Goal driven optimization (GDO), as one of ANSYS DE's components which encompasses three phases (ANSYS 14.0 2011).

Real values of output variables necessary for the first phase of DE as well as for refining and verifying of DE results may be measured through a real experiment (employed especially in the cases when a mathematical expression of a problem is too difficult or impossible) or acquired by means of a numerical model for a real deterministic simulation. In the paper, DE uses a real deterministic simulation which represents a static finite element analysis (FEA). Geometric non-linearity as well as linear changes of active members' lengths (strokes) were included in the FEA calculations. Equilibrium in a geometrically non-linear static FEA is expressed by a formula in global coordinates (Bathe 1996)

$$\left[K_T\right] \cdot \left\{\mathbf{u}\right\} = \left\{\mathbf{F}\right\}, \tag{1}$$

where $[K_T]$ = tangential stiffness matrix (sum of elastic and geometric stiffness matrices); $\{\mathbf{u}\}$ = vector of nodal displacements; and $\{\mathbf{F}\}$ = vector of nodal loading forces.

Optimal strokes of active members were explored according to chosen objectives (i.e. reduction of nodal displacements). Hence, for the inputs were set up strokes of the active members and the outputs of the interest were represented by nodal displacements.

The first phase introduces a design of experiments (DOE), setting up of bounds for input variables and a convenient sampling technique which aim is to achieve an effective covering of the whole search space with as little number of samples (design points) as possible using statistical methods for generating inputs and subsequently obtaining of pertaining outputs by executing a deterministic simulation. Optimal space-filling (OSF) design was utilised for a DOE in the article. OSF is a sampling technique optimised through several iterations, maximising distance among design points in order to achieve a uniform distribution with minimal amount of samples across the entire search space (ANSYS 14.0 2011).

The second phase includes creation of a response surface (RS), defined as an approximation of the real numerical system's response (i.e. a surrogate model or metamodel instead of a computer model for simulation) which basic formulation may be stated as follows (Fang et al. 2006)

$$y\left(\mathbf{x}\right) = f_{MM}\left(\mathbf{x}\right), \quad \mathbf{x} \in \mathcal{D}, \tag{2}$$

where $\mathbf{x}$ = vector of s input variables $\{x_1, \ldots, x_s\}$ in a search space $\mathcal{D}$; and $f_{MM}(\mathbf{x})$ = metamodel (unknown function which has to be approximated).

Once the DOE samples are created and updated via real simulation, RS is created using DOE results. For each output is constructed one RS which is affected by all inputs and so by the other outputs. Accuracy of RSs strongly depend on used DOE scheme. The better is coverage of the search space, the more accurate are generated RSs. A goal for generating RSs is to determine function between inputs and output responses continuously through entire search space by employing statistical metamodels with a satisfactory goodness of fit regarding to real values of outputs. Additionally, several auxiliary numerical simulations may be executed to refine RSs in an efficient way. Such procedure is significantly faster than executing a deterministic simulation of each solution for receiving global responses in a whole search space. For RSs' creation in the article was chosen a Kriging metamodel (i.e. an accurate multidimensional interpolation statistical method) because of its accuracy and auto-refinement capabilities (i.e. adding refinement points to the design domain where they are most needed). Based on the information provided by DOE and RS is one able to achieve effective simulation driven product development. An optimal design of a problem can be found manually by thorough investigation of graphical representation of the RS or automatically using an optimisation technique (ANSYS 14.0 2011).

The third phase is represented by an optimisation of a problem subjected to objectives, constraints and bounds given on inputs using an optimisation algorithm. Since the samples are generated using information gained from RS in contrast to direct solving of each possible input through deterministic simulation, the computational time of the entire optimisation process until a satisfactory solution is found significantly decreases. Optimisation results depends on the quality of RS (ANSYS 14.0 2011).

A multi-objective (MO) decision problem is defined as follows: Given an s-dimensional decision variable vector $\mathbf{x} = \{x_1, \ldots, x_s\}$ in the solution space $\mathbf{X}$, find a vector $\mathbf{x}^*$ that minimises a given set of N objective functions $f(\mathbf{x}^*) = \{f_1(\mathbf{x}^*), \ldots, f_N(\mathbf{x}^*)\}$. Solution space can be restricted by a series of constraints, such as inequalities $g_j(\mathbf{x}^*) \leq b_j$ and/or equalities $h_k(\mathbf{x}^*) = b_k$ for $j = 1, \ldots, J$, $k = 1, \ldots, K$ and bounds $\mathbf{x}^{(L)} \leq \mathbf{x}^* \leq \mathbf{x}^{(U)}$ on decision variables (Konak et al. 2006).

An optimal solution to a MO problem may be formulated by definition: If all objective functions are for minimisation, a feasible solution $\mathbf{x}$ is said to dominate another feasible solution $\mathbf{y}$ ($\mathbf{x} \succ \mathbf{y}$), if and only if, $f_n(\mathbf{x}) \leq f_n(\mathbf{y})$ for all $n = 1, \ldots, N$ and $f_n(\mathbf{x}) < f_n(\mathbf{y})$ for at least one objective function $n = 1, \ldots, N$. A solution is said to be optimal (Pareto optimal) if it is not dominated by any other solution in the solution space. A Pareto optimal solution cannot be improved for any objective without worsening at least one other objective. However, in MO there are more uniformly optimal solutions (Pareto optimal set). Thus, the major goal of a MO optimisation is to identify the best-known Pareto set which represents the Pareto optimal set as much as possible (Konak et al. 2006). For optimisation in the article was chosen MO genetic algorithm, a hybrid variant of elitistic Non-dominated sorting genetic algorithm II (NSGA-II) which is implemented in ANSYS for solving multi-objective optimisation problems (ANSYS 14.0 2011, Deb 2002).

3 EXPLORING FOR AN OPTIMAL SHAPE OF AN ACTIVE TENSEGRITY SYSTEM

DE procedure was utilised for obtaining an optimal shape of active S4 T-pyramid counter-clockwise twisted consisting of 8 nodes and 16 members (12 cables and 4 active struts). From a geometrical point of view a bottom base is of a theoretical size 2×2 m, a top base of 1.414×1.414 m and height of the system is 1.5 m. Three-dimensional (3D) model of explored S4 T-pyramid can be seen in Figure 1.

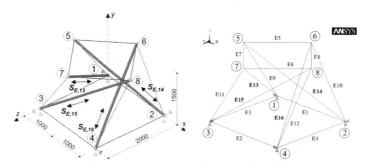

Figure 1.   Three-dimensional solid model of analysed S4 T-pyramid (left), finite element model (right).

Geometrical and finite element model (FEM) of the chosen tensegrity system were created in ANSYS 14.0 Mechanical APDL. For all the cables (E1–E12) were utilised LINK180s (compression-tension 3D line finite elements) with tension only property set up and for all active struts (E13–E16) were applied LINK11s (linear actuators, i.e. 3D line finite elements allowing a linear change of length), detailed description is provided by ANSYS 14.0 (2011). Unit is supported in the nodes 1–4 of the bottom base according to Figure 1.

Chosen properties for cables (Kmeť et al. 2012): stainless austenitic steel 1.4401, cross-sectional area $15.2 \times 10^{-6}\, m^2$ (construction 7×7, 7 wires per each from 7 strands, nominal diameter of $6 \times 10^{-3}\, m$), Young's modulus of 120×10^9 Pa and a prestressing force of 1000 N.

Chosen properties for active members: steel S235, mass of 12.5 kg approximately equal to a mass of a tube with a cross-section ($\emptyset\ 51/4$) $\times 10^{-3}$ m and theoretical length 2.693 m equivalent to the length of an active member and stiffness $1 \times 10^{11}\, Nm^{-1}$. Large value of elastic stiffness was chosen in order to reduce an influence of loading and lengths' modifications of active members on their axial deformation and thereby achieve a more clearly representation of corrected versus non-corrected shape with respect to the initial geometry.

Initial prestressed configuration of active S4 T-pyramid was loaded by its self-weight (the acceleration of gravity $9.81\ ms^{-2}$ was used as a constant) and 8 equal increments of -250 N. Symmetrical as well as asymmetrical vertical load with downward orientation and maximal value $F_y = -2000$ N. Shape of the tensegrity system was corrected for 5 different load cases (LCs). Applied load in each LC is of the same magnitude differing only by location and number of nodal loading forces located in the nodes of the top base 5, 6, 7, 8, seen in Table 1, 2. Once the loading was executed, strokes of active members were incremented by 1/8 of maximal stroke value for each active member independently.

The whole DE procedure was governed using ANSYS 14.0 (more precisely, a combination of Workbench and Mechanical APDL environment) with a following MO formulation:

$$f\left(\mathbf{S}_{E}\right) = U_{y,m}\left(\mathbf{S}_{E}\right) \to 0 \,; \quad m = 5, \dots, 8 \ [m], \tag{3}$$

subject to

$$1 \leq g_{ten,min}\left(\mathbf{S}_{E}\right) = N_{c,min}\left(\mathbf{S}_{E}\right), \ g_{ten,max}\left(\mathbf{S}_{E}\right) = N_{c,max}\left(\mathbf{S}_{E}\right) \leq 20,100 \ [N], \tag{4}$$

$$-1 \geq g_{com,min}\left(\mathbf{S}_{E}\right) = N_{am,min}\left(\mathbf{S}_{E}\right), \ g_{com,max}\left(\mathbf{S}_{E}\right) = N_{am,max}\left(\mathbf{S}_{E}\right) \geq -42,000 \ [N], \tag{5}$$

$$\text{LC } 1 : 0 \leq S_{E,i} \leq 0.01\,; \ \text{LC } 2, \dots, \text{LC } 5 : 0 \leq S_{E,i} \leq 0.05\,; \quad i = 13, \dots, 16\,, [m] \tag{6}$$

where $\mathbf{S}_E$ = design variable vector of strokes $\{S_{E,13}, \dots, S_{E,16}\}$ for the four active members; $U_{y,m}$ = top base's vertical nodal displacements; $N_{c.min}/N_{c.max}$ = minimal/maximal tensile force in the cables; and $N_{am,min}/N_{am,max}$ = minimal/maximal compressive force in the active members.

Equation 3 represents objective functions which objectives are for a reduction of vertical nodal displacements of the top base (nodes 5, 6, 7, 8) until 0 is reached [m]. Equations 4 and 5 introduce constraints given on design variables regarding to limiting tensile and compressive normal forces respectively [N]. Equation 6 represents bounds placed on design variables [m].

Upper values of axial forces 20,100 N and $-42,000$ N for constraints were chosen considering breaking force in cables and buckling resistance of a steel tube (its structural parameters were described earlier in this section) as a limiting axial force for active members respectively.

Firstly, several exploration calculations using GDO were executed to verify capabilities of a loaded tensegrity unit regarding to objectives, constraints and inputs' bounds. Thereafter, a convenient bounds were chosen and a GDO was elegantly established using Workbench as a main environment and Mechanical APDL environment for a batch run. After reading an APDL script (FEM, FEA, input and output variables) used for running a real numerical simulation in a batch mode and setting up of bounds, objectives and constraints, whole DE procedure for finding a satisfactory set of active members' strokes was executed automatically.

As a consequence of searching for optimal strokes of active members was observed that shape of active S4 T-pyramid was unable to completely return into its initial configuration. Nevertheless, vertical nodal displacements at the nodes of the top base 5, 6, 7, 8 were significantly reduced. Since

Table 1. Optimised strokes as well as vertical nodal displacements in the nodes of the interest.

Load case	Loaded nodes	Shape correction	$S_{E,13}$	$S_{E,14}$	$S_{E,15}$	$S_{E,16}$	$U_{y,5}$	$U_{y,6}$	$U_{y,7}$	$U_{y,8}$
			$(\times 10^{-3})$ m							
1.	–	NO	–	–	–	–	−17.8	−17.8	−17.8	−17.8
		YES	5.0	5.5	5.5	5.0	0.0	0.0	0.0	0.0
2.	5, 6, 7, 8	NO	–	–	–	–	−111.2	−111.2	−111.2	−111.2
		YES	16.6	35.0	19.3	21.6	−21.6	−31.4	−16.8	−26.1
3.	5, 6, 7	NO	–	–	–	–	−100.4	−100.5	−100.6	−94.7
		YES	44.6	18.0	1.1	24.2	−27.2	−24.3	−9.4	−3.9
4.	5, 6	NO	–	–	–	–	−87.0	−87.0	−81.4	−81.4
		YES	43.2	2.4	5.1	36.0	−11.8	−5.9	−3.7	−0.6
5.	5	NO	–	–	–	–	−67.7	−62.4	−62.4	−62.4
		YES	2.8	30.5	31.0	8.0	−0.3	1.1	0.6	−0.1

Table 2. Resulting minimal and maximal normal forces in members.

Load case	Loaded nodes	Shape correction	$N_{c,min}$	$N_{c,max}$	$N_{am,min}$	$N_{am,max}$
			N			
1.	–	NO	844 (E1~4)	1404 (E9~12)	−2213 (E13~16)	
		YES	3129 (E3, 4)	5538 (E9, 12)	−8377 (E13, 16)	−8383 (E14, 15)
2.	5, 6, 7, 8	NO	6286 (E1~4)	7831 (E5~8)	−14,709 (E13~16)	
		YES	12,853 (E3)	19,990 (E10)	−33,122 (E13)	−33,816 (E14)
3.	5, 6, 7	NO	4686 (E3)	8050 (E12)	−11,038 (E13)	−13,182 (E16)
		YES	11,445 (E1)	20,094 (E12)	−30,365 (E15)	−32,430 (E16)
4.	5, 6	NO	2971 (E9)	6893 (E12)	−8332 (E13)	−10,438 (E16)
		YES	10,254 (E1)	20,098 (E12)	−28,226 (E14)	−31,560 (E16)
5.	5	NO	1588 (E9)	4795 (E10)	−5263 (E13)	−7296 (E16)
		YES	9098 (E4)	16,848 (E10)	−23,301 (E13)	−25,281 (E15)

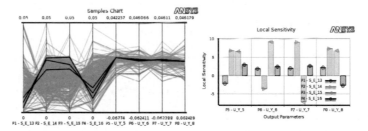

Figure 2. Optimal solutions (left), active members' strokes' sensitivity for LC 5 (right).

normal forces in members increase with number of loaded nodes, load magnitudes and especially when actuators are moving in order to decrease deformations of the system, reduction of vertical nodal displacements was established with respect to limitations given on maximal and minimal normal forces in members (minimal forces' limitations for maintaining initial orientation of forces in members, maximal forces' limitations for maintaining resistance of the system). Note that nodal displacements were reduced each independently on other according to defined objectives with respect to limit forces in members (i.e. top base of the unit is not optimised to be positioned in a horizontal plane). This may cause unidentical nodal displacements reduction in the final corrected shape. Obtained results for both corrected and non-corrected shapes are arranged in Table 1, 2.

Maximal achieved values for horizontal nodal displacements of the top base are as follows: non-corrected shape → $U_{x,6(7)} = -(+)153.2 \times 10^{-3}$ m (LC 2) and $U_{z,5(8)} = +(-)153.2 \times 10^{-3}$ m (LC 2), corrected shape → $U_{x,6} = -64.4 \times 10^{-3}$ m (LC 2) and $U_{z,8} = -57.6 \times 10^{-3}$ m (LC 2).

A slackening of cables or reverse orientation of axial forces in members was not observed during DE. Acquired optimal solutions with the best three candidates with resulting strokes' sensitivity of the active members for LC 5 are shown in Figure 2.

199

4 CONCLUSIONS

Tensegrity systems comprises elastic tensioned and rigid compressed members which interaction defines a self-equilibrated state. When tensegrities are subjected to loading they behave relatively flexibly, hence a question of meeting a serviceability criteria is considerable. The most progressive way for decreasing nodal displacements of tensegrity systems is to implement an active or even intelligent control. In the contribution is presented a shape correction (nodal displacements' reduction) of an active tensegrity pyramid with respect to defined objectives, constraints and bounds on design variables. Utilised DE procedure is not an active or intelligent control (i.e. continuous controlling and governing process) because some steps require user's manual feed (i.e. manually reading an APDL script file, setting up search parameters and running a search process), there is no continuous decision procedure implemented as well as a long search time needed (in the contribution the search time for each load case was in a range of several hours). However, aforementioned DE appears to be convenient for finding of optimal active members' strokes which may be used as learning data for controlling and governing purposes using machine learning methods. With reference to a number of loaded nodes under chosen load magnitude, one was unable to return a tensegrity system entirely into its initial geometry. Vertical nodal displacements in most load cases could not be fully reduced due to rising tendency of axial forces in members when active members are lengthened and limiting resistance's constraints for members. Nevertheless, obtained results showed a significant nodal reduction with respect to defined objectives, constraints and bounds placed on design variables.

ACKNOWLEDGEMENTS

Paper is the result of the Project implementation: University Science Park TECHNICOM for Innovation Applications Supported by Knowledge Technology, ITMS: 26220220182, supported by the Research & Development Operational Programme funded by the ERDF. This research has been carried out in terms of the projects VEGA No. 1/0321/12 and NFP26220120037 Centre of excellent research of the progressive building structures, materials and technologies, supported from the European Union Structural funds.

REFERENCES

Adam, B. & Smith, I.F.C. 2008. Active tensegrity. A control framework for an adaptive civil-engineering structure. *Computer & Structures* 86(23–24): 2215–2223.
ANSYS 14.0 2011. *ANSYS 14.0 Help*.
Bathe, K.J. 1996. *Finite Element Procedures*. New Jersey: Prentice-Hall.
Burkhardt, R.W. 2008. *A Practical Guide to Tensegrity Design*. Cambridge: Cambridge University Press.
Domer, B. & Smith, I.F.C. 2005. An active structure that learns, *Journal of Computing in Civil Engineering* 19(1): 16–24.
Deb, K., Pratap A., Agarwal, S. & Meyarivan, T. 2002. A fast and elitist multiobjective genetic algorithm: NSGA-II. *IEEE Trans Evol Comput* 6(2): 182–97.
Fang, K.-T., Li, R., Sudjianto, A. 2006. *Design and Modeling for Computer Experiments*. Boca Raton: Chapman & Hall/CRC.
Fuller, R.B. 1975. *Synergetics Explorations in the Geometry of Thinking*. London: Collier Macmillan Publishers.
Kmeť, S., Platko, P., Mojdis, M. 2012. Analysis of Adaptive Light-Weight Structures. *Procedia Engineering: Steel Structures and Bridges* 40: 199–204.
Konak, A., Coit, D.W., Smith, A. E. 2006. Multi-objective optimization using genetic algorithms: A tutorial, *Reliability Engineering and System Safety* 91(9): 992–1007.
Motro, R. 2003. *Tensegrity: Structural Systems for the Future*, London: Kogan Page Science.
Singiresu, S.R. 2009. *Engineering Optimization. Theory and Practice, Fourth Edition*. New Jersey: John Wiley & Sons.
Jenkins, B. 2014, *Design Exploration. vs. Design Optimization*. Raleigh: Ora Research LLC.

Advances and Trends in Engineering Sciences and Technologies – Al Ali & Platko (Eds)
© 2016 Taylor & Francis Group, London, ISBN: 978-1-138-02907-1

RC concrete constructions durability estimation according to diachronic model

A.S. Stepanov, A.A. Podzhunas, A.V. Ioskevich & S.E. Nikitin
Peter the Great St. Petersburg Polytechnic University, St. Petersburg, Russia

ABSTRACT: Reinforced concrete is a leading structural building material. This is due to its unique physical and mechanical properties, high raw material base for the production of and relatively simple technology and preparation. However, with the passage of time in reinforced concrete structures evolving process of destruction. The article describes the main causes of these processes. Describes the main types of corrosion of concrete and corrosion classifications. In the second part of the article describes estimation of RC construction durability according to diachronic model. The modern market of construction materials becomes wider by development of building sector. One of the prevailing materials used in construction is RC concrete. It is used both at monolithic construction and at manufacturing of complete prefabricated units at factories. As any constructional material concrete is subject to corrosion. That's why we should deal with present day regulatory framework of designing reinforced concrete and find out by what types of corrosion it is affected.

1 DURABILITY OF BUILDING STRUCTURES

A main part of buildings and structures constructed of concrete and reinforced concrete are exposed the action of corrosive medium during the life, which can cause damage and failure of building structures. It is especially important for industrial buildings where the liquid or gaseous wastes can penetrate into work area. At this time, the development of a quantitative theory of corrosion processes under the action of the concrete variety of corrosive environments and forecast durability of structures are the subject of research (Moskvin, V.M., et al. 1980).

According to SP (Code of practice) 28.13330.2012 "Protection of structures against corrosion", the action of corrosion can be controlled by methods:

– *primary protection method* (considers the aggressiveness of the operating environment in the selection of materials for fabrication of the structures and design requirements by adding inhibitors – agents that reduce the risk of corrosion of the metal) (Morris, W., et al. 2003);
– *secondary protection method*, i.e. surface protection of structure.

Durability can be considered at three levels: material, element and design or object on the whole. Usage of the term durability is not straightforward. The durability of reinforced concrete is expressed through the ability of a material to resist the effects of the environment, reducing its quality. It is possible to distinguish the two most important criteria in Russian and Eurocode regulations:

– sufficiency initial quality characteristics that allow to satisfy the requirements of preserving security and serviceability not below the limits laid down in the rules;
– the economic rationale life, including the minimization of costs (Pukhonto, L.M. 2004).

Current standards SP (Code of practice) 28.13330.2012 "Protection of structures against corrosion. The updated edition of SNIP (Construction Norms and Regulation) 2.03.11-85″ evaluates the degree of aggressive environmental influences based on the assumption that the structure will be operated for 50 years. Another issue is creation of methods of corrosion testing and life prediction of concrete and reinforced concrete elements. Consider the processes occurring in reinforced concrete structures under the influence of aggressive media.

2 CONCRETE CORROSION

It is important to classify of such processes on general grounds to assess the nature and extent of the corrosion process, the aggressive action of various substances in the environment. V.M. Moskvin proposed common classification of corrosion processes that can occur in the concrete interaction with aggressive media. Three main types of corrosion were allocated.

Corrosion of 1 species
To this type belong all corrosion processes that occur in the concrete under the action of liquids, capable of dissolving the components of cement stone. Components of the cement stone dissolved and carried out of the concrete structure. Since most soluble component of the cement stone is calcium hydroxide $Ca(OH)_2$, the corrosion process is usually defined as a process of "leaching" lime. The intensity of the dissolution of CaO (in the form of $Ca(OH)_2$) of the cement stone is determined by the composition of the cement stone, the leaching conditions and, most importantly, the degree of access to the inner surface of the cement stone for water. Resistance and durability of concrete in the form of corrosion 1 increased density increases, the use of concrete and cement the most resistant to the dissolving action of water (pozzolanic and slag cement) (Moskvin, V.M., Ivanov, F.M., Alekseyev, S.N., Guzeyev, Ye.A. 1980).

Corrosion of 2 species
This type of corrosion processes are occurring in the interaction of the components of the cement stone and acid solutions, or certain salts. The destruction of the cement stone is in the surface layers of concrete in contact with aggressive media. The most common example of corrosion 2 types – corrosion under the action of carbonic waters. To protect concrete from damage by the action of acids and acid-salt solutions used concrete or use painting, plastering and lining to protect the concrete surface from contact with aggressive media (Moskvin, et al. 1980).

Corrosion of 3 species
Includes processes in which development occurs in the pores of the concrete accumulation and crystallization of the reaction products with the increase in the solid phase material. Crystallization and other secondary processes create internal stresses, which may cause damage to the concrete structure. This type of corrosion refers to corrosion by reacting with sulphates (SO_4^{2-}). Destruction in this case is due to the growth of salt crystals in the presence of the evaporating surface structure (Moskvin, et al. 1980).

Most harmful to concrete are salts of acids, especially sulfuric acid (H_2SO_4). Because they form in the cement calcium sulfate ($CaSO_4$) and alumina ($Al_2(SO_4)_3$). In particular, calcium sulfoaluminate – "cement bacillus" dissolves easily and increases in size by 2.5 times.

3 REINFORCEMENT CORROSION

Normal usage of reinforced concrete structures is possible if concrete and reinforcement work together. One of the concrete features is to protect reinforcement from corrosion and flame. Corrosion of rebar in concrete often is determined as natural carbonation of concrete.

Moisture, which is penetrating concrete, does positive effect on concrete strength, but when it comes to rebar, it starts to interact with it. As a result reinforcement starts rusting.

Reinforcement corrosion in concrete is particular case of galvanic corrosion. During the process of galvanic corrosion after electrolytic conductor ingression, like pellicle of moisture on the steel surface, it begins.

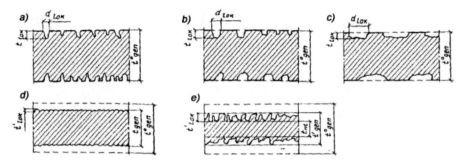

Figure 1. Classification of corrosion destruction (a–local pitting corrosion; b–local holes corrosion; c–local spot corrosion; d–uniform corrosion; e–nonuniform corrosion) from Recommendations…1990.

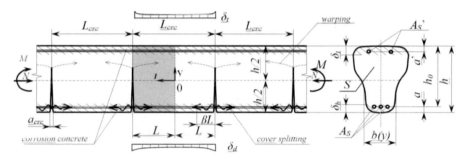

Figure 2. Scheme of corrosion-damaged elements that can be loaded.

Steel divides into cathode and anode area and the anode process begins, in other words begins transition of ions from anode area and their hydration. Here is the chemical equation:

$$Fe^{2+} + 2OH^- = Fe(OH)_2 \text{ and } 4Fe(OH)_2 + 2H_2O + O_2 = 4Fe(OH)_3.$$

$Fe(OH)_3$ is known as rust.

The most influenced factor is concrete moisture. So, when the relative moisture comes to 50% the process reaction rate is maximal (Alekseyev, S.N. 1968, Novichkov, P.I. 2008).

The second way of corrosion trolley is local depassivation of steel, cause of Chlorum ions influence. The Chlorum ions are an intensive stimulant – lead to pitting corrosion of steel.

Rust takes in 2–3 twice as large content, than corroded bars. That content increment around bars causes great tension stress inside the concrete. Resultant micro and macro cracks easing the way of moisture inside, which increase internal stresses exponentially. When they are bigger than concrete strength – destruction starts (Alekseyev, S.N. 1968, Izotov, V.S. 2006, Benin, A.V., Semenov, A.S., Semenov, S.G., Melnikov, B.E. 2012).

Pitting corrosion can cause brittle crack of steel, which especially dangerous in prestressed reinforcement. This kind of corrosion is also dangerous, because Test methods of corrosion state of reinforced concrete structures gets into the concrete at manufacturing stage. Also can get inside during exploitation is in aggressive environment and in construction which isn't protected enough. Classification of corrosion destruction of reinforcement is on the (Fig. 1).

4 DIACHRONIC MODEL OF DEFORMATION A CORROSION-DAMAGED REINFORCED CONCRETE ELEMENTS WITH NORMAL CRACKS

Two-section model considers the stress-strain state of a bended reinforced concrete element with symmetric cross-section and double reinforcement. For transverse bending it is assumed the dominant role of the bending moment M and the longitudinal force N with a negligible effect of shear forces. The system of orthogonal cracks zone is developing in the tension when cracking

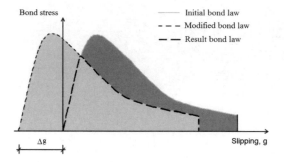

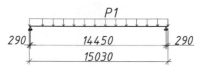

Figure 3. Bond law modifying.

Figure 4. Calculation scheme.

moment Mcrc is reached. Cracks are situated on the approximately equal distance Lcrc = 2L from each other and have approximately equal depth hcrc and width acrc (Fig. 2).

Element is considered as regular system of blocks divided by orthogonal cracks. Blocks are connected with each other by compressed concrete and compressed As' and tensioned As reinforcement. In this approach the stress-strain state calculation of the whole element results to stress-strain state calculation for the symmetric half of the typical S × L block. Two specific sections can be determined: section with crack (l = ±L) and middle section (l = 0).

5 CORROSION INFLUENCE MODELING

Diachronic (greek "dia" – through, "chronos" – time) model is a compilation two-section model and corrosion factors. Diachronic model is open for any corrosion model. Uniform or pitting corrosion is modeled by simultaneous or independent reducing of compressed and tensioned bars area. Different corrosion level is taken into account in section with crack and middle section. Stress-strain state is influencing on aggressive agents penetration velocity inside concrete. As example, concrete degradation is increased in section with crack (δt, Fig. 4). In the same section reinforcement degradation maximum take place due to free access to reinforcement through orthogonal crack (δd, Fig. 2). Bond law modification based on superposition of two factors:

- additional pressure caused by corrosion products which volume, according to different papers, is 2–4 time more than initial steel volume;
- cover splitting due to reinforcement slipping.

This complex physical process is modeled by bond law modifying (Fig. 5) and bond length reducing $L \cdot (1 - \beta)$.

Bond law is modified according to Schlune H. by using an additional reinforcement slipping

$$\Delta g = m \cdot \delta d \tag{3}$$

where δd = reinforcement corrosion depth, m = empiric coefficient. Initial diagram is shifting by slipping axis to negative direction on Δg.

6 CORROSION-DAMAGED REINFORCED CONCRETE BEAM

The calculation of beams pool cover was performed with using of diachronic model. The beam is located in slightly aggressive environment. Slightly aggressive environment for the construction under consideration are a pair of chlorinated water. Material of construction: concrete B55 W6 F75; prestressed cable fittings K7. Figure 4 shows the design scheme of the beam.

Calculation by diachronic model based on stress-strain. Initial data is:

- P1 load (excluding own weight of the beam) is 57.4 kN/m;
- maximum design moment (M) 1700 kN · m;

Figure 5.   Calculation scheme of beam.

Table 1.   The depth of corrosion of concrete and reinforcement.

No	Service life months	Service life years	The depth of corrosion of concrete mm	The depth of corrosion of reinforcement mm
1	12	1	1.9	0.02
2	60	5	4.3	0.12
3	120	10	6.1	0.24
4	240	20	8.5	0.48
5	600	50	13.4	1.19
6	1200	100	18.8	2.39
7	1800	150	22.9	3.58

– cross-section dimensions and reinforcement characteristics (Fig. 5);
– concrete and reinforcement strength and deformation characteristics.

Diachronic deformation model is able to estimate reinforced concrete construction durability maintaining under combined load and corrosion factors. The calculation takes into account the depth of penetration of aggressive environment in the body of the concrete:

$$L(t) = K \cdot tm \qquad (4)$$

where L = the depth of neutralization of concrete; t = time actions of aggressive environment; K, m = coefficients of aggressiveness of the environment (Popesko, A.I. 1996). And take into account the influence of aggressive environment to the reinforcement:

$$\delta(t) = \delta k(t) + \Delta d(t) \qquad (5)$$

where $\delta k(t)$ = corrosion losses stem thickness (depth of corrosion pits) during exposure to corrosive environment t [23]; $\Delta d(t)$ = increasing the diameter of stem due to corrosion products (Popesko, A.I. 1996).

The results of subsidiary calculations are given in the Table 1.

With diachronic model has been calculated beam with different depths of corrosion of the reinforcement and concrete in the respective periods. The results are summarized in Table 2.

Results of calculations are shown graphically in Figure 6.

According to service moment function expected service life is about 75 years. It was shown that it is possible to calculate the reduction properties function of RC elements on different models based on the degradation characteristics of an aggressive environment for various periods.

7   CONCLUSION

Currently there is no assessment methodology of RC structures durability. This is due to the complexity of physical and chemical processes of corrosion in reinforced concrete, which were only touched upon in this article. Actually designed mathematical models describing the destruction of concrete structures do not reflect the full picture of the simultaneous effects of several types of corrosion on the stress-strain state of the structure. The purpose of the further work we see in

Table 2. Relative deformations of reinforcement and concrete.

No	Service life years	Mult (SP 52-101-2003) kN·m	εs, relative deformation of reinforcement	εb, relative deformation of concrete
1	1	2208.0	0.73	0.77
2	5	2208.0	0.75	0.78
3	10	2208.0	0.76	0.79
4	20	2208.0	0.79	0.81
5	50	2208.0	0.89	0.86
6	100	2208.0	1.13	0.99
7	150	2208.0	1.46	1.18

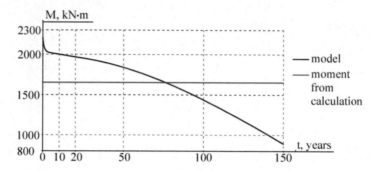

Figure 6. Service moment degradation.

researching a corrosion-damaged RC structures in aggressive environments according to service life calculating model, which is developing in SMiSK SPbSPU.

REFERENCES

Alekseyev, S.N., Ivanov, F.M., Modry, S., Shissl, P. 1990 Dolgovechnost zhelezobetona v agressivnykh sredakh (*Durability of concrete in aggressive environments*).

Belov, V.V., Nikitin, S.E. 2013 *Durability of Reinforced Concrete Elements in Aggressive Environment* (2013) Proceedings of the Twenty-third International Offshore and Polar Engineering.

Benin, A.V., Semenov, A.S., Semenov, S.G., Melnikov, B.E. 2012 *Konechno-elementnoye modelirovaniye protsessov razrusheniya i otsenka resursa elementov avtodorozhnogo mosta s uchetom korrozionnykh povrezhdeniy*. Inzhenerno-stroitelnyy zhurnal.

Fedorov, P.A., Anvarov, B.R., Latypova, T.V., Anvarov, A.R., Latypov, V.M. *2010 O matematicheskoy zavisimosti, opisyvayushchey progress neytralizatsii betona*. Vestnik YuUrGU seriya "Stroitelstvo i arkhitektura".

Izotov, V.S. 2006 *Zashchitnyye svoystva betona po otnosheniyu k stalnoy armature kak funktsiya struktury tsementnogo kompozita (2006)*. Izvestiya KGASU.

Morris, W., Vico, A., Vazquez, M. 2003 *The performance of a migrating corrosion inhibitor suitable for reinforced concrete*. Journal of Applied Electrochemistry.

Moskvin, V.M., Ivanov, F.M., Alekseyev, S.N., Guzeyev, Ye.A. 1980 *Korroziya betona i zhelezobetona, metody ikh zashchity* (Corrosion of concrete and reinforced concrete, methods of their protection).

Popesko, A.I. 1996 *Rabotosposobnost' zhelezobetonnykh konstruktsiy, povrezhdennykh korroziey* (Service-ability of reinforced concrete structures damaged by corrosion).

Pukhonto, L.M. 2004 *Dolgovechnost zhelezobetonnykh konstruktsiy inzhenernykh sooruzheniy* (Durability of concrete structures engineering structures).

Puzanov, A.V., Ulybin, A.V. 2010 *Test methods of corrosion state of reinforced concrete structures*. Magazine of Civil Engineering.

Rozental, N.K. 2007 *Problemy korrozionnogo povrezhdeniya betona*. Beton i zhelezobeton.

Advances and Trends in Engineering Sciences and Technologies – Al Ali & Platko (Eds)
© 2016 Taylor & Francis Group, London, ISBN: 978-1-138-02907-1

Complex modulus and fatigue of recycled asphalt mixtures

J. Šrámek
University of Žilina, Žilina, Slovakia

ABSTRACT: The deformation properties of asphalt mixtures measured by means of dynamic methods and fatigue allow designing pavements which are capable to withstand the expected traffic load. Quality of these mixtures is also expressed through resistance to permanent deformation. Complex modulus of stiffness and fatigue can reliably characterize required asphalt mixture characteristics of the asphalt pavement. The complex modulus (E*) measurements of asphalt mixtures are carried out in laboratory of Department of Construction Management at University of Zilina, utilizing the two-point bending test method on trapezoid-shaped samples. At present, the fatigue is verified on trapezoid-shaped samples and it is evaluated pursuing the criterion of proportional strain at one million cycles (ε_6).

1 INTRODUCTION

Deformation and resistance characteristics and fatigue of asphalt mixtures is of significant importance for asphalt pavement's operational performance. Hitherto, deformation properties assessment was conducted on the principle of dynamic (impact) testing and the fatigue life of asphalt pavement, and additionally, on the principle of resistance decrease or deformation increase in different binders and mixtures characteristics (Decký et al. 2009). Contemporarily, the standard for measurement of complex modulus and fatigue of the asphalt reinforced materials ("mixtures") is used, it provides a better description of the car axle during normal operation at the frequency from 6 to 25 Hz. The complex modulus (E*) measurement of asphalt mixtures is carried out in laboratory of Department of Construction Management at University of Zilina by means of the two-point bending test method on trapezoid-shaped samples. The fatigue is verified on trapezoid-shaped samples and it is evaluated pursuing the criterion of proportional strain at one million cycles (ε_6). The testing equipment and software called KATEMS is used to evaluate fatigue and deformation characteristics.

2 COMPLEX MODULUS AND FATIGUE

Complex modulus is the ratio of tension and deformation at steady, harmonically variable oscillation in consideration of their mutual time shift.

$$E^* = \frac{\sigma_0}{\varepsilon_0} = \left(E_1^2 + E_2^2 \right)^{\frac{1}{2}} \tag{1}$$

Fatigue is the decrease of material resistance by repetitive loading in comparison with resistance by single loading STN EN12697-24.

According to the Slovak dimensioning method, the fatigue is given by

$$S = a - b.\log N_i \tag{2}$$

where a, b – fatigue coefficients; N_i - number of load cycles.

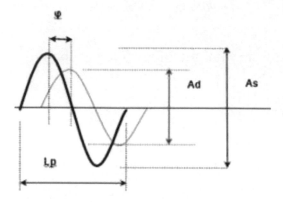

Figure 1.   Measurement of complex modulus.
Lp-length of period; φ-phase-shift
As-amplitude forces
Ad-amplitude deformation

Measurement of the complex stiffness modulus is performed during short-term alternating harmonic loading. It expresses the proportion of maximum amplitudes of excitation tension (σ_0), induced deformation (ε_0) and their phase shift (φ). The sinusoidal stress (force), acts on to the element of linear viscoelastic material, the strain (deformation) varies in time with the same frequency as the inducing stress, but it lags behind by a phase shift. Graphical representation of measurement and complex modulus evaluation is represented in Figure 1 (Schlosser et al. 2013).

3   TESTING EQUIPMENT

The testing of the complex modulus and fatigue were carried out in the laboratory of Department of Construction Management of University of Zilina. Bending tests are widely used for measurement of the stiffness modulus and evaluation of the asphalt paving material fatigue resistance. The two-point bending test on trapezoidal samples is, arguably, the most repeatable and reproducible bending test method detailed in the relevant STN EN 12697 standards. In this test, the sample is mounted as a vertical cantilever. The base is fixed and the top is moved sinusoidally with a constant displacement amplitude. The trapezoidal shape ensures that the maximum values of bending stress and strain occur away from the ends of the samples where the stress is likely to concentrate.

With the Cooper Technology equipment, two trapezoidal samples are tested simultaneously, and the stiffness modulus can be ascertained at a range of frequencies and temperatures. During the fatigue test, the samples are subjected to constant strain amplitude at a selected frequency and temperature until the stiffness modulus decreases to a user-selected target level (normally 50 percent of its initial value).

The equipment (Figure 2) works with constant deviation. Rigid test frame is housed within a cabinet with adjustable temperature ranging from −20°C to +30°C. Frequency can be adjusted within interval of 0,1 Hz to 30 Hz. Precise value of strain amplitude can be set manually. Accurate pre-test displacement can be set by software adjustment of the transducer. The force is induced by two high-precision ±2.5 kN fatigue rated piezoelectric force transducers (Schlosser et al. 2014).

4   MEASUREMENTS RESULTS

The measurements were carried out on the trapezoidal shaped test sample made from recycled asphalt mixture AC 16 L; II.

Figure 2.   Two point trapezoidal bending beam machine.

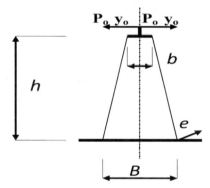

Figure 3.   Scheme of measuring and Dimensions of test samples; B = 56 mm; b 25 mm; e = 25 mm; h = 250 mm for the mixture with D ≤ 14 mm; P_0 – acting force; y_0 – deformation.

Table 1.   Complex modulus of recycled asphalt mixture.

Frequency (Hz)	Complex modulus (MPa)		
5	9520,4	5246,8	4172,2
10	9989,7	6192,6	4462,0
15	10455,8	6611,9	4990,6
20	10583,9	6596,6	5362,3
25	10768,4	7011,0	5821,5
Temperature (°C)	5	15	25

4.1   *Complex modulus*

The measurements were performed on trapezoidal sample according to figure 3 (STN EN 12697-26. 2012).

   The measurement temperature was set to +15°C and frequency was set to 10 Hz. The measurement was conducted in temperatures ranging from +5°C to +25°C. The complex modulus results of the tested recycled asphalt mixture for different temperatures are listed in Table 1 and shown in Figure 4.

209

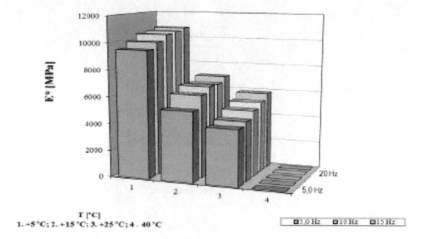

Figure 4.   Diagram of measured complex modulus (E*).

Table 2.   Minimal stiffness, Smin.

Minimal stiffness [MPa]	Category Smin	Minimal stiffness [MPa]	Category Smin
21000	$S_{min\,21\,000}$	4500	$S_{min\,4\,500}$
17000	$S_{min\,17\,000}$	3600	$S_{min\,3\,600}$
14000	$S_{min\,14\,000}$	2800	$S_{min\,2\,800}$
11000	$S_{min\,11\,000}$	2200	$S_{min\,2\,200}$
9000	$S_{min\,9\,000}$	1800	$S_{min\,1\,800}$
7000	$S_{min\,7\,000}$	1100	$S_{min\,1\,500}$
5500	$S_{min\,5\,500}$	No requests	$S_{min\,NR}$

Table 2 shows the classification of mixtures according to values of deformation characteristics (STN EN 13 108-1. 2007). Recycled asphalt mixture is classified as Category $S_{min\,5500}$.

4.2   Master curve

The evaluation is performed by means of master curves – after introducing the gas constant – at the frequency of 3 to 97 Hz. The values are computed according to equation

$$\alpha_T = \exp \frac{\Delta H}{R}\left(\frac{1}{T} - \frac{1}{T_S}\right) \tag{3}$$

where ΔH – the apparent energy activation $(2*10^5\,\mathrm{Jmol}^{-1})$; R – universal gas constant $(8.31434\,\mathrm{Jmol}^{-1}{}^{\circ}\mathrm{K}^{-1})$; T, Ts – temperatures expressed in $^{\circ}\mathrm{K}$ (Ts is the reference temperature)

4.3   Fatigue

The results of the measurements are from samples, which were tested for temperature of $+10^{\circ}\mathrm{C}$ and loading frequency of 25 Hz. The fatigue line is estimated in a bi-logarithmic system as a linear

Table 3. Values of fatigue.

Category	ε_6 [strain $\times 10^6$]	Category	ε_6 [strain $\times 10^6$]
Fat$_{e\ min\ 310}$	310	Fat$_{e\ min\ 100}$	100
Fat$_{e\ min\ 260}$	260	Fat$_{e\ min\ 85}$	85
Fat$_{e\ min\ 220}$	220	Fat$_{e\ min\ 70}$	70
Fat$_{e\ min\ 190}$	190	Fat$_{e\ min\ 60}$	60
Fat$_{e\ min\ 160}$	160	Fat$_{e\ min\ 50}$	50
Fat$_{e\ min\ 135}$	135	Fat$_{e\ min\ NR}$	–
Fat$_{e\ min\ 115}$	115		

Table 4. Fatigue of the recycled asphalt mixtures.

Mixture	T [°C]	F [Hz]	$\varepsilon_6.10^{-6}$ [mikrostrain]	b [–]	r^2 [–]	sN	$\Delta\varepsilon_6$	Category
AC 16 L; II	10	25	96,98	0,9836	0,8985	0,13138	2,91E-09	Fat$_{\varepsilon\ min\ 85}$

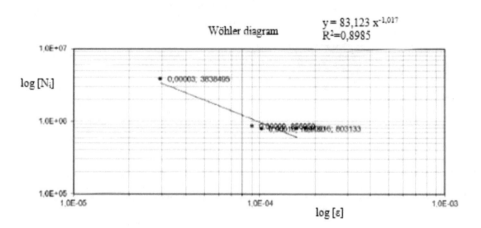

Figure 5. Fatigue – Wöhler's diagram.

regression of fatigue life versus amplitude levels. Using these results, the strain corresponds to an average of 10^6 cycles (ε_6) and the slope of the fatigue line 1/b (STN EN 12697-24. 2012).

The parameters are:

a) ε_6
b) $\Delta\varepsilon_6$
c) slope $1/b$
d) estimated residual standard deviation s_N
e) correlation coefficient r^2

Table 3 shows the mixtures classified according to fatigue values (STN EN 13 108-1. 2007). Recycled asphalt mixture is classified as Category Fat$_{e\ min85}$.

The results of the measurements are listed in table 4 and the example of the fatigue in form of Wöhler's diagram is shown on Figure 5.

5   CONCLUSIONS

It is recommended that the functional test of fatigue for asphalt mixtures is used during their design phase; such test will ascertain the life expectancy of the mixture, i.e. pavement, in consideration given to its operational performance. For the mixture's fatigue resistance assessment, it is important to know the proportional strain at one million cycles ε_6, as well as the trend of fatigue line from Wöhler diagram. The value of slope l/b is very important, If the trend of fatigue line (slope l/b) is different, it will be necessary to consider which mixture is more suitable for particular construction layer and which value is preferable to fatigue of particular asphalt layer. The master curve helps in better evaluation of asphalt layer materials for different temperatures.

Recycled asphalt mixture AC 16 L; II is suitable for local or continuous maintenance repair and rehabilitation of asphalt surfacing.

ACKNOWLEDGEMENTS

This article is created with support of Slovak grant agency VEGA 1/0254/15 Implementation of new diagnostic measurements for the project of the optimize life of roads.

This article is supported by the Slovak Research and Development Agency under the contract No.SUSPP-0005-07.

The research is supported by European regional development fund and Slovak state budget by the project "Research centre of University of Žilina", ITMS 26220220183.

REFERENCES

STN EN 12697-24. 2012: Bituminous mixtures – Test methods for hot mix asphalt – Part 24: Resistance to fatigue.

STN EN 12697-26. 2012: Bituminous mixtures – Test methods for hot mix asphalt – Part 26: Stiffness.

STN EN 13 108-1. 2007: Bituminous mixtures. Material specifications. Part 1: Asphalt Concrete.

Decký, M., Drusa, M., Zgútová, K., Vangel, J., Trojanová, M. Benč, G. & Starší, B. 2009. Design and Quality Control of Earth Structures at Civil Engineering Projects. Scientific monograph. Zilina: BTO print. 479 p.

Schlosser, F. – Mikolaj, J. – Zatkalíková, V. – Šrámek, J. – Ďureková, D. – Remek, Ľ. 2013. Deformation Properties and Fatigue of Bituminous Mixtures. In: Advances in materials Science and Engineering – Mechanical properties and Nondestructive Testing of Advanced Materials, Hindawi Publishing Corporation, Volume 2013.

Mikolaj, J., Remek, L., Life Cycle cost analysis-integral part of road network management system, Procedia Engineering, Volume 91, Pages 487–492, 2014.

Schlosser, F. – Ďureková D. 2014. Rheological properties of asphalt mixtures with additives. In: Advances in materials research.

Advances and Trends in Engineering Sciences and Technologies – Al Ali & Platko (Eds)
© 2016 Taylor & Francis Group, London, ISBN: 978-1-138-02907-1

Determining the mechanical properties of reinforcement coupler systems in Turkey

K. Taskin
Engineering Faculty, Civil Engineering Department, Anadolu University, Eskişehir, Turkey

K. Peker
ERDEMLI Engineering and Consulting Co. Ltd., Beşiktaş, İstanbul, Turkey

ABSTRACT: As the size of structures become larger in the current construction trends, connection of reinforcing rebar for concrete structures will come an important issue for RC structures. Most common rebar connection methods available are the lap splice joint, pressure welded joint, and the mechanical joint methods. When thick rebar's are required to be connected, however, design codes recommends the lap splice joint to be avoided. This is because of potential difficulties in placing concrete in complex rebar area and securing sufficient cover depth, and uncertainty of bond between rebar and concrete. This study will present the results of independent study of mechanical reinforcement splices conducted at the Anadolu University of splicing systems marketed in Turkey. The results of this testing program have shown that mechanical splices are indeed capable of effectively connecting reinforcing bars together and that many are capable of producing bar breakage as the failure mode.

1   INTRODUCTION

Use of mechanical connection systems in reinforced concrete has become increasingly prevalent in Turkey. Mechanical connectors are an alternative to lap and welded splices, and many are capable of developing the full strength of the connected reinforcing bars. There are many advantages for using mechanical connector systems over conventional reinforcing bar lapping. Such examples are overcoming reinforcement congestion problems and convenience when installing precast construction members at sites.

The purpose of this research is to establish the unsuitability of the current Turkish reinforced concrete design standard's (TS500) approach to the performance verification of mechanical connectors.

This paper presents:

- Findings on international trends of mechanical connection systems;
- Recommendations for Turkish reinforced concrete standard review; and
- Indicative test results using newly proposed testing protocol.

2   REVIEW OF DESIGN PROVISIONS FOR MECHANICAL SPLICES

International literature reporting on mechanical connection testing protocols and experimental studies conducted in the United States, Japan, and Europe were assessed to provide recommendations for Turkish reinforced concrete standard. Both static and seismic conditions, in terms of their relevance in the Turkish context, were examined in this literature review.

2.1   Static conditions

Based on assessment of the international literature, a general conclusion was drawn. For use of mechanical connection systems in static conditions, three categories need to be considered, and they are:

- Strength;
- Serviceability limit state; and
- Fatigue loading.

2.2   Strength

The general trend for strength requirement of mechanical connectors in static conditions is that the strength of the connectors must be larger than that of the spliced reinforcing bars. Most reinforced concrete design standard organizations demand an over strength factor to be multiplied to the specified yield strength of the spliced reinforcing bars for the connector's strength requirement. This over strength factor tends to be governed by both safety based on reinforcing bar manufacturer's quality control and on economy considerations. Some organizations are more restrictive than others in this matter, as they require the connector strength to be more than the specified ultimate strength of the spliced reinforcing bars. The basic logic behind the above requirements is that the spliced reinforcing bars must yield and eventually fail before the ultimate failure of mechanical connectors under loading situation, thus avoiding brittle failure of the connectors.

2.3   Serviceability limit state

A number of reinforced concrete design standards recognize possible concrete cracking, which may arise from slip between the spliced reinforcing bar and the mechanical connector, thus constituting a serviceability limit state. It is understood that this slip, which is a permanent or residual deformation, is a matter of manufacturer's quality control on interlock between the spliced reinforcing bar and the connector. It also is understood that the mechanical connection system tends to be more rigid once it is subjected to low stress. The reason for this is the plastic deformation due to bearing within the interlocked system of the reinforcing bar and the connection. Serviceability limit state design in reinforced concrete under static conditions is an important aspect that structural engineers need to consider. Appropriate crack width limits provide an aesthetically sound environment for the public as well as preventing possible corrosion occurrence in reinforcing bars.

2.4   Fatigue loading situation

A few organizations require fatigue testing of mechanical connection systems. This fatigue loading, which is a high number of cycles within the elastic stress range, can affect the mechanical connector's performance. However, assessing fatigue behavior of the mechanical connectors was beyond the scope of the research reported here.

2.5   Seismic conditions

The design provisions for mechanical splices in Turkish code is not enough to design a connection using mechanical coupler in applications.

In British Standard BS 8110: Part 1, the tensile strength of coupled hot-rolled bars has had to be at least 1.15 times the nominal yield stress, f_{sy}. The intention has been that the coupler should be stronger than the bar. In practice the requirement in BS 8110: Part 1 has only meant that the strength of a splice at least equals the required minimum tensile strength of the bars. As a consequence, it has been conceded that it may not be possible to force failure outside a coupler when the bar strength exceeds the minimum. Another requirement in BS 8110: Part 1 has been to ensure that the permanent elongation across the coupler does not exceed 0.1 mm after the bars have been stressed to $0.6f_{sy}$. This elongation arises due to slip of the bars relative to the coupler. It appears likely that the value of 0.1 mm was chosen because it was considered to be sufficiently smaller than the normal upper characteristic limit of 0.3 mm for crack widths in buildings, and does not appear to be supported by the results of research.

In the United States today there are several different codes that these splice manufacturers must meet or at least consider. These codes include the ACI 318 [1995, 1999], the Uniform Building Code [ICBO 1997], the BOCA (Building Officials and Code Administrators) Building Code [1999] and the Southern Building Code [SBCC 1997]. Moreover there are a separate set of code for bridges, principally the AASHTO (American Association of State Highway Transportation Officials) Bridge Design Specification [AASHTO 1998], as well as various state standards for the design of bridges. Lastly there are federal standards written by the U.S. Army Corps of Engineers and other agencies.

Given the complexity of the code landscape in the US, the splice requirements are surprisingly similar. All of the various code bodies in the US follow, to some extent, the basic reinforced concrete code, the ACI 318. Here the traditional requirement for both welded and mechanical splices has been that the splice be capable of developing 125% of the nominal yield strength of the spliced bars.

A discussion of the proper method to reasonably test mechanical splices ensued. So as to provide a simple procedure, an in-air test was selected. While this is not as accurate as an in-member test, it would provide a conservative lower bound for the performance. It was decided to limit the bar size to 26-mm and 32 mm in diameter and to use only Grade 420 steel with a nominal yield of 420 MPa.

The so-called ACI 439 Test Procedure consisted of three separate tests intended to determine how well the spliced bar system performed under simulated seismic loading. A test specimen with a gage length of r approximately 800 mm for these bars, was used to simulate the length of a plastic hinge that would be found in a flexural member. The goal was to determine the response of the specimen to three different types of loading.

Two separate types of tests were conducted to evaluate the behavior of the mechanical splices:

2.5.1 *Monotonic tension test*
The first test performed on all splice assemblies. The specimen was loaded from zero strain up through the 4 percent strain requirement and then on to failure. Each test was performed similar to the tension test specified in ASTM 370A for determining the yield strength of steel [ASTM 1999].

2.5.2 *Stepped cyclic test*
For this test the specimen was started at initial conditions of zero strain under zero load. The splice assembly was loaded in the same manner as in the monotonic load test but only until the average strain across the 20 bar diameter reached 2 percent. At this point in the test the load was reversed and the specimen was unloaded through zero load.

The purpose of compressing the splice assembly was to insure that the splice itself underwent complete unloading in tension before being subjected to the next cycle of tension loading. After being unloaded, the assembly was then cycled four times under a tension load out to 2 percent strain. After this tension-to-compression cycle to 2 percent strain was completed four times, the testing was repeated again at strain values of 2.5, 3.0 and 3.5 percent strain. At each strain level, the testing cycle was conducted four times. After the final unloading at the 3.5 percent strain the assembly then loaded in tension out to failure.

3   TEST SPECIMENS AND RESULTS

20 assemblies (15 of them are Ø32 the others are Ø26)were obtained from each manufacturer participant. These assemblies were divided into two groups. 12 were pulled monotonically to failure, 8 were cycled to 4 percent strain uniformly over 16 cycles and then pulled to failure.

The elongation was measured using Linear Differential Variable Transformers (LVDT's). Two LVDT's were placed at 180-degree intervals around the circumference of the mechanical connection assembly. The purpose of using three measurements taken at 180 degree intervals was to account for any initial lack of straightness of the connector/bar assembly and to insure accurate elongation measurements throughout the testing, both for monotonic and cyclic loading tests.

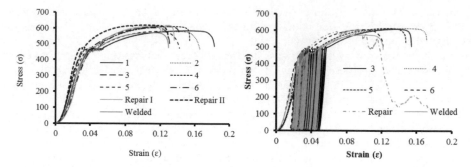

Figure 1.   Stress-strain diagram for Ø32 bar specimens_monotonic and stepped cycling tensile test.

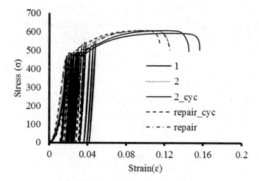

Figure 2.   Stress-Strain diagram for Ø26 bar specimens_monotonic and stepped cycling tensile test.

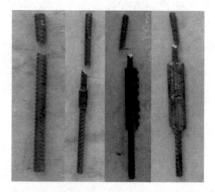

Figure 3.   Failure modes of the Ø26 bar specimens after tests.

The two elongations were then averaged to determine the average overall elongation. This average elongation was divided by the original gage length of the splice/bar specimen to determine the average strain value for evaluation of the mechanical splice. Two other strain measurements were taken on the exterior of the mechanical splice itself and on the bar.

Testing was performed in the Structural Engineering Laboratory at the Anadolu University. Results were reported to the individual participating companies and were used to improve their individual systems as well as a means to develop new design requirements and policies within ACI and Turkish codes.

Typical results are shown in Figure 1–3. It can be seen that the results clearly indicate that the splice is capable of developing a large nonlinear strain in the bars and that system failure is above the 4% acceptance criteria.

Table 1. Summary of company Ø32 Bar Specimens (Tensile Tests).

Test Pattern	Yield σ_y (MPa)	Ultimate σ_U (MPa)	ε_U (%)	σ_U/σ_y	Slip ≤1 mm	Ductility ≥%70ε_U	1.25f_y	4%	test
Monotonic	457*	582	17	1.27	–	11.9	Y	Y	Y
Tensile	455	580	16.5	1.27	0	Y	Y	Y	Y
Test	462	610	14	1.32	0	Y	Y	Y	Y
	460	608	15.4	1.32	0	Y	Y	Y	Y
	465	578	14.1	1.25	0	Y	Y	Y	Y
	465	600	13	1.29	0	Y	Y	Y	Y
	468[Repair I]	615	12	1.31	–	N	Y	N	N
	469[Repair II]	618	12.7	1.32	0	Y	Y	Y	Y
	450[Welded]	570	12.9	1.27	0	Y	Y	Y	Y

Table 2. Summary of company Ø32 Bar Specimens (Stepped CyclicTensile Tests).

Test Pattern	Yield σ_y (MPa)	Ultimate σ_U (MPa)	ε_U (%)	σ_U/σ_y	Slip ≤1 mm	Ductility ≥%70εU	1.25f_y	4%	test
Stepped	461*	609	15.4	1.32	0	Y	Y	Y	Y
Cyclic	462	614	17.0	1.31	0	Y	Y	Y	Y
Tensile	465	613	14.7	1.31	0	Y	Y	Y	Y
Test	467	607	12.1	1.30	0	Y	Y	Y	Y
	467[Rerpair I]	608	–	1.30	–	N	Y	N	N
	461[Welded]	576	12.1	1.25	0	Y	Y	Y	Y

*Reference

What is important in these figures is that the system presented as the "Company KT specimen" exhibited good overall system ductility with failures at strains well in excess of the 4% acceptance criterion. The step cyclic testing results show that the presence of cyclic loading did not reduce the failure ductility significantly and that this particular splice could indeed be expected to maintain load capacity well into the nonlinear regime.

Additional insight can be gained from study of the plots. These different readings permitted observation of where the strain was concentrated during testing. The data that was produced is interesting in that, as shown in Tables 1–2, the strain is not uniform across the splice. It can be seen in both of these figures that there tends to be a difference in strain readings in the splice body, as compared with the overall system. This difference shows where the more flexible parts of the connection are found and where the damage tends to be concentrated.

4 CONCLUSIONS

The purpose of this research was to provide recommendations for use of mechanical connection systems in both static and seismic conditions.

The research was motivated by the facts that the current criteria in Turkish reinforced concrete standard, TS 500 were based on no research, while there had been advancement in this field worldwide in recent years.

New criteria can be used on mechanical connection systems in both static and seismic conditions and also proposed for the upcoming review of Turkish code.

Two criteria were proposed for use of mechanical connection systems in seismic conditions, being strength and serviceability limit state.

The results of this testing program at the Anadolu University have shown that mechanical splices are indeed capable of effectively connecting reinforcing bars together and that many are capable of producing bar breakage as the failure mode; thus indicating that the splice is stronger than the parent bar itself.

REFERENCES

American Concrete Institute. 1999. *Building Code Requirements for Structural Concrete: ACI 318M – 99.* USA

American Concrete Institute – *ACI Committee 318 (1995), Building Code Requirements for Structural Concrete(ACI 318-95) and Commentary – ACI 318R-95*, American Concrete Institute, Farmington Hills, MI.

American Concrete Institute – ACI Committee 318 (1999), *Building Code Requirements for Structural Concrete (ACI 318-99) and Commentary – ACI 318R-99*, American Concrete Institute, Farmington Hills, MI.

American Concrete Institute – ACI Committee 439 (1998), *Standard Specification for Mechanical Reinforcement Splices for Seismic Designs using Energy Dissipation Criteria (ACI-439)*, ACI Technical Activities Committee.

American Society for Testing and Materials (ASTM) (1999), *Standard Test Methods and Definitions for Mechanical Testing of Steel Products (A 370 – 97a), 1999 Annual Book of ASTM Standards, V. 4.02*, American Society of Testing and Materials, West Conshohocken, PA, pp. 166–211.

American Society for Testing and Materials (ASTM) (1999), *Standard Specification for Billet Steel Bars for Reinforcement of Concrete (A 615/A 615 M – 96a), 1999 Annual Book of ASTM Standards, V. 4.02*, American Society of Testing and Materials, West Conshohocken, PA, pp. 308–316.

American Society for Testing and Materials (ASTM) (1999), *Standard Specification for Alloy Steel Bars for Reinforcement of Concrete (A 706/A 706 M – 96b), 1999 Annual Book of ASTM Standards, V. 4.02*, American Society of Testing and Materials, West Conshohocken, PA, pp. 346–350.

American Welding Society (1992), *Structural Welding Code – Reinforcing Steel, D1.4-92, Miami. Building Officials and Code Administrators International (1999)*, Building Code, BOCA International, Country Club Hills, IL.

American Association of State Highway Transportation Officials– AASHTO (1996), *Design Specification for Highway Bridges*, 16th Edition, Washington D.C.

British Standards Institution. 1997. *Structural Use of Concrete: Part 1: Code of Practice for Design and Construction: BS 8110: Part 1: 1997.* UK.

Canadian Standards Association. 1994. *A 23.3 – 94: Design of Concrete Structures. Toronto, Canada*

Department of Transportation. 2001. *Memo to Designers 20 – 9: Splices in Bar Reinforcing Steel.* California.

Department of Transportation. 1999. *Method of Tests for Steel Reinforcing Bar Butt Splices: California Test 670 (CT_670).* California.

German Institute for Construction Techniques. 1998. *Structural Use of Concrete: Design and Construction: DIN 1045.* Germany.

International Conference of Building Officials. 1997. *Uniform Building Code: UBC 1997.* California.

International Conference of Building Officials Evaluation Service (1998), *Acceptance Criteria for Mechanical Connectors for Steel Bar Reinforcement," Acceptance Criteria 133 (AC 133)*, Whittier, CA.

International Conference of Building Officials: Evaluation Service Inc. 2001. Acceptance Criteria for Mechanical Connectors for Steel Bar Reinforcement: ICBO ES AC 133. California.

Patrick M. *Strength and Ductility of Mechanically Spliced Bars.* Centre for Construction Technology & Research. University of Western Sydney, Sydney, Australia.

Raper A.F. 1977. *Evaluation of Mechanical Reinforcing Bar Splice Systems for New Zealand Conditions.* Research and Development Report. Ministry of Works and Development, Wellington, New Zealand.

Splices for Bars (DRAFT) – ISO/CD 15835 – Part 1 & Part 2. Switzerland.

Southern Building Code Congress International (1997), *Standard Building Code*, Birmingham, AL.

Advances and Trends in Engineering Sciences and Technologies – Al Ali & Platko (Eds)
© 2016 Taylor & Francis Group, London, ISBN: 978-1-138-02907-1

Deformation and stress analysis of a concrete bridge due to nonlinear temperature effects

M. Tomko & I. Demjan
Technical University of Košice, Faculty of Civil Engineering, Košice, Slovakia

ABSTRACT: The paper presents a theoretical analysis which models temperature load effects induced on a double-span, double box girder reinforced concrete bridge.

1 INTRODUCTION

In recent years, increased attention has been paid to professional public law consistent with the modelling of structures and the effects of operating loads. The aim of this paper is that the knowledge gained from the study will be subsequently applied to future design methodology, regulations and standards (Kvocak et al. 2013, Kvocak et al. 2014).

It can be concluded that for large-span bridge structures, the incidental influence of long-term effects is progressively becoming a decisive factor when dimensioning bridges. An independent and targeted analysis of individual load effects facilitates the presumption of a good overall design of a bridge subjected to a combination of different effects. Currently, common practice necessitates a more exact analysis of the effects caused by climatic influences, especially in recent years regarding the effect of temperature changes, etc. (Giussani 2009, Kovalakova et al. 2013, Kyong-Ku et al. 2014, Mandula et al. 2002, Pagani et al. 2014, Si et al. 2013, Yun et al. 2014).

In this paper a theoretical analysis of thermodynamic effects was performed on a closed double-box reinforced concrete bridge structure. The modelling and analysis only considered the upper part of the load-bearing bridge structure. This type of bridge was selected to present the results of this study which analysed the worst effects of temperature acting on the bridge structure.

2 TWIN-CHAMBER BRIDGES

The hollow cross-section of a box-chamber girder is characterized by its torsional rigidity even with large spans, it is economical in terms of reinforcement, but is more labour intensive than slab bridges. The construction height of the bridge deck is always greater than their slab counterparts. The cross-section of the box-chamber is often large enough to be used as a double decker bridge, assuming the rigidity is sufficient since additional diagonal reinforcement would obstruct the efficient use of the box-chamber. All structural concrete box girder bridges are monolithically connected, thus forming a whole. The most complete picture of stress at any point of the bridge is obtained if the structure is considered as a whole.

3 BRIDGE SUBJECTED TO NON-LINEAR THERMAL EFFECTS

Addressing the specific calculation of a bridge thermal actions must be preceded by a corresponding analysis of the appropriateness of a particular method, process and simplification of assumptions. Standard EN 1991-1-5 provides guidance for determining the load of buildings and civil engineering

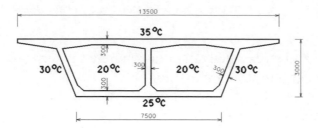

Figure 1.   Cross section of the bridge (mid-span), warming in the vertical direction.

temperature effects that arise as a result of climatic and operating conditions. The Standard presents alternative procedures, values and recommendations for classification for many cases, i.e., national annex. In the case of bridges the National Annex specifies whether the calculations should use the general non-linear or simplified linear temperature components. Standard EN 1991-1-5 is intended for use with EN 1990, additional parts of EN 1991 and EN 1992-1999 for the design of structures.

Standards EN 1991-1-5, EN 1991-2-5, CSN 73 62 03 recognize a linear one-dimensional vertical temperature gradient of the heat transfer rate of the cross-section of a bridge, which is sufficient in certain cases. For significant objects the model is insufficient and instead a non-linear (linearization per parts) of the vertical temperature gradient of heat transfer through the height of the bridge's cross-section is recommended.

Standard EN 1991-1-5 considers heat transfer only in the plane of a bridge's cross-section, i.e., In general, it can be stated that it's the dominant effect of induced temperature. In the study, the heat transfer and the length of the bridge construction is considered to realistically represent the effects of temperature.

The concrete wall of the box-girder profile, a structurally indeterminate element, is subjected to a temperature gradient dependent on individual temperatures between the outside and inside of the wall and from the mutual interaction between the individual walls. Standard EN 1991-1-5 states: 6.1.4.4 Components of the thermal gradient in the walls of the concrete box girder, (1) When designing large concrete box-girder bridges attention must be paid to the possible occurrence of significant temperature differences within those structures between the inner and outer walls.

Each span of the double-span bridge measures 63 m and is mounted on two sliding supports and a single fixed support. The supporting structure of the bridge is formed by a closed profile (Figure 1).

A FEM model of a bridge construction was created for thermal analysis, which focused on the two-dimensional temperature distribution in the bridge construction. Thermal analysis defined the heat transfer through the thickness of each closed wall cross-section and through the length of the bridge construction. The temperature values, which are peripheral, i.e. wall surfaces of various bridge sections are taken from standard EN 1991-1-5.

The temperature of the cross section was subsequently transformed to load conditions, i.e., the bridge structure was subjected to the effects of temperature. A model of the bridge was created and analysed using "structural FEM thermal stress analysis". The FEM analysis considered strain, relative deformation and stress of a double span bridge for a closed single chamber profile created in ANSYS with 38 123 elements and SOLID BRICK volume element 70.

The effects of two-dimensional temperature distribution in the construction of a bridge consisting of a closed profile are considered in three variants:

1) Temperature gradient considers only the vertical direction – warming (Figure 1);
2) Temperature gradient considers only the vertical direction – cooling (Figure 4);
3) The temperature gradient is contemplated in the vertical and horizontal direction – warming (Figure 6).

Some of the results of thermal analysis are shown in Figure 1 to Figure 8.

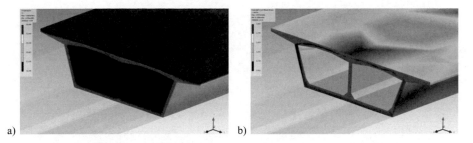

a) b)

Figure 2. a) Heat transfer through the cross-section of the bridge (mid-span no.2), warming in the vertical direction, b) Equivalent stress corresponding to the effects of temperature (mid-span no.2).

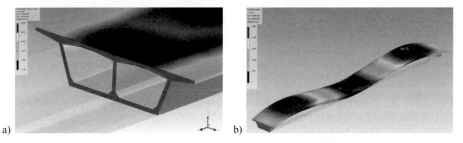

a) b)

Figure 3. a) Deformation of the bridge (δ_z [mm]) in the z direction (mid-span no.2), b) The total deformation of the bridge.

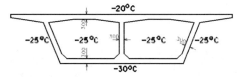

Figure 4. Cross section of bridge (mid-span cross-section), cooling in the vertical direction.

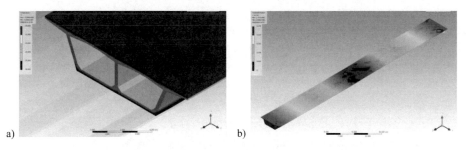

a) b)

Figure 5. a) Heat transfer through the bridge cross-section (mid-span no.2) cooling in the vertical direction, b) The total deformation of the bridge.

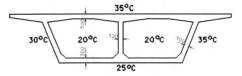

Figure 6. Cross section of the bridge (mid-span cross-section), warming in the vertical and horizontal directions.

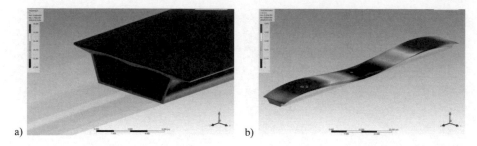

a)   b)

Figure 7.   a) Heat transfer through the cross-section of the bridge (mid-span no.2), warming in the vertical and horizontal directions, b) The total deformation of the bridge.

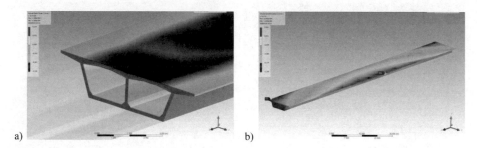

a)   b)

Figure 8.   a) The relative deflection of the bridge (δ_x [mm]) X-axis (mid-span no.2), b) Deformation of the bridge (δ_x [mm]) in the X-axis.

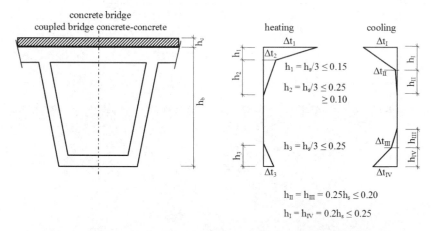

Figure 9.   One-dimensional heat transfer in a concrete box girder according to CSN 73 6203.

4   ANALYSIS OF THE RESULTS

Deformation, relative deformation and axial tension were observed in the "structural FEM thermal stress analysis" of the double-span double box-girder bridge. Standard CSN 73 6203 defines the load effect for the influence of temperature in one direction, i.e. linear influence of the temperature effect on the axis of the chamber wall of the bridge, Figure 9. The temperature applied along the axis of the peripheral walls of the closed box girder as a result of individual effects of temperature in Figure 10.

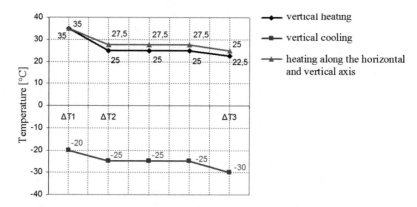

Figure 10. Operating temperature on the axis of the chamber walls of concrete bridges.

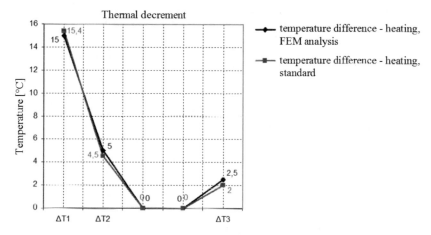

Figure 11. The temperature gradient along the axis of the chamber walls of concrete bridges, warming in the vertical and horizontal direction; a comparison between "structural FEM thermal stress analysis" and standard EN 1991-1-5.

Results of "structural FEM thermal stress analysis" of the two-dimensional heat diffusion in the cross-section of the box girder double-span bridge are compared with values given in EN 1991-1-5, Figure 11. The results presented show that the heat transfer temperature at various points differed in the range of 2.6–25%.

When considering the two-dimensional effect of temperature, the temperature gradient in the vertical and horizontal lines were considered, i.e. that which caused a horizontal bending action on the wall of the closed box girder profile in the horizontal plane of the bridge. The presented results show that the non-linear effect of temperature is spread in two directions affecting the direct stress and deformation of the bridge structure. This fact should be taken into account during the design process, especially in massive bridges of larger spans.

Large span girder bridges should also factor in other conditions and influences that ultimately may affect the results to some degree. The biggest influence on the temperature of the bridge are caused by environmental conditions (air temperature, solar radiation, the radiation emitted from the bridge construction, wind speed), structural layout of the bridge (geometric characteristics, method of storage of the bridge (statically determinate/indeterminate, continuous beam, and so on.)) the shape of the bridge (bridge cardinal direction, curve, etc.), the material of the bridge, etc.

5 CONCLUSION

Knowledge of the temperature effects, the values of operating temperatures, structural static and physical-mechanical properties of the bridge construction are essential for the proper evaluation of the effect of temperature on the bridge structure, therefore, these parameters should also be included in the theoretical computing model, in order to achieve an equivalence between the real behaviour of the bridge construction and the theoretical response of the calculation model. Currently scientific publications, as well as international standards, recommend that significant structures employ spatial models in their calculations in order to consider the real distribution of mass, stiffness of structures and actual loads.

The theoretical approach for calculating the deformation and stress characteristics of a reinforced concrete box-girder beam from the effects of temperature is influenced by the simulation of thermal stress, choice of computational model, but also the conditions imposed on the model.

For practical needs, it would be appropriate if the procedures described in the theoretical modelling of the effect of temperature on the double-box girder is extended to other types of bridge structures and their computational models. Based on the above it can be concluded that the present procedures and the results of the simulation modelling non-linear effects of temperature on the bridge provide useful information for technical construction in practice.

ACKNOWLEDGEMENTS

The paper is carried out within the project No. 1/0321/12, partially founded by the Science Grant Agency of the Ministry of Education of Slovak Republic and the Slovak Academy of Sciences. The paper is the result of the Project implementation: University Science Park TECHNICOM for Innovation Applications Supported by Knowledge Technology, ITMS: 26220220182, supported by the Research & Development Operational Programme funded by the ERDF.

REFERENCES

CSN 73 6203, 1986: *Load bridges*.

EN 1991-1-5: Eurocode 1, 2007: *Actions on structures*.

EN 1991-2-5: Eurocode 1, 2000: *Actions on structures*.

Giussani, F. 2009: The effects of temperature variations on the long-term behaviour of composite steel-concrete beams. *Engineering Structures* Vol. 31, p. 2392–2406.

Kovalakova, M. & Fricova, O. & Hronsky, V. & Olcak, D. & Mandula, J. & Salaiova Brigita, 2013: Characterisation of crumb rubber modifier using solid-state nuclear magnetic resonance spectroscopy. *Road Materials and Pavement Design*. Vol. 14, p. 946–958.

Kvocak, V. & Dubecky, D. 2013: Experimental stiffness verification of composite beams. *Advances in Structural Engineering and Mechanics* (ASEM 13): The 2013 World Congress on Advances, Techno-Press, p. 3171–3178.

Kvocak, V. & Dubecky, D. & Kocurova, R. & Beke, P. & Al Ali, M. 2014: Evaluation and analysis of bridges with encased filler beams. *Civil Engineering and Urban Planning 3*, Taylor and Francis Group, p. 150–153.

Kyong-Ku, Y. & Pangil, Ch. 2014: Causes and controls of cracking at bridge deck overlay with very-early strength latex-modified concrete. *Construction and Building Materials* Vol. 56, p. 53–62.

Mandula, J. & Salaiova, B. & Kovalakova, M. 2002: Prediction of noise from trams. *Applied acoustics*. Vol. 63, p. 373–389.

Pagani, R. & Bocciarelli, M. & Carvelli, V. & Pisani, M. A. 2014: Modelling high temperature effects on bridge slabs reinforced with GFRP rebars. *Engineering Structures* Vol. 81, p. 318–326.

Si, X. T. & Au, F. T. K. & Li, Y. H. 2013: Capturing the long-term dynamic properties of concrete cable-stayed bridges. *Engineering Structures* Vol. 57, p. 502–511.

Yun, L. & Jiantao, W. & Jun, Ch. 2014: Mechanical properties of a waterproofing adhesive layer used on concrete bridges under heavy traffic and temperature loading. *International Journal of Adhesion and Adhesives* Vol. 48, p. 102–109.

Advances and Trends in Engineering Sciences and Technologies – Al Ali & Platko (Eds)
© 2016 Taylor & Francis Group, London, ISBN: 978-1-138-02907-1

Alternative methods for application the earthworks quality control

K. Zgútová & M. Pitoňák
University of Žilina, Žilina, Slovakia

ABSTRACT: This article is about assessing the quality of earth structures in terms of possibilities of using alternative methods. These methods are beneficial to engineering practice with respect to speed and smart testing. Methods described in the article are not yet included in the Slovak technical standards and their application is very rare as they are not verified in practice yet.

1 INTRODUCTION

Today there are many methods for obtaining information about subsoil carrying capacity based on various theoretical approaches. In order to ensure consistency between these methods, it is necessary to determine conditions for measurements, accurately perform comparative measurements using the selected methods, to compare the obtained values, and to determine relationships between them by the statistical method of correlation. If correlations are determined responsibly, it is possible to apply these methods. This would be beneficial for the contractor and the employer due to simple and easy handling, fast results and problem free transport.

2 MEASURING INSTRUMENTS

At our experiment we have used the device Humboltd H4140 and Clegg Model CIST882. For comparison we have used the Light Dynamic Plate LDD100 device – commonly used in the Slovak Republic.

2.1 *WS 32830 – CBR Clegg Impact Soil Tester*

Clegg impact soil tester Model CIST 882 (version with 4.5 kg hammer) is a device suitable for continuous quality assurance monitoring at earth structures – embankments, which can be implemented quickly without interrupting advancing works (Manual). The device (Fig. 1) consists of:

– falling hammer with in-built compaction sensor
– guiding cylinder with integrated base plate and auxiliary handle
– measuring instrument with digital display and connecting cable
– transport and storage case.

The parts are assembled into an easily portable measuring unit. Compacting hammer rises and falls in the vertical guiding cylinder. It falls directly onto the tested material without any pad. The extent to which the soil retards the hammer is defined by the power dependent on compaction of the tested material(Decký et al. 2009). On the falling hammer is located the accelerometer, which is connected with the digital display. The display shows the speed of the hammer descent in the CIV units (Clegg Impact Value).

Figure 1.   Clegg CIST 882.

In order to obtain reliable CIV values it is necessary to perform 5 hammer blows at each tested location. Regardless the tested material, the following measurement procedure must be maintained:

– the tested location must be free of roughness and larger stones (impact hammer falling on the stone can damage the accelerometer)
– prior to testing check if position of the guiding cylinder is vertical. The device is supported by the inner side of the foot. The hammer is elevated up to the line and the button on the screen is pressed. Check whether the first reading shows zero
– make sure that the cable connecting accelerometer and the display is not captured by the falling hammer and watch the value displayed
– without moving the guiding cylinder and pushing the button on the display 4 hammer blows are repeated, in order to obtain final count of 5 blows
– 5 measurements are assessed. The first 2–3 blows should even out small roughness caused by the release of material under the hammer. Further should increase slightly and determine the stiffness of the compacted layer. The device is sensitive to the surrounding shocks, therefore it is necessary to make sure that during the whole process of testing there is no source of vibration or shocks in the radius of at least 15 m from the device, which could adversely affect the final result.

2.2   LDD 100 – dynamic load test

The fundamental of the test is to determine the size of the response (vertical deflection) of the tested half-space to the load by the blow. Deflections are determined from the size of acceleration detected by the accelerometer. In accordance with the User manual (LDD 100) from the retraction of plate is calculated the deformation modulus M_{vd} according to the formula:

$$M_{vd} = \frac{F}{d \cdot y_{el}} \cdot (1 - \mu^2) \qquad (1)$$

where F = power [N]; μ = Poisson's ratio [–]; y_{el} = size of the elastic deflection under the centre of the loading plate [mm] and d = diameter of the loading plate [mm].
Measuring by the Light dynamic plate is performed as follows:

– measuring plate is placed on the surface of the tested soil which must be flat and the plate shall touch down on its entire surface. If there are small unevenness it is required to smooth the tested location by the sand.
– connecting cable is connected to the plate by the connector

Figure 2.   HumboldtH-4140.

– impactor is installed on to the plate and the locking pin locks off the weights. Weights are lifted
  and are latched.
– at the beginning there is one impact, which secures the plate touch down on the surface. After
  the rebound, the weight is retained in the raised position and locked.
– evaluation unit is switched on (and the printer) and the measurement is prepared
– measuring impacts are performed, after the rebound the weight is always retained in the raised
  position and locked
– connector is disconnected from the plate, weights is locked in the bottom position and the device
  can be moved to the next tested location.

2.3   H-4140 – geogauge Humboldt – Dynamic method using the principle of mechanical impedance method

The fundamental of the test is to determine the mechanic impedance of the tested layer. It measures
the pressure transmitted to the layer surface and the resulting surface speed as a function of time
(Zgútová et al. 2012).

The device during one measurement cycle generates electromagnetic energy at the level of 25
frequencies (range 100 Hz to 196 Hz). Measuring interval lasts 75 seconds and the gauge conducts
7400 measuring cycles (Fig. 2). Output parameters for each measurement interval are:

– *modulus of elasticity* – in the original literature referred to as a Young's modulus in accordance
  with Hooke's law for isotropic, linear elastic material. For the resolution referred to as E_H.
– *stiffness* – measured in the range from 3 MN.m^{-1} to 70 MN.m^{-1}

According to the manual the EH value is determined according to the formula:

$$E_H = \frac{F}{1,77R.\delta}(1 - \mu^2) \tag{2}$$

where $E_H =$ the dynamic modulus of elasticity of the soil determined by the Humbolt device at
one phase of the test [MPa]; $\mu =$ Poisson's ratio [–]; $R =$ outer diameter ring-shaped foot [mm] and
$\delta =$ deflection [mm].

3   FIELD OF TESTING

Prior to experiment the testing field had to be prepared. In autumn 2012 the top soil layer was
removed and on the right side begun the construction of the railway stand. On the left side was

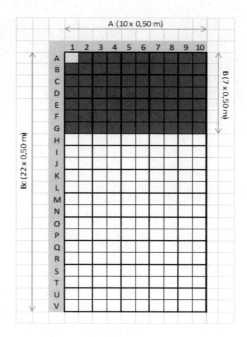

Figure 3.   The most homogeneous third of the testing field.

in spring 2013 removed the further layer of soil/depth 0.3–1.2 m/to even the terrain. The soil sample (100 kg) was taken and laboratory analysis was performed at the Geotechnical Department Laboratory. In the testing field corners were constructed footings, carrying beams which are arranged in the longitudinal direction of the testing field. On beams were placed moving cross-beams. Corner footings served to outset the testing field. The size of the tested field is 11 × 5 m. Prior to commencing experiment it was necessary to adjust minor unevenness of the testing field. Larger stones were removed, depressions were filled, upraised locations were aligned and adjusted locations were compacted (Zgútová et al. 2012). Visual inspection during the works determined that the test field is not homogeneous, and the middle part is located more coarse material, therefore the initial measurements were carried out by using the apparatus LDD 100 and Clegg CIST 882.

On the testing field was marked the grid with the square size 50 cm and on 25.4.2013 were performed measurements at each square. In total 220 initial measurements were taken with each device. From the measured data, we created a surface graph for each of the apparatus used, and from the results we determined that the most homogeneous part is the upper third (Figure 3), with the dimensions 3.5 × 5 m. Further measurements were conducted only on this part. Laboratory analysis was performed on sample taken from the upper third of the testing field (Ďureková et al. 2015).

The tests carried out on the sample:

– Sieve analysis
– Hydrometer test
– Determination of plasticity
– Determine the limits of fluidity
– Proctor Standard

Evaluation determined that it is a clay with medium plasticity F6 – CI.

Clay subsoil is quite frequent in Slovak geology. In some cases unsuitable soil is replaced, or subsoil improvement is applied. Such soil can be also used at layered embankments. Such embankments require increased attention to the control of compaction, moisture content and the degree of consistency.

Table 1. Humboldt H4140.

Consistency	Moisture interval	Date	Formula	Correlation
Hard consistency	8%–9.99%	27.5.'14 26.5.'14	$E_{vd} = -0.0007E_H{}^2 + 0.3192E_H - 2.4842$	$R = 0.8254$
	10%–11.99%	19.6.'13	$E_{vd} = 0.0004E_H{}^2 + 0.2309E_H - 2.3927$	$R = 0.8091$
	12%–13.99%	2.7.'13 4.10.'13	$E_{vd} = 0.0038E_H{}^2 - 0.2597E_H + 11.896$	$R = 0.9127$
Firm consistency	14%–15.99%	9.10.'13	$E_{vd} = -0.0029E_H{}^2 + 0.8185E_H - 31.701$	$R = 0.8691$
	16%–17.99%	20.5.'13 9.5.'14 22.5.'14 23.5.'14	$E_{vd} = 0.0051E_H{}^2 - 0.5027E_H + 21.048$	$R = 0.819$
	18%–19.99%	29.5.'13	$E_{vd} = 0.0006E_H{}^2 + 0.062E_H + 4.6911$	$R = 0.8963$

Measurements procedure:

– prior to commencement we sampled the soil from which we determined the laboratory humidity (dry top layer, we have removed at thickness of about 2 cm to avoid distortion of the actual value of humidity)
– as the first were performed measurements by the Clegg CIST 882. At measurements by this device there are clear imprints in soil. When using the next apparatus the exact tested location was easily determined. Thus it was possible to statistically evaluate a pair testing. The device was placed on the testing location according to the grid, the distance between measuring points was approximately 50 cm.
– further measurements we carried out by the Humboldt H4140, the device was placed as close as possible to the imprint of the Clegg device (or imprint of Clegg was located in the middle of the Humboldt ring), in a way that the imprint would not interfere into the measuring ring of the device
– last measurements were performed by the device LDD 100, because at measurements by the H4140 it was necessary to fill the cavities under the plate by sand in order to achieve better contact with the measured material. The LDD was placed on the same spots.
– After completing the measurements, we aligned the minor surface soil disruptions incurred by measurements, particularly from devices Clegg CIST 882 and LDD100, in order to create ideal conditions for the next day measurements.

Obtained values were continuously during measurements manually recorded because the Clegg CIST 882 and Humboldt H4140 do not have the function to store the data.

4 CONCLUSIONS AND RESULTS

At evaluation was taken into account the impact of climate and soil conditions. Comparative measurements had to be carried out always within the interval of several hours in order not to alter the consistency and the moisture content of soil, which has a great impact on the soil behaviour. Obtained values were evaluated statistically for both devices. (Tabs 1,2).

The results show that it is possible to consider the interchangeability of these devices in practice, which would lead to more efficient quality control.

Obtained correlations are suitable basis for the introduction of alternative methods in to the quality assurance process in Slovakia. Their use in combination with the static loading test simplifies, speed up and enhance the quality assurance process at earth structures.

Table 2. Clegg CIST 882.

Consistency	Moisture interval	Date	Formula	Correlation
Hard consistency	8%–9.99%	27.5.'14 26.5.'14	$E_{vd} = 0.0301CIV^2 + 1.7836CIV - 3.1223$	$R = 0.7972$
	10%–11.99%	19.6.'13	$E_{vd} = 0.3982CIV^2 - 5.5153CIV + 32.927$	$R = 0.879$
	12%–13.99%	2.7.'13 4.10.'13	$E_{vd} = 0.0751CIV^2 + 0.0629CIV + 7.1015$	$R = 0.777$
Firm consistency	14%–15.99%	9.10.'13	$E_{vd} = -0.046CIV^2 + 3.6876CIV - 12.543$	$R = 0.8962$
	16%–17.99%	20.5.'13 9.5.'14 22.5.'14 23.5.'14	$E_{vd} = 0.2697CIV^2 - 2.634CIV + 14.6$	$R = 0.8437$
	18%–19.99%	29.5.'13	$E_{vd} = 0.1683CIV^2 - 0.8741CIV + 9.4097$	$R = 0.8916$

ACKNOWLEDGEMENTS

This article is created with support of Slovak grant agency VEGA 1/0254/15 Implementation of new diagnostic measurements for the project of the optimize life of roads.

This article came to existence thanks to support within the frame of OP Education for project Support of quality of education and research for area of transport as an engine of economics, (ITMS: 26110230076), which is cofinanced from sources of European Social Fund.

The research is supported by European regional development fund and Slovak state budget by the project "Research centre of University of Žilina", ITMS 26220220183.

REFERENCES

Decký, M., Drusa, M., Zgútová, K., Vangel, J., Trojanová, M. Benč, G. & Starší, B. 2009. Design and Quality Control of Earth Structures at Civil Engineering Projects. Scientific monograph. Zilina: BTO print. 479 p.

Ďureková, D., Šrámek, J. & Danišovič, P. 2015. Non-destructive verification methods of eartworksquality. Magazine SSaM 2/2015, Bratislava: Weltprint. In process of reviewing.

Zgútová, K., Decký, M., Šrámek, J. & Ďureková, D. 2012. Non-destructive method of controlling compaction of earthworks using device Humboldt. Conference Testing and Quality in Civil Engineering.

Part B: Buildings and structures, Water supply and drainage, Construction technology and management, Materials and technologies, and Environmental engineering

Advances and Trends in Engineering Sciences and Technologies – Al Ali & Platko (Eds)
© 2016 Taylor & Francis Group, London, ISBN: 978-1-138-02907-1

Building shells made up of bent and twisted flat folded sheets

J. Abramczyk
Department of Architectural Design and Engineering Graphics, Rzeszow University of Technology, Poland

ABSTRACT: The work outlines possibilities of shaping roof forms made up of flat profiled sheets transformed freely into shell shapes during their assembling to shell directrices. A unidirectional folding allows each sheet fold to be a structural member, which may be bent, twisted or sheared. The use of the above folding for building covers requires solving many issues including related to the restrictions referring to: a) shape changes of the folds that tend to preserve the rectilinearity of their axes, b) strength and buckling changes, c) modelling these changes. A form diversification of all folds of an individual shell resulting from various curvatures and a mutual position of its directrices is a phenomenon, which enables us to create innovative, original and diversified shell structures.

1   INTRODUCTION

1.1   *Flat folded steel sheets transformed into shells*

A flat folded sheet is characterized by orthotropic properties resulting from its unidirectional folding. Its great intentional bending and torsional transformations into a shell shape can be easily realized at right angles to the directions of its folds (Figure 1a, b) (Abramczyk 2011, Reichhart 2002) opposite to the directions of the fold axes, where deformations are small.

Crosswise stiffness of each fold is smaller than its longitudinal stiffness by about one order of magnitude, so that it is possible to deform the fold shape from flat into shell by applying very small transversal force. The easiness of these crosswise deformations results from the fact that: a) flat walls (flanges and webs) of each fold can be easily bent or twisted out of their planes, b) the cross sections of each fold are open (Figure 2).

The intentional transformation the shape changes of the fold from planar into spatial while its assembling to the shell directrices is called an initially deformation. Only such a type of deformation of the shell fold carried out before a useful load is considered in the paper. If a technique of assembling the fold in the shell assures a freedom of its crosswise width increments, then its deformation is called a free deformation. In the case of such a deformation, the initial effort of the fold is possibly small, so its resistance to dead load does not decrease excessively and the

Figure 1.   The intentional transformation: a) an experimental shell; b) a structure composed of two individual shells.

Figure 2. Twisted and bent flanges and webs of a numerical model of the transformed sheet.

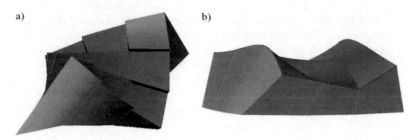

Figure 3. Discontinuous and continuous shell structures.

transformed fold is able to work as a structural member (Reichhart 2002). The proper manner of assembling the fold ends to the directrices, ensuring free deformation of each fold in the shell, enables us to omit a bimoment and bending-torsional moment of a freely deformed thin-walled fold in its stress calculations.

The free deformation is realized to locate the fold in a shell before the useful load. We may assume big displacements and small strains of its walls, so we should expect nonlinear geometrical and strength characteristics of the deformed fold for its big deformations (Abramczyk 2014a).

In addition, the deformation usually causes a diversification of cross section shapes on the length of each fold. The neutral axes of all shell folds tend to preserve their rectilinearity during the free deformation (Abramczyk 2011). Thus, it is possible to shape many various forms of individual shells made up of the freely deformed sheets connected to one another with their longitudinal edges, however, they have to be rectilinear (Figure 1b) (Reichhart 2002).

1.2 *Compound shells*

In order to shape a curvilinear form, many individual rectilinear shells have to be put together with their transversal edges into a ribbed or discontinue structure (Figure 3a, b).

It is worth stressing that the form of the shell made up of the flat sheets transformed into shell shapes is achieved as a result of assembling crosswise ends of all folds to shell directrices (Abramczyk 2014b), so the form is dependent on: a) curvatures and a mutual position of the shell directrices, b) geometrical and physical properties of the folded sheet, c) a mutual position of the folds and directrices, d) the technique of the fold assembly.

2 THE OBJECTIVE OF THE RESEARCH

2.1 *The aim and the concept*

The main objective of the paper is to present the usage of chosen geometrical and strength properties of the flat folded sheet transformed freely into a shell shape for shaping building shells characterized by original forms. The dependencies of a type and degree of the free deformation on the shape and strength changes of each fold of the flat sheet with regard to its work as a structural member are also sought. Each shell fold is taken as an open beam (Reichhart 2002).

To obtain the freely transformed folds, the adequate technique of the fold assembling to the shell directrices has to be applied. In this case, the longitudinal force $N(x)$, shearing forces $Q_y(x)$, $Q_z(x)$, bending moments $M_y(x)$, $M_z(x)$, the bimoment $B(x)$ and bending-torsional moment $M_\omega(x)$ (Maguncki & Ostwald 2005) are equal to zero.

The bending moment M_x causing crosswise fold bent in relation to the fold direction is small and we can assume $M_x \cong 0$.

The shear stress (Maguncki & Ostwald 2005) linearly varies across the fold wall thickness as follows:

$$\tau_{sv} = \frac{M_{sv}(x)}{J_t} t \tag{1}$$

where t – thickness of fold's walls, $M_{sv}(x)$ – the Saint-Venant's torsion moment; J_t – geometric stiffness for the Saint-Venant's torsion of the cross section.

The Saint-Venant's torsion moment defined at the cross section of the fold, and presented in (Obrebski 2005), can be modified as follows:

$$M_{sv} = G \cdot J_t \cdot \frac{d\theta}{dx} \cdot \beta_M(\alpha) \tag{2}$$

where G – module of elasticity of the fold material; $\theta(x)$ – angle of rotation of the fold cross section with respect to the x axis; $\beta_M(\alpha)$ – function of the twist angle α, resulting from free and big deformation of each shell fold.

The maximum shear stress, given in (Jastrzębski & Mutermilch & Orłowski 1985), for rectangular thin-walled cross section can be written in the modified form:

$$\tau_{max} = G \cdot t \cdot \frac{d\theta}{dx} \cdot \beta_t(\alpha) \tag{3}$$

where $\beta_t(\alpha)$ – function of the twist angle α, resulting from free and big deformation of the fold.

Finally, we the strength condition (Bazant & Cedolin 1991) for an open-profile thin-walled fold is as follows (Maguncki & Ostwald 2005):

$$\tau_{sv}\sqrt{3} < \sigma_{al} \tag{4}$$

where σ_{al} – allowable stress should be much greater than the initial effort if the fold is expected to carry useful load.

In spite of many significant restrictions associated with the shape changes of the freely deformed fold, which are described in the next parts of the paper, the possibility of shaping diversified and original shell forms is great due to various types and degree of the free deformations resulting primarily from the curvatures and mutual position of the shell directrices. The restrictions result from: a) open cross sections, b) the requirement that the deformations have to preserve the recti-linearity of the neutral axis and the possible small effort of each transformed fold so that the fold could be a structural member of the shell and capable of carrying the dead loads. In this case, an effective shell stress should be very small in relation to the allowable stress.

2.2 *Critical analysis*

Many researchers (Davis & Bryan 1982, Nilson 1962) from outside the Rzeszow centre have pointed out that: a) the possibilities of shaping diversified forms are firmly restricted, b) it is impossible to shape the shells different from sectors of cylindrical surfaces, conoids or hyperbolic paraboloids by means of the deformed flat folded sheets on account of the difficulties related to the assembly of the folds to the shell directices and the preservation of their rectilinearity. They usually reduce the designed forms by an initial acceptance of a representative of one of the above simple surfaces. Thus, during the assembly the fold widths are adapted to the calculated positions of the rulings of the initially accepted surfaces by additional stretching or compressing crosswise fold ends relative to the directrices. Such an action is ineffective because it causes the shell fold strength to increase.

Adam Reichhart overcame the restrictions resulting from the specificity of the fold shape changes during a freely torsional deformation (Reichhart 2002). He elaborated an algorithm which makes it possible: a) to calculate the positions of the rectilinear rulings of the proper ruled surface modelling free and innovative shell form with taking into account the character of the shape changes of the freely twisted folds, b) to obtain an effective adaption of the fold shapes to the curvatures and mutual position of the shell directrices without using additional forces by stretching or compressing the crosswise fold ends.

He defined two kinds of the free deformations: a) a crosswise bending deformation called a free bent deformation, b) a torsional deformation called a free twist deformation.

He also very generally defined a free bending-torsional deformation as causing the fold to be bent and twisted. However, he didn't defined: a) the way of determination of the degrees of the fold bent and twist, b) a way of creating of a model of such a deformed fold, c) the way of dimensioning of the freely bent and twisted folds. He replaces the models of the freely bent and twisted fold with the models of the freely twisted folds making capital of the neutral surface of each shell fold. In addition, his method uses the models reduced to peculiar, central sectors of a peculiar type of ruled surfaces known as right ruled paraboloids. In some cases, such an assumption appears incorrect because the shaping precision is inaccurate.

Such a defect does not appear during the free bend and twist deformation defined by Jacek Abramczyk. Each model of such deformed fold is created by means of: a) a helix corresponding to the fold contraction; b) straight lines perpendicular to the helix referring to the fold's neutral axes. Two models, created this way, of the consecutive freely bent and twisted folds allows us for creating a smooth connection of the models of the consecutive shell folds in the shell instead of discontinuous displacement of two sectors of right ruled paraboloids. In this case, the shaping precision increases as well as the cooperation of two adjacent folds in the shell can be modelled.

3 NUMERICAL RESEARCH

3.1 *The assumptions*

In order to investigate into the shape and effort changes of the freely bent and twisted profiled sheet, its numerical model and border constrains were defined.

The profile T $85 \times 0.88 \times 5100$ mm is modelled as a sum of flat rectangular sides referring to the central surface of this profile. Its numerical model is created by applying a thickness of 0.88 mm to these sides.

The freedom of its crosswise width increments was assured strictly because the sheet was suspended with flexible long bars to a stiff frame (Figure 2).

The flat profiled sheet was subjected to the true free bend and twist deformation while the numerical investigations carried out by means of the ADINA application (Bathe 2006).

3.2 *The early research*

The results of the researches associated with this virtual model and theoretical analyses carried out on this model can be only fragmentary compared with results of the experimental researches performed earlier for a different purpose in the laboratory hall of the Rzeszow centre.

In the case of the experimental research (Wasilewski 2013), each consecutive fold was deformed after stiff fastening of the previous fold to the shell directrices, so the assembled fold is not able to deform freely (Figure 4a). The values of the fold twist angle are shown on the axis of abscissae but the values of the maximum effective stress in the bottom flange at the crosswise fold end are shown on the axes of ordinates.

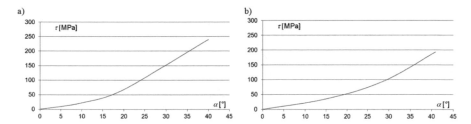

Figure 4. The relationship between the twist angle and the effective stress of: a) the not completely freely bent and twisted experimental fold; b) the numerical model of the freely bent and twisted fold.

3.3 *The obtained results*

The analogous relationship between the fold twist angle and maximum effective stress in the bottom flange at the crosswise fold end, obtained during numerical research is presented in the diagram (Fig 4b).

The results obtained during both researches are coincident as regards the tendencies but they are different from themselves as regards their values. The difference is the effect of the diversified border conditions of the sheets deformed during both examinations.

Each freely bent and twisted sheet contracts at half its length and extends identically at both crosswise ends (Figure 1a). Its edge model is often created by means of the neutral axes of its folds, directrices supporting its both crosswise ends and its line of contraction.

Its cross sections slightly deplane out of their planes. The central cross sections of all shell folds are located in accordance with a shell contraction.

Each fold tends to preserve the perpendicularity between its neutral axis and the plane of its cross sections. Preservation of this perpendicularity is difficult at half its length. Therefore, the fold deformation should be free.

4 THE ANALYSE ON MODELLING THE FREELY BENT AND TWISTED SHEETINGS

Properties of helical surfaces meet the requirements presented earlier and referring to a satisfying measurement description of the freely bent and twisted folds of the designed shell. The line of striction of the helical surface is a helix. Its rulings are straight lines binormal to the helix. The osculating planes of the helix correspond to the planes of the cross sections of the shell fold.

The parametric conditions of the helical surface are as follows:

$$x = a \cdot \cos(u) + \frac{b \cdot \sin(u)}{\sqrt{a^2 + b^2}} \cdot v \tag{5a}$$

$$y = a \cdot \sin(u) + \frac{-b \cdot \cos(u)}{\sqrt{a^2 + b^2}} \cdot v \tag{5b}$$

$$z = b \cdot u + \frac{a}{\sqrt{a^2 + b^2}} \cdot v \tag{5c}$$

where a – the radius of the helix being the line of striction of the helicoid; b – the spiral lead of the helix; and u, v – two independent variables of the helicoid.

If we take the zero value of the variable v, then we will obtain the parametric conditions of the helix.

The dependence of the curvature of the helix of the helical surface on the twist angle α of the fold is presented in Figure 5a. The dependence of the spiral lead of the helix of the helical surface on the twist angle α of the fold is shown in Figure 5b. The dependence between the twist angle α and the variable β_t of the considered fold profile is presented in Figure 5c.

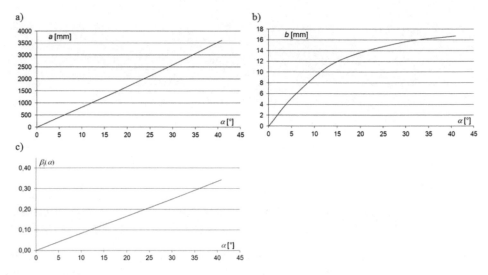

Figure 5. The characteristics of the helicoid formed from the freely bent and twisted folds.

5 CONCLUSIONS

The geometrical properties of helicoids surfaces enable the author to describe the shape changes of the folded sheets connected to each other with their longitudinal edges as a flat strip and, next, freely bent and twisted into shell shapes to obtain an individual shell.

This particular kind of ruled surfaces modelling the shells made up of the freely bent and twisted folds was distinguished on the basis of the shape and strength changes observed during the numerical research and experimental research carried out in the Rzeszow centre.

The innovative method of shell shaping creates new possibilities of wider application of diversified forms of the building shells made up of the freely transformed profiled sheets. This way shaped shell is characterized by a possibly small initial effort and original shell forms.

REFERENCES

Abramczyk, J. 2011. *An influence of shapes of folded sheets and their directrices on the forms of the building covers made up of these sheets*. Rzeszow: Doctoral thesis.

Abramczyk, J. 2014a. Some examinations on behavior of profiled sheets freely twisted in a shell. *Lightweight Structures in Civil Engineering; Proc. XX Intern. Sem. of IASS Polish Chapter: 6–14*. Warsaw: MICRO-PUBLISHER-C-P Jan B. Obrebski.

Abramczyk, J. 2014b. Principles of geometrical shaping effective shell structures forms. *Journal of Civil Engineering, Environment and Architecture* XXXI (61), 5–21.

Bathe, K.J. 2006. *Finite element procedures*. Cambridge: MA Klaus-Jurgen Bathe.

Davis, J.M. & Bryan, E. R. 1982. *Manual of stressed skin diaphragm design*. London: Granada.

Gioncu, V. & Petcu, D. 1995. Corrugated hypar structures. *Proc. Intern. Conf. IASS/LSCE: 637–644*. Warsaw: MICRO-PUBLISHER-C-P Jan B. Obrebski.

Jastrzębski, P. & Mutermilch, J. & Orłowski, W. 1985. *Wytrzymalosc materiałów*. Warszawa: Arkady.

Maguncki, K. & Ostwald M. 2005. *Optimal design of selected open cross sections of cold-formed thin-walled beams*. Poznan: Publishing House of Poznan University of Technology.

Obrebski, J. B. 1991. *Cienkoscienne prety proste*. Warsaw: Publishing House of Warsaw University of Technology.

Reichhart, A. 2002. *Geometrical and structural shaping building shells made up of the transformed flat folded sheets*. Rzeszow: Publishing House of Rzeszow University of Technology.

Wasilewski, A. 2013. *Wpływ krzywizny powloki z blach na jej sztywnosc*. Rzeszow: Diploma thesis.

Advances and Trends in Engineering Sciences and Technologies – Al Ali & Platko (Eds)
© 2016 Taylor & Francis Group, London, ISBN: 978-1-138-02907-1

Sulphate removal from mine water – precipitation and bacterial sulphate reduction

M. Balintova, S. Demcak & M. Holub
Civil Engineering Faculty, Košice, Slovakia

ABSTRACT: The Slovak Republic is a country with long mining traditions. Nowadays in Slovakia exist a few localities with prevailing acid mine drainage generation conditions (either in the form of direct outflow from mine or leachate from tailing ponds). High concentrations of heavy metals, sulphates and low pH are limiting for many various treatment technologies in these acidic waters. The paper deals with methods to remove sulphate from water by precipitation and bio-chemical technology with the aid sulphate reducing bacteria. Acid mine drainage, from the flooded Pech shaft (Smolník), contained a sulphate concentration of about $3200\,mg.L^{-1}$. Part of the sulphate was removed by precipitation. Sulphate reducing bacteria was used for the residual $1200\,mg.L^{-1}$ of sulphate. After applying sulphate reducing bacteria the concentration of sulphates in solution was decreased by 30%.

1 INTRODUCTION

Sulphates can cause various kinds of problems for environment and human health depending on its concentration and on the earth alkaline cations (Balintova and Petrilakova, 2011). The World Health Organisation (WHO) has established a $250\,mg.L^{-1}$ maximum tolerable level of sulphate in water, but many countries have recommended lower values. That is also the case for Slovakia, where the allowable concentration of sulphates in purified wastewaters is under $250\,mg.L^{-1}$ – Government Regulation no. 269/2010. The largest producers of wastewater containing sulphates are the paper industry, textile industry, metallurgical industry and mining. Acid Mine Drainage (AMD) is a big problem in the mining industry which is formed after terminating mining activity. It causes surface water pollution, which may impact aquatic life as well as the whole ecosystem (Younger et al., 2002). Methods that are used to remove sulphates are physicochemical (precipitation, membrane process, ion exchange and sorption) (Benatti et al., 2009; Mansoura et al., 2009) and biological chemical (sulphate reducing bacteria) (Luptakova et al., 2013).

AMD are formed during the weathering of sulphide minerals under oxidation conditions (in contact with water, atmospheric oxygen, micro minerals and aerobic bacteria) (Montero et al., 2005; Xie et al., 2009). AMD acidic effluents are characterized by a very low pH value of about 2 and above the limit concentrations of heavy metals (Cu, Zn, Cd, As, Mn, Al, Pb, Ni, Ag, Hg, Cr and Fe), toxic elements and sulphates (Kontopoulos, 1998). High levels of AMD cause acidification and metal contamination of large areas of land, water and damage the health of wildlife (Robinson-Lora and Brennan, 2009). Importantly, once AMD has formed, it can remain for hundreds of years, and it is generally difficult and costly to control (Alakangas et al., 2013). For these reasons, it is urgent to develop novel approaches for the efficient control of AMD production.

The aim of this article was to study sulphate removal from mine water in Smolnik using sulphate reducing bacteria (SRB). High concentrations of heavy metals, mainly iron, have a negative impact on the degradation process of bacteria. This element was removed from AMD by oxidation and sorption (Kusnierova et al., 2014). SRB reduces the sulphide to hydrogen sulphide via a respiration

Table 1. Properties of real sample of AMD (input concentration).

Type	Fe_{total} [mg.L^{-1}]	SO_4^{2-} [mg.L^{-1}]	pH
AMD	270.0	2200	4.0

process. This process is time limited but has economical benefits and high level of wastewater treatment can be reached (Luptakova et al., 2012).

2 MATERIAL AND METHODS

2.1 *Acid mine drainage samples*

Real sample of AMD is containing a high level of sulphates and iron concentration (Table 1). For experimentation the sample was collected from the abandoned subsurface mine in Smolnik (Slovakia).

2.2 *Precipitation and oxidation*

High concentrations of iron cations require the pretreatment of AMD samples before the experiment. For this reason iron cations were removed by oxidation using of 31% H_2O_2. Subsequent precipitation with a 0.1 M NaOH adjusted pH on input value. The following reaction takes place during the experiment:

$$2Fe^{2+} + H_2O_2 \rightarrow 2Fe^{3+} + 2OH^-$$ (1)

The solution was filtered and concentrations of iron cations and sulphates were determined. Precipitate was dried in a laboratory oven at a temperature of $105 \pm 1°C$ and tested by FTIR method.

2.3 *Sulphate reducing bacteria*

The process of reduction by sulphate reducing bacteria (SRB) requires an optimal pH ≈ 7 and it is negatively affected by the presence of heavy metals. For removal of the remaining heavy metals and adjustment pH used an inorganic sorbent SLOVAKITE (adding 1 gram per 100 millilitres) (Holub et al., 2014). The solution was filtrated after 24 hours sorption. The concentration of heavy metals and sulphates in the filtrate was determined by colorimeter DR890 (HACH LANGE, Germany).

A bacterial culture of SRB (*Desulfovibrio sp.* and *Desulfotomaculum sp.*) was isolated from potable mineral water (Gajdovka spring, Kosice-north, Slovakia), which was also used for the experiments (Luptakova and Kusnierova, 2005). It was selected from the mixed cultures by Postgate's method (Postgate, 1984) and modified using the dilution method (Karavaiko et al. 1988).

For the experiments 200 mL of the solution was prepared. The composition of this solution is described in Table 2. The control solution is marked "K" which was without bacteria and "SRB" indicates the solutions with bacteria. K1 and SRB1 represent the nutrient medium enriched about sulphates due to the increase of sulphate concentration in the solution. K2 and SRB2 represent the nutrient medium without sulphates (Luptakova et al., 2013; Postgate, 1984).

The control solutions K1 and K2 without sulphate bacteria and the solutions SRB1 and SRB2 which contain the mentioned bacteria were used for determining the biochemical reduction. Solution SRB2 contained the reached sulphates of matrix medium.

Table 2. Compositions of solutions which were used in experiment with SRB.

Indication of solution	Composition
K1	100 ml treated AMD, 100 ml nutrient medium (with sulphates)
SRB1	100 ml treated AMD, 80 ml nutrient medium (with sulphates), 20 ml SRB
K2	100 ml treated AMD, 100 ml nutrient medium (without sulphates)
SRB2	100 ml treated AMD, 80 ml nutrient medium (without sulphates), 20 ml SRB

Table 3. Concentrations of AMD sample (input concentrations and concentrations after iron precipitation).

	Concentrations [mg.L^{-1}]	
	SO_4^{2-}	Fe_{total}
Input data (pH $= 4.0$)	2200	270
After iron oxidation and precipitation (pH $= 3.8$)	1200	0.03

2.4 Apparatus and instrumentation

A Colorimeter DR890 (HACH LANGE, Germany) with appropriate reagents was used to determine the dissolved Fe_{total} concentration.

Spectromom195 instrument for nefelometric method was used to measure the sulphate concentration during the sulphate reducing bacteria experiment. The absorbance of the sample was measured at a wavelength of 490 nm.

pH values were determined by pH meter inoLabph 730 (WTW, Germany) (pretreatment AMD experiment).

IR spectra were measured by Alpha FT-IR Spectrometer with ALPHA's Platinum ATR single reflection diamond ATR module (Bruker, Germany).

Also c, % ion removal, was calculated using the following equation:

$$c = \frac{(c_0 - c_e)}{c_0} 100\%$$

(2)

where c_0 is the initial concentration of appropriate ions (mg.L^{-1}), c_e is equilibrium concentration of ions (mg.L^{-1}).

3 RESULTS AND DISCUSSION

3.1 Pretreatment AMD

Pretreatment of AMD was important for furtherer experiments with SRB. The primary objective was to remove the high concentration of iron cations from AMD due to their negative influence on the subsequent experiments. The iron cations were precipitated by hydrogen peroxide to form an insoluble solid. This part of sulphates was also removed. It can be supposed that a dependence exists between sulphates and iron in water solution.

The removal efficiency for iron was 99.99% and sulphate 45.45%. Input concentrations and concentrations after iron precipitation are shown in Table 3.

Low values of pH after oxidation of iron (Table 3) and the presence of heavy metals (mainly Cu, Al, Mn, Mg and Zn) in AMD can lead to the inhibition of the process with bacteria. For heavy

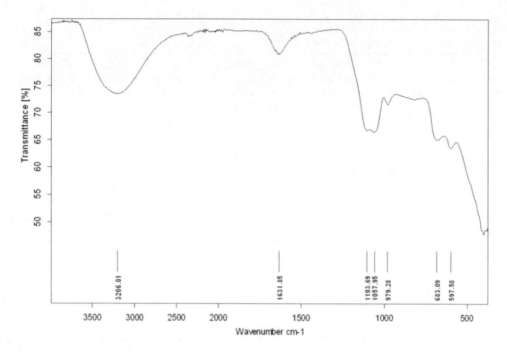

Figure 1.   Infrared spectra of precipitate after iron oxidation.

metal removal from the solution after pretreatment AMD inorganic sorbent SLOVAKITE was used. After 24 hours of sorption the pH of mixture was of 7.1.

The presence of iron (III) hydroxy sulphate in the precipitate was confirmed by infrared spectrum measurement (shown in Figure 1). The IR spectrum includes a broad OH-stretching band centred in area at $3200 \, cm^{-1}$. Intense bands at 1100, and $1060 \, cm^{-1}$ reflect a strong splitting of the wave number $\nu 3 \, (SO_4)$ fundamental due to the formation of a bidentate bridging complex between SO_4 and Fe. This complex may result from the replacement of OH groups by SO_4 at the mineral surface through a ligand exchange or by the formation of linkages within the structure during nucleation and subsequent growth of the crystal. Related features due to the presence of structural SO_4 include band at 680 and $600 \, cm^{-1}$ that can be assigned to $\nu 4 \, (SO_4)$ (Pallova et al., 2010).

3.2   *Sulphate reducing bacteria*

For the nutrient medium without sulphates (K2, SRB2) the initial concentration in solution was about $1200 \, mg.L^{-1}$. Whereas for the nutrient medium enriched with sulphates (K1, SRB1), the concentration was increased from $1200 \, mg.L^{-1}$ to $1950 \, mg.L^{-1}$.

As it is clear from Figure 2, the significant decrease of sulphates concentration occurred in both cases after the ninth day. This fact can be explained by the lack of culture medium, sodium lactate, consumed by bacteria and/or the final product of degradation, hydrogen sulphide, inhibited the process of reduction of sulphates. These factors caused the end of the mentioned process. In both cases, because of too high an initial concentration of sulphates related to the prepared nutrient medium, the process of sulphates treatment was ineffective. The final concentration of sulphates in SRB2 at the end of the experiment was of $853 \, mg.L^{-1}$ which represents a decrease of less than 30%.

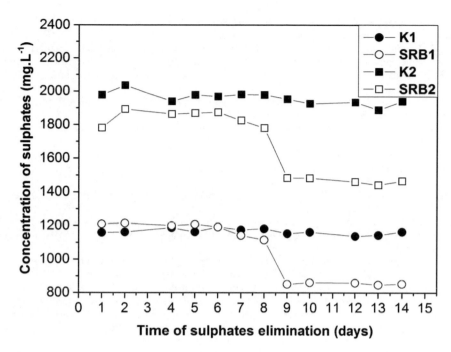

Figure 2.   Elimination of sulphates from water solutions by the sulphate reducing bacteria (matrix with and without sulphates).

4   CONCLUSION

Removal of sulphates from wastewater is an actual problem and most methods are used are economically and technologically difficult. One modern, cheap and effective technology use sulphate reducing bacteria for sulphate removal. However the biological removal of sulphates in AMD depends on more factors as concentration of heavy metals and value of pH.

Partially decreasing of the concentration of sulphate anions (45.45%) was obtained together with iron precipitation by the oxidation of Fe^{2+} to Fe^{3+}.

The biological part observed a dependence between sulphate removal and its concentration in an aquatic solution. Sulphate reducing bacteria spent all nutrient medium (lactate sodium) and the process of reduction was stopped. Hydrogen sulphide (product of degradation) inhibited reduction too. Produced hydrogen sulphide can be used to precipitate heavy metals from the mine water or to the production of elemental sulphur.

ACKNOWLEDGEMENT

This work has been supported by the Slovak Grant Agency for Science (Grant No. 1/0563/15).

REFERENCES

Alakangas L., Andersson E., Mueller S., 2013, Neutralization/prevencion of acid rock drainage using mixtures of alkaline by-products and sulfidic mine wastes, Environ. Sci. Pollut. Res. 20, pp. 7907–7916.
Balintova M., Petrilakova, A., 2011, Study of pH Influence on Selective Precipitation of Heavy Metals from Acid Mine Drainage, Chemical Engineering Transactions, Vol. 25, pp. 1–6.

Benatti C.T., Granhen Tavares C.R. & Lenzi E., 2009, Sulfate removal from waste chemicals by precipitation, Journal of Environmental Management, vol. 90, issue 1, pp. 504–511, 2009.

Holub M., Balintova M., Demcak S., Pavlikova P., 2014, Application of various methods for sulphates removal under acidic conditions, In: SGEM 2014: 14th International multidisciplinary scientific geoconference: GeoConference on Ecology, Economics, Education and Legislation: conference proceedings: vol. 2, pp. 39–45, ISBN 978-619-7105-18-6.

Karavaiko G.I., Rossi G., Agate A.D., Groudev S.N., Avakyan Z.A., 1988, Biotechnology of metals – manual, Centre for International projects GKNT, Moscow, Russia.

Kontopoulos A., 1998, Acid mine drainage control. In: Effluent treatment in the mining industry, University of Concepcion, Chile, p. 57.

Kusnierova M., Prascakova M., Nowak A.K., Gorazda K., Wzorek Z., 2014, Biogenic catalysis in sulphide minerals weathering processes and acid mine drainage genesis, In Acta Biochimica Polonica, vol. 61, no. 1, p. 33–39. (1.389 – IF2013). ISSN 0001-527X.

Luptakova A., Kusnierova M., 2005, Bioremediation of Acid Mine Drainage by SRB, Hydrometallurgy, 77, pp. 97–102.

Luptakova A., Ubaldini S., Macingova E., Fornari P., Giuliano V., 2012, Application of physical-chemical and biological-chemical methods for heavy metals removal from acid mine drainage, Process Biochemistry, vol. 47, No. 11, pp. 1633–1639.

Luptakova A., Kotulicova I., Macingova E. , Jencarova J., 2013, Bacterial elimination of sulphates from mine waters. In: Chemical Engineering Transactions, vol. 35, pp. 853–858.

Mansoura, C., Lefèvreb, G., Pavageaua, E.M., Catalettea, H., Fédoroff, M. & Zannac, S., 2009, Sorption of sulfate ions onto magnetite, Journal of Colloid and Interface Science, vol. 331, issue 1, pp. 77–82.

Montero, S.I.C., Brimhall, G.H., Alpers, C.N., Swayze, G.A., 2005, Characterization of waste rock associated with acid drainage at the Penn Mine, California, by ground-based visible to short-wave infrared reflectance spectroscopy assisted by digital mapping. Chem. Geol., 215, pp. 453472

Pallová Z., Kupka D., Achimovičová M., 2010, Metal mobilization from AMD sediments in connection with bacterial iron reduction, Mineralia Slovaca, vol. 42, pp. 343–7.

Postgate J.R., 1984, The sulphate-reducing bacteria, Cambridge University Press, Cambridge, United Kingdom.

Regulation of the Government of the Slovak Republic no. 269/2010, which lays down the requirements for achieving good condition of waters (in Slovak) <http://www.zbierka.sk/sk/predpisy/269-2010-z-z.p-33654.pdf> accessed 8.4.2015.

Robinson-Lora M.A., Brennan R.A., 2009, Efficient metal removal and neutralization of acid mine drainage by crab-shell chitin under batch and continuous-flow conditions, Bioresour. Technol., 100, pp. 5063–5071.

Xie X.H., Xiao S.M., Liu J.S., 2009, Microbial communities in acid mine drainage and their interaction with pyrite surface. Curr. Microbiol., 59, pp. 71–77.

Younger P.L., Banwart S.A., Hedin R.S., 2002, Mine Water: Hydrology, Pollution, Remediation, Environmental Pollution, Vol. 5, Kluwer academic Publishers, Dordrecht, Netherlands.

Advances and Trends in Engineering Sciences and Technologies – Al Ali & Platko (Eds)
© 2016 Taylor & Francis Group, London, ISBN: 978-1-138-02907-1

Arsenic removal from water by using sorption materials

R. Biela & T. Kučera
Institute of Municipal Water Management, Brno, Czech Republic

ABSTRACT: The article deals with the occurrence of arsenic in water, the health risks and the possibilities of arsenic removal from water by using sorption materials. In this work a laboratory experiment was carried out with water from an underground source with the increased content of arsenic. This water was filtered through a selected sorbents GEH and CFH 0818. The results show that these sorption materials can reduce the arsenic content below the over-limit values just after 2.5 minutes to a tenth of the limit value prescribed by the Regulation of the Ministry of Health 252/2004 Sb. valid in the Czech Republic. Furthermore, it has been proved that the sorption materials also have an effect on the removal of iron, manganese and turbidity from water.

1 INTRODUCTION

Arsenic is a widespread element which occurs in the environment in organic and inorganic forms. It is a toxic semi-metal found in four allotropic modifications: yellow, brown, black and grey. It mainly occurs in the environment in the form of sulphides (arsenicopyrite $FeAsS$, realgar As_4S_4, auripigment As_2S_3) and it is a frequent component of various types of soil and rock. It occurrence in rocks is mainly in the form of components of nickel, cobalt, silver, gold and iron ores, and it is sometimes found as a trace element in many coal deposits. Inorganic arsenic is usually found in water as a result of rock scouring and weathering, through wastewater and atmospheric deposition. It is very often a common component of groundwater as well as surface water. It has a strong ability to accumulate in river sediments (Pitter 2009).

Arsenic is strongly poisonous and long-term consumption of water with low As concentrations results in chronic diseases. Its toxicity is to a great extent dependant on the degree of oxidation. As(III) compounds are approximately five times up to twenty times more toxic than As(V) compounds. Its carcinogenic effects have been confirmed.

The most significant anthropogenic sources of arsenic are mining and ore industries, tanneries, applications of certain insecticides and herbicides, fossil fuel burning, wood preserving agents, power plant fly ash leachate and metallurgical industry (Pitter 2009), (Gray 2008).

2 OCCURRENCE OF ARSENIC IN WATER

Arsenic is a relatively common element occurring naturally in ground water. It is found both in the organic and inorganic forms. Under various redox conditions, arsenic is stable in the oxidation states of $+5$, $+3$, -3 and 0. As^{III} is stable and prevailing in water rich with oxygen and, on the contrary, the prevailing form in slightly reduction environment (e.g. in groundwater) is As^V. Pentavalent arsenic is found as AsO_3^{3-}, $HAsO_4^{2-}$ and $H_2AsO_4^{-}$. Trivalent arsenic occurs as $As(OH)_3$, $As(OH)_4^{-}$, AsO_2OH^{2-} and AsO_3^{3-}. The dominant bond of the pentavalent and trivalent arsenic is determined by the pH value of the water environment (Ramakrishna 2006).

As^{III} oxidation to As^V takes place chemically and biochemically. Fast and efficient reaction is ensured by chlorination, if chloramines are used, the reaction decelerates. Oxidation using oxygen dissolved in water is very slow, acceleration can be ensured by catalysing with copper compounds.

As^V reduction to As^{III} is relatively easy, through the addition of iron sulphate or sulphides. This reaction is very frequent in the hypolimnion of water reservoirs and lakes. Again, reduction using oxygen dissolved in water in anoxic conditions is very slow. Therefore, reaching an equilibrium takes quite a long time in still water and As^{III} can also be found in oxic conditions in the epilimnion and, on the contrary, As^V can be found in anoxic conditions in the hypolimnion. Depending on the water composition, the oxidation and reduction times reach tens of days (Pitter 2009).

The usual concentrations of arsenic in groundwater and surface water range from units up to tens of micrograms per litre. A natural value of arsenic concentration in groundwater is considered as $5\,\mu g.l^{-1}$. Mineral water in the Karlovy Vary springs in the Czech Republic contains approx. $150\,\mu g.l^{-1}$ of arsenic on average. The Glauber III spring in Františkovy Lázně contains approx. $800\,\mu g.l^{-1}$, it is the so-called "arsenic mineral water". Arsenic concentrations in sea water usually range between $1\,\mu g.l^{-1}$–$9\,\mu g.l^{-1}$. Wastewater discharged from large laundries contains arsenic in concentrations reaching up to $100\,\mu g.l^{-1}$. Extraordinarily high concentrations can be found in mining water in the vicinity of arsenic ore deposits (even over $1000\,\mu g.l^{-1}$). The average concentration of arsenic in drinking water in the Czech Republic is around $2\,\mu g.l^{-1}$ (Vosáhlo 2012), and the Regulation issued by the Ministry of Health of the CR No. 252/2004 Sb. permits the highest permissible arsenic content value in drinking water at $10\,\mu g.l^{-1}$.

3 ARSENIC HEALTH RISKS

Arsenic toxicity depends on its oxidation state. Trivalent arsenic is almost 60 times more toxic than pentavalent arsenic. Both these forms are more acutely toxic than most organic forms of arsenic (Vaishya & Gupta 2011). Organically bound arsenic is no particularly dangerous to people as it is excreted in urine after ingestion fast and almost without changes. Dissolved arsenic in inorganic form is quickly absorbed after ingestion and is to a large extent discharged only after the detoxication in the liver in a period of ca. 4 days, with the first step being methylation and formation of mono- and dimethylarsenic acids, which are even more toxic than inorganic arsenic. The degree of arsenic adsorption in humans in skin contact is not precisely known; however, experimental studies confirm low arsenic absorption through the skin during washing and an external arsenic bind in hair and probably also the skin (Pomykačová et al. 2010).

In some countries, arsenic is the most important toxic metal contained in groundwater and drinking water (Macrae et al. 2007). It is estimated that 226 million people (Murcott 2013) worldwide are exposed to the effects of arsenic in drinking water. Mass cases of chronic arsenic poisoning were observed for example in India, Taiwan and Bangladesh. As concentrations in groundwater used in these countries for drinking purposes reach up to $3.7\,mg.l^{-1}$, most groundwater in Bangladesh contain arsenic concentrations above $0.05\,mg.l^{-1}$, with a maximum of about $1.0\,mg.l^{-1}$. It is estimated that the groundwater is consumed for drinking purposes in Bangladesh by approx. 20 million inhabitants (Pitter 2009).

In terms of time effects, the effects of arsenic on health if ingested orally may be divided into acute and chronic. Acute arsenic poisoning resulting in death occurs if drinking water containing $60\,000\,ppb.l^{-1}$ of arsenic is consumed. The drinking of potable water containing 300–$30\,000\,ppb$ As.l^{-1} causes stomach and intestine irritation, nausea, vomiting and diarrhea. There is a drip in red and white blood cells accompanied by symptoms of fatigue, disrupted intestinal walls, burning sensation in palms and feet. Chronic effects occur during long-term oral ingestion, causing chronic poisoning, which is manifested by skin changes, in particular coarse skin on palms and feet, warts and changes in the vascular system (Koppová & Fletcher 2009). Even low concentrations during chronic effects may result in cancer (skin cancer, urinary bladder cancer, kidney cancer) and heart disorders. People drinking water containing $3\,\mu g.l^{-1}$ on a daily basis are 1:1000 more likely to contract urinary bladder or lung cancer. This likelihood is 3:1000 (Chuanyong & Xiaoguang 2009) at a concentration of $10\,\mu g.l^{-1}$.

The World Health Organisation (WHO) and the US Environmental Protection Agency (US EPA) both consider arsenic as a confirmed human carcinogen. As the mechanism behind the effects of arsenic in carcinogenesis is still not absolutely clear, there is an uncertainty over the effects of low arsenic doses. For the time being, it is not possible to decide whether there is an absolutely safe arsenic dose (Pomykačová et al. 2010).

4 ARSENIC REMOVAL FROM WATER

There are many methods to remove heavy metals from water. Currently, the most frequently used method is sorption using a granulated medium based on iron oxides and hydroxides. This is a selective, undemanding, economically acceptable and highly efficient method which can reduce arsenic concentration in water below the limit of $10\,\mu\mathrm{g.l^{-1}}$. The function is based on the principle of irreversible chemisorption of removed arsenic. The most common adsorbents include:

– GEH;
– Kemira CFH;
– Bayoxide E33.

GEH (Granulated Eisen Hydroxide) was developed by the Berlin University, Department of Quality Control in order to remove arsenic and antimony from water. It is produced by the German company GEH-Wasserchemie GmbH. The treatment technology consists of the contaminant adsorption to granulated ferric hydroxide (GEH sorbent) in a reactor through which the treated water flows (GEH Wasserchemie GmbH 2011). It is imported to the Czech Republic by Inform-Consult Aqua s.r.o, Příbram.

CFH adsorbent was developed by the company Kemira in Finland. It is a granulated medium based iron hydroxide oxide. The advantage of these materials consists in easy handling and almost no requirements for material storage. This material can be washed by water as well as air (Kemwater ProChemie s.r.o. 2012). It is imported to the CR by Kemwater ProChemie s.r.o., Bakov nad Jizerou. There are 2 types of this materials available in the market marked at CFH 12 and CFH 0818, differing in granularity – see Table 1.

Bayoxide is a granular medium based on iron oxide. It was developed by the company Severn Trent in cooperation with Bayer AG. The arsenic removal system is titled SORB 33. The advantage of this system is the removal of As^{III} and As^{V} together with iron and manganese removal. The reported water treatment ability is given at the arsenic content of $11–5000\,\mu\mathrm{g.l^{-1}}$ and iron content of $50–10\,000\,\mu\mathrm{g.l^{-1}}$ (Severn Trent Service 2014).

5 EXPERIMENTAL ARSENIC REMOVAL

The objective of the experiment was to compare the arsenic removal efficiency using two selected adsorption materials, i.e. GEH and CFH 0818. Properties of these filtration materials are provided in Table 2.

Two columns with an internal diameter of 4.4 cm were set up for the purposes of the experiment. The adsorption medium was filled onto the drainage layer made of glass beads so as to avoid clogging the control valves due to loose filtration material. The layer of the filtration medium in

Table 1. Granularity of filtration materials Kemira CFH (Kemwater ProChemie s.r.o. 2012).

Kemira CFH 12		Kemira CFH 0818	
Dispersion [mm]	Presence [%]	Dispersion [mm]	Presence [%]
2–0.85	92.7	2–0.5	97.6
<0.85	5.9	<0.5	2.4
>2	1.4	>2	0

Table 2. Adsorption material properties.

Parameter	Unit	GEH	CFH 0818
Chemical composition	–	$Fe(OH)_3 + \beta Fe\text{-}O\text{-}OH$	Fe-O-OH
Particle size	mm	0.2–2	1–2
Density	$g.cm^{-3}$	1.25	1.12
Specific surface	$m^2.g^{-1}$	250–300	120
pH working range	–	5.5–9	6.5–7.5
Grain porosity	%	72–77	72–80
Colour	–	dark brown – black	brown – brown red
Material description	–	humid grainy	dry grainy

Table 3. Raw water analysis.

Raw water

t [min]	pH –	T [°C]	Turbidity [ZF]	Fe [mg.l^{-1}]	Mn [mg.l^{-1}]	As [µg.l^{-1}]
0	7.01	12.7	12.4	4.47	0.454	26

Table 4. Analysis after filtration through the GEH adsorbent.

GEH

t [min]	pH –	T [°C]	Turbidity [ZF]	Fe [mg.l^{-1}]	Mn [mg.l^{-1}]	As [µg.l^{-1}]
2.5	7.08	13.7	2.79	0.16	0.063	1
7	7.05	13.6	1.96	0.16	0.052	1
15	7.30	13.6	2.06	0.16	0.027	<1

the GEH material was 0.55 m, and in the CFH 0818 material it was 0.58 m. For the purposes of filtration, we respected the conditions prescribed by the manufacturers of the adsorption materials. As these were new filtration materials, it was necessary to soak them in water for as minimum of 45 minutes before use as instructed by the manufacturers so as to release the residues from production.

The filters were then flushed bottom up using drinking water from the Brno public water supply system. To ensure the correct set-up of the flushing and the filtration itself, a flow meter with a throttle jet for fine flow rate control was used.

Water with an elevated arsenic concentration was extracted from a groundwater source, the Janovice borehole, in the vicinity of the town of Přelouč in the Pardubice Region in the Czech Republic. The raw water analysis is provided in Table 3.

During filtration, the flow rates changed according to the required retention time, which was 2.5 minutes, 7 and 15 minutes and the efficiency of arsenic removal from water was observed along with the performance of the adsorption materials removing elements that form integral parts of groundwater resources, being iron and manganese.

The results of the analysis after filtration through the GEH and CFH 0818 adsorbents are shown in Table 4 and Table 5. To measure pH, a digital pH meter was used as it can also measure the temperature of liquid. Turbidity was measured by a portable turbidimeter. A spectrophotometer was used to determine the concentrations of iron, manganese and arsenic.

Table 5. Analysis after filtration through the CFH 0818 adsorbent.

CFH 0818

t [min]	pH –	T [°C]	Turbidity [ZF]	Fe [mg.l^{-1}]	Mn [mg.l^{-1}]	As [µg.l^{-1}]
2.5	7.86	13.7	2.86	0.20	0.018	1
7	7.94	13.7	2.48	0.14	0.022	1
15	8.06	13.6	1.94	0.09	0.018	<1

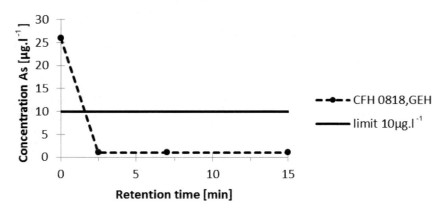

Figure 1. As removal efficiency using sorption materials.

The tables and charts in Fig. 1 indicate that both materials are excellent in removing arsenic and achieve identical sorption performance. Even at the shortest retention time (2.5 min), the arsenic concentration in treated water was below the limit of the highest permissible value as per Regulation 252/2004 Sb., valid in the Czech Republic. Longer retention times did not have any major influence on the arsenic removal. If the retention time was 15 minutes, the measured concentration was below one; however, the precise value cannot be measured by the instrument. Given the arsenic removal velocity, this is a contact filtration.

The experiment also determined that the applied sorption materials remove iron and manganese from groundwater. The GEH material eliminates iron right after the contact down to a value of 0.16 mg.l^{-1} and after a longer contact time, the removed iron concentration does not change. To remove manganese below the limit value for drinking water it is necessary to ensure filtration with a contact time longer than seven minutes. The CFH 0818 material eliminates iron depending on the time. The iron concentration values drop with the increasing filtration period. Manganese is removed in contact independently of the retention time to a concentration value of ca. 0.02 mg.l^{-1}. The applied sorption materials are also efficient for water turbidity elimination as shown in Figure 2.

6 CONCLUSION

Laboratory tests of arsenic removal from water were conducted as part of the specific research carried out at the Institute of Municipal Water Management, Faculty of Civil Engineering, Brno University of Technology in Brno. The results show that the GEH and CFH 0818 sorption materials can reduce the arsenic content below the over-limit values just after 2.5 minutes to a tenth of the

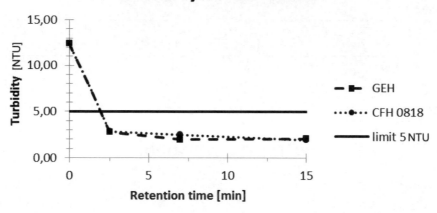

Turbidity elimination

Figure 2.   Efficiency of turbidity elimination by using the sorption materials.

limit value prescribed by the Regulation of the Ministry of Health 252/2004 Sb. valid in the Czech Republic. Furthermore, it has been proved that the sorption materials also have an effect on the removal of iron, manganese and turbidity from water.

REFERENCES

Chuanyong, J. & Xiaoguang, M. 2009. Nanoparticles for Treatment of Arsenic. *Nanotechnologies for Water Environment Applications:* 116–136.
GEH Wasserchemie GmbH. 2011. GEH 102 Arsenicentfernung. GEH Wasserchemie GmbH, Osnabrük. http://www.geh-wasserchemie.de/files/datenblatt_geh102_de_web.pdf
Gray, N.F. (2nd edition) 2008. *Drinking Water Quality. Problems and Solutions.* Cambridge: Cambridge University Press.
Kemwater ProChemie s.r.o. 2012. Kemira CFH12, CFH0818. Kemwater ProChemie s.r.o., Kosmonosy.
Koppová, K. & Fletcher, T. 2009. Hodnotenie zdravotných rizík z príjmu arzénu, odhad vplyvu expozície arzénu na riziko rakoviny obličiek, močového mechúra a bazocelulárneho karcinómu kože. *Sborník konference Pitná voda.* Trenčianské Teplice: 3–10.
Macrae, J. et al. 2007. Isolation and Characterization of NP4, Arsenicate-Reducing Sulfurospirillum from Maine Groundwater. *Environmental Engineering*: 81–88.
Murcott, S. 2013. Arsenic contamination: a worldwide call to action. *Water 21 – Magazine of the International Water Association*: 15–18.
Pitter, P. (4th edition) 2009. *Hydrochemie.* Praha: VŠCHT Praha.
Pomykačová, I. et al. 2010. Problematika arsenu v pitné vodě v České republice. *Sborník z X. ročníku konference Pitná voda 2010.* České Budějovice: W&ET Team, 145–150.
Ramakrishna, D. et al. 2006. Iron Oxide Coated Sand for Arsenic Removal: Investigation of Coating Parameters Using Factorial Design Approach. *Pract. Period. Hazard. Toxic Radioact. Waste Manage:* 198–206.
Severn Trent Service. 2014. Bayoxide – E 33 – Arsenic Removal Media. Severn Trent Service, Washington. http://www.environmental-expert.com/products/bioxide-e33-arsenic-removal-media-15343.
Vaishya, R. & Gupta, S. 2011. Arsenic(III) Adsorption by Mixed-Oxide-Coated Sand: Kinetic Modeling and Desorption Studies. Toxic Radioact. Waste 15, SPECIAL ISSUE – *Contaminant Mixtures: Fate, Transport and Remediation*: 199–207.
Vosáhlo, J. 2012. *Hodnocení kvality vody v úpravně vody Mokošín.* Master thesis. Brno: University of Technology.

The paper was prepared under the solution of the grant project on special research at BUT in Brno titled "Efficiency monitoring of water treatment processes in microcontamination elimination" (FAST-S-15-2701).

Advances and Trends in Engineering Sciences and Technologies – Al Ali & Platko (Eds)
© 2016 Taylor & Francis Group, London, ISBN: 978-1-138-02907-1

The consequences of shifting single wagon consignments to the road infrastructure and its environmental impact

V. Cempírek, I. Drahotský & P. Novák
Jan Perner Faculty of Transport, Pardubice, The Czech Republic

ABSTRACT: This paper contains the conclusions of a study, which was conducted for a rail cargo carrier. One of the main activities of the carrier is the transportation of single wagon consignments as well as the set of more complete wagon loads. It is the transport of the cargo loaded in one carriage or in a more wagons (maximally five wagons) with single consignment note (way-bill). This business segment is for business entity economically unprofitable because revenues do not cover the total costs incurred for this activity. If this business segment was repealed, the entire transport volume of goods would be transferred to road cargo. The main aim of this study was to analyze the single consignments (i.e. one complete wagon loads) as well as complete wagon loads transportation. The analysis is focused on the period from 2008 to 2013.

1   INTRODUCTION

National cargo carriers in most of European countries face the inefficiency in the segment of single wagon consignments. The share of these consignments on the total transport volume is almost the same, approx. 30%–35%. Thus the share of the complete train loads reaches 65%–70% by these cargo carriers. The rail cargo liberalization enabled the entry of dozens of new cargo carriers (private companies) into the transport market. Is causes the drop of the volumes of the former national carriers in the segment of the complete train loads and they record a loss then.

The data mentioned in the table imply that, in case of the share of single consignments reaching approximately 35%, these are huge volumes which would be shifted to the road cargo. The share of railway cargo reaches 18% on the total transport market in the EU countries, road cargos share is 75%.

Table 1.   Transport volume in chosen European countries in thousands of tons.

Country	2008	2009	2010	2011	2012	2013
Germany	371,298	312,087	355,715	374,737	366,140	373,738
Spain	26,906	21,087	21,578	23,899	24,903	24,949
France	108,536	86,126	85,045	91,789	87,539	88,989
Italy	95,810	76,336	84,435	91,811	88,505	87,960
Hungary	51,543	42,277	45,794	47,424	46,884	–
Austria	121,579	98,887	107,670	107,587	100,452	95,449
Poland	248,860	200,819	216,767	248,606	230,878	232,596
Slovakia	47,910	37,603	44,327	43,711	42,599	48,401
Switzerland	69,864	61,848	63,989	65,038	60,270	64,999

Source: http://ec.europa.eu/eurostat/

Table 2. Basic indicators describing the single consignments transport development.

Years	Transport volume [th. net]	Transport performance [th. netkm]	Ø transport distance [km]
2008	28,951	5,337,557	184
2009	22,130	3,977,358	180
2010	26,937	4,589,737	170
2011	28,647	5,043,450	176
2012	25,803	4,447,198	172
2013	16,950	4,400,000	259
Total	149,418	27,795,300	191

Source: ČD Cargo, authors

Based on the European Commission's White paper Roadmap to a Single European Transport Area – Towards a competitive and resource efficient transport system and it's 10 goals for a competitive and resource efficient transport system 30% of the road cargo over 300 km has to be shifted to other transport modes such as rail and waterborne transport by 2030 and more than 50% of road cargo exceeding 300 km has to be shifted by 2050. The recommendation should be realized with the support of green corridors for cargo. The document supports multimodal transport and single consignments, stimulation for inland waterborne ways integration into the transport system and ecological inventions in cargo promotion. The terms concern introduction of new vehicles and watercrafts, eventually their modernization, too.

The following paper shows the found out results and the consequences of the shift of the single consignments segment to the road cargo.

2 SINGLE CONSIGNMENTS ASSESSMENT

Single consignment is considered as the cargo loaded in one carriage or max. in 5 carriages (the transport is not realized as the complete train load) and the consignment is attached by one consignment note (way-bill). The development of these consignments is shown in the Table 2. The data from all the transport is included (national, export, import and transit consignments).

The highest drop in the area of single consignments was recorded in year 2009, the drop was 23.5% in comparison with year 2008. The average rise of 25.6% in comparison with the year 2009 was recorded in following years 2010 and 2011. The drop of 12.8% in comparison with the year 2011 and more than 1% in comparison with 2012 is estimated preliminary again in 2013. The estimation of the average transport distance reaches 259 km, this distance is longer of 47% in comparison with previous years.

The share of single consignment on the total transport volume is 36% in average. It proves that this is indispensable segment, it is essential to save it.

The total loss of single consignments in years from 2009 to 2013 reaches approx. 5.05 bn. CZK, it means that the average annual loss is approx. 1.01 bn. CZK.

3 COMPLETE TRAIN LOADS ASSESSMENT

The share of complete train loads transport on the total transport volume reaches 64% in average. It is important segment on the transport market. All the data from the transport are included (national, export, import, transit).

The biggest drop of the transport volume in the segment of complete train load consignments was recorded in 2009. The drop reached 8.7% in comparison with the year 2008. The rise of 8.0% in comparison with 2009 was recorded in following years 2010 and 2011. The drop of 2.5% in

Table 3. Economic indicators of the single consignments.

Years	Transport charges [th. CZK]	Complete costs [th. CZK]
2008	5,641,340	–
2009	4,185,252	6,060,367
2010	4,452,109	6,268,320
2011	4,765,597	5,600,922
2012	4,404,912	4,595,200
2013	4,200,000	4,530,000
Total	27,649,210	27,054,809

Source: ČD Cargo, authors

Table 4. Basic indicators describing the complete train loads transport development.

Years	Transport volume [th. net]	Transport performance [th. netkm]	Ø transport distance [km]
2008	48,350	8,202,080	170
2009	40,627	7,074,848	174
2010	42,564	7,019,831	165
2011	50,003	8,822,629	176
2012	47,541	8,580,076	180
2013	40,300	7,800,000	193
Total	269,385	47,499,464	147

Source: ČD Cargo, authors

Table 5. Economic indicators of the complete train loads.

Years	Transport charges [th. CZK]	Complete costs [th. CZK]
2008	7,250,669	–
2009	6,461,302	6,947,066
2010	6,315,375	5,968,480
2011	7,438,664	6,015,136
2012	7,228,556	5,393,035
2013	6,300,000	4,550,000
Total	40,994,566	28,873,717

Source: ČD Cargo, authors

comparison with 2011 was recorded in 2012 and the drop of 15.2% in comparison with 2012 is expected in 2013. The average transport distance of the complete train loads is 193 km (it is prolonged of 11.6% in comparison with the average in years from 2008 to 2012).

The profit of approx. 3.2 bn. CZK was gained by the complete train loads transport in years from 2009 to 2012. The average annual profit reaches then 780 mn. CZK.

The outcome of the analysis for the single consignments and complete train loads shows the average annual loss of 0.399 bn. CZK (1.179 – 0.780).

4 FINAL ASSESSMENT FOR SINGLE CONSIGNMENTS

The shift to the road cargo would be conducted in case of termination of the single consignment transport in rail cargo. Simplified calculation enables to count the economic costs for the deterioration of road infrastructure.

Road tractor with semi-trailer causes the road deterioration of approx. 15.824 CZK/km. The average transport distance of the single consignments on the rail infrastructure is 191 km. Transport volume in net tons in 2013 was 16.95 mn. nt by single consignments.

This volume would be carried by approx. 770 th. of semi-trailers according to the formula (1):

$$Nsemi-trailers = \frac{Qnt}{CWsemi-trailer} \quad \text{[semi-trailers]} \tag{1}$$

where $N_{semi-trailers}$ = the number of semi-trailers [semi-trailers]; Q_{nt} = single consignments transport volume [nt]; $CW_{semi-trailer}$ = capacity weight of the semi-trailer [t].

$$Nsemi-trailers = \frac{16.95 \; mn. \; nt}{22 \; t}$$

$$Nsemi-trailers = 770 \; th. \; of \; se \; mi-trailer \; s$$

Annual covered distance is approx. 129 mn. km (assuming the use of the semi-trailers on the level of 80%).

Using the formula (2) to count the deterioration of road infrastructure is numbered:

$$Psemi-trailers = Lannual \times Psemi-trailer \; deterioration \quad \text{[CZK]} \tag{2}$$

where $P_{deterioration}$ = road infrastructure deterioration [CZK]; L_{annual} = annual covered distance of the semi-trailers [mn. km]; $P_{semi-trailerdeterioration}$ = road infrastructure deterioration caused by the semi-trailer [CZK/km].

$$Pdeterioration = 129 \; mn. \; km \times 16 \; CZK/km$$

$$Pdeterioration = 2.064 \; bn. \; CZK$$

The formulas (3) and (4) meant for counting of the CO_2 emissions in case of shifting the consignments to the road cargo:

$$Consdiesel = Lannual \times \varnothing Consdiesel \quad \text{[l]} \tag{3}$$

where $Cons_{disel}$ = total consumption of the diesel [l]; L_{annual} = annual covered distance of the semi-trailers [mn. km]; $\varnothing \; Cons_{diesel}$ = estimated average consumption of the diesel oil [l/km].

$$Cons \, disel = 129 \; mn. \; km \times 0.35 \; l/km$$

$$Consdisel = 45.2 \; mn. \; l$$

$$TPCO_2 = Cons \, diesel \times QCO_2 \quad \text{[t]} \tag{4}$$

where TP_{CO2} = total production of CO_2 [t]; $Cons_{diesel}$ = total consumption of diesel oil [l]; Q_{CO2} = the amount of CO_2 created when burning 1 liter of diesel oil [t] = 2.64 kg CO_2.

$$TPCO_2 = 45.2 \; mn. \; l \times 0.00264$$

$$TPCO_2 = 119.328 \; t \; CO2$$

The above mentioned value characterizing the deterioration of road infrastructure and the calculation of total CO_2 production are only some negative consequences of the shift of the single consignment system to the road cargo. It shall be mentioned that there are existing disproportions between current toll system applied only on chosen roads in the Czech Republic and the rail network usage charge where using any part of the whole rail network is subject to pay the charge. Another negative aspect of the shift to the road cargo is the high accident rate in road transport. This issue is far beyond the extent of this paper.

5 CONCLUSION

The researchers carried out the analysis of the single consignments transport that means one complete wagon loads as well as the set of more complete wagon loads, focused on the period from 2008 to 2013. The average transport distance of the single consignment was counted and related deterioration of the road infrastructure when shifting these consignments to the road cargo was counted afterwards. The quantification of the emission production from road traffic when shifting single consignments to the road cargo was counted based on the above mentioned data. Because of the lack of available needed data from other national carriers the comparison with other European countries couldn't be carried out.

Recommendations for single consignments efficiency:

a) Reducing of the stations with forwarding right for single consignments (Germany decreased the amount of stations by 20% according to the plan MORA C in the year 2003). Analogic procedure was done in Switzerland and Poland in the year 2012, the reducing was also by 20%.
b) Predict knot stations for forwarding of the single consignments and to set the preferential tariff for the operation to and from these stations.
c) Set up the handling of line sections according to the real needs, eventually consign the handling these line sections to the private operators.
d) To handle the occasionally used loading/unloading stations according to the customer's needs, but with covering all the incurred costs by the customer.
e) To adjust efficiently the amount of local connecting trains designed for haulage of the single consignments to the seasonal fluctuations according to the local commodities.
f) To adjust the amount of through cargo trains that carry the single consignments between key shutting yard stations during the week to reach the highest possible payload. The optimization task in this case is quite complicated, because the savings rising from cancellations of the insufficiently loaded trains have to be compared to the costs of the standstill of wagons in the shutting yard stations especially in case of foreign wagons.

Implementation of EU Transport policy conclusions requires a higher level of cooperation between land transport modes. Therefore, it is not desirable to transfer single consignments from rail cargo to road cargo. EU supports the construction of multimodal logistic centers for conventional shipments, i.e. for palletized cargo shipments. Such systems take advantage of rail cargo for bulk transport over long distances. For the collection and distribution (transportation to the first and last mile) the flexible road transport is being used.

REFERENCES

Becker, Klaus G. 2014. Handbuch Schienengüterverkehr. *DVV Media Group GmbH* 1. Hamburg. ISBN 978-3-7771-0458-4.
Cempírek, V. 2012. Nový systém pro horizontální překládku. *Logistika* 1/2012: 16. Economia Praha. ISBN 1211-0957.
European Commission. Eurostat statistics. Available from: <*ec.europa.eu/eurostat/web/transport/data*>
Institut Jana Pernera, o.p.s. 2013. *Optimalizace technologických procesů ČD Cargo, a.s.*. Praha.

Kortschak, B. 2007. Rangieren Abschaffen! *Sonderheft 131 ZEV Glasers Annalen Tagungsband SFT Graz*: 50–55.

Kortschak, B. 2010. Rangieren ist überholt. *Deutsche Verkehrs-Zeitung (DVZ)* 64 (37) vom 27: 6 März.

Kortschak, B. 2011. Rangieren abschaffen, Einzelwagenverkehr retten! *Internationales Verkehrswesen* (63) 6/2011: 30–33.

Kortschak, B. 2012. Einzelverkehr wettbewerbsfähig gestalten: Rangieren Abschaffen! *Sonderdruck aus dem Jahresbuch für Controlling und Rechnungswesen*. LexisNexis. ISBN 978-3-7007-5126-7.

Ministry of Transport. 2013. *Transport Yearbook*. Department of transport Strategy & Transport Research Centre. ISBN 1801-3090.

Široký, J. et al. 2012. *Transport technology and control*. Brno: Tribun EU. ISBN 978-80-263-0268-1.

Advances and Trends in Engineering Sciences and Technologies – Al Ali & Platko (Eds)
© 2016 Taylor & Francis Group, London, ISBN: 978-1-138-02907-1

Determination of diagnostic action intervals for planned interval maintenance

N. Daneshjo
Faculty of Business Economics, University of Economics, Košice, Slovakia

P. Beke
Civil Engineering Faculty, Technical University of Kosice, Košice, Slovakia

ABSTRACT: Initial frequency of diagnostics functions for the given object is determined on the basis of the service man experience. According to the first measurements at single functions the time course of the monitored diagnostics quantity or the parameter $T_D(t)$ ascertained. Maximum increase of the parameter is compared with the critical value of the parameter. The critical value of the parameter corresponds to an important failure in our case. On the basis of this comparing the new frequency of diagnostics functions is determined so that probability of crossing the parameter of the relevant failure x_p up to further function performing is 5 percent.

1   INTRODUCTION

We divide the maintenance according to character of performed works on the preventive maintenance (solicitude about means) and on restoration of damaged parts realized by repair or exchange.

The aim of the effective maintenance programme is to prevent from deteriorating object reliability and to maintain durability and service life of parts in the specified boundaries. The maintenance level should have such the value that total costs on working of the object are at minimum level.

We do not know when the important failure sets in at the given functions. Therefore it depends on suitable choice of maintenance works optimum interval. The following maintenance advances were taken in practice:

a) Planned periodical maintenance:

The solicitude in narrower conception does not include in diagnostics activities. On one hand it simplifies the maintenance process but on the other hand a risk of unforeseen failures increases. It is suitable precisely to know the middle time up to the object failure and also the density of failures distribution probability at this method of maintenance.

On the basis we choose the firm time intervals of maintenance works with optimum length of the interval (T_{opt}). The time intervals between single maintenances must be so short that the probability of premature failure development is at minimum level in the given interval. On the other hand a large density of maintenance works can overprice considerably the object operation.

This maintenance advance is called also maintenance with hard time intervals (HTL – Hard Time Limit). Such a maintenance is performed e.g. at exchange of motor oil after the stated time of operation.

Some important parts are restored by exchange irrespective of their failure in some branches with increased claims on safety (air-traffic companies).

b) Planned periodical maintenance connected with diagnostics of object technical state:

Likewise as in previous case besides maintenance functions regular diagnostics functions are planned. It is a higher form of maintenance because it is demanded an analysis of state from a service man and the maintenance is performed in the case of necessity only. This advance saves costs on the maintenance but on the other hand demands diagnostics functions and higher claims on qualification.

c) Planned interval maintenance:

We do not know the density of failures distribution probability or the given density presents a trouble incidence already from beginning of object operation in this case.

2   DIAGNOSTIC ACTION INTERVALS FOR PLANNED INTERVAL MAINTENANCE

Initial frequency of diagnostics functions for the given object is determined on the basis of the service man experience. According to the first measurements at single functions the time course of the monitored diagnostics quantity or the parameter $T_D(t)$ ascertained. Maximum increase of the parameter is compared with the critical value of the parameter. The critical value of the parameter corresponds to an important failure in our case. On the basis of this comparing a new frequency of diagnostics functions is determined so that probability of crossing the parameter of the relevant failure x_p up to further function performing is 5 percent.

The wear is intense in the phase of running-in because produced areas that are in touch are not still adapted mutual. The contact of the areas is transmitted on unevenness peaks of surfaces roughness profile. This unevenness is by mutual effect plastically deformed or worn.

After the phase of running-in every part has still the so-called reserve on wear. This reserve is gradually exhausted and it is shown outwardly by limited function of the given mechanical nodal point. More functional surfaces can create the whole function of the nodal point in some cases. Then malfunction is shown by sum wear of all surfaces. The reserve is not given by intense of wear but allowed clearance of the mechanical nodal point. This clearance determines an important failure.

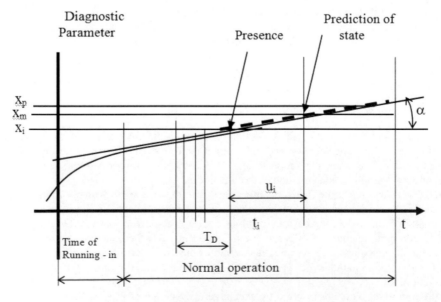

Figure 1.   Example of wearing course in normal operation.

Then the relevant failure can be found still in normal phase of wear in some cases. A forecast of optimum time for function of maintenance is simply computable in this case. It is a linear dependence:

$$u_i = \frac{x_m - x_i}{tg\,\alpha} \tag{1}$$

Precision of calculation will depend from precision of determination of wear trend directions. We must have a record of one measurement at least in normal phase of operation.

After the normal phase of operation the phase of intense deteriorating operation follows. The wear is more intense in this phase because of growing dynamic forces at increased clearances but it can be also because of material surface layer fatigue. Its critical state of wear interferences up to this phase then the calculation of the optimum time for maintenance has more complicated character.

It is necessary to perform a function of maintenance at ascertaining an important failure that is given by the value x_m.

T_D – time interval in that diagnostics functions were performed,
t_i – time of the last diagnostics function,
x_p – critical state of parameter (important failure),
x_m – the state of the parameter after achievement it, it is necessary to perform maintenance (monitored failure),
x_i – the measured state of the parameter on time t_i,
u_i – the optimum time for maintenance performing.

$$tg\,\alpha = \frac{dx}{dt} \tag{2}$$

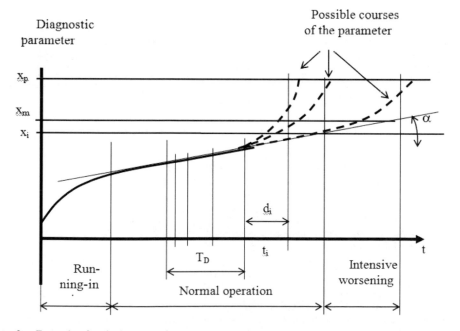

Figure 2.   Example of typical course of wear.

Frequency of diagnostics function:

$$f_i = \frac{A}{(x_m - x_i)^B} \left(\frac{dx}{dt}\right)^C + D \qquad (3)$$

Time interval of following diagnostics function:

$$d_i = \frac{1}{f_i} \qquad (4)$$

A, B, C, D – empiric, non-dimensional coefficients choice so that is 95% certainty that in this interval it does not come to the important failure.

The stated interval d_i can be further adapted according to following economical calculation on the basis of verified results:

$$k_d = \frac{n_d}{n_u + n_p} \qquad (5)$$

n_d – specific costs on application of diagnostics function
n_u – specific costs on maintenance performed by correction of important failure
n_p – specific costs on shutdown periods at correction of the important failure

If the coefficient k_d is larger than we can lengthen the interval d_i, we accept by the higher risk that it comes to the important failure but the total costs on operation will be lower. If the coefficient k_d is smaller than we shorten the interval d_i. In the case, $k_d = 1$ we do not change the interval d_i.

3 ECONOMICAL JUSTIFICATION OF MAINTENANCE AND TECHNICAL DIAGNOSTICS UTILIZATION

We meet in current life also with such reliable products that do not demand maintenance and diagnostics during all time of life if they are used by prescribed method. It is an ideal solution from the standpoint of the user but it is less such systems. The preventive maintenance also the diagnostics has mostly an important significance.

We must know the reliability of the given product respectively system for an estimate of maintenance use or technical diagnostics economic advantages. In many cases, it is very difficult to determine the reliability.

There are following reasons:

- The system reliability was not monitored till now then there are not statistical records.
- Determination of the system reliability is exacting on time and costs.
- Information about reliability is unavailable for public from commercial reason.
- The system reliability could not be verified because it goes about a new system.

It goes out often in practice from assumptions and experiences of experts only at economical evaluation. Also this information has a high value and often is sufficient for putting suitable technical diagnostics and maintenance.

Experiences from evolutionary works of product have an important significance. We predestine preventive maintenance works for some parts e.g. lubrication of bearings and design of lubricating passages, oil gauge in gear box etc at design of product in essence. Structural solution of these nodal points is such one that they enable the simplest performing maintenance works.

It is possible to judge the reliability of object (product, manufacturing system) from some standpoints practically in stage of its arising already.

We classify the reliability from this standpoint as follows:

- Reliability of design (quality of design).
- Reliability of working (quality of conformance).
- Reliability in working.

Economical access rests in essence in it that we determine the list of probable failures and we can compute costs for their removing.

Simultaneously we have available also the estimate of probability density for occurrence of single failures also other available information about reliability.

3.1 Costs connected with failure at proposed use of new maintenance and diagnostics

U_i – costs on preventive maintenance of the element i

C_i – probability figure of failures during the stated period if new maintenance is used ($C_i < B_i$)

D_i – costs on use of technical diagnostics (installation, price)

E_i – probability number of failures during the stated period if new technical diagnostics is used ($E_i < B_i$)

n – number of monitored parts that we decided to take into consideration.

Economical advantage sets in after fulfilling the following unevenness:

$$\sum_{i=1}^{n} A_i B_i > \sum_{i=1}^{n} (A_i C_i + U_i)$$ (6)

For introducing the new technical diagnostics:

$$\sum_{i=1}^{n} A_i B_i > \sum_{i=1}^{n} (A_i E_i + D_i)$$ (7)

In the case that it goes about relevant systems where it is possible to come to endangering human lives or high material damages, the given system must have unconditionally secured protection by the maintenance also the diagnostics. It is not question whether to use a protective system but it is the question which protective system has be used in this case. The above-mentioned access to the economical solution is in strong measure depended from right information. Firstly, an overall analysis of possible failures must be made. It is possible to secure through expert experiences only.

The right determination of the coefficients B and C is a further problem. The coefficients are also dependent from human factor. If serious data about failures of the system are not available then it is possible to estimate these coefficients only. There are, however, the cases when also at $B_i = 1$ (one relevant failure during the stated period) it is sufficient on the economical introducing the new maintenance or the diagnostics.

4 CONCLUSION

Division in company is considered usually to be only a service department, which is spending a large amount of financial means, whereas its real benefit for the company is debatable. However, without a well-organized and effectively managed maintenance cannot be realized the main objective of every company, i.e. creation and increasing of profit, which is a result of reliable operation of production machines, transport equipment and other machinery that are used for producing or delivery of quality products and services in the right time and with a suitable price. Optimisation of processes in company requires also a standardisation of maintenance approaches, background papers and information base, which offers a complex overview about maintenance.

Standardisation of activities enables to obtain early and correct information about the necessary maintenance, about its duration, idling or about consumption of sources. It is useful to elaborate a working methodology in order to perform complex activities, which are requiring a special treatment or safety rules. If there is established a unified model of machinery and equipment, it enables fast orientation and overview. The real records about operation are useful for optimisation and for correct tuning of maintenance system, as well as for analysis and statistics.

ACKNOWLEDGEMENT

The article was prepared in the framework of the research project VEGA 1/0582/13 supported by the Scientific Grant Agency of the Ministry of Education and Sciences.

REFERENCES

Al Ali, M. 2013. Analýza vplyvu rozmiestnenia výstuh na odolnos' tenkostenných profilov tvarovaných za studena. *Ocel'ové, kompozitné a drevené nosné konštrukcie a mosty: 38. aktív pracovníkov odboru ocel'ových konštrukcií* 38(1): 5–10.
Daneshjo, N. 2012. Pohl'ad na diagnostiku, údržbu a spol'ahlivos' strojov a ich význam v letectve. Košice: TU.
Ižaríková, G. & Džoganová, Z. 2014. Analysis of the psychoacoustic parameter by the method weighted sum of the order. *Interdisciplinarity in theory and practice* 2(5): 86–90.
Knežo, D. 2013. Calculation of parameters of nonlinear regression function for mathematical model of diffusion. *Transfer inovácií* 15(28): 187–189.
Kravec, M. 2014. Application of multidimensional statistical methods in steel industry. *International Journal of Interdisciplinarity in Theory and Practice* 2(3): 70–78.
Šeminský, J. 2012. Metodológia navrhovania a povaha technického diela. *Strojárstvo extra* 2(5): 16/1–16/3.

Advances and Trends in Engineering Sciences and Technologies – Al Ali & Platko (Eds)
© 2016 Taylor & Francis Group, London, ISBN: 978-1-138-02907-1

Diagnostics reliability of machines and manufacturing systems

N. Daneshjo & M. Kravec
Faculty of Business Economics, University of Economics, Košice, Slovakia

P. Beke
Civil Engineering Faculty, Technical University of Košice, Košice, Slovakia

ABSTRACT: A division in company is considered usually to be only a service department, which spends a large amount of funds, whereas its real benefit for the company is debatable. However, the main goal of every company, i.e. creation and increasing of profit, which is a result of reliable operation of production machines, transport equipment and other machinery that are used for producing or delivery of quality products and services in the right time and with a suitable price cannot be realized without a well-organized and effectively managed maintenance. Optimisation of processes in a company requires also a standardisation of maintenance approaches, background documents and information base which offers a complex overview about the maintenance.

1 INTRODUCTION

Each of the mentioned components can be equipped with own counter as an indicator of wear-out; further a plan of operation site together with designation of the maintenance points has to be defined. Documentation has to involve relevant schemes, pictures, lubrication planes and other technical details concerning used equipment.

It is necessary to create an effective system for scheduling of maintenance interventions in order to ensure a stable operation of machines and machinery during planning of maintenance. In other words: by means of the planned maintenance to maintain machines in such a state, which avoids unexpected breakdowns or idling.

The prevention plays a very important role in the maintenance process. It creates a base for effective maintenance. Another considerable factor is reliability. Reliable machine works without a failure during required time interval. If we talk about reliability, we think about strategy, methodology and programs developed in order to optimise manufacturing systems. A global approach to the reliability integrates all the above-mentioned initiatives into one complex system, which is built on serious basic components.

Usually it is possible to identify certain typical changes of machine or machinery parameters, which occur before a real failure appears for example:

- Uncommon intensity of mechanical vibration.
- Higher temperature.
- Increased noisiness.
- Reduced tolerance of products.
- Worsened lubrication.
- Other appearances.

It can be stated that the technical diagnostics is a multi-parametric system and the diagnostic process has to be performed complexly, taking into consideration all the presented facts. The technical diagnostics is a measuring process which is based on monitoring of various physical characteristics during current operation, whereas any change of these parameters is a relevant indicator of an undesirable dynamic change inside of machine.

It is necessary to clear certain questions at machines diagnostics solution. It is suitable to know the answers on the following questions for sensible application of diagnostics methods in technical practices:

1. Which object (product, manufacturing system, machine....) requires technical diagnostics for reliability or safety increase?
2. How can a relevant failure arise?
3. What methods and means of technical diagnostics do we know to reveal in time the development of a failure state and to avoid to the relevant failure by?
4. How do we prove to defend economically the need of this diagnostics?

The answers on these questions create a born structure of the object "Elements of machines diagnostics". In general, it is possible to define the diagnostics as the determination, the control of the object technical state from the standpoint of a failure occurrence.

According to the Slovak technical standard STN 01 0105 we define the technical diagnostics as a branch busying with methods and means of ascertaining objects technical state. At the same time we understand the technical diagnostics as the diagnostics realized with non-destructive methods and without dismounting. The same standard states that the technical state of the object is the object state determining, its ability to exert demanded functions at stated conditions of its application.

2 CLASSIFICATION OF TECHNICAL DIAGNOSTICS

The suitable technical diagnostics with following measures resulting from real state is one from the decisive implements for increase of plants service ability. The technical diagnostics is defined as the process at that the topical technical state of objects is detected on the basic of objective evaluating symptoms determined with measuring technique means. Exerting diagnostics is possible with regard on a phase of machine or device life.

The aim of the diagnostics is not a measurement but disclosing machine failures in the state without dismounting. We can prevent so to failures and to realize repairs and to lower costs on the maintenance. It does not need to go about degrading factors only devaluing a material structure and deteriorating operation of the machinery at complicated technical machineries. It is suitable to classify the technical diagnostics from following points of view:

A. According to the degree of diagnostic system automation:

a) Semi-automatic diagnostics.
b) Automatic diagnostics.

Every diagnostic system is a regulation system in essence. The aim of the regulation system is to reach minimum a deviation regulated parameter from the ideal course.

Adaptive systems of management also coincide here. The principle difference from current regulation systems rests in it that the diagnostics device does not emit correction orders into management system for reaching optimum conditions of work but let correction measures only.

This measure rests in emitting warning report for service (alarm) or in blocking work of some object system parts so that it does not go to growth of damages. Then the aim of the technical diagnostics is to prevent the relevant crossing critical state of the parameter. In some cases, the boundary between regulation and diagnostics is very narrow, e.g. at classifying wasters during automatic gauging on manufacturing line.

In this case, the undesirable deviation is also the failure simultaneously therefore we coordinate such a system to the technical diagnostics and we mark as systems of checking in general.

There are also systems where a correction measure is a change of working conditions but not a blocking of system. It is possible to say that the machine has begun to work on forced regime.

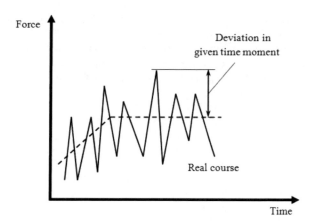

Figure 1.   Example of process regulation.

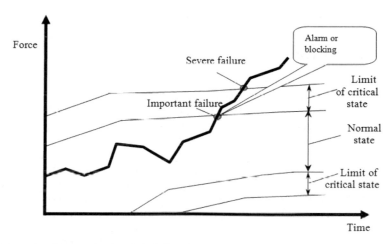

Figure 2.   Example of processional technical diagnostics.

After finishing work of course it is necessary to remove the cause of the failure. They are the cases where blocking of system presents higher costs than work in forced regime.

If a man is in feedback chain in the system of the technical diagnostics, e.g. for data collection from single devices then the system is only partly automated or semi-automated in opposite case system is fully automated.

In this regard, the technical diagnostics is overlapped with the maintenance because the data collection and evaluation of diagnostics object state precedes maintenance works in most cases. These diagnostics measures are coordinated to maintenance works in practice.

B. According to the time of technical diagnostics performing:

a) Processional diagnostics.
b) Out of process diagnostics.

The process diagnostics is performed during diagnostics object work. Out of process diagnostics is performed then when the diagnostics object is beside operation. Thus, it is performed before beginning function or after termination of object function.

If it is system working diagnostics then such a system is marked as on-line.

In a case that the system is stopped from operation for failure then test diagnostics performed on such system is called off-line. The diagnostics also is marked off-line if the state of device is evaluated on portable diagnostics device. The diagnostics on-line is the diagnostics of elements that are not dismounted or disconnected from the whole of device in electrical repair practice. The device, however, can be beside in operation. The diagnostics off-line is then the diagnostics of disconnected parts.

Both entrances of diagnostics have advantages and disadvantages. The diagnostics on-line is advantageous from the little laborious and fast determination state point of view. It is inconvenient from high claims on diagnostics intelligence and a percentage of diagnostics covering. The diagnostics covering expresses a measure of detecting and failure locality. On the other hand the off-line diagnostics is advantageous from regard of low claims on diagnostics means. It is sufficient the low intelligence of diagnostics. The disadvantage rests in necessity of uncoupling function parts from the whole. Assembling and dismounting of elements put high demands on time also means. Diagnostics covering is 100 % however.

C. According to the method of information obtaining about a technical state of object:

a) Non-testing methods (physical methods).
b) Testing methods (functional methods).

It is preferential division of technical diagnostics methods on physical and functional in professional literature. We can coordinate the functional methods to the testing methods and the physical methods to the non-testing methods. The definitions referring the physical and functional methods are meanwhile ambiguous. So that it is hard to decide what method it is in many practical cases. Only the choice dynamics output quantities (signals) are watched mostly during its function (operation process diagnostics) at the non-testing (physical) method from the given diagnostics object. We always suppose that this object does not have operation failure in the beginning at the non-testing method and that we disclose in time arise of the important failure by the used method.

Following operation temperature of bearing can be an example of the non-testing method. The limits of upper operation temperature are stated in the simplest case of evaluation. After crossing this limit we can judge on arise of the important failure that it is necessary to be solved by the diagnostics or maintenance interference. We subject the watched quantities to the mathematical analysis in a more complicated case. The diagnostics parameter (diagnostic index) is its result. The diagnostics parameter on difference from the diagnostics quantity need not be drawn on time dependence in a general case. It is possible also to border with certain limits its change that always runs in time. E.g. the occurrence of the certain spectral line in frequency spectrum of the monitored signal can be a diagnostics parameter. The time between single diagnostics functions (time of measurement) together with time on evaluation and the diagnostics interference must be shorter than the assumed time between the last registered deviation of quantity and its development into the relevant failure at the right non-testing method. Distinguishing ability of measurement corresponds to every digital measurement.

It has to be adequate to changes of a measured quantity without escape of information about a technical state of object (Figure 3). At analogous technique the distinguishing is given by measuring chain exactness.

Sampling or sampling frequency determines a density of diagnostics functions. In general a level division of diagnostics functions intervals need not be equal. The high sampling frequency secures more information simultaneously but costs on measurement are higher. The highest possible sampling frequency is limited by the necessary time of value registration. It is the time necessary for performing and record of standard value coincidence with the measured value. The distinguishing states the minimum difference between the measured values.

The higher is distinguishing, the more precise is the value determination. Of course, costs also grow with distinguishing increase. At the testing (functional) method, the given diagnostics object

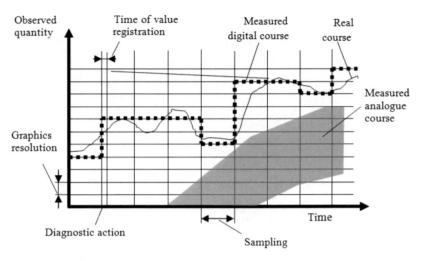

Figure 3. Technical state of object.

is submitted to the exactly stated tests. The response of object is measured at it. The diagnostics system supplies on entrances of the diagnostics object beforehand-defined testing signals.

Their physical size occurs also in operation conditions and has influence on function of the object. The testing signal is caused by a functional change of the object. The correspond output signals are monitored on outputs. We do not know beforehand to say whether this object has or does not have an important failure at the testing method. If, of course we know this then the test serves on location of the failure. Testing diagnostics is performed mostly in out of process diagnostics.

The structural diagnostics can be the testing or the non-testing method in certain sense. It goes out from the assumption that every machine must regularly error-free to work if it consists from error-free parts mutually error-free switched over. Only static quantities are characteristic for structure of part in certain sense. Structure diagnostics firstly goes out from the measurement of static quantities as dimensions, shape and surface roughness. Or we evaluate also inside structure of part material in certain sense. Finally, the switchover between object parts is checked.

A peculiarity of structure diagnostics is that a function is not self-checked by a physical quantity that rises at operation but other physical quantity that as a rule has no influence on a function during operation. Most frequently it is a light as a porter of information about technical state of an object. From this point of view we can consider the method that is testing. Then a testing signal is a quantity that does not cause a functional change of the object. On the other hand these quantities can be measured also during work of the device as physical quantities that are changed by the working conditions influence. From this point of view we could consider as a non-testing method.

The technical state of the machine can be judged on the basis of static and dynamic quantities. In general, it applies that if all static quantities are in order in all parts of the machine then also dynamic quantities must be in order during operation of the machine. A choice of suitable diagnostics quantities depends on their availability and stated value for the concrete object.

It is necessary to know an internal structure and function of a testing object on evaluation of testing signals.

After analysis and optimisation of testing steps we can test practically by two methods:

1. Right responses on single testing steps are laid in a memory of computer. Output values are compared with the memory (Figure 4).
2. Right responses on testing steps are obtained so that at the same time we load two equal objects but we know about one that it is without failure. Output values are compared with the object without failure (Figure 5).

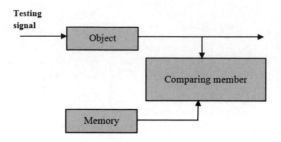

Figure 4.   Testing steps.

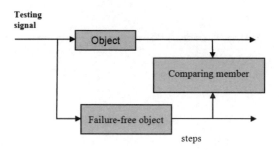

steps

Figure 5.   Testing steps.

3   CONCLUSION

It is suitable to make evident some notions in the following sections for clearing elementary tasks of reliability. The technical diagnostics has the narrow continuity with the maintenance of machines and devices and is one from methods of securing total reliability of objects. The reliability is an important part of the object quality. The final aim of all technical effort is a satisfaction of people need. The most general notion expressing this need is the quality.

The satisfaction of people need has its philosophical limits as the need of people is various and variable over time. We can however receive a simple agreement that the best satisfaction of need it is possible to reach through high quality of products.

ACKNOWLEDGEMENT

The article was prepared in the framework of the research project VEGA 1/0582/13 supported by the Scientific Grant Agency of the Ministry of Education and Sciences.

REFERENCES

Al Ali, M. et al. 2012. Analysis of the Initial Imperfections Effect on the Thin-Walled Cold-Formed Compressed Steel Members. *Communications: Scientific Letters of the University of Zilina.* 14(4): 83–87.
Daneshjo, N. et al. 2011. *Diagnostics, maintenance and reliability of machines manufacturing systems.* Germany – Dr. Enayat Danishjoo.
Fabian, M. & Boslai, R. 2011. "Kulatý roh" jedna ze základních konstrukcí povrchového modelování. *IT CAD.* 21(1): 46–49.
Knežo, D. 2014. Inverse transformation method for normal distribution and the standard numerical methods. *International Journal of Interdisciplinarity in Theory and Practice.* 2(5): 6–10.
Sedláková, A. & Al Ali, M. 2011. Kompozitné materiály na báze ľahčených betónov. *Chemické listy.* 105(16): 445–447.
Šeminský, J. 2014. Present trends in designing of technical systems. *Applied Mechanics and Materials.* 460(1): 73–80.

Advances and Trends in Engineering Sciences and Technologies – Al Ali & Platko (Eds)
© 2016 Taylor & Francis Group, London, ISBN: 978-1-138-02907-1

Methodological approach of teaching sustainability concepts to civil engineering students

P. Ganguly & K. Škrlantová

Institute of Technology and Business, České Budějovice, Czech Republic

ABSTRACT: Sustainability is often overlooked in favor of other technical or humanitarian subjects in many civil/construction engineering programs. This has been noted not only by UNESCO but also by many eminent researchers in engineering education. To help address this issue, this paper presents approaches to introduce civil engineering students to sustainability issues by using a conceptual model based on the emerging issues of sustainability in the construction industry.

1 INTRODUCTION

About forty three years after the Stockholm Conference, on the Human Environment, after 21 years of the first international conference on sustainable construction in Tampa and after the end of the Decade of Education for Sustainable Development (DESD) in 2014, we may note that in certain universities of the USA, Europe, Australia and some other countries the engineering education for sustainable development (EESD) has taken deep root through the curriculum design and subsequent realization. Education is held to be central to sustainability as both are inextricably linked. However, the distinction between education and education for sustainability is enigmatic for many and when it relates to EESD, the problem is more pronounced. This issue has been stressed during the World Conference on Education for Sustainable Development in November 2014 in Japan. As the decade for education for sustainable development closes down, the UNESCO has come up with the Global Action Program on ESD (GAP), which has two objectives and five priority areas (UNESCO, 2014). Two of the following priorities directly relate to the universities: (1) integrating sustainability practices into education and training on environment issues (whole-institution approaches and (2) Increasing the capacity of educators and trainers.

The role of the university in creating a sustainable society is being emphasized not only by the UNESCO, GULF (Global University Leaders Forum) and ISCN (International Sustainable Campus Network) but also by several professional societies around the globe. The ASCE (American Society of Civil Engineers) looks forward to 2025 and believes that civil engineers will act as master builders, stewards, innovators and integrators, managers of risk and uncertainty, and leaders in shaping public policy. Attaining of this belief will ask for review the present curriculum of civil engineering programs in the universities as the starting point of the process to educate the students toward the future profile of the civil engineering and the necessity of imparting them with the required knowledge and skills. Notwithstanding all necessary skills, the future graduates should have sustainability literacy and skills that will be of absolute importance. Therefore, in the present day scenario, most of the universities will have to go through a process of self searching of what is missing in civil engineering programs and to find out the best ways to modify the curriculum according to their particular situation and the ultimate goal of equipping the future graduates with the necessary sustainability skills that will be demanded in the future by the employers.

2 METHODOLOGICAL APPROACH TOWARDS TEACHING

2.1 *Evaluation of awareness of stakeholders*

Stakeholders of program are usually the academics and students of the particular program and likely enough the teachers from many other disciplines. Other .important stakeholders are future employers from the industry and society, representatives from professional association of civil engineers and representative of the accreditation commission at the national level. We recommend to have well prepared symposium, which should present views of each group of stakeholders on the relevant questions. It might be helpful to proceed with a questionnaire survey among the teachers and students on the relevant issues. The questionnaire survey will be followed by an academic workshop where relevant academics will participate and take part in open discussion on several key questions on the proposed program/courses. The issues may be concept of sustainable development, learning and skill developed through the teaching and learning process, degree of imparting sustainability skills through present curriculum and teaching-learning process, needs of developing new courses with holistic approach to sustainable development, proposed pedagogical approach for developing sustainability skills and others (Desha, 2007). After overall analysis of the survey data and the workshop deliberation, the results would be summed up towards further tasks related to sustainability teaching to civil engineering students. It will be required by the initiators to prepare themselves to acquaint the top and middle level of university management on the sustainability issues and the needs to incorporate those in future programs. The essential information should contain development of sustainability issues, future profile of graduate engineers and required steps at university, faculty and departmental level. Evaluation of this group of persons may be done by a workshop supported by background literature and documents). If it will be possible to convince this group of persons, then the output may be: (1) University sustainable development policy (2) Strategy of incorporating sustainability issues in programs (1st cycle, 2nd cycle and 3rd cycle) as appropriate (3) Green light for drawing Action plans, (4) Membership of international or national network of institutions and enterprises on sustainability issues, like ISCN and GULF.

2.2 *Assessment of graduate outcomes and program objectives*

Graduates of the program, which teaches sustainability issues in civil engineering, will have ability to design a system, component, or process to meet desired needs within realistic constraints such as economic, environmental, social, political, ethical, health and safety, manufacturability, and sustainability and ability to function in multidisciplinary teams. These abilities may be imparted if the educational objectives are tied with those. Moreover, the program should lead the students to an understanding of professional and ethical responsibility, the broad education necessary to understand the impact of engineering solutions in a global, economic, environmental, and societal context and recognition of the need for, and an ability to engage in life-long learning.

The program educational objectives should be student-focused and will be intended to provide each student with the necessary skills required to practice civil engineering. Moreover, the graduates will develop the sustainability skills to tackle unfamiliar problems and demonstrate an ability to understand, formulate, analyze, design, and provide solutions in the field of civil engineering. These should be established by demonstrating professional responsibilities while addressing social, cultural, economic, sustainability, and environmental considerations in the solution of engineering problems while working in integrated multi-disciplinary teams and appreciating the value of multiple perspectives in engineering problem solving. They must have the ability to explain, communicate and defend their solutions effectively using graphic, verbal, and written techniques.

2.3 *Assessment of existing program and courses*

A broad assessment of the existing program will be required t in the light of the student outcomes planned for and the first task is to find out the proportion awarded to each group of subjects like social science, management, economics, structural/design, technology, water resources, sustainability, energy supply etc. Division of courses and their placement in each group depends on the practices and the structure of the present curriculum. However, in the institutions using European or American credit system, there might a limitation of total credits for first cycle and second cycle programs. Therefore, measuring in total credits and their division for each year will be the logical choice and then to find out the percentage covered by each group of subjects. The next thing is to investigate what is the content of sustainability in each course, which courses may be put together, which courses are to be innovated to reflect the sustainability issues etc. This might require a SWOT analysis for each of the course and the time required could be quite long and will require interdisciplinary team to assess the sustainability aspects of the program and the courses while respecting the requirements of imparting sustainability skills to the students within the length of the of program.

2.4 *Program/course design*

The future teacher will be necessarily at a cross road while designing the program/course. He has to respect multiple conditions, like scarcity of the resources, academic requirements of the university, accreditation requirements etc. Usually, as literature indicates, there are different ways adopted by the universities or program designers depending on the diversified understanding of the program objectives and the student outcomes. The oldest way was to introduce some environment specific subjects, then, we witness the system of introducing the approach by choosing an integrated approach, which basically aimed to introduce the various principles of sustainable building to the respective civil engineering subjects, like in subjects like construction material, construction physics or energy management of building, civil engineering technology (Nossoni, G. 2014). This is definitely commendable understanding of the necessity of resource efficiency and energy efficiency. Other way is to integrate the courses with design/research projects. The Authors feel that perhaps the best way is, if possible, to design the program with three basic aspects, like introducing the basic theory of sustainable development and the ways to achieve that, taking up sustainability issues in relevant engineering subjects and finally application of the knowledge in design/research projects. The need of knowledge and skills for economic and social sustainability may be furnished by academics of relative subjects. If space is not available due to high pressure from the core subject areas, then this may be solved by offer of elective subjects (Sinnott & Thomas, 2012).

2.5 *Pedagogical approach*

Learning and teaching need to be at the very heart of the relationship between teachers and students of the program/course. The program/course designer should integrate the affective domain and the cognitive domain of learning and will use a combination of lecture based learning, (LBL) Student-based learning (SBL), problem-based learning (PBL), experiential learning (ELT), and the emerging use of appreciative inquiry (AI) to enhance the learning experience of the participants. As they are no novice in the area of tertiary education, so lecture based learning will be limited. Rather learner-based learning (LBL) will be preferred as LBL includes a learning environment that fosters mutual helpfulness and freedom of expression and it increases learner commitment to the learning process through involvement in planning and active participation in the learning experience while building on prior experiences and knowledge. Validation of individual learner will be possible through a group evaluation by students, defense of project, continuous evaluation through tests and a full written examination with oral test. An attempt should be made to employ set of principles on validation. A set of common principles for validation may be organized according to six main themes; purpose of validation, individual entitlements, responsibilities of organization and stakeholders, confidence and trust, impartiality and credibility and legitimacy.

3 EXPERIENCE FROM COURSE UNDER ERASMUS INTENSIVE PROGRAM

3.1 The course organization

The project, "Learning Sustainable Building Principles" had been realized under the Erasmus Intensive Program. The pinnacle of the project was to conduct two weeks intensive course which was realized from 03.08.2014 to 16.08.2014 with the participation of 28 students and 17 teachers from 8 universities from 5 countries (Turkey, Slovakia, Latvia, Lithuania and the Czech Republic). The location of delivering the was the premises of the Institute of Technology and Business in the town of České Budějovice in Czech Republic The choice of the partner universities were made on the basis of the fact that they have established civil or architecture department but have not yet started to teach sustainability issues after a basic review of the curriculum.

The students were first cycle students and they had been selected by each university on the basis of their English proficiency, good academic standing and interest in the subject. During selection, the universities were asked to create a gender balance among selected teachers and students. Teacher selection was done on the basis of English proficiency and ability to lecture topics on sustainability issues, help students in the project work and able to realize the pedagogical approach of the course (ERASMUS, 2013–2014).

3.2 Formulation of the intensive course structure

Before applying for the European grant, the course structure was consulted with the partner universities by the applying/coordinating institution (Institute of Technology and Business) and was mutually agreed on the understanding that in case of approval the fine tuning of the course will be done. After the approval of the project, intensive consultations were carried out and a final course structure was adopted. Those consultations led to certain modification of the structure and scheduling of the lectures, examination, evaluation pattern etc. The most important change is incorporation of student project work to greater extent, formation of multidisciplinary and international team and giving more emphasis on objective testing, and project presentation and evaluation, examination and evaluation by interdisciplinary and international team of teachers. This included also the selection of appropriate teacher responsible for respective lectures. The course had been divided into two parts, lecture and design project under the supervision of the international teaching group.

Final intensive course structure had the following parameters: 10 days teaching-learning activities (8 hours a day out of which 50% lectures and 50% design project), one whole day excursion in a LEED certified building (a hotel) to note the sustainability features, assessment of students by a final written test (70 points) and on the quality of presentation and defend of the project by the project team (30 points), awarding of 5 credits for successful students who obtain 70 or more points and classification as per ECST).

3.3 Intensive course content

It was agreed among the partners, that the course will be designed in an integrating manner and will be oriented to sustainable building principles. Due to very short time (160 teaching hours) available for the course, there were only roughly 80 hours for lecturing. Hence the following lecture topics were taken up: Origin of sustainability and overview of development of concept of the sustainable development, National Strategy of sustainable development and its impact on building industry, Sustainable town planning, To Learn Sustainability Principles from Vernacular, Sustainable building principles, Materials for sustainable design, Reconstructions of historical buildings according to principles of sustainability, Material selection for sustainability, Traditional Hammams buildings, Passive design for sustainability, Local building materials, produced with low energy consumption, and their usage for construction of Buildings for Producing Agricultural Products, Sustainable construction materials, Recycling the waste into construction material, Why build Green? Water saving. Merits of Green building, Hydropower for Sustainable Development,

Storm water management, Water efficiency in sustainable building, Waste aspects in sustainable building design, Waste aspects in sustainable building design, Passive solar design principles for buildings, Energy efficiency, Building energy and indoor environmental simulation.

3.4 *The design projects*

The project teams were formed as international and multidisciplinary teams. Moreover, gender balance has been observed to the extent it was possible. Due to different knowledge base, varied point of views and focused area of each participant made interesting multidisciplinary teams. Small conflicts between both disciplines (architect vs. engineer) opened many interesting questions and supported a process of proposing a solution of designing the sustainable building. Participants in each group required good leadership and teamwork. All teams were also subjected to teacher supervision (each group had two teachers from two countries and as far as possible from different specialization) and functioned as advisors of the respective project team. The advisers/supervisor have been so assigned that each group has the possibility to have intensive consultation in the particular direction in the multifarious areas of sustainable building issues. Three projects have been done by the students. The basic idea was to submit proposals of sustainable building (three types of building) at the campus of the host institution. The types of buildings were: laboratory, student centre and kindergarten and the proposals should respect the sustainability issues. Students were asked to give stress on resource and energy efficiency of the construction. Keeping in view their academic level (2nd or 3rd year students) and the short time (80 contact hours for project work), it is possible to conclude that they had done commendable work.

3.5 *Assessment of intensive course*

It was the intention to accommodate the possibilities of the universities while drawing the structure of the course. This project had been a test case with the selected target groups to find the possibilities of planning and realizing a course. It has brought out several strong and weak points. For example; sustainability skills demand the ability to work in multidisciplinary and international teams. This has been achieved even with various degree of language proficiency. One of the weak points of the course is that the course has not addressed to social sustainability and economical aspects of sustainability. This was known from the beginning and the major reason was time constraints of the course. One of the greatest achievements of the course is sustainability skills can be imparted at all level of studies and with case studies and problem oriented explanation and design go a far way to develop the skills than simply reading and committing to memory the theoretical topics.

3.6 *Recommendations*

On the basis of on-site experience, dialogue with colleagues and knowledge of the university education, the authors wish to make certain recommendations. Firstly, the universities in all countries should act as a change agent to bring in a future sustainable society and use adequate tools, such as, (Sustainable development policy, vision of curriculum and research needs for the future graduates, assessment of sustainability understanding among the employees and students, organize curriculum modification to impart future skills needed by the graduates at all level etc. Authors being engineers will also like to stress on engineering education research which unfortunately has not yet attracted the sufficient attention of engineering educators in most of the EU28, particularly the countries which became EU Member after the year 2000.

4 CONCLUSIONS

Although the AGENDA 21 came out in 1992, the sustainability had never got much exposure to be mainstreamed into the educational system of future professional engineers (United Nations, 1992).

Although almost all countries in the globe had prepared a document like Sustainable development strategy, yet most of the universities do not have a sustainable development policy and an action plan. The present trend of introducing teaching sustainability issues in civil engineering education programs in universities of the USA, Australia and a few countries of the EU is the result of accreditation requirement of some professional institutions in some countries.

Public and private universities are not under the control of the State. But the quality analysis is based on the attitudes of the accreditation commission, which may or may not approve the program which attracts the students and consequently less funding to run the organization. If the accreditation commission will require as student outcome "sustainable skills", only then the university may move. This is not a vision; it is the absence of vision. If anybody will be interested to introduce sustainability, the person requires to stat with the assessment of the stakeholders and this can be done by a sustainable skill survey for different groups of stakeholders either by a questionnaire survey and a workshop to conduct qualitative research and focus groups. This may lead to various conclusions which may work as the stepping stone of curriculum assessment to find out the sustainable issues already incorporated in the instruction of the civil engineering students at all levels of study program and then an attempt may be made to fill up the necessary gaps. Usually, the program designers look basically on the environmental sustainability aspects, as has been done in the project "Learning Sustainable Building Principles" It is also necessary to count with the lack of required proficiency of the colleagues or unwillingness to test something new. Therefore, consideration should be given to organize reeducation of faculty members to develop sustainability teaching skills. Sustainability does not have only one pillar (environmental sustainability) but also the, social and economic pillars. Present program structures may not offer enough space to include all those issues, this is a serious problem and should be taken care of by revaluing the program or through association of students with community based projects or design projects or by research projects. Introduction of sustainability issues in curriculum by right methods to impart sustainable skills for future graduates is a time taking process but the path should be taken as early as possible.

REFERENCES

Desha, C.J.K. et al. 2007. The importance of sustainability in engineering education: A toolkit of information and teaching material, *In Engineering Training and Learning Conference*, September 12–13, 2007, Centre for Environmental Systems Research, Griffith University, Australia. Available at: http://www.naturaledgeproject.net/documents/icdpaper-final.pdf (cited on 30. 1. 2015).

Erasmus IP project "Learning Sustainable Building Principles". 2013–2014. Instate of Technology and Business, České Budějovice, Czech Republic.

Nossoni, G. 2014. An innovative way to teach sustainability in Civil engineering material class, *In 121st Annual ASEE Conference*, June 15–18, 2014, Indianapolis, USA. Available at: http://www. asee.org/public/conferences/32/papers/8759/view (cited on 10. 2. 2015)

Sinnott, D. & Thomas, K. 2012, Integrating Sustainability into civil engineering Education: curriculum development & Implementation, *In 4th International Symposium for engineering education*, July 2012, University of Sheffield. Available at: http://isee2012.group.shef.ac.uk/docs/papers/paper_13.pdf (cited on 18. 3. 2015).

UNESCO Roadmap for implementing the education for sustainable development. 2014, Paris, France. Available at: http://unesdoc.unesco.org/images/0023/002305/230514e.pdf (cited on 14. 3. 2015).

United Nations Agenda 21 (Promoting Education, Public Awareness and Training – chapter 36), 1992, New York, USA. Available at: http://www.unep.org/Documents.Multilingual/Default.asp?DocumentID= 52&ArticleID=4415&l=en (cited on 15. 1. 2015).

Advances and Trends in Engineering Sciences and Technologies – Al Ali & Platko (Eds)
© 2016 Taylor & Francis Group, London, ISBN: 978-1-138-02907-1

Development of an indoor navigation system using NFC technology

Ľ. Ilkovičová & A. Kopáčik
Slovak University of Technology in Bratislava, Department of Surveying, Bratislava, Slovakia

ABSTRACT: Indoor navigation (navigation in indoor environment) presents a new challenge in the field of personal navigation systems. Although there are a lot of options how to create a navigation system, but only few of them satisfy the requirements of users in terms of accuracy. Systems based on Near Field Communication (NFC) technology provide accurate positioning also in indoor environment, where other navigation techniques (e.g. GNSS) are not available. The advancement of smartphones, wireless networking and NFC technologies have opened up a new solution of indoor navigation. This paper presents an NFC and OS Android based indoor positioning system developed and implemented at the Department of Surveying, Faculty of Civil Engineering SUT in Bratislava to help the students and the visitors. NFC technology allows quick and easy transfer of the information in the navigation system by contact of a mobile device and an NFC tag anywhere in the building.

1 INTRODUCTION

In recent years, we are the witnesses of rapid increase of activities aimed at the field of navigation technologies, including the navigation in indoor environment. The indoor navigation differs from the traditional navigation in open areas by impossibility of usage of Global Navigation Satellite System (GNSS) for the position determination in the buildings. Due to this fact, a new innovative approach is necessary, and it opens new possibilities for the solution of the mentioned problematic. Achievement of the required accuracy of the position, the correct determination of the floor in the building, the usage of elevators, etc. often can be a problem in indoor navigation. The research activities in the field of indoor navigation are focused on the development and improvement of navigation techniques, which are using different types of signals, e.g. WLAN (Han, 2014), (Stook, 2012), (Panyov, 2014), ultrasound systems, inertial measurement systems (Woodman, 2007), Bluetooth (Halberg, 2003), (Bekkelien, 2012), or Radio Frequency Identification (RFID) technology (Nakamori, 2012), (Ting, 2011). For the users are more usable and friendly the solution, which not require the installation of expensive components. One of the most suitable solution is the usage of smartphones, which are owned by the large part of the population, and the people have them with yourself almost at any time. An example of a low-cot system is a localization system based on QR codes (Gule, 2014), (Rahul, 2013).

The paper presents a localization system based on RFID technology using Near Field Communication (NFC) tags. For automation of the position determination process NFC tag reader was programmed for OS Android. The determination of the current position of the user is done by the identification of an NFC tag with known coordinates, when the smartphone is able to start the communication with the NFC tag in their neighborhood. The position of the user is shown in a spatial model of the building and the trajectory to the chosen destination is displayed of the user's smartphone. The operation of the system was tested at the Faculty of Civil Engineering, Slovak University of Technology in Bratislava. The system described in the paper becomes the subsystem of the information system of the faculty in the future.

2 THE THEORETICAL BASIS OF THE LOCALIZATION SYSTEM

According to the worldwide trend of automation of systems and services, an integral part of SMART buildings are the navigation (localization) systems operating in the indoor environment of the given building. A navigation system allows for the users easier and more efficient use of the building environment. In addition to the normal operation, navigation systems increase the safety of the building in crisis situation (navigation to the emergency exits, navigation of the deployed rescue teams). Positioning systems are simplified forms of navigation systems, which instead of real time navigation allow the determination of the current position of the visitor considering to the environment of the building. Besides the position, they can provide additional information necessary for faster orientation and planning of the visit (information about the offices, the personal in these offices, etc.).

Positioning system, working on the bases of localization of the users using NFC tags, is based on the identification of these tags, positioned on the characteristic places in a building, by the technology of RFID. The NFC tags placed at these locations allow the users to determine their actual position in the map, resp. in the virtual spatial model of the building with the required accuracy. It is possible due to the fact that the position of the NFC tag is given exactly (with required accuracy) and user's position is determined according to the position of this tag. The NFC tag have to contain the information, with which we are able to identify the user's location (it may be coordinates or an URL address). Beside the information about the position, the NFC tag can contain additional information about the object and environment where is the user currently situated.

The requirements of every localization system is as follows: simple actualization of data (the ability to update only a chosen part of the system as needed), the possibility to place the map of the building (floor plan) on a local webserver to make the application more independent. Very important is the ability to add additional data to the chosen part of the indoor environment (e.g. photographs, contacts, etc.) and to add to the whole system a kind of interactivity according to the requirement of the operator (owner) of the building.

The most appropriate equipment to install a positioning system is a smartphone, especially because of its technological parameters, such as screen size and simplicity and intuitiveness of the control (for a smartphone is not a problem to run multiple applications at once). In the following parts of this paper, is designed positioning system running on Android OS, which is currently used by most of software developers for smartphones. The advantage of OS Android is not only high number of free apps, but also the fact, that it is an "open" operating system (anyone can write applications using freeware compilers). This property allows the modifications of the designed system by the users.

It was designed online concept of the positioning system, which is characterized by its undemanding storage space, since the building model is stored on an external server. To identify the NFC tag, it was developed JAVA application, by which is the user redirected to a web server and is identified its location on the map. The map can be controlled interactively, the additional information associated with each object can be used, and so on.

3 NFC TECHNOLOGY

NFC is the abbreviation for Near Field Communication. It is the international standard for contactless exchange of data. In contrast to a large range of other technologies, such as wireless LAN and Bluetooth, the maximum distance of two devices is 0,1 m. Like Bluetooth and WiFi, and all manner of other wireless signals, NFC works on the principle of sending information over radio waves. Through NFC data is send through electromagnetic induction between two devices (Fig. 1). NFC works on the bases of tags, it allows you to share some amount of data between an NFC tag and an Android powered device or between two Android powered devices. Tags have various set of complexities. The data stored in the tag can be written in a variety of formats, but Android APIs (Application Program Interface) are based around a NFC standard called as NFC Data Exchange

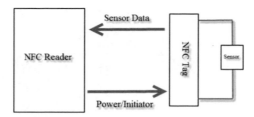

Figure 1. Data transmission using NFC technology (idlehandsproject.com, 2015).

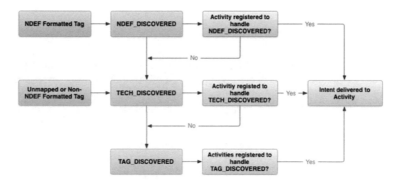

Figure 2. Flowchart of the tag identification (developers.android.com, 2015).

Format (NDEF). The transmission frequency for data across NFC is 13.56 MHz, and data can be sent at either 106, 212 or 424 kilobits per second, which is quick enough for a range of data transfers from contact details to swapping pictures, songs and videos.

4 APPLICATION NFC_APP

One of the specific solution of above proposed positioning system is the application NFC_APP. The application is developed in JAVA and designed for OS Android for use in smartphone Samsung Galaxy S4. Fig. 2 shows a flowchart of the application for identifying NFC tags using a smartphone that supports NFC technology (developers.android.com, 2015). The following flowchart describes process, when the NFC tag is close to smartphone.

NFC tags system works in Android with three type of filters (NDEF_DISCOVERED, TECH_DISCOVERED, TAG_DISCOVERED). An action NDEF_DISCOVERED is triggered, when the NFC tag is defined in NDEF format (this format is recommended by the official Android developer's website). An action TECH_DISCOVERED is triggered if the tag does not support NDEF, but another known technology. An action TAG_DISCOVERED is started, if chip technology could not be detected (developers.android.com, 2015).

As mentioned, not every device has support for NFC technology, and on the beginning of the application it must be verified.

```
if (mNfcAdapter == null) {
        // Stop here, we definitely need NFC
        Toast.makeText(this, "This device doesn't support NFC.",
Toast.LENGTH_LONG).show();
        finish();
        return;
    }
```

Figure 3.   Procedure of the localization.

In the given part of the code is verified the presence of the support of the NFC technology in the system. For this purpose was created a class mNfcAdapter, through which is possible to communicate with the hardware in order to verify the supporting technology.

It is also necessary to verified, whether is the NFC technology in the device enabled (on).

```
if (!mNfcAdapter.isEnabled()) {
        mTextView.setText("NFC is disabled.");
    } else {
        mTextView.setText(R.string.explanation);
    }
```

To gain the access to the NFC hardware, the requirement must be declared in the application manifest.

```
<uses-permission android:name="android.permission.NFC" />
<uses-feature
    android:name="android.hardware.nfc"
    android:required="true" />
```

After the tag identification is the user automatically redirected to the external server, where is located the interactive model of the building (Fig. 3). The positioning system was tested at the Slovak University of Technology in Bratislava at the Department of Surveying. The plan of the department (one floor in the block A of the building of the Faculty of Civil Engineering) was created in AutoCAD software. Subsequently, it was created a webserver, where was stored a floor plans, and where have been programmed additional information (redirection to the university information system, sending email to the teacher, show schedule, see the website of the department) and the name of the teacher who seats in the office (Fig. 4). This kind of information is displayed after clicking on each office, also with the route to the searched office. The database of other additional information, which are displayed by clicking on the icons, and which we also added to the map, contains for example also a phone number, teacher's schedule, and automated redirection to sending mail or to the website of the department (Fig. 4).

After determination of the user's position and displaying the map, the user can find all the necessary information about a person or an office he is looking for by clicking on one of the offices. Thus user is able to find the most optimal way from his current position to the searched office, or person (Fig. 4). Created positioning system is based on free access to all of its components, what is its indisputable advantage. This advantage will be reflected not only on the availability of the components, but also on the price. This low-cost system is fully interconnected to the information system and database STU in Bratislava, and includes specific services, what is a great tool not only for new students as well as visitors of the university.

As mentioned, the usage of the NFC technology is very variable. In addition NFC tags is this technology used also in mobile accessories, printers, ID cards or leaflets, but the maximum usage

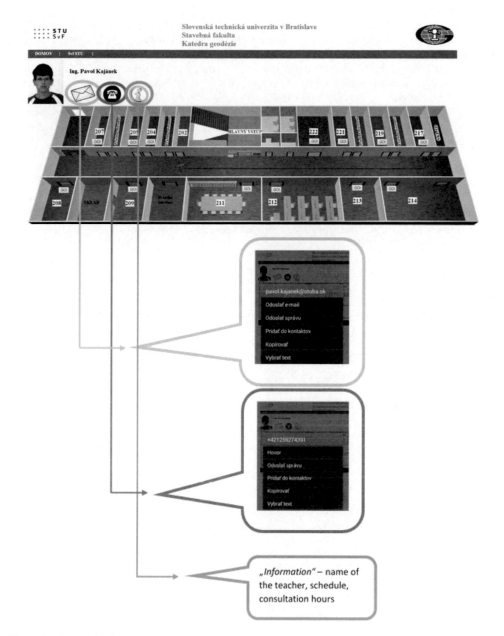

Figure 4.   Interactive icons.

the experts see in contactless payments. The question remains, of course, security and expansion of NFC around the world, but today we can say that this is the technology of the future.

5   CONCLUSION

Since there is no universally applicable navigation system for indoor environment of intelligent buildings, and such systems are not yet an ordinary part of smartphones, the topic of indoor navigation using smart devices is still a big challenge. Creating of positioning system using NFC technology is an easy way to locate the users in the areas of buildings. At the Department of Surveying on SUT in Bratislava was launched a pilot project to test the localization system using QR codes, which was partly automated with the determination of the position with NFC technology. Java application with integrated NFC reader named NFC_APP was created. The system works online, so all the data needed for interactive positioning are placed on a created webserver. The system described in the paper, would be improved in the future, enabling interactive navigation in real time.

ACKNOWLEDGEMENT

"This publication was supported by Grant Project VEGA 1/0445/13 Indoor navigation of equipment, funded by the Scientific Grant Agency of the Ministry of Education, science, research and sport of the Slovak Republic and the Slovak Academy of Sciences."

REFERENCES

Bekkelien, A. 2012. *Bluetooth Indoor Positioning [online]*. Geneva: University of Geneva. [cit.28.2.2015], availabble on: http://tam.unige.ch/assets/documents/masters/bekkelien/Bekkelien_Master_Thesis.pdf. pp. 49

Gule, K. 2014. A Survey-Qr Code Based Navigation System for Closed Building Using Smart Phones. In *International Journal for Research in Applied Science and Engineering Technology (IJRASET)*. 2(4): 442–445.

Han, D. et al. 2014. Building a Practical Wi-Fi-Based Indoor Navigation System. In *Pervasive Computing, IEEE*. 13(2): 72–79.

Hallberg, J., et al. 2003. Positioning with Bluetooth. In *Telecommunications, ICT 2003. 10th Intenational Conference (Volume:2). 23 Feb.–1 March 2003*. Danvers: IEEE.

Nakamori, E., et al. 2012. A New Indoor Position Estimation Method of RFID Tags for Continuous Moving Navigation Systems. In *International Conference on Indoor Positioning and Indoor Navigtion, 13–15th November 2012*. Sydney: University of New South Wales.

Panyov, A. et al. 2014. Indoor Positioning Using Wi-Fi Fingerprinting, Pedestrian Dead Reckoning and Aided INS. In *2014 International Symposium on Inertial Sensors and Systems Proceedings, 25–26 Feb. 2014*. Danvers: IEEE.

Rahul Raj, C.P., et al. 2013. QR code based navigation system for closed building using smart phones. In *Automation, Computing, Communication, Control and Compressed Sensing (iMac4s), 2013 Interntional Multi-Conference. 22–23 March 2013*.

Stook, J. 2012. Localization with Wi-Fi Fingerprinting: towards Indoor Navigation on Smartphones. In *8th International Symposium on Location Based Services, 16–18 november 2012*, Munchen: TU Munchen.

Ting, S.L., et al. 2011. The Study on Using Passive RFID Tags for Indoor Positioning. In *International Journal of Engineering Business Management*. 3: 9–15.

Woodman, O. 2007. An introduction to inertial navigation [online].[cit. 28.2.2015], available on: http://www.cl.cam.ac.uk/techreports/UCAM-CL-TR-696.pdf. University of Cambridge, Computer Laboratory. ISSN 1476-2986. pp 37.

http://www.idlehandsproject.com/research-wireless-passive-sensor-data-with-nfc-transmission/, 28.3.2015

www.developers.android.com, 28.3.2015

Advances and Trends in Engineering Sciences and Technologies – Al Ali & Platko (Eds)
© 2016 Taylor & Francis Group, London, ISBN: 978-1-138-02907-1

The method of ice field strength reduction in port Sabetta on Yamal peninsula

I.D. Kazunin & A.I. Alhimenko
St. Petersburg State Polytechnical University, St. Petersburg, Russia

M.U. Nikolaevskiy
Morstroytechnology" Ltd, St. Petersburg, Russia

ABSTRACT: The construction of port Sabetta is being carried out on the Yamal peninsula in Arctic north conditions. It has been designed to solve the following tasks: to provide the transloading of hydrocarbon crude from Yamal gas condensate field, and transport natural gas and oil by sea to other countries. However there are difficulties for navigation in port Sabetta water area due to the problem of its freezing in severe Arctic climate. Three ways of solving this problem have been suggested nowadays: mechanical destruction of the ice sheet (by means of ice-breakers), heat exposure (using bubbler devices), ice melting by warm water. The paper presents a new method of ice sheet destruction, which suggests decreasing heat conductivity of upper water layers. To achieve this it is possible to use a sort of blanket made up of separate polyethylene spheres (buoys) closely located to each other covering the harbour area, and to allow their shift during ship passage. In addition, these elements will make pores in ice field reducing its strength and allowing the ship to break ice easily. Using the buoys will allow to cut significantly energetic and, consequently, economical costs for pilotage in north conditions.

1 INTRODUCTION

The construction of north sea-port Sabetta (port coordinates: latitude 71°16′ north, longitude 72°04′ east, registration number A-18) (fig. 1), located at Yamal peninsula (The Russian Federation) is designed for the transloading of hydrocarbon crude from Yamal gas condensate field to provide the transfer of natural gas and oil by sea transport to the countries of Western Europe, North and South America and Asia-Pacific region.

One of the problems to solve in the port construction is providing optimal conditions for ship navigation, arrival at terminal and mooring to perform cargo-handling operations. The port water area to be protected from freezing is Swa = 111 ha (it was calculated considering the need for turning basin, operational waters and other constructions according to standard regulations) (Minin, 2015) (fig. 2).

2 CURRENT METHODS OF ICE CONTROL

Keeping harbour areas ice-free in northern regions has been carried out for a fairly long time. A number of methods to provide navigation in harbour areas have been suggested so far. These can be divided into two groups: mechanical destruction of the ice sheet and heat exposure (Bogorodskiy et al., 1983).

Mechanical destruction of the ice sheet primarily suggests pilotage with the help of icebreakers. Icebreakers destroy ice sheet breaking it into separate small ice floes which allows ships of the

Figure 1.   Port Sabetta location. Port Sabetta 3D model. Berths for ship mooring, LNG plant "LNG-Sabetta" (photo: Novatek).

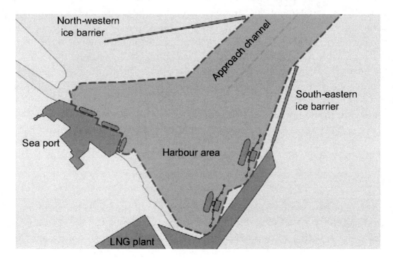

Figure 2.   Port Sabetta. Top view (photo: skmost).

appropriate ice class to follow them in wake. However, such pilotage allowing navigation of a single ship or a caravan is rather expensive, and channels formed after ship passage freeze quickly which requires establishing new shipping lanes each time. It should be noted that ice sheet formed on freezing with inclusions of previously separated ice floes is significantly stronger and requires more energy expenditures for the next passage (Alhimenko et al., 2014).

Examples of using heat exposure for ice sheet are mainly confined to the use of bubbler installations (based on the idea of transfer of relatively warm lower water layers to the upper ones) (Carstens, 1977), as well as warm water supply (Bogorodskiy et al., 1983). In particular, nowadays the combination of warm water supply with bubbler device has been found successful in Helsinki Vuosaari harbour ice control: Huachen Pan and Esa Eranti (Huachen et al., 2009) offered a system which allows using the heat of thermal electric power station (TPS) circuit for harbour area warming.

Energy consumption assessment of three main methods in ice control is performed in relation to energy costs per 1 m^3 of ice destruction:

– assessment of icebreaker energy cost has been performed according to its ship power plant capacity ($N_{spp} = 95$ MW) spent on the passage of $L = 1$ m of ice with the thickness $h = 1$ m at the speed of $V_s = 1$ m/s:

$$Q = \frac{L}{V_s} \cdot N_{spp} = \frac{1}{1} \cdot 95 = 95 MJ$$

– assessment of bubbler device energy cost has been performed including warm water energy (+40°C) and 30% of ice destruction by bubbling:

$$Q = 0,7 \cdot (c_v \cdot m \cdot \Delta T + m \cdot L) = 0,7 \cdot (2100 \cdot 1 \cdot 40 + 1 \cdot 2100) = 270 MJ$$

– assessment of ice-melting energy has been performed using the method of warm water discharge without mixing:

$$Q = c_v \cdot m \cdot \Delta T + m \cdot L = 385 MJ$$

3   ICE THICKNESS ANALYSIS IN PORT SABETTA

In port Sabetta conditions it is important to provide just the possibility of passage for ships of the appropriate ice class and their mooring. Single ice floes can remain on the surface if they do not prevent navigation. Thus, the task of ship pilotage can be transformed to the ice sheet reduction rather than its total destruction. This will allow cutting the energy costs on pilotage significantly in ice sheet conditions.

For energy cost rate assessment, it is necessary to calculate the heat balance of the system (Donchenko et al., 1987), (Kozlov, 2000), which seems to have specific features resulting from the offered construction.

Figure 3 shows heat balance diagram considering the main heat flows from the bottom, water, air and heat input from ground waters. With the decrease in environmental temperature, the water is passing heat into the air (q_{air}). Once the temperature of upper water layers has fallen to zero and lower, ice formation begins. The water itself in addition to its own heat receives the heat from the bottom (q_b) and ground waters (q_{gw}). Besides, due to the river flow (W_f) energy dissipation of the heat flow occurs (q_{wt}) which also affects the water temperature (T_w) and its ability to resist freezing.

The equation for heat balance of the system (fig. 3) considering the assumption that energy disbalance will be converted into ice cover formation (ρ_{ice} – ice density (kg/m^3), L – specific heat of ice melting (J/kg), (h_{ice} – ice thickness (m)) appears as following (Yureneva et al., 1976):

$$dE = \frac{dMi}{dt} = \rho_{ice} \cdot L \cdot \frac{dh_{ice}}{dt} = \sum_{i=1}^{n} q_i \qquad (1)$$

where heat flows appear as following:

$$\sum_{i=1}^{n} q_i = q_{air} + q_{wt} + q_{qw} + q_b \qquad (2)$$

According to standard relations, heat input from the water into the air through ice, snow cover and buoy constructions can be written as:

$$q_{air} = \alpha \cdot (T_w - T_{air}) \qquad (3)$$

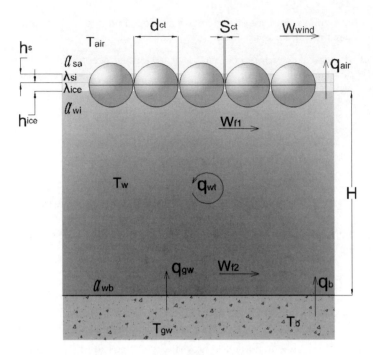

Figure 3.   Heat balance diagram of harbour area protected by buoys.

where $\alpha = \dfrac{1}{\frac{1}{\alpha_{sa}} + \frac{h_s}{\lambda_s} + \frac{h_{ice}}{\lambda_{ice}} + \frac{1}{\alpha_{wi}}}$

Heat input from the bottom to the water:

$$q_b = \alpha_{wb} \cdot (T_b - T_w) \qquad (4)$$

Heat input from ground waters to the water:

$$q_{qw} = C_v \cdot (T_{gw} - T_w) \qquad (5)$$

Water friction depending on flow type, buoy construction and energy dissipation of turbulent flow:

$$q_{qw} = \rho \cdot \varepsilon \cdot H \qquad (6)$$

where ρ – water density (kg/m^3), ε – energy dissipation of the flow, H – depth (m).
Furthermore:

$$\varepsilon = g \cdot W_{f1} \cdot I \qquad (7)$$

where W_{f1} – river flow rate (m/s), g – gravity acceleration (m/s^2), i – bottom slope (cm/km).
Heat exchange is taken as steady-state, where heat transfer coefficient α (W/(m^2*deg)) and heat conductivity coefficient λ (W/(m*deg)) are presented as functions (Elema, 2013),

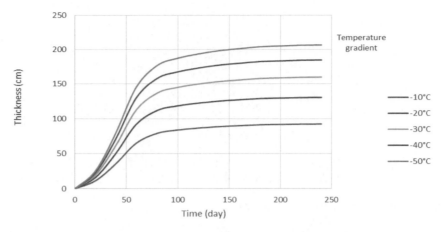

Figure 4. Design chart of changing in ice thickness on the surface of the water area protected by the buoys depending on the temperature gradient between fresh water and environmental temperature.

(Yureneva et al., 1976):

$$\alpha_{sa} = f\left(W_{wind}; d_{ct}; S_{ct}\right)$$

$$\lambda_{s} - f\left(h_{s}; d_{ct}; \lambda_{ct}; S_{ct}\right)$$

$$\lambda_{ice} = f\left(h_{ice}; d_{ct}; \lambda_{ct}; S_{ct}\right) \qquad (8)$$

$$\alpha_{wi} = f\left(W_{f1}; d_{ct}; S_{ct}\right)$$

$$\alpha_{wb} = f\left(W_{f2}\right)$$

$$q_{wt} = f\left(W_{f1}; d_{ct}; S_{ct}\right)$$

where W_{wind} – wind rate (m/s), d_{ct} – buoy diameter (m) and taken based on ice thickness in calculation area (in Sabetta water area ice thickness is about 1 m), S_{ct} – distance between the buoys (m), h_{s} – snow thickness (cm) (in our case for calculation we will take it about 5 cm), h_{ice} – ice thickness (cm), W_{f1} – river flow rate at the surface (m/s), W_{f2} – river flow rate at the bottom (m/s).

The analysis of formulas (1), (2) results in differential equation for the rate of ice thickness increase depending on external effects:

$$\frac{dh_{ice}}{dt} = \frac{1}{\rho_{ice} \cdot L} \cdot \left(q_{air} + q_{wt} + q_{qw} + q_{b}\right)$$

It is necessary to solve the equation (9) subject to (8), where heat flow intensity between the bottom and the water is regarded taking into account that the temperature at the bottom tends to zero (the proximity to the permafrost) and peculiar features of port Sabetta (Kirillov, 2012): $W_{wind} = 6{,}8$ m/s, $d_{ct} = 2$ m, $\lambda_{s} = 0,1$ W/(m*deg), $\lambda_{ice} = 2{,}2$ W/(m*deg), $T_{air} =$ from-8°C to-48°C, $T_{w} \approx +2°C$, $T_{gw} = 0°C$, $S_{ct} = 0,1$ m, $h_{s} = 5$ cm, $\lambda_{st} = 80,4$ W/(m*deg), $W_{f1} = 0,1$ m/s, $W_{f2} = 0,04$ m/s, $i = 4$ cm/km.

As a result, for the accepted snow thickness $h_{s} = 5$ cm relations of ice formation rate are following:

The chart (fig. 4) presents changing in ice thickness formation basing on the solution to equation (9) depending on the temperature gradient of fresh water $T_{w} = +2°C$, environmental temperature ranging from $T_{air} = -8°C$ to $-48°C$, and buoy material – alloy steel.

The analysis of equation (9) (with and without snow cover) shows that the presence of soft snow (5 cm in the given case) leads to the reduction of total ice thickness within the taken period.

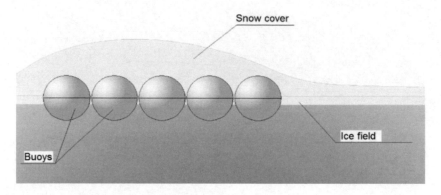

Figure 5. Diagram of snow and buoys interaction.

However, the snow becomes compacted over the time, and its heat conductivity rises, and therefore ice thickness increases. To avoid the decrease in heat conductivity of snow in dry winter conditions at the port area it is possible to set a device for artificial snow generation to cover the fairway protected by buoys, and having this snow retained between the buoys, thus to maintain low heat conductivity of soft snow and to reduce ice thickness.

4 CONCLUSIONS

Heat balance analysis (9) revealed that the most effective methods to reduce ice thickness h_{ice} are:

- Changing ice heat conductivity by means of metal or polyurethane buoys (filled with air) included in ice field structure
- External isolation of ice cover and buoys with natural or artificial snow (natural isolator) generated by special devices located at the harbour entrance (fig. 5)

Therefore the production of an isolating covering layer in port Sabetta water area will allow to prevent it from total freezing and provide safe navigation at the lower energy costs compared to methods existing nowadays.

REFERENCES

Alhimenko, AI, et al (2014). "Hydrotechnical constructions of sea ports," *LAN*, 427.
Bogorodskiy, VV, Gavrilo, VP, Nedoshivin, OA (1983). "Ice destruction. Methods, technical facilities," *Gidrometeoizdat*, 232.
Carstens, T (1977). "Maintaining an Ice-Free Harbour by Pumping of Warm Water the Heat budget," *Proceedings of the POAC Conference,* 1, 26–30.
Donchenko, RV (1987). "Ice regime of rivers SSSR," *Gidrometeoizdat*, 242.
Elema, VA (2013). "Engineering thermodynamics and heat transfer," *Novorossiysk*, 312.
Ilyushin, AA and Lenskiy, VS (1959). "Strength of materials," *Fizmatlit*, 373.
Huachen, P, Esa E (2009). "Flow and heat transfer simulations for the design of the Helsinki Vuosaari harbour ice control system," *Journal Elsevier,* 304–310.
Kirillov, VV (2012). "State ecological expertise in Sabetta area on the Yamal peninsula," *Moskva,* 52.
Kozlov, DV (2000). "Ice of freshwater habitats and water streams," *MGUP*, 263.
Minin, MV (2014). "Construction of sea port Sabetta," *arctic-info.ru,* 1.
Yureneva, VN and Lebedeva, PD (1976). "Heat engineering manual," *Moskva*, 1, 743.
Yureneva, VN and Lebedeva, PD (1976). "Heat engineering manual," *Moskva*, 2, 897 p.

Advances and Trends in Engineering Sciences and Technologies – Al Ali & Platko (Eds)
© 2016 Taylor & Francis Group, London, ISBN: 978-1-138-02907-1

3D CFD model of a V-notch weir

V. Kiricci & A.O. Celik
Department of Civil Engineering, Anadolu University, Eskisehir, Turkey

M.A. Kizilaslan & E. Demirel
Department of Civil Engineering, Eskisehir Osmangazi University, Eskisehir, Turkey

ABSTRACT: CFD-Computational Fluid Dynamics models provide comprehensive information about the fluid flow around water structures. A CFD solution allows obtaining almost all of the flow parameters for any desired time interval and location in the flow domain. The accuracy of these models is enhanced when they are supported and verified by experimental studies. However, it is not always possible to measure all of the parameters by physical experiments due to complex flow conditions and limited measurement techniques. In some cases, experience and fluids mechanics knowledge together with verification via gross flow parameters when the turbulence resolving scales are not significant may be sufficient and even superior compared to empirical design tools. In this study, flow characteristics of a V-notch weir is modeled by CFD. Experimental verification is also provided and the results are discussed within the context of design procedure and performance of the specific weir type that is being investigated.

1 INTRODUCTION

Hydraulic characteristics and flow behavior of the several types of weirs has been subject to many studies. In addition to laboratory experiments, successful CFD analysis was conducted. Researches have focused on different flow conditions and geometric structures of the channel and the weir. Rombust et al. (2014) revealed using CDF results that the discharge calculation for weirs by common formulations can be insufficient for different flow conditions. Hoseini et al. (2013) studied the determination of discharge coefficient of rectangular broad-crested side weir in trapezoidal channel by CFD and physical model experiments. They achieved a good agreement between results for different Reynolds, Froude numbers and channel slope values. Sarker et al. (2004) compared the CFD and experimental results of free surface profiles on a rectangular board crested weir and concluded that the flow structure around a rectangular board crested weir cannot be defined by 1-dimensional inviscid flow theory. Aydin (2012) modelled free surface flow over a triangular labyrinth side weir in subcritical turbulence flow conditions by using CFD and compared the results with experimental studies. It was shown that CFD analysis can be a trusted source for hydraulic investigations of weirs in particularly for the turbulent flow conditions. Hargreaves et al. (2007) emphasized the importance of the chosen turbulence model for the sensitivity of the CFD results. The studies conducted on this subject have clearly shown that CFD method is a reliable and powerful tool to investigate flow on weirs.

This study presents a CFD study of V-notch weir. The aim is to obtain a verified numerical model of a weir flow generated in a laboratory flume. The major goal of the current efforts of ongoing research is to obtain a 3D model sufficient to explore the weir coefficient, particularly the variations in this coefficient with flow conditions. This work is supported by a laboratory experiment campaign which is also planned to be presented in the same conference.

2 3D CFD MODEL

In this study, a commercial CFD solver ANSYS CFX V.14 is used for the 3D analysis of the turbulent flow upstream of a triangular shaped weir. CFX code is capable to solve many types of flow conditions including multiphase, turbulent, free-surface flow. Open channel flow is a quite challenging issue in fluid dynamics field and should be modeled carefully. This is due to the well known identification issue with the free surface and its variation. CFX uses volume of fraction (VOF) function to track and capture free surface in multiphase flow solutions. VOF function basically represents the distribution of the volume of the fluids in the flow domain and takes values between "1" and "0" according to the volume of the fluid in each discrete cell (Lv X et al., 2009). This study utilizes the VOF function to resolve the details of free surface variations. Below the model being discussed is presented in detail.

2.1 *Geometry and mesh*

CAD geometry and mesh of the model created by ANSYS DesignModeler module and model geometry is formed in the same scale as the physical experiments. Triangular weir is located in the middle of the channel. Isometric view of the geometry and the boundaries of model are shown in Figure 1.

Meshing process is one of the most critical stages of a CFD solution. Mesh type and size of mesh elements are directly related with the accuracy and stability of the solution. That is, the grid system has to meet certain requirements (Juretic et al., 2010). In addition to some control criteria for the mesh quality, it is an undeniable requirement to have sufficient knowledge on physics of the problem to judge the most appropriate mesh type and size. Geometric characteristics of mesh elements is one of these requirements and can be expressed by skewness parameter (You et al., 2006). Grid quality depends on the equiangular structure of the mesh elements (note that this is only relevant to cfd studies). If the number of isogonic elements is high in the domain, the mesh quality will be high as well and the skewness value will be low. Another process required for a more accurate CFD solution is providing the mesh independency (Roache, 1997). Certainly for the most cases, refining the mesh increases the accuracy of the solution but an uncontrolled mesh refinement may result in unnecessarily high computational load. For this reason, a size analysis should be applied to determine the optimum mesh element dimensions (Kirkgoz et al. 2008). The maximum mesh size can be determined in which the value of any selected control variable does not change for a finer mesh in the successive solutions from coarse to fine grid configuration

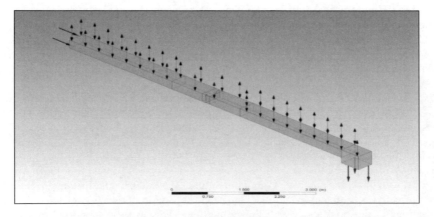

Figure 1.   CAD geometry and boundaries of the model.

(Filonovich et al., 2013). Area average shear stress at the bottom of the channel is selected as the control variable for the mesh independency study and solutions are repeated for four different mesh sizes. It is observed that the value of the average shear stress at the bottom of the channel remains within an acceptable limit around 0.015 m and finer mesh sizes. Details of the mesh independency study are given in Figure 2 and Table 1. The size determined using this approach is employed in the actual model which yields the mesh properties given in Table 2.

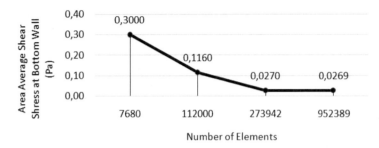

Figure 2.   Change of control parameter according to mesh sizes.

Table 1.   Mesh independency process.

Type	Size (m)	Elements	Nodes	Area Averaged Wall Shear (Pa)
Structured Coarse	0.050	7680	10143	0.300
Structured Medium	0.020	112000	126315	0.116
Structured Fine	0.015	273942	299600	0.0270
Structured Finer	0.010	952389	896000	0.0269

Figure 3.   Mesh view of the model. Refinement (inflation) near the weir is emphesized with the inset figure.

Table 2.   Details of the used mesh.

Number of Elements	1988919
Number of Nodes	549582
Average Skewness	0.287

In addition to the mesh size considerations for the flow domain in general, inflation layers are applied near the bottom and side wall regions of the channel to resolve viscous effects and boundary layer where considerably high changes in the gradients of flow parameters take place. View and details of the mesh used in solutions is given in Figure 3 and Table 2.

2.2 *Setup and boundary conditions*

CFX allows both steady state and transient analysis options. In steady state solutions, flow characteristics do not change with time after the initial unsteady behavior of the flow reached to a steady state condition (convergence). It should be noted that, it cannot always be possible to have a steady solution due to the unsteady nature of the flow type. In this case, a time independent solution can be expected as observed in the physical experiment. Therefore a steady state analysis type is used. Boundary conditions are defined to reflect the exact physical experiment conditions. Initially, a uniform velocity inlet type is used to provide an uninfluenced flow development as in the experiment and a static pressure outlet is used at the end of the channel. As seen in Figure 4. channel outlet is a connected tank allowing a gravity driven flow near the section which otherwise would be defined as a pressure outlet. This type of an outlet at the end of channel does not dictate the downstream water level. Top of the channel is defined as atmospheric pressure opening to ensure the mass balance of air in the solution domain. Both, side and bottom walls of the channel are defined as "no slip" smooth adhesive wall. Side walls of the channel are made of glass and bottom is made of plexiglas material. Adhesive contact angle of water between glass and plexiglas is used as $0°$ and $71°$ respectively to capture the adhesive effects. For a more realistic approach, buoyancy is activated and inhomogeneous multiphase option selected to solve the velocity fields for water and air phases separately. Surface tension is also activated and continuum surface force is applied (ANSYS CFX Solver Theory Guide).

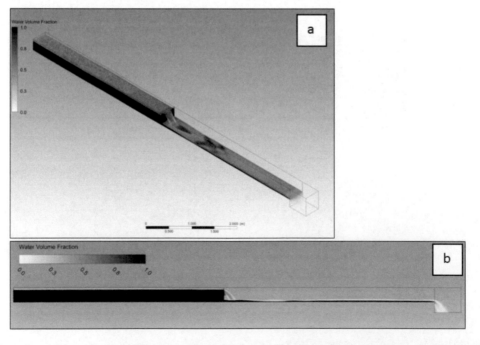

Figure 4.   a) Isometric view of the water volume fraction; b) Side view of the water volume fraction, result for Q = 8 lt/s.

High resolution advection scheme and turbulence numeric is used for computations of momentum, continuity and turbulence kinetic energy equations. Shear Stress Transport (SST) turbulence model is preferred for the analysis. Standard k-ε model is a widely used turbulence closure in engineering applications and very effective in terms of computational load. Otherwise, SST turbulence model uses both k-w model near the wall at boundary layer and standard k-ε model at free shear flow region to cope with problems of turbulence viscosity computations in k-w model. Additionally, SST turbulence model provides more accurate approach for the separated flow conditions and negative pressure gradient zones. While one of the aims of this ongoing research is to determine the turbulence closure that fits the best to the problem in hand, the details provided in this paper is not concerned with the best turbulence model.

3 RESULTS AND CONCLUSION

CFD solutions were repeated for different discharge conditions (9, 8, 6 and 5 lt.s^{-1}) This study presents the results from only 8 lt/s inlet discharge as the similar trends are obtained in all runs. The first important finding to be reported is that the free surface levels, (flow depth upstream of the weir at the uniform section, h) of CFD and experiment are quite consistent with each other. h was found to be 24.5 cm both from CFD and experiments respectively (note the estimated measurement error in the experiments, $\Delta h = \pm 1$ mm). The velocity (time average) contours are given in Figure 5 for 8 lt.s^{-1} run.

Experimental verification is provided using the time-averaged velocity profiles of CFD and experimental findings at a centerline vertical direction 2.5 m upstream of the weir. The comparison is given in Figure 6. The max error here (deviation of CFD results from the experimental results) is reported to be 16% excluding the velocity measurement point near the bottom. While such an error margin may not be acceptable for a fundamental turbulence study, it is believed that the CFD results can be useful in the design procedure of such water structures. Experimental details and more verification results will be presented in the accompanying article presented in the same conference.

A 3D CFD model of a flow with free surface interacting with a hydraulic structure was presented in detail. The results indicate that the water structure design procedure can take advantage of CFD models with experimental gross flow parameter verification.

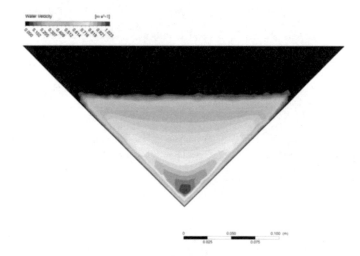

Figure 5. Water velocity distribution at the triangular weir.

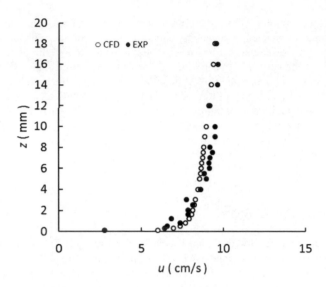

Figure 6.   Comparison of velocity profiles obtained from CFD and experiments.

REFERENCES

ANSYS CFX Solver Theory Guide.

Aydin M. C. (2012). CFD simulation of free-surface flow over triangular labyrinth side weir. Advances in Engineering Software. 45, 159–166.

Filonovich M. S., Azevedo R., Rojas-Solórzano L. R., Leal J. B. (2013). Credibility analysis of computational fluid dynamic simulations for compound channel flow. Journal of Hydroinformatics. 15.3, 926–938.

Hargreaves D. M., Morvan H. P., Wright N. G. (2007). Validation of the volume of fluid method for free surface calculation: the broad-crested weir. Engineering Applications of Computational Fluid Mechanics. 1(2), 136–146.

Hoseini S. H., Jahromi S. H. M., Vahid M. S. R. (2013). Determination of Discharge Coefficient of Rectangular Broad-Crested Side Weir in Trapezoidal Channel by CFD. International Journal of Hydraulic Engineering. 2(4), 64–70.

Juretic F., Gosman A. D. (2010). Error Analysis of the Finite-Volume Method with Respect to Mesh Type. Numerical Heat Transfer, Part B: Fundamentals: An International Journal of Computation and Methodology. 57, 414–439.

Kirkgoz, M. S., Akoz, M. S., and Oner, A. A. (2008). Experimental and theoretical analyses of two-dimensional flows upstream of broad crested weirs. Can. J. Civ. Eng., 35(9), 975–986.

Lv X., Zou Q. P., Reeve D. E., Zhao Y. (2009). A novel coupled level set and volume of fluid method for sharp interface capturing on 3D tetrahedral grids. J Comput Phys. 229(7), 2573–604.

Sarkar M.A., Rhodes D.G. (2004). Calculation of free-surface profile over a rectangular broad crested weir. Flow Measurement and Instrumentation 15, 215–219.

Roache P. J. (1997). Quantification of uncertainty in computational fluid dynamics. Annu. Rev. Fluid. Mech. 29, 123–60.

Rombust P., Tralli A., Verhaart F., Langeveld J., Clemens F. (2014). Applicability of CFD Modelling in Determining Accurate Weir Discharge: Water Level Relationships. 13th International Conference on Urban Drainage, Sarawak, Malaysia.

You D., Mittal R., Wang M., Moin P. (2006). Analysis of stability and accuracy of finite-difference schemes on a skewed mesh. Journal of Computational Physics. 213(1), 184–204.

Advances and Trends in Engineering Sciences and Technologies – Al Ali & Platko (Eds)
© 2016 Taylor & Francis Group, London, ISBN: 978-1-138-02907-1

Construction philosophy "Light Building"

B. Kovářová
Faculty of Civil Engineering, Department of Technology, Mechanization and Construction Management,
Brno University of Technology, Brno, Czech Republic

ABSTRACT: Light Building combines the latest technology, materials, procedures and safeguards for investors. It frees the construction of the classic "heavy" and "wet" processes of concrete and masonry as well as all the complexity and potential mistakes. Light Building brings investors lightness, quality, speed, unique thermal properties and many other added values. The part of the concept is also ensuring the high quality of the materials and technologies. Selecting and examining individual components of the building can be achieved to the maximum extent extraordinary guarantees. To achieve above mentioned structural insulated panels are used. In paper is described history of it and current possibilities, which is Light Building based on. Typical example is included.

1 HISTORY OF SIP'S

The building technology SIPs (structural insulated panels) is gradually taking hold in our construction market. It was developed in the USA, it was first used for the construction of a residential building in 1952. Intensive work on the development of SIPs was since 1935, but as starting of use SIPs panels is considered to year 1952 where mentioned building architect Alden P. Dow used panels with a core of expanded polystyrene on. (Liška 2011) The basis of SIP technology is therefore a panel made by sticking large-scale wood-based panels at the core of the thermal insulator. The condition is sufficient core strength because the panel does not contain any reinforcing elements. Therefore, the SIP's panel also sometimes referred to as layered insulated panel without reinforcing ribs. From the static point of view SIP panel is always considered as a whole, which needs to be emphasis during the preparation of the project, since it is not possible in any way weaken, e.g. with grooves for distribution. Piping and wiring in this case must be kept in cavities of spaced walls (ANSI/APA 2009).

Using the SIPs panel construction gives the wide application of the external walls, over the ceiling structure to the roof structure. It is thus the basis for the application of prefabrication in the construction of energy-efficient buildings. It is also the basis for Light Building.

Light Building as a term is used for the opposite of conventional construction. Easy, fast, quality and cheap to build, easy to use and save money on energy. Get rid of classic heavy and wet processes during construction (Liška 2012) Light Building term isn't worldwide used at the moment, but exactly describes philosophy of its construction.

2 PREFABRICATION IN WOODEN CONSTRUCTIONS

The usage of prefabrication allows you to move generally complex and laborious processes from site to equipped factories with consequent acceleration and improvement of construction, increase productivity, and in particular transformation "building" to "assembly." (Liška 2009).

Generally divided into prefabrication plain prefabrication (plain structural components of different sizes and levels of completion) and 3D prefabrication (building or multipurpose cell spatial structures segments). (TimberFrame Design)

Figure 1.   Isometric diagram of the module and position.

Prefabrication however, also has its disadvantages. With increasing degree of prefabrication it decreases variability of buildings. The versatility of the use of prefabricated elements is also limited. With increasing dimensions of prefabricated elements decreases rapidly ease of transport and handling both in the factory and in the final assembly site.

The above limits have been optimized within the production of SIPs. The result is a base panel width of 1200–1500 mm. Panel thickness is inferred from the desired heat resistance. According to the manufacturers it differ sheath material and methods of mutual connection panels. The most commonly used as the sheathing OSB panels, plywood, cement-wood chips board or fiber cement. (DimensionCanada Network)

The SIPs with these dimensions and structure are easy to handle and, if necessary, also easily divisible directly on site during the assembly of the object.

The disadvantage of this structural system is its somewhat higher purchase price in comparison to traditional building materials. This disadvantage is compensated by lower operating costs and better use of space, as due to the material and technical characteristics it is possible to use lower wall thickness while maintaining or achieving better parameters in comparison to traditional building materials used for extensions of objects. Due to the use of only one type of building material for all types of load bearing structures are simultaneously significantly reduced the overall life cycle costs of the object. (VareaModul)

Among the disadvantages of the structural system can also include the need for additional insulation against fire. In the basic configuration due to applied building system and details of connections can be considered in case of fire resistance only with cladding panel. The most commonly used SIPs panels have a coat of OSB, a required fire protection without additional modifications do not. That is achieved by tiling the plasterboard fire protection in combination with mineral insulation systems oriented from the outside of the object. Sometimes the required fire resistance reaches also installing spaced walls with fire insulation. This solution is mostly used in combination with the implementation of piping and wiring in the cavity of spaced wall. (Structural Insulated Panels)

Another limiting factor is the need for the construction of a certified firm, which is equipped, trained for the proper implementation of this type of construction. This ensures the quality of the construction and the use of quality and for the usage of certified materials and processes.

3 PLANAR AND 3D PREFABRICATION

As mentioned above, one aspect of a classification is its degree of prefabrication.

Plain prefabrication using SIPs panels represents the possibility of producing large panels with prepared openings for windows and doors. The advantage of this system is the rapid assembly of the object at the site. The disadvantage of this system is the "sensitivity" to the dimensional tolerance and quality of the foundation structure. Using a panel format mentioned above, i.e. up to width 1250 mm, largely eliminates the above deficiency, but at the cost extension of the assembly building and increasing labor intensive at construction site. (Europanel domy)

Generally, planar prefabrication based on SIP's prefabricated elements represents an increase in the efficiency of building production compared to traditional construction materials. On the other hand, it should be noted that finishing work remains in comparison with conventional construction basically in the same range.

3D prefabrication, on the other hand, represents a significant increase in the efficiency of the assembly work using factory pre-finished building cells, which are built in into the final work. These spatial elements are usually equipped and have surface finishes, sanitary ware, etc., so that on site only takes place assembly of these cells into the overall construction works. The ratio of finishing work is to the maximum possible extent eliminated and moved to the factory. Finishing work is to be carried out also from certified materials and certified procedures while respecting SIP technology (e.g. not grooving performed as mentioned above).

Prefabrication on the one hand speeds up the assembly process at the site, on the other hand, places increased requirements on the precision of foundations and accuracy of the position of utilities for the object.

Prefabrication is one of the cornerstones of the philosophy Light Building Construction.

4 3D PREFABRICATION – THE BASIS OF LIGHT BUILDING

The term "light company," originating in Japan, simply means a minimum of materials to create maximum added value. The same principle applies in the construction industry. For example, a house is built of heavy brick or concrete and is insulated with light insulation. We can also build a house of a layered structure, which includes an insulator, which takes over the static function. We save energy, which would cost production of bricks or concrete. It increases so the internal space of the house thanks to thinner walls, thus saving time and money. This house is about the idea, about the needs of individual investors, therefore about the architecture, design and new approaches. Light Building is exactly the new approach. Within the Light Building combines planar and spatial prefabrication, as described above. Panels for floors, walls and roofs are a standard element of the building system. Spatial prefabricated elements for these houses are 3D cores. The core is always equipped with electric switchboard for the whole house. Within the core, the house is connected to sewerage systems, water and electrical connections. The core is supplied in the form of a toilet with boiler, sink, washing machine connection, or as a fully equipped bathroom with an area of 7.5 square meters. Always surfaces are finished equipped with all fixtures and fittings. Equipment of core is dependent on the type of house and customer's wishes. Usage of this core, however, places additional requirements on the layout of the object: the kitchen must be due to the connection located as adjacent room to the core. (Liška 2012)

Other technologies for application in the context of philosophy Light Building is not much. More are those that do not belong to Light Building. These houses have minimal heat loss, so it pays for their heating to use heating convectors or underfloor electric heating. To deal with hot-water systems, gas connections, or solid fuels is completely unnecessary and expensive. In houses Light Building is consistently used forced ventilation with heat recovery.

The construction is carried out so that the first is to build a foundation for building and prepare the shaft with a connection of all installations. To the substructure is fitted in the right position core and connected with installation at the shaft. And only now around the core build a house. The basic construction element of these objects is therefore prefabrication as a core around which to base planar prefabrication "finishes" residential house. (Europanel ®)

It is important to realize that not every wooden structure is based on the philosophy Light Building; Light Building mainly represents a new view of the construction of houses based on a combination of spatial and planar prefabrication in order to maintain the maximum benefits of both types of prefabrication.

Figure 2.   Placing ground screws (Liška 2011/11).

Figure 3.   Installation of core (Liška 2011/11).

5 EXAMPLE OF IMPLEMENTATION OF LIGHT BUILDING HOUSE

In the following example shows typical implementation process of Light Building house. For the example was chosen the family house with ground screws and foundation beams.

Building site was Janov (Hřensko, Czech Republic). Chosen house has built up area of $79\,m^2$, converted space of $390\,m^3$. Needed construction time was 2 months (October, November 2011). Total costs 2,5 Mio CZK w/o VAT.

On site construction begins by placing screws and foundation beams.

Following installation of the above core with installations on the foundation beams to the exact position. This activity requires great accuracy. Inaccuracies in position ultimately may impact at floor coverings.

Figure 4.   Use of prefabrication (Liška 2011/11).

Figure 5.   Final family house (Liška 2011/11).

It follows enclosures around core structures of the building. This phase uses mostly planar prefabrication, although there may also be building elements for accelerating spatial prefabrication, as shown in the following figure. It is not a typical usage of the mentioned spatial prefabrication, because the cell does not have finished surfaces.

Completion of the object has been held in a conventional manner planar prefabrication and finishing work.

6 CONCLUSION

Like any new construction technologies even Light Building brings to the process of building a number of advantages. The most important here include accelerate and improve their own construction process by transferring a part of the production process from site to central manufactures with repeatable high accuracy and maximum completion of spatial prefabricates.

Each technology has its disadvantages, of course, which at Light Building technology primarily we count increased requirements for the precision of foundation of a building and for the precision on location of shaft with connection of the object. As a disadvantage to some extent can be also mark a plan requirement of adjacent kitchens to core.

Despite the above disadvantages Light Building technology represents a significant advance in understanding the use of prefabrication of low-energy buildings based on wooden structures. Its application brings significant financial savings within the life cycle of the object and reduces the environmental impact of the production of building materials and own build process.

REFERENCES

ANSI/APA PRS-610.1. Standard for Performance-Rated Structural Insulated Panels in Wall Applications. 12/2009
Byvaj lacno. 2014. *Europanel.* [Online] Available from: http://www.byvajlacno.sk [Accessed 21st March 2015]
DimensionCanadaNetworks. N.A. [Online] Available from: http://www.dimensioncanada.ca [Accessed 26th January 2015]
Europanel ®. N.A. [Online] Available from: http://www.europanel.cz [Accessed January 2015 – March 2015]
Europanel domy. N.A. [Online] Available from: http://www.montovanedomy.cz [Accessed March 2015]
Liška, L. 2009. Univerzálny stavebný systém pre drevostavby – technológia SIPs. *Stavebné materiály* (11/2009): 28–31
Liška, L. 2011. SIPs technologie konstrukčních izolovaných panelů. *Realizace staveb* (5/2011): 14–17
Liška. L. 2011/11. Private archive.
Liška, L. 2012. Nový směr ve výstavbě úsporných domů – Light Building. http://www.lighbuilding.cz
Structural Insulated panels. N.A. [Online] Available from: http://sipbuilding.wordpress.com [Accessed February 2015]
TimberFrame Design. N.A. [Online] Available from: http://www.timberframedesign.net [Accessed 27th January 2015]
VareaModul Nízkoenergetické domy. N.A. [Online] Available from: http://www.vareamodul.cz [Accessed 14th February 2015]

Advances and Trends in Engineering Sciences and Technologies – Al Ali & Platko (Eds)
© 2016 Taylor & Francis Group, London, ISBN: 978-1-138-02907-1

Company size impact on construction management documents processing and using

M. Kozlovska, M. Spisakova & D. Mackova
Faculty of Civil Engineering, Technical University of Košice, Košice, Slovakia

ABSTRACT: Nowadays nobody doubts about sense and content of the construction project preparation. Additionally construction project preparation presents an essential assumption for the trouble free progress of the execution phase of construction. An emphasis is on the active using of the construction project preparation documents, which are used for the observing of the contract conditions, budget and required quality of the construction project. The aim of this paper is to investigate (using chi-squared test) the impact of company size to construction management documents processing and using in Slovak construction companies. This investigation is created through results of research done by Institute of Construction Technology and Management at the Faculty of Civil Engineering, Technical University in Kosice.

1 INTRODUCTION

Construction management documents, required for construction procurement, referred to construction processes which are determined by costing, scheduling, and planning based on individual construction activities involved to documentation of construction project (Kozlovská et al., 2014). A first key step is identification of these activities. The construction industry, distinct from many other industries, has well-designed methods for decomposing the complexity of major multi-year projects into fundamental work items and tasks for use in detailed cost and schedule tracking. (Winch, 2002). There are different definitions of construction procurement. Moshini (1993) described the construction procurement as a process involving a sequence of decisions and/or actions that a client engages in as soon as the need to acquire a new facility arises. Alarcon et al. (1999) defined construction procurement as the process used to supply equipment, materials and other resources required to carry out a project. Charvat (2000) noted that construction procurement is the process through which the client brings together the team and resources needed to translate project plans into physical reality. The International Standard Organization's document on construction procurement also defined construction procurement as a process which creates, manages and fulfils contracts relating to the provision of goods, services and engineering and construction works or disposal, or any combination thereof. From these definitions according Ibem et al., there appears that construction procurement is a process involving a series of activities and steps through which clients acquire specified goods and services related to engineering and construction works within a given period of time, cost and agreed terms (Ibem et al., 2014).

We can divide construction procurement into two groups – public or private. Both of them it should be processed construction management documentation. This paper is focused on survey of this documentation processing and using in Slovak construction companies and identifies influences of the company size on the level of documents processing. The impact of this factor on mentioned documents processing and using, according to available sources, was not the subject of previous researches.

The governmental funding in the construction sector is provided through the public procurement. According to Transparency international, construction works is subject with the largest share of total volume of public procurement in Slovak republic. The public procurement should reflect transparency, economy, efficiency and the absence of discrimination, and should respect ethical and moral values (Juszczyk et al., 2014, Kozik et al., 2013, Radziszewska-Zielnia, 2010). Given that, the public procurement is a process which aims to ensure the proper functioning of state bodies by choosing the most efficient, transparent and cost-effective contracts and suppliers financed from public funds. The public procurement has a specific procedure of implementation. The governments provide funding for a broad range of projects and programmes covering areas such as regional and urban development, employment and social inclusion, agriculture and rural development, maritime and fisheries policies, research and innovation and humanitarian aid. The funding for a regional and urban development can be considered as a significant effect to construction industry. The construction works present 40 % out of total volume of public contract in Slovakia (tender.sme.sk, 2015).

In Slovakia, the public procurement is governed by Act 25/2006 Coll. of Laws on Public Procurement where are transposed legislation of EU. We can indirectly deduce from the Act's provisions that procurement should reflect transparency, economy, efficiency and the absence of discrimination, and should respect ethical and moral values. The Act applies to orders for the supply of goods, carrying out of works, provision of a variety of services, tenders, and the administration of public procurement. Within the application of public procurement law, there should be a proper definition of an order and the object of the contract, whether that is goods, services or works (Taby, 2015). Construction works present all work associated with the construction, alteration, modification or removal of structures. The requisites of documents are determined by the Act 254/1998 Coll. of Laws on Public works which identifies the conditions for the preparation, assessment and quality of public works.

Public authorities may use different criteria when evaluating tenders – for example they may select according to the lowest price offered or they may use other criteria. In the latter case, each applicant should be informed of the different weighting given to the different criteria (for example price, technical characteristics and environmental aspects).

For any public procurement exceeding the thresholds must to be published contract notice or notice of a design contest and notice of the contract award announcing the results of the public procurement.

Public authorities may only begin evaluating tenders after the deadline for submission has expired. If the tenderer has submitted a tender, he has the right to be informed as soon as possible whether or not you have won the contract. If he has not been selected, he are entitled to a detailed explanation of why your tender was rejected. The public authority must observe strict confidentiality regarding the exchange and storage of your data.

According the requirements of Public Procurement Law, the main criteria for bids evaluation is the lowest offered price. On the other hand, procurer also assesses the documents aimed to preparing of construction.

The aim of this paper is to investigate the impact of company size to construction management documents processing and using in Slovak construction companies. The price offered by tenderers is not the subject of this research.

Based on the analyses of public procurement procedure (determined by Public Procurement Law) and particular documents for construction preparing (determined by Public Works Law), construction projects documents were analysed.

Tenderer has to processes the *concept of construction procedure* where is exactly demonstrated the construction procedure and ability of constructor to perform the construction (Baskova & Krajnak, 2013, Mesaros & Mandicak, 2015). The tenderer has to process also the *list of machinery and its technical specification* (Gasparik & Gasparik, 2010) needed for the realization of construction works. The needed machinery has to be in accordance with the proposed construction processes, construction schedule and required quality of construction works.

The *schedule of construction works* with the critical path describes a time aspects of construction process. It determined the sequence of processes and define the processes which are on critical path, ie. affect the overall construction period. Considering schedule, the procurer assesses the ability to meet the deadline specified in the by the business.

The aspect of building site is describes by the *proposal of building site equipment and organization* in terms of Slovak legislation. It is aimed to the designed clear drawing, localization of sanitary site equipment and office site equipment, information about the electricity consumption for site, proposal of storage space site, way of construction waste disposal, condition of occupational safety and health, way of ensure site fire protection, schedule of site equipment removal and important contacts of company (Strukova & Kozlovska, 2013). The proposal of building organization provides the comprehensive overview of building site organization which is used for the assessment of way of construction methods and site preparation.

The *quality plan for construction realization* is document processed by quality manager in terms of the Act 254/1998 Coll. of Laws on Public works and also according the standard STN EN ISO 9001. There is defined the inspection and test plan for particular construction process, type and measures of tests. Based on the quality plan the procurer can assess the compliance the required and offered quality of construction works what presents quality aspects of construction (Kozlovska et al., 2014).

The aim of this paper is to investigate the impact of company size to construction management documents processing and using in Slovak construction companies. This investigation is created through results of two researches done by Institute of Construction Technology and Management at the Faculty of Civil Engineering, Technical University in Kosice. The first study (chapter 2) of research finds out the impact of company size to complexity of documents for construction within public procurement. The second study (chapter 3) of research investigates and verifies (using chi-squared test) the impact of company size to selected construction management documents creating and using. The second study (a larger number of samples – 256) was used to confirm the results of the first study (a smaller number of samples – 11).

2 FIRST STUDY OF DOCUMENTS ASSESSMENT FOR CONSTRUCTION WITHIN PUBLIC PROCUREMENTS

This study is focused on investigation of the company size impact to complexity of documents for construction within public procurement. The subjects of study are eleven tenders for public procurement. Tenders are made for four projects – reconstructions of comparable type (historical buildings) and volume of construction works. Five small and six medium companies are public procurement tenderers in this study.

After examination of project documentation of each project within assessment in this study it was necessary to get information about project documentation of four reconstructions.

Tenders for four reconstruction projects were assessed in this study. The subject of the first reconstruction (Project No. 1) is mainly exterior changes because of the building aesthetics. The object of this solution is mainly replacement of all windows and doors on the façade which are largely unsatisfactory and dilapidated condition. Repairs related to the exchange of external and internal windows sills, exchange wooden shingles on the roof, including flashings, dismantling plumbing products and repair chimneys. The largest part of the second reconstruction (Project No. 2) is the roof insulation, which is in a dilapidated condition. Doors will be also replaced and wider ramp will improve immobile readers' services. It will be completely repaint the entire interior equipment and repaired the cracks in the walls. The subject of Project No. 3 reconstruction is windows and doors replacement and complete reconstruction of the social facilities. For all objects also take place painting, floor replacement and reconstruction of interior plaster, roofing replacement and insulation of attic space. In all basements will be a new waterproofing. South side of the building will be insulated. Project No. 4 includes the façade reconstruction and renovation of the castle wall.

Subsequently, eleven tenders for public procurement were examined. There were two tenderers for the Project No. 1 and three tenderers for all of the others three projects. The documents complexity of eleven tenders was evaluated.

The task of this study was to find out the impact of company size to complexity of documents for construction within public procurement which are defined in chapter 1. On the basis of analysis of construction project document submitted for public procurement, it was compiled table and presence or absence of each document was recorded in this table for each tenderer and the results are shown in table 1.

Table 1. Company size impact to complexity of documents for construction within public procurement.

Project	Project No. 1		Project No. 2			Project No. 3			Project No. 4		
Tenderer	No. 1	No. 2	No. 3	No. 4	No. 5	No. 6	No. 7	No. 8	No. 9	No. 10	No. 11
Company size	S	M	S	S	S	M	M	M	S	M	M
Number of complete documents	11/13	13/13	13/13	9/13	11/13	10/13	12/13	13/13	11/13	11/13	13/13

S – small-sized company (up to 50 employees)
M – medium-sized company (up to 250 employees)

Table 1 provides an overview of company size impact to complexity of documents for construction within public procurement documents. It can be concluded that only four (three processed by medium-sized company) of the eleven companies provided complete documentation for construction within public procurement. One document was missing in one tender, four applicants were missing two documents, one tenderer missing three documents and one missing four documents.

According to company size of tenderers and results of complexity assessment we can say hypothesis that the larger companies process and use the construction documents by higher level. Considering the sample size is not enough for the statistical verification of this hypothesis, the hypothesis will be verified in the second study (chapter 3) with the larger sample size.

3 SECOND STUDY OF CONSTRUCTION MANAGEMENT DOCUMENTS PROCESSING AND USING IN SLOVAK CONSTRUCTION COMPANIES

Since 2003 Institute of Construction Technology and Management at the Faculty of Civil Engineering, Technical University in Kosice have done a research of construction management documents processing and using in Slovak construction companies through questionnaire survey. The questioning is the most widely used method of research. The basic method of research is questionnaire to respondents. A questionnaire was focused to find out the level of processing and using in Slovak construction companies depend on the company size.

This research analyses data collected in three years – 2003, 2008, 2013. Respondents (experts) were asked by personal questioning to answer questions relating to construction management documents. In 2003 were interviewed 48 respondents, 159 respondents in year 2008 and 59 respondents in year 2013.

This study is part of mentioned research and analyses processing of three selected documents of construction management documents depend on the company size:

- schedule of construction works (question Q1),
- quality plan for construction realization (question Q2),
- proposal of building organization (proposal of storage space site) (question Q3).

The questionnaire mentioned above contained questions related to processing of these documents and the respondent had scale of possible answers: every time, sometimes and never documents are processed. The share of answers (in percent) for all three questions depending on company size is shown in figure 1.

Using chi-squared test (significance level $\alpha = 0.05$) it was investigated influence of the company size to construction management documents processing. A null hypothesis was: The processing and using of documents is independent on the company size. The share of answers (in percent) for all three questions depending on company size is shown in table 2. Consequently chi-squared test was done and the results are shown in table 3.

As we can see in table 3, p-value of chi-squared test is significantly lower than significance level $\alpha = 0.05$, therefore we reject null hypothesis and accept alternative hypothesis: *Processing and using of construction management documents is dependent on the company size.*

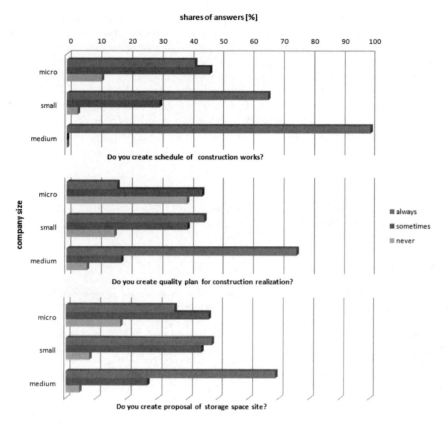

Figure 1. The share of answers for questions Q1, Q2, Q3 depending on company size.

Table 2. Shares of answers for questions Q1, Q2, Q3 [%].

	always [%]			sometimes [%]			never [%]		
	Q1	Q2	Q3	Q1	Q2	Q3	Q1	Q2	Q3
micro	42	16	35	47	44	47	11	39	18
small	66	45	48	30	39	44	4	15	8
medium	100	76	69	0	18	27	0	7	4

Table 3. Results of chi-squared test.

	p-value
Q1	$1.054 \cdot 10^{-16}$
Q2	$1.895 \cdot 10^{-16}$
Q3	$1.286 \cdot 10^{-16}$

4 DISCUSSION AND CONCLUSION

The construction process is a very complex system. Primarily, the successfully completed project must be carefully planned and controlled also during the construction process. The aim of this paper was to investigate the impact of company size to construction management documents processing and using in Slovak construction companies. This investigation was created through results of two researches done by Institute of Construction Technology and Management at the Faculty of Civil Engineering, Technical University in Kosice. The first study of research found out the impact of company size to complexity of documents for construction within public procurement. The second study of research investigated and verified (using chi-squared test) the impact of company size to creating and using of selected construction management documents. The hypothesis, that company size has impact on processing and using construction management documents was accepted based on the results of studies and statistic verification. It can be concluded, that larger companies process and use the construction documents by higher level.

ACKNOWLEDGEMENTS

The article presents a partial research result of project VEGA – 1/0677/14 *"Research of construction efficiency improvement through MMC technologies"*.

REFERENCES

Alarcon, L.F., Rivas, R. & Serpell, A. 1999. Evaluation and improvement of the procurement process in construction projects. *Proceedings IGLC-7, 26–28 July 1999*, University of California, Berkeley, USA.

Baskova, R. & Krajnak, M. 2013. An analytical approach to optimization of the costs for construction work. In *International Multidisciplinary Scientific GeoConference Surveying Geology and Mining. Albena, 16–22 June 2013.* Sofia: STEF92 Technology, Bulgaria.

Charvat, W.C. 2000. *Construction Procurement in the Architect's Handbook of Professional Practice*, New York.

Gasparik, J. & Gasparik, M. 2010. Machine selection optimizing method for building processes with software suport. In *27th International Symposium on Automation and Robotics in Construction, Bratislava, 25–27 June 2010.* Elsevier B.V.

Ibem, E.O. & Laryea, S. 2014. Survey of digital technologies in procurement of construction projects. *Automation in construction* 14: 11–21.

International Organization for Standard (ISO), 2010. Construction Procurement-Part 1: Process, Methods and Procedures, ISO, Geneva, Switzerland.

Juszczyk, M., Kozik, R., Lesniak, A., Plebankiewicz, E. & Zima, K. 2014. Errors in the preparation of design documentation in public procurement in Poland. In *Procedia Engineering Creative Construction Conference 2014, 21–24 June 2014.* Elsevier Ltd.

Kozik, R. & Plebankiewicz, E. 2013. Bid documentation in public procurement in Poland. *Organization, technology and management in contruction* 5: 712–719.

Kozlovska, M., Strukova, Z, & Tazikova, A. 2014. Integrated assessment of buildings quality in the context of sustainable development principles. *Quality Innovation Prosperity* 18 (2): 100–115.

Mesároš, P. & Mandičák, T. 2015. Information systems for material flow management in construction processes. In *IOP conference series: materials science and engineering, Tomsk, 15–17 October 2014.* IOP Publishing Ltd.

Moshini, R.A. 1993. Knowledge-based design of project-procurement process. *Journal of computing in civil engineering* 7: 107–122.

Radziszewska-Zielnia, E. 2010. Methods for selecting the best partner construction enterprise in terms of partnering relations. *Journal of Civil Engineering and Management* 16: 510–520.

Rules and procedures, available on: http://europa.eu/youreurope/business/public-tenders/rules-procedures

Statistics of Transparency international, available on: http://tender.sme.sk

Strukova, Z. & Kozlovska, M. 2013. Environmental impact reducing through less pollution from construction equipments. In *International Multidisciplinary Scientific GeoConference Surveying Geology and Mining. Albena, 16–22 June 2013.* Sofia: STEF92 Technology, Bulgaria.

Taby, M. 2013. Public procurement in the Slovak Republic, available on: http://spectator.sme.sk/c/20048881/public-procurement-in-the-slovak-republic.html

Winch, G. 2010. *Managing Construction Projects*. John Wiley & Sons.

Advances and Trends in Engineering Sciences and Technologies – Al Ali & Platko (Eds)
© 2016 Taylor & Francis Group, London, ISBN: 978-1-138-02907-1

Testing of concrete abrasion resistance in hydraulic structures

A. Kryžanowski & T. Birtič
Faculty of Civil and Geodetic Engineering, University of Ljubljana, Ljubljana, Slovenia

J. Šušteršič
IRMA Institute, Ljubljana, Slovenia

ABSTRACT: In the paper the phenomenon of abrasion of concrete in hydraulic structures is analysed, which is caused by the abrasion process resulting from the sediment carried by water. In detail the adequacy of the laboratory procedure for definition of the level of abrasion of concrete in the standard ASTM C 1138 hydraulic structures was investigated. Concrete composites with different mechanical properties were analysed within the research programme. The basic composition of concretes was modified by additives of cement binder, steel fibres and PP fibres, and rubber aggregate. High abrasion resistance of concrete was achieved with the addition of rubber aggregate. To this effect, a measuring system was designed based on laser triangulation with linear illumination of the surface measured. The system allows for contactless, quick and precise measuring of average wear and other topological characteristics dependent on the type of the tested concrete.

1 INTRODUCTION

In hydraulic structures the term 'abrasion' means the process of disintegration of exposed concrete surfaces, resulting from loads arising from sediment transport (Kryžanowski 1991 and 2009). The rate of disintegration of the concrete surface largely depends on the transport capacity of water and the manner of solid matter transport (Kryžanowski 2009, Mikoš 1993). The protection of structures against abrasion damage is realised by protective linings made of abrasion-resistant materials, together with the appropriate structural solutions. The development trend in this area is moving towards finding appropriate technical solutions and analyses of material suitability, taking into consideration the following criteria: (1) high resistance to physical and chemical processes of the action of the stream, (2) availability of the materials used, (3) feasibility and economy of the project, (4) minimal maintenance costs, and (5) durability of the design solutions. The latter is of vital importance from the viewpoint of reducing maintenance costs and ensuring normal operational readiness of the structures at the same time (Kryžanowski et al. 2009, Jakobs et al. 2001).

There is no general criterion for defining abrasion resistance when designing concretes in hydraulic structures. Usually, abrasion resistance of concretes is assessed based on a set of parameters that define the single mechanical properties of concretes, such as: compressive strength, tensile strength, aggregate strength, use of special cements, modulus of elasticity, water/cement (w/c) ratio, surface polishing, concrete cure, cement additives (fly ash, fibres), connected with investigating methods that more or less realistically simulate abrasive processes (Kryžanowski 2009). In this study, particular attention was paid to the use of granular rubber to replace a proportion of the gravel aggregate in order to improve the concrete's abrasion resistance. The addition of rubber aggregate changes the mechanical properties of the concrete, by reducing its compressive and bending strength, but increasing its toughness. This means that concretes with rubber aggregate are used in the production of paving blocks where increased resistance to wear of exposed surfaces is required (Toutanji 1996, Eldin et al. 1993). The problem of concrete's abrasion resistance was given special attention in the construction of the Sava River hydropower plants (Šušteršič et al. 2004). The abrasion resistance tests of the concretes used on the Sava River demonstrated that ASTM C1138 (underwater method) was the most appropriate test method (Kryžanowski et al. 2012).

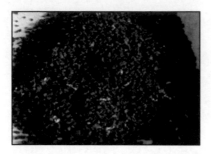

Figure 1. Rubber aggregate of cubic shape.   Figure 2. Rubber aggregate of elongated shape

2 PREPARATION OF CONCRETE COMPOSITION

Concrete mixtures were prepared by using Portland cement with 30% of slag, type: CEM II/A-S 42.5 R, which is in accordance with the SIST EN 197-19 standard. The aggregate was obtained by separating the natural crushed gravel from the alluvial Quaternary filling of the Sava River at the site. Fractions 0–4, 4–8 and 8–16 mm were used. Six samples of different concrete composition were intended for test purposes (Table 1): The C1 composition is adopted as control composition, which is basically the same as the composition of abrasion resistant concrete built in the spillways of the first plant of the Lower Sava River cascade. The nominal maximum gravel of 8 mm was adopted in the C2 composition and all further modifications. The C2 composite with minor modifications was used in the spillway of the second plant of the Lower Sava River cascade.

With the PC1 composition, representing the initial composition for all further modifications, (1) the mineral additive ($SiO_2 > 90\%$) and super plasticizer were replaced by polymeric binder (styrene-butadiene polymer with dry portion in dispersion $45.6 \pm 0.3\%$); (2) the proportion of steel fibres was replaced by doubling the polypropylene fibres ($L = 10\,mm$, $\varnothing 30 \sim 40\,\mu m$); (3) the proportion of the finest fraction (0–4 mm) was replaced by rubber aggregate. In the study, rubber aggregate was used, which is the end product of recycling scrap vehicle tyres following two different processes:

- Rubber aggregate obtained by mechanical crushing of scrap vehicle tyres with a characteristic cubic grain shape, similar to the usual crushed mineral aggregate. Rubber aggregate fractions of 0–4 mm and 4–6 mm, mostly of uniform composition, were used in the study (Fig. 1).
- Rubber aggregate obtained by grinding the tyre tread when renewing tyres (i.e. retreading), with a typically elongated shape and a distinctly fibrous structure of the rubber parts. In the study, we used 0–4 mm rubber aggregate fractions, with a very uneven texture (Fig. 2).

In the PC1 composition, 10% by volume of mineral aggregate with the 0–4 mm fraction was replaced by rubber aggregate of elongated shape, and, similarly, in the PC2 composition, 10% by volume of mineral aggregate was replaced by the rubber aggregate of cubic shape. In the PC3 composition, the rubber aggregate percentage by volume was doubled to 20%, with partial replacement of the mineral aggregate of fractions 0–4 mm and 4–6 mm. In the PC4 composition, the mineral aggregate was fully replaced by rubber aggregate of fraction 0–6 mm. We also added stone meal due to a lack of fine aggregate of fraction 0.5–1 mm. The w/c ratio in the composites did not vary considerably (Tab. 1). The compositions of concretes were prepared in the laboratory mixer with a vertical shaft and with a volume of 75 dm^3. Immediately after the mixing, the fresh concrete properties, such as temperature, slump, air content and density, were determined, following the standard SIST-EN procedures (Tab. 2). The properties of the hardened concrete were proven using the standard investigation methods. The average values of investigation results of the hardened concrete are given in Table 2.

Table 1. Concrete mixture proportions (kg per m^3 of concrete).

		C1	C2	PC1	PC2	PC3	PC4
Cement		440	450	450	450	450	450
W/C		0.39	0.41	0.41	0.41	0.41	0.41
Super plasticizer		3.23	3.31				
Polymer				90	90	90	90
Mineral supplement		22	22.5				
Steel fibers		40	40				
PP Fibers		0.5	0.5	1	1	1	1
Gravel	0–4 mm	818	1168	888	888	888	
aggregate	4–8 mm	226	477	514	514	390	
	8–16 mm	633					
	0–0.1 mm						337
Rubber	0–4 mm				59		380
aggregate	0–6 mm				59	118	38

Table 2. Results of investigations of fresh and hardened concrete.

Concrete composition			C1	C2	PC1	PC2	PC3	PC4
Concrete temperature		[°C]	13.5	16.6	12.7	14.0	14.5	19.8
Slump t = 0		[mm]	125	90	185	100	100	10
Slump t = 30 min		[mm]	80	60	125	70	85	/
Density		[kg/m^3]	2372	2315	2069	2137	2027	1546
Air content		[%]	3.0	5.4	9.5	6.1	7	8.9
Compressive strength	2 days	[MPa]	26.6	31.4	14.0	20.7	15.6	2.9
	28 days		61.4	65.4	27.2	32.1	24.3	6.2
Modulus of elasticity		[GPa]	36.1	31.8	19.0	22.4	17.8	3.3
Wear	28 days	[mm]	3.65	4.59	3.73	3.59	1.27	0.09
	400 days		2.15	2.75		2.92	1.51	0.17

3 PROGRAMME OF THE RESEARCH WORK

3.1 *Testing the wear of concrete specimens*

The study was performed in accordance with the standard ASTM C 1138 method (Fig. 3); namely, the literature contains an abundance of comparisons and results that the evaluation of our study results could be based on. The result of the test is the average depth of wear expressed by the average volume of wear on the surface of the specimen in the duration of the test (Liu, 1981). In our study the method of measuring the specimen's surface wear was upgraded using a continuous three-dimensional measurement based on laser triangulation and an advanced measurement analysis. The three-dimensional measuring system is based on the principle of laser triangulation, where the laser projector illuminates the surface along a line, while the camera aligned at a triangulation angle takes images of the illuminated surface. The shape of the surface is reflected in the laser contour curvature on the captured image. The entire surface is scanned while the profilometer rotates around the turntable axis (Fig. 4). The surface wear is calculated as the difference of depth images of the specimen surface before the test, and the specimen surface in the individual time increments of testing. The three-dimensional illustration of the eroded surface gives many more possibilities of interpreting the results of abrasion resistance measurements than the standard procedure, which only allows for qualitative interpretations of the damage to the surface based on photo documentation – such as position, depth, and the severity of wear (Kozjek et al. 2015).

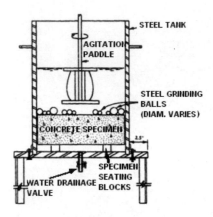

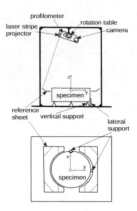

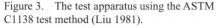

Figure 3. The test apparatus using the ASTM C1138 test method (Liu 1981).

Figure 4. Experimental system for 3D measuring of concrete wear using laser triangulation (Kozjek et al. 2015).

The abrasion resistance test using the ASTM C 1138 test method was performed at 28- and 400-day ages of specimens. At the 28-day age, the wear was tested using only the standard procedure for all test compositions. In wear measurements at the 400-day age, we modified the standard protocol by shortening the prescribed time interval of measurements from 12 hours to 3 hours in the first 36 hours of test duration; then the interval was extended to 6 hours until the prescribed total 72 hours of test duration. In this way we wanted to minimise the possibility of error in measurements, and to condense the measurements in the initial phase of testing when the wear dynamics is largest. In this phase, the concrete specimens' wear by the standard procedure was upgraded by using a laser profilometer to measure surface wear. This part of wear tests was not performed for PMC1 specimens due to technical difficulties.

• study, we used 0–4 mm rubber aggregate fractions, with a very uneven texture (Fig. 2).

3.2 Measurement results and interpretation

Figure 5 shows that all 28-day age rubberized concretes possessed excellent abrasion resistance; comparable results were obtained in the C1 composition, and poorer results in the C2 composition. By increasing the share of the rubber aggregate, the resistance to wear increases significantly, and in the PC4 composition it falls within the measurement accuracy limits. The abrasion resistance of conventional concretes significantly improves with the age of specimens: C1 and C2 compositions show comparable resistance. Somewhat less pronounced is the improvement of abrasion resistance in the PC2 concrete, where by increasing the percentage of rubber aggregate the abrasion resistance slightly decreases with the specimen age.

By densifying the time interval of measurements we also analysed the dynamics of wear progression. Figure 6 shows a distinct trend of wear growth in the initial 3 hours of test duration in all compositions, which, however, later decreases and, finally, stabilises after approx. 36 hours. This trend is best reflected in the compositions with high rubber aggregate percentages. It is less pronounced with conventional concretes, where the initial phase of growth, which lasts 6 hours with the C1 composition, is followed by a less pronounced phase of fading, but then, after 36 hours, there is another small upward trend.

One of the reasons for using a laser profilometer in abrasion resistance tests is to check the accuracy of wear measurements using the standard test by weighing the specimens. Figure 7 show that the measurement results obtained by using the standard method and those using the laser profilometer do not differ significantly. Notable deviations between the measurements' results were recorded in the PC4 composition, where the severity of wear was within the measurement tolerances (up to 15 g or 0.1 mm). In this case, laser profilometry measurements are more credible (0.005 mm accuracy of measurements).

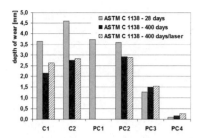

Figure 5. Comparison of results of measuring abrasion resistance of concretes.

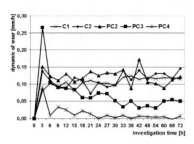

Figure 6. Dynamics of wear progression.

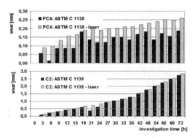

Figure 7. Comparison of the results according to the standard test method and laser measurements.

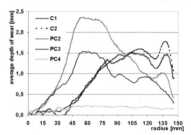

Figure 8. Surface wear profile of the specimen after test completion.

Based on monitoring the progression of concrete surfaces' wear using three-dimensional measurements we prepared typical cross-section profiles with average rates of wear as a function of distance of the observation point from the axis of the specimen. The figure shows two characteristic wear profiles (Fig. 8):

- The mixes with the addition of rubber aggregate revealed a narrow scour hole at the border of the impact area of the agitating paddle, of a width of around 30 mm, which steeply and evenly lowered towards the axis and the rim of the specimen. The specimen is eroded all over the surface. The scour hole is most pronounced in the PC3 composition and it considerably decreases with increasing the percentage of rubber aggregate content. In the PC4 composition, a characteristic scour hole shape is detected; however, the values are considerably lower than those found in other compositions.
- In compositions with mineral aggregates (C1 and C2) a wide scour hole at the rim of the specimen occurs, in a form of a 60-mm wide coil. Towards the axis, the rate of surface damage gradually decreases and is negligible in the operation area of the agitation paddle.

As seen in Figure 8, surface wear is not uniform, but rather it changes with the distance from the center of the specimen. This can be explained with the operation mechanism of the abrasion resistance test using the ASTM C1138 test method. Due to the rotating agitation paddle in the container a potential vortex occurs as a combination of rotational and irrotational flows. Rotational flow occurs in the core of the vortex, limited by the operation area of the agitation paddle, while irrotational flow occurs outside of the vortex's core. The water's velocity in the core of the vortex increases directly proportionally to the distance from the axis of the container towards the rim, reaching its maximum at the outer rim of the container. In the area of the irrotational flow, the water's velocity decreases inversely proportionally to the distance from the core of the vortex towards the rim of the container where it reaches its minimum.

The occurrence of various types of erosion damage was not specifically addressed in this study. We have identified a similar phenomenon already in previous studies, but only with concretes with extremely solid mineral aggregate, where the scope and type of erosion damage of the surface were similar to those found with rubberised concrete mixes. The use of laser profilometry will allow us to study the phenomenon in greater detail in future research. The studies will focus on studying the impact of changing the strength characteristics of rubberised concretes on abrasion resistance, and the possibility of interpreting concrete resistance from the aspect of the shape of surface damage.

4 CONCLUSIONS

As part of constructing the HPP chain on the Sava River, various studies were carried out to define the adequate test method and concrete mixes to achieve abrasion resistance of concrete lining used for spillway structures. High-performance concretes were analysed by using stone aggregate and the modified composition by adding rubber aggregate, which considerably improved abrasion resistance. The abrasion resistance tests of the concretes used on the Sava River demonstrated that ASTM C1138 (underwater method) was the most appropriate test method.

We studied the abrasion resistance of rubberized concretes by using two types of rubber aggregates: crushed rubber aggregates as the product of mechanical crushing of scrap vehicle tyres, and rubber chips obtained as waste product in the retreading of tyres. We used various rubber aggregate percentages (from 10% to 20%) in proportion to the mineral aggregate, and the mineral aggregate was also fully replaced by the rubber aggregate. The concrete mixes with the addition of rubber aggregate had above-average abrasion resistance based on the comparable characteristics of usual high-performance concretes.

In the study the method of measuring the specimen's surface wear was upgraded using a continuous three-dimensional measurement based on laser triangulation and an advanced measurement analysis. The three-dimensional illustration of the eroded surface gives many more possibilities of interpreting the results of abrasion resistance measurements than the standard procedure. This was the preferred method for measuring low rate of wear, where the standard procedure by weighing the specimen is not credible enough due to the lower accuracy of measurements. Testing the specimens' surface wear revealed that the shape of eroded surface and the deterioration process varied according to the aggregate used, which could not be established using standard measurements. The application of laser profilometry allows for a more accurate interpretation of the deterioration process with the possibility of analysing the impact of individual material parameters on the dynamics of the concrete surface deterioration process.

REFERENCES

Eldin, N.N. & Senouci, A.B. 1993. Rubber-tire particles as concrete aggregate. Cement Concrete Aggregates. Vol. 15 1: 74–84.
Jakobs, F., Winkler, K., Hunkeler, F. & Volkart, P. 2001. *Betonabrasion im Wasserbau*, VAW, 168. Zürich: ETH.
Kryžanowski, A. 2009. Abrasion resistance of concrete on hydraulic. Structures. *PhD thesis*. University of Ljubljana.
Kryžanowski, A. 1991. Analyses of concrete resistance to deterioration due to water action. *MSc thesis*. University of Ljubljana.
Kryžanowski, A., Mikoš, M., Šušteršič, J. & Planinc, I. 2009. Abrasion resistance of concrete in hydraulic structures. *ACI mater. j.*, 106 4: 349–356.
Kryžanowski, A., Mikoš, M., Šušteršič, J., Ukrainczyk, V. & Planinc I. (2012). Testing of concrete abrasion resistance in hydraulic structures on the Lower Sava River. *Strojniški vestnik – Journal of Mechanical Engineering*, vol. 58, no. 4, p. 245–254.
Kozjek, D., Pavlovčič, U., Kryžanowski, A., Šušteršič, J. & Jezeršek, M. 2015. Three-dimensional characterization of concrete's abrasion resistance using laser profilometry. *Strojniški vestnik – Journal of Mechanical Engineering*, in print.
Liu, T.C. 1981. Abrasion resistance of concrete. *Journal Proceedings – ACI*, vol. 78, no. 5, pp. 641–350.
Mikoš, M. 1993. Fluvial abrasion of gravel sediments, *Acta hydrotechnica* 11 10: 107 pp. University of Ljubljana.
Šušteršič, J., Kryžanowski, A., Planinc, I., Zajc, A., Dobnikar, V., Leskovar, I. & Ercegovič, R. 2004. Technical report: Performance of concrete exposed to underwater abrasion loading (in Slovenian). Ljubljana: IRMA.
Toutanji, A.H. 1996. The use of rubber tire particles in concrete to replace mineral aggregates. Cement & Concrete Composites. Vol. 18: pp. 135–139.

Advances and Trends in Engineering Sciences and Technologies – Al Ali & Platko (Eds)
© 2016 Taylor & Francis Group, London, ISBN: 978-1-138-02907-1

Assessment of the effect of polymer binder on some properties of repair mortars

L. Licholai, B. Debska & M. Jedrzejko
Rzeszow University of Technology, Rzeszów, Poland

ABSTRACT: One of the most popular methods of the reinforced concrete structure repair is filling in concrete losses with different kinds of repair mortars. Usually these are polymer mortars or mortars modified with polymers. This article presents the results of investigation on some chosen properties of a repair system based on epoxy resin modified with poly(ethylene terephthalate) glycolisate (PET). The article demonstrates how some important characteristics of repair mortars change with the increase in polymer content. The composites analysed were compared with standard cement mortars and also with the results obtained by other authors. The flaws and assets of each of the repair systems were discussed.

1 INTRODUCTION

Concrete is a universal construction material widely used in the building industry all over the world. It is a composite obtained by mixing a binder (cement), water and aggregate as well as possible admixtures or additives. After curing it shows high compressive strength accompanied by low tensile strength. Because of its low chemical strength, concrete is susceptible to smaller or greater damage. Owing to its versatility it is applied in different environments, but due to this it is often exposed to the effect of various factors of destructive nature. The ensuing damage can badly affect the durability of objects, safety of their functioning and their aesthetic qualities. Concrete as a construction material is cheap and easy to use, but its repair during its lifetime is complicated and costly. Maintenance and repair of concrete will always be of importance as concrete structures are expected to last very long. Concrete repair has become almost a kind of universal challenge. Even the present creation of high utility concretes – high quality and ultrahigh quality ones – does not necessarily mean that a long-lasting material has been obtained. Thus the necessity of repair will also involve new generation concretes (Czarnecki & Emmons 2002). The matter of the guarantee of repair effectiveness and consequently of long term serviceability of the object remains open – in particular with regard to the choice of the material. A good decision in this respect is one of the basic conditions for a successful repair. There are many methods of repair and repair materials. The variant to be applied should guarantee good compatibility – that is, the material to be repaired and that used for the repair should be fully compatible.

Standard cement mortars (CM) can be used for repairs (Horszczaruk & Brzozowski 2014, Mallat & Alliche 2011). However, the most often used repair systems are polymer-cement mortars (PCM) obtained by adding a polymer, oligomer or monomer to a cement mixture. Composites own their unique properties to polymers. Some interesting examples of their effective application for repairs and a review of the factors that may determine the quality of the combination can be found, among other things, in the following works: (Beushausen et al. 2014, Ahmad et al. 2012, Luo et al. 2013, Brien & Mahboub 2013, Dawood & Ramli 2014, Łukowski 2007, Gruszczyński 2011, Al-Zahrani et al. 2003, Afridi et al. 1995, Saccani & Magnaghi 1999, Schulze & Killerman 2001, Garbacz et al. 2006, Momayez et al. 2005, Mirza et al. 2014. An alternative solution is resin concretes (PC) which do not contain cement but are obtained by mixing synthetic resins or monomers with aggregate and then curing the resin binder (Mirza et al. 2014, Jung et al. 2015,

Dębska & Lichołai 2012 a, b). Here, the content of polymers is much higher than in polymer-cement concretes. That is the reason why resin concretes are rather rarely used for repairs. They are mostly applied for special purposes when, for example, an object repaired must be back in use in a few days and even hours (e.g. early reopening of a runway). Such concretes are also applied to repair constructions made of upgraded strength concretes, in case of exposure to chemical aggression (e.g. electrolytic tanks in the copper industry) or pipes in micro tunneling (trenchless underground water-pipe network laying) (Czarnecki 2010). Very high mechanical strength and good chemical resistance of resin concretes ensure improved duration of joints made of such materials. However, the application of materials of appropriate durability does not necessarily mean lasting repair effects (Czarnecki & Emmons 2002). Moreover, synthetic resins used in polymer concretes in large quantities are rather expensive. Besides, using only materials very similar to traditional concrete mixtures for repairs does not guarantee success and deprives the builder of the significant benefits of new materials. One of them is undoubtedly the epoxy mortar suggested in this paper. It was obtained by partially substituting resin binder with glycolisate based on poly(ethylene terephthalate) waste and propylene glycol. The mortar shows very high values of strength parameters, very good chemical resistance, low water absorbability and good adhesion. Moreover, the modifier applied reduces production costs and allows rating this composite as environmentally friendly. The article presents the results of strength tests, water absorbability tests and concrete substrate adhesion tests for the modified epoxy mortar. The parameters were compared with the results described in literature obtained for other repair systems whose polymer content is 0% and 15% by weight relative to their cement weight.

2 EXPERIMENTAL DETAILS

2.1 Materials

The resin mortar was fabricated using Epidian 5 epoxy resin (Organika Sarzyna S.A.), Z-1 hardener (Organika Sarzyna S.A.), standard sand of 0–2 grain size and poly(ethylene terephthalate) glycolisate based on propylene glycol. Mortar samples without modifier were also made.

2.2 Mix proportions

The composition of the mortar (resin/aggregate ratio (Z/K) and the percentage content of glicolisate PET) was designed based on the literature data and the author's own experience gained from the previous preliminary experiments (Dębska & Lichołai 214, Lichołai & Dębska 2014). Determinations were made for the composite which earlier showed the best values of the checked characteristics, i.e. for a mortar with a modifier at 9.0% and 0.25 resin/aggregate ratio Z/K. Samples of mortars without the modifier were also made for comparison.

2.3 Sample preparation

The mortar that was prepared was put in $40 \times 40 \times 160$ mm steel moulds, specifically shaped to be used later in further experiments to determine tensile strength at bending. After the flexural strength tests, there were obtained six beam halves which were later used to carry out compressive strength tests. Three $60 \times 60 \times 5$ mm samples of mortars were prepared for water absorbability tests. The mortar for adhesion tests was arranged as described by the PN-EN 1542 specification, in a form of 5 mm thick layer, on blocks of standard concrete of C30/37 strength grade. Concrete of this grade was selected with regard to the results quoted in the literature (Łukowski 2007). Later, the experimental results obtained were compared with them. The specimens were left under laboratory conditions for seven days to cure.

2.4 Testing procedure

Tests for tensile strength at bending f_f as well as for compression strength fc were conducted in testers supplied with appropriate inserts, on standard beams, following the PN-EN 196-1:2006 specification.

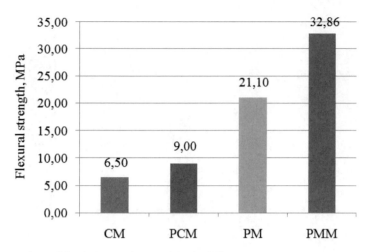

Figure 1.   Comparison of flexural strength test results for different repair systems.

The determination of adhesion by f_h tearing was performed according to the PN-EN 1542 specification. The method involved tearing off steel rings glued to the surface of the epoxy mortar with an epoxy two-component glue. The area to be investigated was outlined by an appropriate spot drilling of the surface. The investigation was conducted with a Dyna Pull-off Tester making use of steel, 20 mm thick disks of 50 mm in diameter.

The absorbability tests were done following the PN-EN ISO 62:2008 specification.

3   RESULTS AND DISCUSSION

3.1   *Flexural strength*

The results for epoxy mortar significantly differ from the other ones, which was shown in Figure 1.

Flexural strength for epoxy composite (PM) reached 21.1 MPa. The use of the PET modifier (PMM) improves it by 50%, up to 32.86 MPa. While comparing these parameters with those of a standard cement mortar (CM) one can see a five-time increase in the flexural strength determined. The polymer-cement mortar (PCM) also shows higher flexural strength in comparison with standard cement mortar. It equals 9 MPa.

3.2   *Compressive strength*

The results of the compressive strength test for epoxy mortars and cement mortars with different contents of polymer (0–15% by weight) are shown in Figure 2. The highest compressive strength was obtained for the epoxy mortar modified with PET glycolisate (PMM). It equaled 116.08 MPa. The epoxy mortar without modifier (PM) had compressive strength of 97 MPa. The polymer-cement mortar (PCM) showed compressive strength at a level of 70 MPa. The lowest value, 33 MPa, was achieved by the cement mortar (CM). Also the polymer-cement mortar (PCM) shows lower compressive strength compared to epoxy mortars.

The lowering of compressive strength may be in this case due to the delay of the cement hydration process as a result of the formation of a polymer film reducing the access of water (Al-Zahrani et al. 2003).

3.3   *Absorbability*

Comparison of absorbability test results for various repair systems is shown in Figure 3. Cement mortar (CM) shows the highest absorbability that equals 6.8%. An addition of the polymer reduces

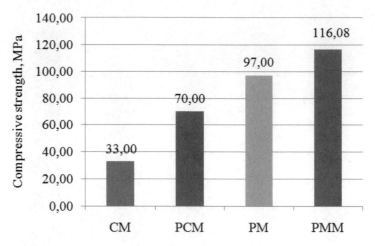

Figure 2. Comparison of compressive strength test results for various repair systems.

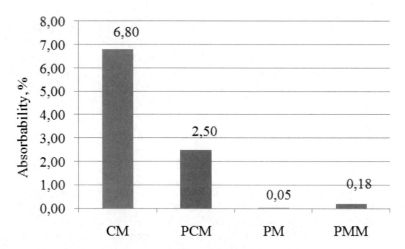

Figure 3. Comparison of absorbability test results for various repair systems.

that value to 2% in the case of polymer-cement mortars (PCM). Epoxy mortars show the lowest absorbability. For resin mortar with an addition of glycolisate PET (PMM) absorbability equals 0.18%, and in the case of a repair without the modifier (PM) only 0.05%.

3.4 *Adhesion*

During the adhesion test one tear-off model was observed. This model of tear illustrates cohesion damage in the layer of the repaired substrate. In such a case the value of adhesion is higher than that determined. The correct adhesion value would be obtained if the damage occurred at the interface of the repair and repaired materials.

Results of concrete substrate adhesion tests for epoxy mortar and for epoxy mortar modified with S3 glycolisate are shown in Table 1.

During the adhesion test for the epoxy mortar, in the case of all the measuring points, destruction occurred in the concrete. This means that the adhesion of epoxy mortar with a modifier as well as without it is higher than the tensile strength of concrete itself and is contained within a range of 1.34 MPa–3.78 MPa (Table 1). This adhesion is comparable with that obtained without an addition of a modifier. Also the thickness of the mortar had only a marginal effect on the test result.

Table 1. Results of adhesion tests for epoxy resins.

Strength class for concrete	No. of measurement	Tear model	Tearing off strength, N	Adhesion, N/mm^2
Results for a 5 mm thick repair mortar without modifier/with modifier				
30/37	1	A	5570/6110	2.84/3.11
	2	A	6270/5700	3.19/2.9
	3	A	7410/4860	3.78/2.48
	4	A	5150/2990	2.62/1.52
	5	A	6370/3590	3.25/1.83

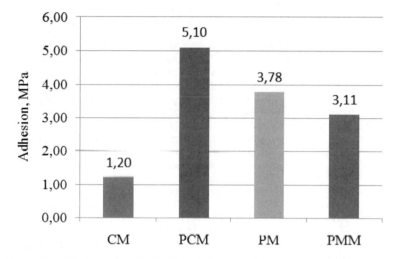

Figure 4. Comparison of adhesion results for three different repair systems.

Unfortunately, the investigation failed to determine the real adhesion of the repair mortar. That was due to the insufficient tensile strength of the concrete used for the experiment.

Figure 4 lists the maximum results obtained during the adhesion test for three different repair systems.

The lowest adhesion values are characteristic of a standard cement mortar (1.2 MPa). For the polymer-cement mortar adhesion was 5.1 MPa. For epoxy mortars, the PET modifier has no considerable effect on the characteristic discussed. The results obtained for resin composites do not represent the real adhesion. It can only be said that the adhesion is higher than 3.78 MPa.

4 CONCLUSIONS

For the application scope of repair materials it is important that the repair systems be characterized by high strength, low absorbability and good adhesion to the substrate under repair. A comparison of these parameters for mortars containing different amounts of polymers allows reaching the following conclusions:

- Resin mortar and modified resin mortar show the highest values of flexural and compressive strength values while preserving very low absorbability.
- The highest adhesion values are characteristic of polymer-cement mortar, but the results can hardly be compared with those obtained for resin mortars because of too weak concrete substrate.

REFERENCES

Afridi, M.U.K. et al. 1995. Water retention and adhesion of powdered and aqueous polymer-modified mortars. *Cement and Concrete Composites* 17: 113–118.

Ahmad, S. et al. 2012. Use of polymer modified mortar in controlling cracks in reinforced concrete beams. *Construction and Building Materials* 27: 91–96.

Al-Zahrani, M.M. et al. 2003. Mechanical properties and durability characteristics of polymer and cement-based repair materials. *Cement & Concrete Composites* 25: 527–237.

Beushausen, H. et al. 2014. The influence of superabsorbent polymers on strength and durability properties of blended cement mortars. *Cement & Concrete Composites* 52: 73–80.

Brien, J.V. & Mahboub, K.C. 2013. Influence of polymer type on adhesion performance of a blended cement mortar. *International Journal of Adhesion & Adhesives* 43: 7–13.

Czarnecki, L. & Emmons, P. H. 2002. Naprawa i ochrona konstrukcji betonowych. *Polski Cement*. Kraków.

Czarnecki, L. 2010. Polymer concrete. *Cement Lime Concrete* 2: 63–85.

Dawood, E.T. & Ramli, M. 2014. The effect of using high strength flowable system as repair material. *Composites: Part B* 57: 91–95.

Dębska, B. & Lichołai, L. 2012. Badanie możliwości wykorzystania modyfikowanych zapraw epoksydowych w procesach naprawczych betonów. Cz.1. Oznaczanie właściwości wytrzymałościowych, absorpcji wody oraz odporności chemicznej. *Zeszyty Naukowe Politechniki Rzeszowskiej. Budownictwo I Inżynieria Środowiska* 59: 149–160.

Dębska, B. & Lichołai, L. 2012. Badanie możliwości wykorzystania modyfikowanych zapraw epoksydowych w procesach naprawczych betonów. Cz.2. Oznaczenie przyczepności. *Zeszyty Naukowe Politechniki Rzeszowskiej. Budownictwo I Inżynieria Środowiska* 59: 161–168.

Dębska, B. & Lichołai, L. 2014. A study of the effect of corrosive solutions on selected physical properties of modified epoxy mortars. *Construction and Building Materials* 65: 604–611.

Garbacz, A. et al. 2006. Characterization of concrete surface roughness and its relation to adhesion in repair systems, *Materials Characterization* 56: 281–289.

Gruszczyński, M. 2011. Zastosowanie nowoczesnych betonów specjalnych na przykładzie falochronu wyspowego w porcie Gdynia. *Architektura Czasopismo Techniczne Politechniki Krakowskiej* 11: 265–272.

Horszczaruk, E. & Brzozowski, P. 2014. Bond strength of underwater repair concretes under hydrostatic pressure. *Construction and Building Materials* 72: 167–173.

Jung, K.-Ch. et al. 2014. Evaluation of mechanical properties of polymer concretes for the rapid repair of runways. *Composites: Part B* 58: 352–360.

Jung, K.-Ch. et al. 2015. Thermal behavior and performance evaluation of epoxy-based polymer concretes containing silicone rubber for use as runway repair materials. *Composite Structures* 119: 195–205.

Lichołai, L. & Dębska, B. 2014. The multidimensional response function exemplified by epoxy mortars: Looking for the global extreme. *Archives of Civil and Mechanical Engineering* 14: 644–675.

Luo, J. et al. 2013. Bonding and toughness properties of PVA fibre reinforced aqueous epoxy resin cement repair mortar. *Construction and Building Materials* 49: 766–771.

Łukowski, P. 2007. Polimerowo-cementowa zaprawa naprawcza o podwyższonej przyczepności do betonu. *Awarie budowlane 2007, XXIII Konferencja naukowo-techniczna, Szczecin- Międzyzdroje 23–26 maja 2007.*

Mallat, A. & Alliche, A. 2011. Mechanical investigation of two fiber-reinforced repair mortars and the repaired system. *Construction and Building Materials* 25: 1587–1595.

Mirza, J. et al. 2014. Preferred test methods to select suitable surface repair materials in severe climates. *Construction and Building Materials* 50: 692–698.

Momayez, A. et al. 2005. Comparison of methods for evaluating bond strength between concrete substrate and repair materials. Cement and Concrete Research 35: 748–757.

Saccani, A. & Magnaghi, V. 1999. Durability of epoxy resin-based materials for the repair of damaged cementitious composites. *Cement and Concrete Research* 29: 95–98.

Schulze, J. & Killermann, O. 2001. Long-term performance of redispersible powders in mortars. *Cement and Concrete Research* 31: 357–362.

Advances and Trends in Engineering Sciences and Technologies – Al Ali & Platko (Eds)
© 2016 Taylor & Francis Group, London, ISBN: 978-1-138-02907-1

Liquid-filled cavity in basic insulation glass unit

J. Lojkovics, D. Kosicanova & R. Nagy
Technical University of Košice, Civil Engineering Faculty, Institute of Architectural Engineering,
Department of Indoor Technologies and Building Services, Košice, Slovakia

ABSTRACT: Our theoretical part of research focuses on designing a modern active façade transparent element, which can utilize incident solar energy. The system is energy efficient due to liquid absorbing flowing heat in the window cavity. Currently, full scale model of window without glass unit is prepared. Other components of system such as glass unit with measuring points, heat exchanger, pumps, volume storage tank and measuring equipment are designed to complete the system and then measuring in laboratory conditions may start.

1   INTRODUCTION

1.1   *Liquid-filled cavity in window – principles*

According to Chow et al (2011), a water-flow window carries a water circuit that allows a stream of clean water to flow upward within the entire space between two glass panes. Positioned at the external wall this works as an advanced window device able to reduce solar transmission. In this way, not only the excessive solar heat is absorbed by the flowing water stream, the indoor solar heat gain is also effectively reduced. From the window, the heat-carrying water flows to the water-to-water heat exchanger and preheats the cold feed water stream. When hot water is in need, this pre-heated water stream will be brought up to the required temperature by gas/electrical heater. The water-filled glazing system works with either natural or mechanical water circulation, hence consumes no energy except pumping power where required. Besides, the outer and inner glazing can be both clear glass panes at low cost, and maintaining excellent psychological and daylight performance.

1.2   *Possibility of applications of liquid-filled window*

There is an energy-saving potential in applications with these systems, especially in buildings with large glass surfaces which make use of hot water like gyms, swimming pools, hotels. Non-residential buildings are very suitable for these façade elements, because of the way that they use and consume hot water, electricity for HVAC systems etc.

2   CAVITY AND LIQUID MATERIAL PROPERTIES

2.1   *Glass properties*

Construction of window counts with basic Insulation Glass Unit (IGU) with triple insulation glass. Increased mechanical loads caused by presence of liquid in the window cavity lead us to use tempered glass in all three positions A, B, C as shown in Figure 1. In effort to increase energy collected from solar irradiance, a metallic layer on glass pane is supposed to be used in position 1 or 4. This thesis was presented in the previous papers.

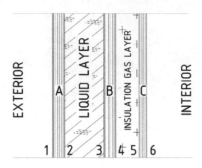

Figure 1.   Detail of glass panes in window.

Table 1.   Liquid properties.

Properties	Water at 20°	Ethylene-glycol at 20°	Air at 20°
Density	$9.9778\ 10^2$	$1.1156\ 10^3$	1.2047
Dynamic Viscosity	$9.7720\ 10^{-4}$	0.0213	$1.8205\ 10^{-5}$
Kinematic Viscosity	$9.7937\ 10^{-7}$	$1.9119\ 10^{-5}$	$1.5111\ 10^{-5}$
Specific heat	$4.0764\ 10^3$	$2.3865\ 10^3$	$1.0061\ 10^3$
Conductivity	0.6048	0.2499	0.0256
Thermal diffusivity	$1.4868\ 10^{-7}$	$9.3891\ 10^{-8}$	$2.1117\ 10^{-5}$
Thermal expansion coefficient	$3.4112\ 10^{-3}$	$9.3891\ 10^{-8}$	$3.4112\ 10^{-3}$
Prandtl number	6.5870	203.63	0.7156

2.2 *Liquid properties*

Due to ambient conditions we consider to use mixture of water and ethylene-glycol as a usually used fluid in common technical applications. This mixture requires anti-freeze and specific optical properties. Table 1 shows relevant physical properties at 20°C. Difference between water and ethylene-glycol is observed, even in Prandtl number, importance of which is described in part 3.1 of following chapter.

3   CONVECTION IN PRIMARY CIRCUIT OF WINDOW – CAVITY CONVECTION

3.1 *Geometry of cavity*

As Figure 2 shows, there are 12 built-in sensors S1–S12. These sensors provide us the vertical temperature field in liquid layer in the cavity. This is the most important point in further supply pipe improvement. Dimension of window as primary circuit is 1.750×1.150 m.

3.2 *Dimensionless criteria – Reynolds number*

The dimensionless term Reynolds number, Re, is universally employed in the correlation of experimental data on frictional pressure drop and heat and mass transfer in convective flow. According to Lighthill 1970, Reynolds number represents the ratio of force associated with momentum (ρu^2) to force associated with viscous shear ($\rho u/L$). Alternatively it may be regarded as a measure of the ratio of turbulent energy production per unit volume ($\rho u^3/L$) to the corresponding rate of viscous dissipation ($\eta u^2/L^2$). Below a lower critical value of Reynolds number flow is laminar, above a higher critical value flow is turbulent. Between these values the flow is in so called 'transition'. The higher critical value is strongly dependent on upstream conditions, Reynolds observed values

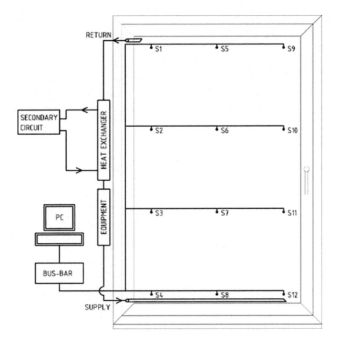

Figure 2. Schematic view of system.

of between 11,800 and 14,300 for water in bell mouth tubes of a few centimeters diameter. The lower critical value is less sensitive and is usually quoted simply as the critical Reynolds number. Its value for smooth circular pipes and tubes is approximately 2,000.

From the above we note that value of Re can be described as:

$$R_e = \frac{dv\rho}{\eta} \tag{1}$$

where d = characteristic dimension; v = fluid velocity; ρ = fluid density and η = fluid viscosity.

3.3 *Dimensionless criteria – Grashof number*

According to Hewitt (1994), Grashof number, Gr, is a nondimensional parameter used in the correlation of heat and mass transfer due to thermally induced natural convection at a solid surface immersed in a fluid. It is defined as:

$$G_r = \frac{gl^3 \beta \Delta T}{v^2} \tag{2}$$

where g = acceleration due to gravity, l = representative dimension, β = coefficient of expansion of the fluid, ΔT = temperature difference between the surface and the bulk of the fluid, v = kinematic viscosity of the fluid.

The significance of the Grashof number is that it represents the ratio between the buoyancy forces due to spatial variation in fluid density (caused by temperature differences) to the restraining force due to the viscosity of the fluid. Since Reynolds number, Re, represents the ratio of momentum to viscous forces, the relative magnitudes of Gr and Re are an indication of the relative importance of natural and forced convection in determining heat transfer. Forced convection effects are usually insignificant when $Gr/Re^2 \gg 1$ and conversely natural convection effects may be neglected when

Gr/Re$^2 \ll 1$. When the ratio is of the order of one, combined effects of natural and forced convection have to be taken into account.

3.4 Dimensionless criteria – Prandtl number

According to Hewitt (1994), Prandtl number, Pr, is a dimensionless parameter representing the ratio of diffusion of momentum to diffusion of heat in a fluid.

$$P_r = \frac{v}{\alpha} \tag{3}$$

where v = momentum diffusivity and α = thermal diffusivity. The Prandtl number can alternatively be expressed as:

$$P_r = \frac{\mu c_p}{k} \tag{4}$$

where μ = absolute or dynamic viscosity, c_p = specific heat capacity and k = thermal conductivity. Prandtl number is a characteristic of the fluid only and it reaches values for gases Pr = 0.7–1.0, for water Pr = 1–10, liquid metals Pr = 0.001–0.03, oils Pr = 50–2000. Ethylene glycol is considered as oil in this example, and the value of Prandtl number at 20°C reaches value Pr = 203.6.

3.5 Dimensionless criteria – Rayleigh number

According to Hewitt (1994) & Bejan (2004), Rayleigh number is associated with buoyancy of natural convection. Rayleigh number is a function of Grashof number, which describes the relationship between buoyancy and viscosity within a fluid and Prandtl number, which describes relationship between momentum diffusivity and thermal diffusivity. Thus the Rayleigh number itself may be viewed as the ratio of buoyancy and viscosity forces times the ratio of momentum and thermal diffusivities:

$$Ra_x = Gr_x \, Pr \tag{5}$$

where Gr_x = Grashof number and Pr = Prandtl number.

According to Jicha (2001) at natural flow for vertical surface we can observe in both laminar and turbulent flow. It is represented by value of Rayleigh critical number Ra_x. If $Ra_x < 10^9$ the flow is laminar, $Ra_x > 10^9$ the flow is turbulent [Jicha 2001]. It is necessary to say, that the fluid properties Pr, v, α and β are evaluated at the film temperature which is defined as:

$$T_f = \frac{T_s + T_\infty}{2} \tag{6}$$

where T_s = Surface temperature (temperature of the wall) and T_∞ = Quiescent temperature (fluid temperature).

3.6 Dimensionless criteria – Nusselt number

According to Hewitt (1994), Nusselt number, Nu, is the dimensionless parameter characterizing convective heat transfer. It is basically defined as:

$$Nu = \frac{\alpha L}{k} \tag{7}$$

where α is convective heat transfer coefficient, L is representative dimension (e.g., diameter for pipes), and k is the thermal conductivity of the fluid. Nusselt number is a measure of the ratio between heat transfer by convection (α) and heat transfer by conduction alone (k/L).

Table 2. Correlations for Nusselt number in rectangular enclosures for various ranges of Ra and Pr.

Fluid	Ra_s Range	Pr Range	H/S Range	Nu_S
Gas	$< 2 \times 10^3$	–	–	$Nu_S = 1$
Gas	$2 \times 10^3 - 2 \times 10^5$	0.5–2	11–42	$Nu_S = 0.197\, Ra_s^{0.25}\,(H/S)^{-1/9}$
Gas	$2 \times 10^5 - 2 \times 10^7$	0.5–2	11–42	$Nu_S = 0.073\, Ra_s^{0.3333}\,(H/S)^{-1/9}$
Liquid	$< 2 \times 10^3$	–	–	$Nu_S = 1$
Liquid	$10^4 - 10^7$	1–20,000	10–40	$Nu_S = 0.042\, Ra_s^{0.25}\,Pr^{0.012}\,(H/S)^{-0.3}$
Liquid	$10^6 - 10^9$	1–20	1–40	$Nu_S = 0.046\, Ra_s^{0.3333}$

4   SPECIFIC CONDITIONS OF NATURAL CONVECTION IN ENCLOUSURES

4.1   *Internal free convection*

In all the natural convection applications discussed above, heat is exchanged between a surface and a surrounding fluid of infinite extent. In this section, we focus attention on natural convection inside enclosed spaces, where heat is transferred between two surfaces separated by a fluid. Applications occur in building design, double-pane windows, solar collectors, cryogenic containers, electronic equipment etc.

According to Kaminski (1986), the rectangular vertical enclosure is considered. The length, L, is very large compared to the height, H, and spacing, S, therefore, we may approximate the geometry as two-dimensional. The left wall is at temperature T_1 and the right wall is at T_2, where $T_1 > T_2$. The top and bottom surfaces are perfectly insulated. Heat is transferred from the hot to the cold surface by natural convection of the fluid within the enclosure. The character of the buoyancy-induced flow depends on the Rayleigh number defined, for this case, as:

$$Ra_S - Gr_s\, Pr = \frac{g\beta\varrho^2(T_1-T_2)S^3}{v^2}\, Pr \qquad (8)$$

where g = acceleration due to gravity, S^3 = representative characteristic dimension, distance between two parallel surfaces in this case, β = coefficient of expansion of the fluid, ΔT = temperature difference, v = kinematic viscosity of the fluid.

If Rayleigh number is very low, $(Ra < 1708)$, viscous forces are much larger than buoyancy forces and no fluid motion occurs. This may happen, for example, if the spacing, S, is small or the viscosity, μ, is large. At these low Rayleigh numbers, heat transfer is by pure conduction in the fluid between the walls, and Nusselt number reduces to $Nu = 1$. At larger Rayleigh number, fluid near the hot left wall rises and fluid near the cold right wall sinks. A large rotating fluid cell promotes heat transfer between the plates. As Rayleigh number increases further, the boundary layers on right and left walls become thinner, and counter-rotating cells appear in the corners. At very high Rayleigh numbers, the flow becomes transient and turbulent.

According to Kaminski (1986), the Nusselt number and heat transfer rate equation to be used with these correlations is:

$$Nu = \frac{\alpha S}{k} \qquad (9)$$

$$Q = \alpha A(T_1 - T_2) = \alpha HL(T_1 - T_2) \qquad (10)$$

where H = height of cavity, L = length of the cavity.

According to Jicha (2001), there are other correlations describing internal free convection in vertical cavity as shown in Table 3. Nusselt number represents an average value, also the restrictions are required. For cavity correlations, all fluid properties are evaluated at the average surface temperature $T = (T_1 + T_2)/2$. L is in this case the distance between hot and cold walls.

Table 3. Correlations for Nusselt number in rectangular vertical cavity by Jicha (2001).

Average Nusselt Number	Restrictions
$Nu_L = 0.18((Pr/Pr + 0.2)Ra_L)^{0.29}$	$1 \leq (H/L) \leq 2$
	$10^{-3} \leq Pr \leq 10^5$
	$10^3 \leq Ra_L \leq (Ra_L\ Pr/0.2 + Pr)$
$Nu_L = 0.22((Pr/Pr + 0.2)Ra_L)^{0.28}(H/L)^{-1/4}$	$2 \leq (H/L) \leq 10$
	$Pr \leq 10^5$
	$10^3 \leq Ra_L \leq 10^{10}$
$Nu_L = 0.42\ Ra_L^{1/4}\ Pr^{0.012}(H/L)^{-0.3}$	$10 \leq (H/L) \leq 40$
	$1 \leq Pr \leq 2 \times 10^4$
	$10^4 \leq Ra_L \leq 10^7$

5 CONCLUSIONS

Due to different properties of flowing liquid in the cavity, it is necessary to establish values of Rayleigh number. These values represent type of flow that can be laminar, transient or turbulent one. In vertical enclosures the values of Rayleigh number reach value for water $Ra = 1.4769\ 10^6$, $Ra = 2.6134\ 10^9$ for ethylene-glycol, respectively. These values indicate non-laminar flow based on condition $Ra \gg 2000$ for both liquids. Using ethylene-glycol increases the ratio of the energy transferred by convective heat transfer.

Due to the closer cavity width (glass panes distance $S = 0.022$ m), a heat conduction transfer is assumed to prevail over the energy transferred by convective heat transfer. As mentioned above if they are close together, so that S/L is small, the boundary layers join and the flow is fully developed at the exit. Natural convection in a cavity depends on the boundary conditions at the walls.

There is a scope for further research of this issue and applying more appropriate correlations, based on very specific geometric and material conditions in the window, which will continue in further publication.

ACKNOWLEDGEMENTS

This work was funded by project KEGA 052TUKE-4/2013 Využitie virtuálneho laboratória pri navrhovaní energeticky efektívnych budov.

REFERENCES

Bejan, A. 2004. Convection heat transfer. 3rd ed. Hoboken, N.J.: Wiley.
Chow, T.T., Hunying, L., Zhang, L. 2010. Innovative solar windows for cooling demand climate. Solar energy Materials & Solar Cells, vol. 94, pp. 212–220.
Chow, T.T., Hunying, L., Zhang, L. 2011. Thermal Characteristics of water-flow double-pane window. International Journal of Thermal Sciences, vol. 50, pp. 140–148.
Hewit, G., Shires, G. L. and Bott, T. R. 1994. Process Heat Transfer. Boca Raton, FL.: CRC Press.
Jicha, M. 2001. Přenos tepla a látky. Brno, CZ.: Vysoké učení technické v Brně, Fakulta strojního inženýrství.
Kaminski, D.A., Prakash, C. 1986. Conjugate natural convection in a square enclosure: effect of conduction in one of the vertical walls. International Journal of Heat and Mass Transfer, vol. 29, pp. 1979–1988.

Advances and Trends in Engineering Sciences and Technologies – Al Ali & Platko (Eds)
© 2016 Taylor & Francis Group, London, ISBN: 978-1-138-02907-1

Potable water savings by use of rainwater in school building

G. Markovic & Z. Vranayova
Technical University in Košice, Faculty of Civil Engineering, Department of Building Services,
Košice, Slovakia

ABSTRACT: Stormwater management is quite a new topic in Slovakia. There is neither a legal framework nor standards or guidelines on how to implement sustainable stormwater management techniques. Rainwater harvesting as a form of source control measures could contribute to sustainability in stormwater management as well, by supporting potable water conservation and sustainability in water management in general. Rainwater harvesting also lowers the amount of runoff entering a stormwater sewer and helps decrease water consumption of potable water.

1 INTRODUCTION

Rainwater harvesting is not the only part of source control measured in SWM, it is also feasible way of controlling water consumption and supporting qualitative and reasonable water use for different purposes. One of the objectives of WFD (2000) is to promote sustainable water use (Ocipova, 2009), based on the long-term protection of available water resources and we can say that RWH contributes to this objective.

There are numerous techniques, approaches and models known around the world on how to support sustainable stormwater management, especially in urban areas, where the stormwater can cause significant damage (Lash et al., 2014; Yu et al., 2014). The aim is to manage stormwater as close to its source as possible which is also called source control covers number of measures. Rainwater harvesting as a form of source control could contribute to sustainability in stormwater management, by supporting potable water conservation and sustainability in water management in general. There are at least two very important facts which need to be considered when dealing with rainwater management. The first is the increasingly changing climate, resulting in short term, but more intensive precipitation on one hand and increasing droughts in some countries on the other. The second fact is ever increasing urbanization over the years which has changed the natural water processes and increased urban runoff significantly. These facts have influenced urban drainage and it is assumed that they will influence it even more in the future.

Rainwater represents high quality alternative source of water for buildings supply but of course, we must take into the account the designated purpose for rainwater use. For example, there is big skepticism about the quality of rainwater in our country (Ocipova, 2007, Botka, 2011; Poliak, 2011). However, there are numerous research and studies around the world about rainwater quality which shows sufficient quality for the purpose of flushing toilets, irrigation, cleaning and of course as drinking water after necessary treatment (Despins et al., 2009; Schets et al., 2010; De Busk and Hunt, 2013). Our own research of rainwater quality demonstrates thatrainwater meets quality standards for the purposes of collection, storage and re-use, as well as for the purposes of rainwater infiltration (Ahmidat et al., 2013).

2 RAINWATER – ALTERNATIVE WATER SUPPLY SOURCE FOR BUILDINGS

The capacity of the storage reservoir has to correspond to its intended consumption (toilet, irrigation, laundry, kitchen, bathroom, etc.) and investment exigency (Bose, 1999).

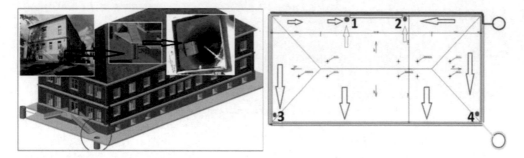

Figure 1.   Location of infiltration shafts near building PK6, ground plan of PK6 roof.

Another factor influencing the capacity of the storage reservoir is needed rainwater supply. The rainwater supply calculated annually for dry spells will require a bigger storage reservoir than is necessary for the case of stable rainfalls. Lower rainwater regions will require an alternative water supply. If rainwater storage reservoir represents the only drinking and service water supply, investment exigency factor will not play such an important role and storage reservoir capacity will have to be big enough to ensure ample water supply (Mitchell et al., 1997; Coombes et al., 2002). Following these conditions, rainwater storage reservoir capacity will be determined by:

- required water supply
- rainfall quantity and quality
- rooftop area

The intensity of rainfall in combination with rooftop area will determine the maximum accumulation rate. If water quantity is not insufficient, area for capturing needs to be increased or water consumption needs to be minimized. There are many ways of influencing water consumption in a positive way for example flushing toilets with an economic flushing system (with smaller and bigger flush), economic water tap etc. If these steps do not lead to minimized water consumption, water has to be gained also from alternative sources (Hlavatá, 2005).

Effective use of rainwater from capturing rainwater for further reuse represents one of the alternatives of saving precious potable water. Every building has the potential to be used for capturing rainwater. Surface area or surface for capturing rainwater is determined by the roof structure or other compact surface from which rainwater is drained by drainpipes into a storage reservoir.

The most common type of buildings used for capturing rainwater is the family house. Theoretical yearly profit of rainwater depends on precipitation and roof area for rainwater capturing. Rainwater is – Slovakia only used as service water.

3   EXPERIMENTAL RESEARCH IN THE CAMPUS OF TU KOŠICE

The project APVV SUSPP-0007-09 with title "Increasing rainwater management efficiency for the purpose of energy demand minimization" has been implemented at the Faculty of Civil Engineering, Technical University of Kosice. It has been devoted to the investigation of quantity and quality of rainwater and its infiltration. The resources that provide us with information about the quality and quantity of rainwater are located at the campus of Technical University of Kosice. First is a rain gauge located on the roof of University Library (Zeleňáková et al., 2011,2014).

The research takes place near building PK6 on campus. All rainwater runoff from the roof of this building is flowing into two infiltration shafts (Figure 1). The roof area of the PK6 building is 548,55 m^2.

Both infiltration shafts are located on the east side of the building. The shafts are realized from concrete rings with an outer diameter of 1000 mm. The measuring devices that provide us with

Figure 2.   Measurement devices – Data unit M4016 in shaft A, measurement flume with ultrasonic level sensor in shafts.

information about the volume of incoming rainwater from the roof of the building PK6 and also information about the quality of rain water are located in these infiltration shafts (Markoviè and Vranayová, 2013).

3.1   Measuring devices

The data acquisition system, respectively a control/data unit for generating measurement data, is the M4016 data unit, which is situated in infiltration shaft A (Figure 2). Infiltration shaft B, devices located in this shaft, are also connected to the control unit. Registration and control unit equipped the M4016 includes universal data logger, telemetric station with built-in GSM module, programmable control automat and multiple flow meter if a M4016 is connected to an ultrasonic or pressure level sensor (Figure 2) (Markoviè et al., 2014). Under the inflow, respectively rain outlet pipe in the shaft, there are measurement flumes for metering the inflow of rainwater from the roof of the building in both infiltration shafts. Rainwater from the roof of the building PK6 is fed by rainwater pipes directly into measurement flumes, which are placed under the ultrasonic level sensor which transmits data of the water level in the measurement flumes to the data unit M4016 (Figure 2). The unit M4016, from which the signal is transmitted, has an ultrasonic level sensor with a preset of up to 14 equations or the most used sharp crested weirs. Flow rate is calculated from the relationship between water level/flow rates. The purpose of our measurements is to calculate the instantaneous and cumulative flow, from the water level used by predefined profile – Thomson weir.

3.2   Possible use of harvested rainwater in real conditions at the university campus of Kosice

Total rainfall represents the theoretical amount of rainfall in mm, falling on a surface of interest. Totals of rainfall depend on specific locations. The average of yearly totals of rainfall is about 770 mm/year in Slovakia.

As already mentioned above, the research is done in the area of Kosice – at the campus of Technical University of Košice. The average value of annual precipitation for the Košice-city is 628 mm/year. Average precipitation rainfall in Kosice varies from 28 mm in January (winter) to 86 mm in June (summer).

Taking into account the 548 m^2 PK6 roof area, theoretical potable water savings according to yearly precipitation from the years 1900–2010 (figure 3) for the city of Kosice is shown in figure 4 and represents the potential volume of rainwater for a building's water supply.

The rain gauge which is located at the campus of the Technical University of Košice provides us with information about rainfall intensities. These intensities are compared with real volumes of rainwater inflow to the infiltration shaft mentioned above. Measured values of rainfall during our

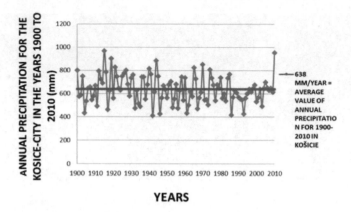

Figure 3. Annual precipitation for the Kosice in the years 1900 to 2010.

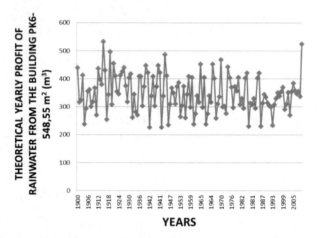

Figure 4. Theoretical yearly profit of rainwater from the building PK6 according to yearly precipitation from the years 1900–2010 for the city of Kosice.

research are shown in figure 5. We use a recording heated rain gauge for year round measuring. There are known unheated rain gauges as well used for a limited part of year when the temperatures aren't so low. Heated rain gauge is used for measuring liquid precipitation (rain) and solid precipitation (snow) as well.

Initial measurements started and continue to be in use in infiltration shaft A since March 2011, when the inflow of rainwater runoff from (212 m^2) of roof of the building PK6 began to measured. In March 2012, the research was extended to measurements of rainwater quantity in infiltration shaft B. It provides us with data regarding rainwater quantity from the entire roof area (548,55 m^2) of PK6 building. Figure 6 represents theoretical volumes of rainwater for 548,55 m^2 of roof according to formula 1 and measured rainfall while figure 7 represents a comparison of theoretical volumes and real measured volumes of rainwater for the roof area of 548,55 m^2 of PK6 building for possible use in this building. (Note: august 2012 without data due to equipment failure).

4   CONCLUSION

The widespread opinion in our country is that the rainwater harvesting system should provide full coverage of water demand for a specific purpose in building. But the main goal of this article and

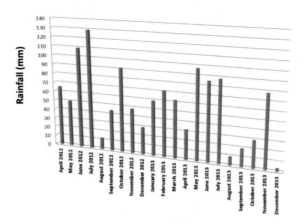

Figure 5.   Measured values of rainfall during our research.

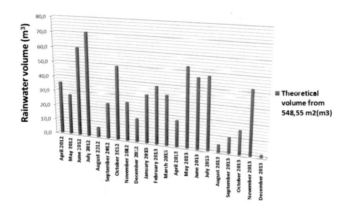

Figure 6.   Theoretical volumes of rainwater from roof of PK6 building.

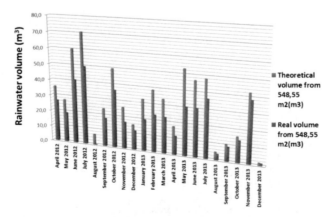

Figure 7.   Theoretical volumes of rainwater from roof of PK6 building and real volumes of rainwater from roof of PK6 building during research.

rainwater harvesting system is to show rainwater as an alternative source of water and avoid of wasting precious drinking water which can be replaced by rainwater. From an environmental point of view the most volume of drinking water is wasted for the purpose of – flushing toilets (Gabe et al., 2012; Ghimire et al., 2012). When we consider with 6 liters per one flush of toilet, figure 6 represent the number of possible flushes during our research and also represents the volume of drinking water which could be replaced by rainwater for this purpose. The total amount of rainwater measured during our research from March 2011 to December 2013 is $552{,}57\,\mathrm{m^3}$. Measured real volumes of rainwater are lower than theoretical volume of rainwater for all the calculated months. It's given by calculating with a coefficient of runoff according Slovak standard STN 73 6760 which is equal to 1 and of course, includes losses such as evaporation, absorbency of roofing materials and losses during water runoff.

ACKNOWLEDGEMENTS

This work was supported by VEGA 1/0202/15- Sustainable and Safe Water Management in Buildings of the 3rd. Millennium and APVV-SK-CZ-2013-0188 Lets Talk about the Water – An Essential Dimension of Sustainable Society of the 21st Century.

REFERENCES

Bose K. 1999 *Dešt'ová voda pro záhradu a dum (Rainwater for the garden and house)*, HEL, Ostrava

Botka M. 2011 Skúsenosti s návrhom a realizáciou ZTI v objekte (Experience in the design and implementation of plumbing in the house) In: *16. Inter conference Sanhyga*. Piešt'any, pp. 55–63.

Coombes P.J., Kuczera G., Argue J., Kalma J.D. 2002 An evaluation of the benefits of source control measures at the regional scale. *Urban Water*, vol. 4 (4), pp. 307–320.

DeBusk K. M., Hunt W. F. 2014 Impact of rainwater harvesting systems on nutrient and sediment concentrations in roof runoff, *Water Science & Technology: Water Supply*, Vol 14 No 2, pp. 220–229.

Despins Ch., Farahbakhsh K., Leidl Ch. 2009 Assessment of rainwater quality from rainwater harvesting systems in Ontario, Canada, *Journal of Water Supply*, Vol 58 No 2, pp. 117–134.

Hlavatá H. 2005 The analysis of changes of atmospheric precipitation in Košice in the years 1900 – 2005, Bratislava: SHMI, branch office Košice.

Lash D., Ward S., Kershaw T., Butler D., Eames M. 2014 Robust rainwater harvesting: probabilistic tank sizing for climate change adaptation, *Journal of Water and Climate Change*, No 4, pp. 526–539.

Markoviè G., Kaposztasova D., Vranayová Z. 2014 The Analysis of the Possible Use of Harvested Rainwater and its Potential for Water Supply in Real Conditions, *WSEAS Transactions on Environment and Development*. Vol. 10, p. 242–249.

Mitchell V.G., Mein R.G., Mcmahon T.A. 1997 Evaluating the resource potential of stormwater and wastewater; an Australian perspective, *Australian Journal of Water Resources*, vol. 2(1), pp. 19–22.

Ocipova, D. (2007) *Risk of Legionellosis and thermical disinfection/* Riziko infekcie Legionellou a termická dezinfekcia.In: Topenářství instalace. Vol. 41, no. 8 (2007), p. 49–51. – ISSN 1211-0906.

Ocipova, D. (2009): *Experimentálny výskum vplyvu teploty v zásobníku z hl'adiska mikrobiologického znečistenia.* In: SANHYGA 2009. – Bratislava : SSTP, 2009 S. 157–164. – ISBN 9788089216291.

Jasminska, N.: *Design of heating and hot water supply system like a power of low-temperature heating utilization in combination with solar collectors.* In: Budownictwo o zoptimalizowanym potencjale energetycznym Vol. 11, no. 1 (2013), p. 34–43 ISSN: 2299–8535.

Poliak M. 2011 The use of rainwater, rainwater tanks – practical experiences and recommendations for design. In: *16. International conference Sanhyga. Piešt'any*, October 2011

Schets F.M., Italiaander R., Van den Berg H.H.J.L., De Roda Husman A. M. 2010 Rainwater harvesting: quality assessment and utilization in The Netherlands, *Journal of Water and Health*, Vol 08 No 2, pp. 224–235.

The Water Framework Directive 2000/60/EC.

Yu H., Huang G., Wu Ch. 2014 Application of the stormwater management model to a piedmont city: a case study of Jinan City, *Water Science & Technology*, Vol. 70 No 5, pp. 858–864.

Advances and Trends in Engineering Sciences and Technologies – Al Ali & Platko (Eds)
© 2016 Taylor & Francis Group, London, ISBN: 978-1-138-02907-1

Effect of shrinkage-reducing admixture on the properties of cement composites

K. Matulova & S. Uncik

Slovak University of Technology in Bratislava, Faculty of Civil Engineering, Bratislava, Slovakia

ABSTRACT: Cement composites are one of the most durable materials, however, development of cracks adversely affects its durability and functionality. One of the greatest weaknesses of these materials is shrinkage and the formation of shrinkage cracks. This paper presents the results obtained in studying the effect of various doses of shrinkage reducing admixture (SRA) on properties of cement paste. A drying shrinkage, compressive strength, basic rheological properties (consistency) and density of prepared mixtures were measured in this study. The obtained results showed that shrinkage reduction due to the SRA products was rising with increasing dose of admixture without deterioration of workability. Also, there was not achieved significant impact of SRA on development of compressive strength of samples. These results give a good basis for use of the admixture to solve the problems of excessive shrinkage of concrete structures.

1 INTRODUCTION

Shrinkage of cement composites generally means reducing the volume of cement matrix due to chemical and physical processes occurring in these materials. Shrinkage causes stress in their structure, which can lead to cracks or deformations of construction. These volume changes are often attributed to drying of the concrete over a long time period, although recent observations have also focused on early age or plastic drying problems. Shrinkage of concrete takes place in two distinct stages: early and later ages. The early stage is commonly defined as the first day, while the concrete is setting and starting to harden. After hardening of concrete, the water which has not been consumed by cement hydration leaves the pore system, in case there is not a balance between the moisture content in the composite and the surrounding environment. While drying, capillary forces are formed in the pore system together with the surface tension of water causing reduction of the pore and material volume. This is known as drying shrinkage, which is substantially independent from water evaporation intensity, which depends on the temperature of the environment, relative humidity and air velocity (Bajza & Rousekova 2006). Shrinkage intensifies especially with the increasing water-cement ratio, content of fine particles and water and increasing of share of cement paste. Some structures, such as slabs and walls, are especially prone to crack due to the large surface exposed to air-drying. The basic technological measure to reduce shrinkage is the curing of fresh concrete. In some conditions, however, the appropriate curing of concrete, reduction of water cement ratio or the use of suitable additives do not necessarily guarantee prevention of cracking. Control of cracking may be done by appropriate reinforcement. The reinforcement however, does not reduce shrinkage but can keep cracks from widening. Shrinkage Reducing Admixtures (SRA's) provide a method to reduce the strains caused by drying shrinkage and subsequently drying shrinkage stresses. Advantages to this method are that with the exception of the added SRA, concrete mixture proportions and mixing remain relatively unchanged. Also, no expansion is induced with shrinkage reducing admixtures. Thus, the drying shrinkage can be reduced for essentially any currently used concrete. The purpose by which the reduction in shrinkage occurs is thought to be from the reduction in surface tension of the pore water solution (Berke et al. 2003). The reduction in capillary tension may decrease the concrete volume changes due to air drying or internal self-desiccation. Thus, the effectiveness of a SRA depends on porosity and stiffness of concrete (Ribeiro et al. 2003).

In order to evaluate the efficiency of SRA product on cement paste, four sets of mixtures were prepared with the same materials but with different dose of SRA. The experimental study was based on measurement of consistency and density of the fresh cement pastes and the drying shrinkage, weight loss, density and compressive of the hardened cement pastes.

2 MATERIALS AND MIXTURE DESIGNS

Tests were done using four mixtures of cement pastes containing cement CEM I 42.5 R, in accordance with STN EN 197-1, with the following properties: compressive strength after 2 days 30.0 MPa, after 28 days 53.6 MPa, initial setting time 150 min, setting time 190 min, normal density 29.6%, soundness (expansion) by Le Chatelier 5.0 mm. To reduce shrinkage of cement pastes, shrinkage–reduction admixture (SRA) was used in 1%, 2% and 3% from dose of cement. The SRA is made up combination of these alcohols: 2-ethylpropan-1,3-diol, prophyli-dyntrimethanol and 5-ethyl-1,3-dioxan-5-methanol with a density of $1010 \, kg/m^3$, pH 10–11. All pastes had a w/c ratio $= 0.4$.

3 MEASURING METHODS

Cement pastes were mixed in a normalized laboratory mixer. After that, the properties of the fresh pastes were determined: consistency of the paste (determined by a flow test on the Haegermann flow table) and density. Samples to determine free shrinkage were $40 \times 40 \times 160$ mm beams with glass contacts in front. Samples to determine compressive strength were cylinders with a height of 50 mm and a diameter of 50 mm. Samples to determine shrinkage were cured one day (24 hours) in a humid environment and after demolding of samples the initial measurement of the length were done. Further samples were kept in the laboratory conditions with a temperature of about $20 \pm 2°C$ and a relative humidity about $55 \pm 5\%$. Drying shrinkage of samples was determined as loss of length compared to the initial length of sample determined after 1 day of curing in a moist environment. Shrinkage was measured after 2, 3, 4, 7, 14, 28 and 56 days of curing in the laboratory environment. The length measurement was done using a Graf-Kaufmann device. The weight loss was measured on the same prisms used for the shrinkage measurements. The compressive strength of hardened pastes was measured on under water-cured cylinders after 1, 7, 28 and 90 days.

4 RESULT AND DISCUSSION

4.1 *Consistency and density*

The measured values of consistency and density of fresh cement pastes for each mixture are given in Table 1.

It can be seen that the addition of SRA improves consistency of cement pastes and flow diameter ranged from 210.5 to 217.0 mm.

Density of fresh pastes which contains SRA was slightly lower than the density of the reference paste and ranged from 1860 to 1880 kg/m^3.

4.2 *Shrinkage and weight loss*

Measured values of shrinkage (ε), respectively swelling and weight loss (Um) of hardened cement pastes are given in Figure 1 (respective Figure 2) and Figure 3.

A volume change of reference paste was the largest in the early days of drying. Over time, curing shrinkage gradually decreased. The total value of shrinkage of reference paste after 56 days was

Table 1. Consistency and density of fresh pastes.

Dose of SRA (%)	Consistency (mm)	Density (kg/m³)
0	208.5	1 890
1	210.5	1 880
2	216.0	1 880
3	217.0	1 860

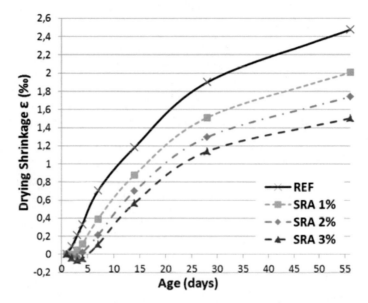

Figure 1. Drying shrinkage of reference sample and the samples with 1%, 2% and 3% doses of SRA.

slightly above 2.4‰. As can be seen an applications of the SRA had significant effect on volume changes of pastes.

Figure 1 shows that with higher dose of shrinkage-reducing admixtures the overall shrinkage decreased. At the higher dose of SRA (2%, 3% by weight of cement), the length of the samples in the first 4 days were practically stable, respectively in one case (3%) experienced modest expansion (Fig. 2). At 56 days, the shrinkage of the SRA 3% was almost 40% less than the reference mixture. The reduction in shrinkage due to SRA products is attributed to the reduction of capillary tension on concrete pores. These results are consistent with the results of other authors, who also achieved reduction of shrinkage with SRA admixture (Berke et al. 1997, Collepardi 2005, Pease 2005, Ribeiro et al. 2003).

Figure 3 shows weight loss. As can be seen, there is slightly higher weight loss in the case of higher dose of SRA like reference cement mixture. These differences are no so significant what mean that the addition of SRAs probably does not affect the drying process.

4.3 Compressive strength and density of hardened pastes

The compressive strength (fc) and density results for hardened cement pastes are given in Figure 4 and Figure 5.

Figure 4 shows the compressive strength of the four cement mixtures. It can be seen that the addition of SRAs didn't have significant effect on the compressive strength of cement pastes.

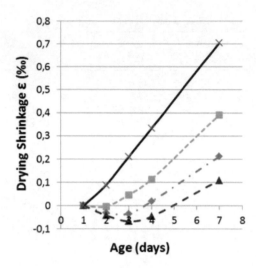

Figure 2.   First 7 days of drying shrinkage of reference sample and the samples with 1%, 2% and 3% doses of SRA.

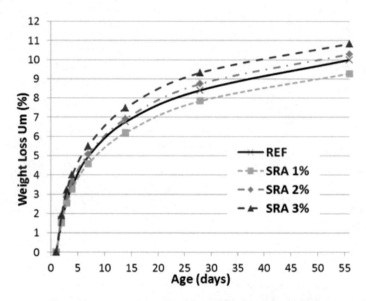

Figure 3.   Weight loss of reference sample and the samples with 1%, 2% and 3% doses of SRA.

A low reduction of compressive strength (15%), in compare with reference mixture, was obtained at 28 days for the mixture with the highest dose of SRA – 3%. On the other hand, 90 days results showed slight increase in compressive strength of samples with addition of SRA. The opinions of other authors about effect of shrinkage-reducing admixture on the strength characteristics of cement composites are different. The literature reports cases in which the measurements showed a reduction in the compressive strength (about 10–25%) by using SRA, but also increase. Significant impact on the results had the treatment conditions of composite (Berke et al. 2003, Collepardi 2005, Pease 2005, Ribeiro et al. 2003).

The bulk densities of hardened cement pastes are not significantly different. The initial bulk density of pastes with a higher content of SRA additives are slightly lower, which corresponds well with the results of compressive strengths. As is evident in Table 1, SRA additive improves the

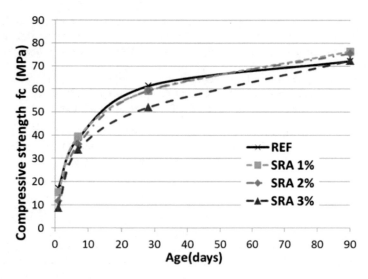

Figure 4. Compressive strength of reference sample and the samples with 1%, 2% and 3% doses of SRA.

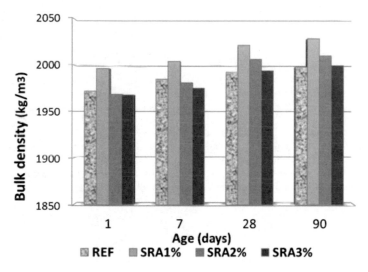

Figure 5. Density of hardened cement pastes with 0%, 1%, 2% and 3% doses of SRA.

workability of cement pastes, but also prolongs the formation of a solid structure of cement matrix. This fact was reflected by reducing the initial densities, as well as compressive strengths of hardened pastes. Over hydration of cement stone, this difference compared and 90-days compressive strengths of pastes with the addition of SRA were already slightly higher than the compressive strength of the reference paste.

5 CONCLUSION

Based on the results presented in this paper, the following conclusions can be drawn:

- the results showed that SRA additive improves the workability of cement pastes, but also prolongs the formation of a solid structure of cement matrix,

- the compressive strength reduction due to addition of SRAs was higher with higher dose in the early stages of solidification but gradually the differences were reduced, compared to the reference mortar, and a long-term compressive strengths of pastes with the addition of SRA were even slightly higher,
- the addition of SRAs had a positive effect on tested cement pastes. Reducing of shrinkage was increased with increasing dose of SRA, without deterioration of workability,
- the most significant effect of SRA was achieved at 3% dose of admixture.

These results give a good basis for use of the admixture to solve the problems of excessive shrinkage of concrete structures. Complementary tests are being done to understand better the effect of this chemical admixture on concrete pore structure and other properties of cement composites.

ACKNOWLEDGMENTS

This article was created with the support of the Ministry of Education, Science, Research and Sport of Slovak Republic within the Research and Development Operational Program for the project "University Science Park of STU Bratislava", ITMS 26240220084, co-funded by the European Regional Development Fund.

REFERENCES

Bajza, A. & Rousekova, I. 2006. *Technology of concrete*. Bratislava: Jaga Group. ISBN 80-8076-032-2.

Berke, N.S. & Dallaire, M.P. & Hicks, M.C. & Kerkar, A. 1997. New developments in Shrinkage-Reducing Admixures. In: *Superplasticizers and Other Chemical Admixtures in Concrete. Rome, 1997*: 971–978.

Berke, N.S. & Li, L. & Hicks, M.C. & Bae, J. 2003. Improving Concrete Performance with Shrikage-Reducing Admixtures. In: *Superplasticizers and Other Chemical Admixtures in Concrete. Ottawa, 2003*: 37–50.

Collepardi, M. 2005. Effects of Shrinkage Reducing Admixture in Shrinkage Compensating Concrete under Non-Wet Curing Conditions. In: *Cement and Concrete Composites* 27(6): 704–708.

Pease, B.J. 2005. *The Role of Shrinkage Reducing Admixtures on Shrinkage, Stress Development, and Cracking* (*Master thesis*). Indiana: Purdue University.

Ribeiro, A.B. & Carrajola, A. & Goncalves, A. 2003. Effectiveness of Shrinkage-Reducing Admixtures on Different Concrete Mixtures, In: *Superplasticizers and Other Chemical Admixtures in Concrete. Ottawa, 2003*: 299–310.

Accuracy evaluation of rapid method for concrete's frost resistance determination

S. Nikolskiy & O. Pertseva
Peter the Great St.Petersburg Polytechnic University, St. Petersburg, Russia

ABSTRACT: In this research it has been analytically proved that using of concrete's residual strain as a measure of damage instead of decreasing of tensile strength increases an accuracy of material's frost resistance by the time of freeze-thaw cycling. Also it has been experimentally shown that ratio of relative decreasing tensile strength to residual strain in direction perpendicular to compression is assumed to be independent on values tensile strength and residual strain for a given concrete and on the ways of achieving them during mechanical or freeze-thaw cycling. Taking this into account patented methods for estimation of concrete's freeze-thaw resistance as per values tensile strength and ε received after freezing and thawing cycles of some specimens and their postliminary failure by linear compression was substantiated.

1 INTRODUCTION

Frost resistance of concrete is an ability of water-saturated concrete specimen to maintain repeated standard thermo cycles without noticeable damages. Different types of water pressure cause concrete's freeze-thaw deterioration, such as hydraulic and osmotic pressure (Rønning, 2001), capillary pressure (Pukhkal et al., 2015) and other types of water influence according to existing freeze-thaw resistance theory. Internal damages by frost deterioration are of great importance for the external strength of concretes (Liisma et al., 2013). In order to determine the concrete mix composition, it is necessary to take into account freeze-thaw resistance. In order to keep comfortable micro-climate inside of structure (Vuksanovic et al., 2014) measurement of durability of concrete is of grave importance (Soldatenko, 2011). The Worldwide experience offers a vast number of ways for determination of durability of the concrete structures (Garanzha et al., 2014), but accordingly to the European standard (EN 206:2013, 2013), (Swedish Standard, 2005) there four main methods to determine the concrete frost resistance: Slab test, CDF, CIF-Test and Cube-Test. These test methods contain the following steps: curing and preparing the specimens, pre-saturation of the specimens and their thermo cycling. The test liquid simulates a deicing agent contains 3% of NaCl weight and 97% weight of (demineralized) water in case of the freeze-thaw test and deicing salt resistance and demineralized water to test the freeze-thaw resistance of concrete respectively. Scaling of the specimens is measured after a well defined number of freeze-thaw cycles and leads to an estimate of the resistance of the tested concrete against freeze-thaw damage (RILEM Technical Committee, 1996). The test methods however differ in terms of their procedures and conditions (RILEM Technical Committee, 2004). Also CIF test shows determination of internal damage by measuring the relative dynamic modulus of elasticity (taking into account ultrasonic transit time) (Bunke, 1991).

There are two different standard types of methods for determining the freeze-thaw resistance of concrete- basic (GOST 10060 2004) and reference (GOST 10060 2004) in the Russian Federation. Considerable spread of values of strength for concrete specimens (variation coefficient $\rho \approx 17\%$) in constant conditions of preparing and testing determines random spread of choosing average values $\bar{R}$ in the range of $\Delta \bar{R} / \bar{R} = \pm 3\,\rho\,\sqrt{n}$ where n is the volume of bath of specimens. In this case for proving the significance of relative decreasing $\bar{R}$ on 0,05...0,15 it is necessary to test more that

$\bar{R}/\bar{R} = 0,068\,\bar{R}$ 50 specimens and support and confidence figure is equal 0,95. Therefore, main disadvantages of the basic method are high labor input and small operability by virtue of the fact that duration of the basic thermo cycle is equal 4,5 at least and F $\gg$ 50.

One of the existing reference methods is a Dilatometric rapid method of determining the freeze-thaw resistance of concrete (GOST 10060, 2004). This method is a prototype for the method, which has been offered by me. In this method concrete's freeze-thaw resistance is determined by the maximum relative difference of volume deformations of the tested concrete and standard specimens in accordance with tables provided in standard specification (GOST 10060, 2004) taking into account concrete's type, its form and the size of specimens.

However, the results from the tables provided in state standard specification are acceptable only for Portland cement concrete and slag Portland cement concrete without surface-active additives (PEAHENS), such concretes are used extremely seldom now. Now lot of new concretes are investigated, tested and used, for example, nanomodificated concrete (Ponomarev, 2007), high-strength concrete (Barabanshchikov et al., 2007), concrete on the basis of fine-grained dry powder mixes (Belyakova et al., 2012) etc. Moreover, there are a vide scope of modern concrete mixtures, such as light-weighted concrete (Vatin et al., 2014a), vibropressed structures (Vatin et al., 2014b), hight- performance concrete (Ponomarev et al., 2014), concretes with additives (Korsun et al., 2014), (Akimov et al., 2015). These are also porous materials and, therefore, should be tested for frost-resistance. It is also worth mentioned, that type of cement, especially its hydration retarder (Skripkiunas et al. 2013), influences freeze-thaw resistance of concrete most (Nagrockiene et al., 2014). In order to obtain new tables long labour-consuming experiences, which imply using, basic methods are needed (Dikun, 2005).

2 PROJECT'S OBJECTIVES

The aims of this paper are to obtain the dependence between relative decreasing in strength by rate of strain and substantiate the new method for determination of concrete frost resistance. The offered method belongs to non-destructive methods, which provide appropriate results as direct (Köliö et al., 2014). Using of measure of damage D by the time of freeze-thaw cycling is offered for removal of disadvantages. In accordance with experimental data (Kuncevich, 1983), (Nikolskaya & Nikolskiy, 2009) dependence $D = Ac^q$ is well approximated, where: A = D (when c = 1) – value of D after first thermo cycle; c – number of thermo cycles; q – constant of material.

In accordane with this statement, evaluation of maximal measurement errors (ΔF and $\Delta F/F$) by the time of determination F is possible with help of dependence:

$$\delta R = A * c^q \,, \tag{1}$$

where $A = \delta R_1$ (when c = 1) – decreasing of strength after first thermo cycle and δR – decreasing of original strength R. The first consequence from (1) covers to constant of material q:

$$q = \{\ln(\delta R_c / \delta R_1)\}/(\ln(c)) \,, \tag{2}$$

After differentiation (3):

$$q\frac{\Delta F}{F} + \{\ln(F/c)\}\Delta q = \frac{\Delta R}{R} + \frac{\Delta \delta R_c}{\delta R_c} \Rightarrow$$

$$\Rightarrow \pm \frac{\Delta F}{F} - \frac{1}{q}\left\{\left(\frac{\Delta R}{R}\right) + \left(\frac{\Delta \delta R_c}{\delta R_c}\right)\right\} + \{\ln(F/c)\}\left(\frac{\Delta q}{q}\right) = 0 \,, \tag{3}$$

where $\Delta R/R$, $\Delta \delta R/\delta R_c \gg \Delta R/R$ and $\Delta q/q$ are maximal relative measurement errors of determination original value R, decreasing R after c thermo cycles or values q correspondingly. By the time of using non-destructive method (Nikolskiy, 2010) $\Delta R/R = 0,03$ or $\Delta R/R = 0.068$, when we used bathes from 25 specimens. So, $\Delta \delta R = 2 \Delta R$.

Therefore, $\Delta F/F$ is depend on measurement errors of strength $\Delta R/R + \Delta \delta R_c / \delta R_c$ with accumulation factor $1/q$ and maesurement error of $\Delta q/q$ with accumulation factor $\ln(F/c)$. If $q = 1$ ($\Delta q = 0$), that we will get measurement error of calculation F by $\delta R = \delta R_1 c$ from (3):

$$\pm \frac{\Delta F}{F} = \left(\frac{\Delta R}{R}\right) + \left(\frac{\Delta \delta R_c}{\Delta R_c}\right), \tag{4}$$

where $\Delta R/R \approx 0,03$; $\Delta \delta R_c = 2\Delta R \approx 0,06R$; $\Delta R_c \leq |\delta R/R| = (0,05 \ldots 0,15) R$.

After differentiation (2):

$$\frac{\Delta q}{q} = \frac{1}{\ln(\delta R_c / \delta R_I)} \left\{ \frac{\Delta \delta R_c}{\delta R_c} + \frac{\Delta \delta R_I}{\delta R_I} \right\}, \tag{5}$$

where $\Delta \delta R_c = \Delta \delta R_1 = 2\Delta R$; $\delta R_c = \delta R_1 c^q$; $\Delta \delta R_1 / \delta R_1 \gg \Delta R/R$. After plugging (5) in (3):

$$\frac{\Delta F}{F} \approx \frac{1}{q} \frac{\Delta \delta R_c}{\delta R_c} \left\{ 1 + \left(\ln\left(\frac{F}{C}\right)\right) \left(\frac{1 + c^q}{\ln(c)}\right) \right\}. \tag{6}$$

Accumulation factor $\frac{1+c^q}{\ln(c)}$ (when $q = 1$) has a minimum when $c = 4$ and is numerically equal 3,6. When $F = 40$ and $c = 4$ $\ln(F/c) = 2,3$; $\left(\ln\left(\frac{F}{c}\right)\right) \left(\frac{1+c^q}{\ln(c)}\right) \approx 8,3$, and $\frac{\Delta F}{F} \approx \frac{9,3}{q} \frac{\Delta \delta R_c}{\delta R_c}$.

While $\Delta \delta R_c = \delta R_c$ (when c is too small), it is $\Delta F/F \approx 9,3/q$. The result proved that contribution of measurement error of q in $\Delta F/F$ is more in 8 times than contribution of measurement error of R. It's the case, when measure of damage is a decreasing of strength. To decide this problem using of rate of strain $\varepsilon = \delta(l)$, where $l \approx 100\,mm$ – basc length, $\delta l = l_c - l$ – the base deformation and l_c – base length after "c" thermo cycles, is offered. But for that it is necessary to determine dependence between decreasing in strength and rate of strain. So, this value is defined as $Z = (\delta R/R)/\varepsilon$.

3 FORMATTING THE RESULTS

To determine this dependence concrete mix was tested. Mix contained a portlandt cement (12%) of brand 400, sand (25%), granite crushed stone 5 …20 mm (56%) and water (7%). 108 samples cubes with an edge of 150 mm were prepared. There points of intersection of diagonals of opposite sides were spaced far apart from each other (to 1,5 mm). Samples hardened 28 days in water at the room temperature, and then 60 days in damp sand at 18 …26°C. From these 108 samples, 8 samples were cycled alternately thermally and mechanically, and 100 samples were used for a frost resistance assessment by a basic method. At realization of a basic method l distance between points of each sample was measured by means of a tool microscope before tests for frost resistance or for durability. Rate of strain ε was counted as δ/l. Changing of distance δ between points at realization of a method (Nikolskiy, 2009) was carried out on $20 \pm 2°$C before and after cycling by means of a bracket with variable base and a measuring head of hour type (the price of division of 1 micron).

Threshold loading L_0 of a water-saturated sample was determined by a way (Nikolskiy & Vorontsova, 2011), registering the acoustic issue (AI) by means of the AF-15 device at cyclic loading and unloading of a sample to zero. In the first experience load of L was brought to 11 t; in the absence of acoustic issue in the course of the end of unloading value of L was increased by 5% and so until at the end of unloading there was no AE. For accepted an average of L two last cycles. The limit of long durability R_0 was found as the relation L_0 to the average area of two loaded sides. Dependence z was determined as $(L_1 - L_2)/L_1 \varepsilon_1$, where L_1 is a maximal value L in the first loading and L_2 – in the second loading (Figure 1). Results of this experiment are given

Table 1. Results of experimental determination dependence between R and ε after 20, 42, 84 and 105 cycles (thermocycles and mechanical cycles).

No of specimen		1	2	3	4	5	6	7	8	Average
Origin value R_0, [MPa]		9,45	16,8	22,7	23,4	24,3	24,3	25,0	30,1	22,0
After 20 cycles	R_0, [MPa]	9,18	16,2	22,2	22,7	23,5	23,9	24,1	29,4	
	$\left(\frac{\delta R}{R}\right)\cdot 10^5$	2857	3571	2203	2991	3292	1646	3600	2326	2811
	$\varepsilon\cdot 10^5$	216	210	204	206	204	200	202	190	204
	z	13,23	17,00	10,80	14,92	16,14	8,23	17,82	12,24	13,8
After 42 cycles	R_0, [MPa]	9,03	16,15	21,9	22,3	23,4	22,9	23,9	20,4	
	$\left(\frac{\delta R}{R}\right)\cdot 10^5$	4482	3869	3524	4701	3704	561	4440	3222	4213
	$\varepsilon\cdot 10^5$	406	452	404	408	394	396	410	394	408
	z	11,04	8,56	8,72	11,52	9,40	14,55	10,73	8,18	10,32
After 84 cycles	R_0, [MPa]	8,10	14,7	19,1	21,2	21,0	23,0	22,6	26,7	
	$\left(\frac{\delta R}{R}\right)\cdot 10^5$	1429	893	1586	940	1244	1230	1253	1130	1213
	$\varepsilon\cdot 10^5$	1206	1217	1188	1196	1201	1207	1196	991	1175
	z	11,85	5,31	13,35	7,86	10,36	9,31	7,36	11,04	9,55
After 105 cycles	R_0, [MPa]	7,75	14,2	18,6	20,1	21,1	20,9	21,4	26,4	18,8
	$\left(\frac{\delta R}{R}\right)\cdot 10^5$	1799	1547	1806	1410	1317	1399	1440	1229	1488
	$\varepsilon\cdot 10^5$	1582	1440	1435	1534	1561	1459	1297	1196	1439
	z	11,37	10,74	12,59	9,19	8,43	9,59	11,10	10,28	10,40
$\bar{z}$ for specimen		11,40	10,76	11,04	10,73	10,21	10,30	11,05	10,34	$\bar{z}=10,73$

		S^2	S	ρ
R – average value of the long-term strength in conditions of compression, δR – decreasing R after specimen cycling,	z	1,94	1,394	0,134
ε – rate of strain in the direction, perpendicular compression; $z=(\delta R/R)/\varepsilon$; zk-value z after 105 cycles; $\bar{z}$ – average value z for specimen; $\bar{\bar{z}}$- average value for $\bar{z}$; S2 –sampling variance; S-mean square deviation; ρ- variation coefficient	$\bar{z}$ for specimen	0,183	0,428	0,04

Table 2. Comparison of results by non-destructive method and standard method.

Average value of:	Non-destructive method	Basic method
$\delta R/R$	0,1488	0,161
ε	0,01439	0,01448
z	10,4	11,1

in the Table 1. Dependence z is near to linear. Results by non-destructive method were compared with results by standard method (Table 2).

4 RESULTS

Results by non-destructive method differ 6,3% from results by basic method. Also this dependence is not delicate to changing some thermo cycles to mechanical cycles. So, value of z is constant and dependence of relative decreasing in strength is linear. After mathematical manipuleations it becomes clear, that using ε as a measure of damage is more accurate than using δR, in approximately in 35 times.

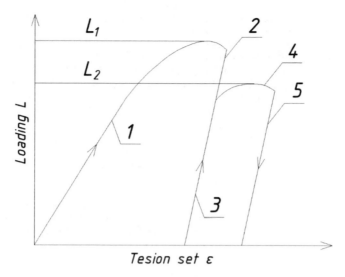

Figure 1. This is a shedule of specimen loading in conditions of monoaxial compression after thermocycles: 1 - line of first loading; 2 – line of first unloading; 3 – line of second loadind; 4 – line of second loading; 5 – line of second unloading; ε – relative longitudinal tension set; L_1 – extreme loading of the first loading; L_2 – extreme loading of the second loading.

After taking into account theese results new method for determination of concrete frost resistance was offered by application for a patent (Nikolskiy & Pertseva, 2014).

REFERENCES

Akimov, L. & Ilenko, N. & Mizharev, R. & Cherkasin, A. & Vatin, N. & Chumadova, L. 2015. Influence of Plasticizing and Siliceous Additives on the Strength Characteristics of Concrete. *Applied Mechanics and Materials* 725–726: 461–468.

Barabanshchikov, Yu.G. & Semenov, K.V. 2007 Increasing the plasticity of concrete mixes in hydrotechnical construction. *Power technology and engineering* 41(4): 197–200.

Belyakova, E.A. & Kalashnikov, V.I. & Kusnetsov, Y.S. & Tarakanov, O.V. & Volodin, V.M. 2012. Next generation concrete on the basis of fine-grained dry powder mixes. *Magazine of Civil Engineering* 34(8): 47–53.

Bunke, N. 1991. Schriftenreihe des Deutschen Ausschusses für Stahlbeton. *Prüfung von Beton – Empfehlungen und Hinweise als Ergänzung zu DIN 1048* 422: 12–15.

Dikun, A.D. et al. 2005. *Construction Materials* 52(8): 55–56.

EN 206:2013 2013.

Garanzha, I. & Vatin, N. 2014. Analytical methods for determination a load capacity of concrete-filled tubes under axial compression. *Applied Mechanics and Materials* 633–634: 965–971.

GOST 10060 2004.

Köliö, A. & Rantala, T. & Lahdensivu, J. & Nurmikolu, A. 2014. Freeze-thaw resistance testing of concrete railway sleepers. *5th International Conference on Concrete Repair; proc.: 533–539*. Belfast: CRC Press.

Korsun, V. & Vatin, N. & Korsun, A. & Nemova, D. 2014. Physical-mechanical properties of the modified fine-grained concrete subjected to thermal effects up to 200°C. *Applied Mechanics and Materials* 633–634: 1013–1017.

Kuncevich, O.V. 1983. *Betony vysokoj morozostojkosti dlja sooruzhenij Krajnego Severa.* Leningrad:Stroyizdat.

Liisma, E., Raado, L.M. 2013. Internal and external damages of concrete with poor quality of coarse limestone aggregate. *CESB 2013 PRAGUE – Central Europe Towards Sustainable Building 2013: Sustainable Building and Refurbishment for Next Generations; proc.: 1–4.* Prague: Klokner institute.

Nagrockiene, D. & Girskas, G. & Skripkiunas, G. 2014. *Construction and Building Materials* 66(9): 45–52.

Nikolskaya, T.S. & Nikolskiy, S.G. 2009. *III international conference "Populjarnoe betonovedenie"; proc., St. Petersburg, 27 February – 2 March 2009.: 35–43.*

Nikolskiy, S.G. et al. 2009. RU Patent 2305281.

Nikolskiy, S.G. et al. 2010. RU Patent 2380681.

Nikolskiy, S.G. & Pertseva, O.N. 2014. RU Application for a patient 2014116713 since 24, May, 2014.

Nikolskiy, S.G. & Vorontsova, E,A, 2011. *XL Week of Science; proc., St. Petersburg, 6–11 December 2011.* St. Petersburg: SpbSPU: 344–430.

Ponomarev, A.N. 2007. *Construction and Building Materials* 24(4): 22–25.

Ponomarev, A. & Knezevic, M. & Vatin, N. & Kiski, S. & Ageev, I. 2014. Nanosize scale additives mix influence on the properties of the high performance concretes. *Journal of Applied Engineering Science* 12: 227–231.

Pukhkal, V. & Murgul, V. & Kondic, S. & Zivkovic, M. & Tanic, M. & Vatin, N. 2015. The study of humidity conditions of the outer walls of a "Passive house" for the climatic conditions of Serbia, City Nis. *Applied Mechanics and Materials* 725–726: 1557–1563.

Rønning, T.F. 2001. *Freeze-Thaw Resistance of Concrete? Effect of Curing Conditions, Moisture Exchange and Materials.* Trondheim: The Norwegian Institute of Technology Press: 416.

Skripkiunas, G. & Nagrockiene, D. & Girskas, G. & Vaičiene, M. & Baranauskaite, E. 2013. *Procedia Engineering* 57: 1045–1051.

Soldatenko, T.N. 2011. Model identifikatsii i prognoza defektov stroitelnoy konstruktsii na osnove nechetkogo analiza prichin ih poyavleniya. *Magazine of Civil Engineering* 25: 52–61.

Swedish Standard 2005, *Concrete testing – Hardened Concrete-Frost Resistance,* SS 137244, Sweden.

Vatin, N. & Gorshkov, A. & Nemova, D. & Gamayayunova, O. & Tarasova, D. 2014. Humidity conditions of homogeneous wall from gas-concrete blocks with finishing plaster compounds. *Applied Mechanics and Materials* 670–671: 349–354.

Vatin, N.I. & Pestryakov, S.S. & Kiski, S.S. & Teplova, Z.S. 2014. Influence of the geometrical values of hollowness on the physicotechnical characteristics of the concrete vibropressed wall stones. *Applied Mechanics and Materials* 584–586: 1381–1387.

Vuksanovic, D. & Murgul, V. & Vatin, N. & Pukhkal, V. 2014. Optimization of microclimate in residential buildings. *Applied Mechanics and Materials* 680: 459–466.

RILEM Technical Committee. TDC, CDF Test, Test Method for the Freeze-Thaw-Resistance of concrete with sodium chloride solution, RILEM TC 117-FDC Recommendation, Germany, 2001. 27p.

RILEM Technical Committee. TDC, CIF Test, Test Method of frost resistance of concrete, RILEM TC 176 Recommendation, Germany, 2004. 17p.

Advances and Trends in Engineering Sciences and Technologies – Al Ali & Platko (Eds)
© 2016 Taylor & Francis Group, London, ISBN: 978-1-138-02907-1

A vanished settlement in the Ore Mountains – the creation of 3D models

J. Pacina & J. Havlicek

Czech Technical University in Prague, Faculty of Civil Engineering, Prague, Czech Republic

ABSTRACT: The settlement in the Czech–German borderland in North-West Bohemia (the Czech Republic) was affected by the political situation after World War II. Most of the German speaking inhabitants were deported to Germany and the region (especially the Ore Mountains) has not been repopulated since WWII. The aim of this paper is to present the methods for identifying, modelling and visualizing an abandoned/vanished settlement using geoinformatics methods – old maps analyses, LIDAR and UAV surveys and old-photos based 3D modelling. All of these methods offer tools to preserve the current state of the (almost) vanished settlement and to visualize the settlement that has almost disappeared. All of these outputs are accessible in the internet environment and are very important for preserving the culture heritage of this region.

1 INTRODUCTION

The world's landscape has been affected by human activity in many ways. During the past hundred years land-use has greatly changed; for example cities have grown and every year more land is used by industry. A very specific situation is in the region of North-West Bohemia (the Czech Republic). This region used to have a normal Central European settlement structure with compact towns and villages regularly spread over the area; surrounded by fields and forests.

The political situation in the 1930s and 1940s had a huge impact on the settlement of this region traditionally inhabited by Germans or German-speaking people. In 1938, the entire region of Sudetenland, based on the Munich Pact, was annexed to Germany. Many people with Czech nationality moved to the Protectorate of Bohemia and Moravia. After the Second World War, the majority of the German-speaking inhabitants were deported to Germany. The region was re-populated by Czechoslovak citizens but it was never populated as before 1938. This especially applies to the border-mountains (the Ore Mountains in this region) with rough living conditions and a rising border-military zone which led to a large number of extinct settlements.

The main aim of this paper is to introduce the possibilities of identifying, modelling and visualizing these vanished settlements in the Ore Mountains. In the mountain areas, there are still recognizable vanished settlement structures in contrast to the coal basin, where the landscape is completely changed by open-pit mining (Pacina & Cajthaml, 2014). This is an important part of the cultural heritage we need to preserve for further generations.

The area of interest is situated in the cadastre of Hora Sv. Šebestiána (Ústí nad Labe region, Chomutov district) directly on the border with Germany. This town used to be an important center of the region. In the nearby valley there was a very sophisticated system of watermills used primarily for milling grain and for sawing wood from the surrounding forests. There used to be a railway running to Reitzhein (neighboring town in Germany) as well with a magnificent bridge arching over the valley.

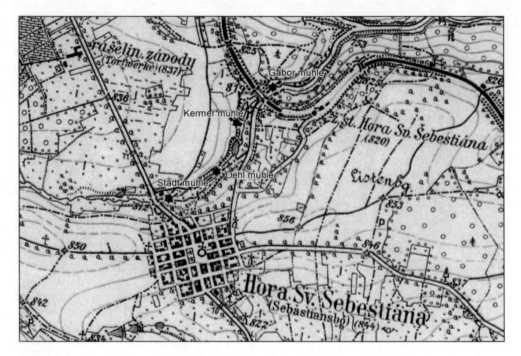

Figure 1.   Location of the water-mills on processed 3rdP Military Survey maps (1938).

Based on historical facts[1], there were altogether four water-mills situated in the valley – see Figure 1. The extinction of all of the water-mills (there are more vanished water-mills in the mountain region) was caused by the political situation after WWII. The borderlands were depopulated and there was no need for such a high amount of saw-mills. Another influencing factor was the technological development of replacing water powered saw-mills with electrical machinery.

2   IDENTIFICATION BASED ON AERIAL IMAGERY

Reconstruction and analysis of extinct settlement requires different data sources. Here we work with maps, aerial imagery, LIDAR, direct surveyed data and much more. The base data for the analysis are old maps. Maps of different types, scale and age help to describe our area of interest. Aerial imagery has been available for our area of interest since 1953. Current technological advances provide the opportunity to collect our own aerial imagery using Unmanned Aerial Vehicles (UAV). Data collected in a large scale using UAVs contained much more detail than current data obtained from commercial sources. LIDAR data (Light Detection and Ranging) offer a precise view of the current morphologic structure. All of these methods were tested in a nearby location Jilmová (Ulmbach) and is described in detail in (Pacina & Hola, 2014). The 3D settlement reconstruction is performed in 3D modelling software using old archived building plans, old photographs and postcards.

The identification of the mills residuals was performed on the current orthophoto available as a WMS service from the Czech Office for Surveying, Mapping and Cadastre. This orthophoto, with a spatial resolution of 25 cm/pixel, wasn't as detailed as required; thus a new survey was performed.

_____

[1] Facts collected in local historical archives.

Figure 2.   Flight plan and stabilized Ground Control Points in the area of interest.

Figure 3.   Water-mill residuals identified in a derived orthophoto and Digital Surface Model.

In spring 2014, a UAV was used to create an orthophoto of the area together with a detailed Digital Surface Model (DSM). Based on the size of the area of interest and the elevation differences (deep valley) this method was rejected. According to these facts a new method using a small aircraft for Small Format Aerial Photography was developed (Aber & et al., 2010). Many studies (Cardenala & et al., 2004; Chandler & et al., 2005; Quan, 2010) have proved that the classic compact cameras preciseness for close-range (aerial and earthbound) photogrammetry is, in comparison with professional aerial cameras, sufficient for the given tasks (DSM creation, orthophoto).

This method uses the same gimbal (camera stabilization) as a drone. A special mount allowing the placement of the gimbal into the aircraft and shooting time-lapse images was developed. A Sony Nex7 camera, in combination with a 16–55 mm f/3.5–5.6 lens, was used in the test survey. The average flight altitude was 300 m, flight speed 90 km/h and the images were shot every second which assured about 90% image overlap. The imagery side-lap was estimated to be 70–80% assuring the resulting model complexity. Spatial resolution of the produced raster DSM and orthophoto may

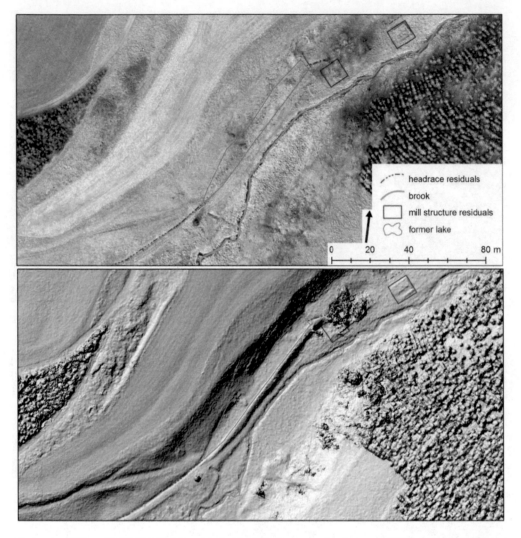

Figure 4.   Water-mill (Oehl Mühle) residuals identification on orthophoto and DSM derived from aerial imagery.

be up to 5 cm/pixel. The test area is fully covered with ground control points (GCP) measured by RTK GPS. The mission flight plan together with an example of the stabilized GCPs is presented in Figure 2. The images were processed using PhotoScan, Agisoft LLC and using the Structure from Motion algorithms (Verhoeven, XX).

3   3D MODEL CREATION

The identification of the former settlement and water-mill technology residuals (lakes, headraces) was performed using the orthophoto and DSM derived from the aircraft imaging mission. A field survey was performed as well. The identified objects are presented in Figure 3 and the details of Kermer mühle in Figure 4. The 3D model created from the aerial images is accessible at https://skfb.ly/BUY7.

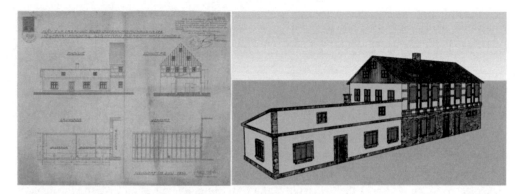

Figure 5.   Oehl mühle 3D reconstruction visualized in SketchUp (right) and the original structure drawing from 1930.

Figure 6.   Final visualization of the reconstructed structures – Kermer mühle presented in ArcScene.

Modelling the vanished landscape and structures is important for preserving the culture heritage as this is the only way how the no longer existing buildings and landscape can be viewed online (Duchnova, 2015) or printed on a 3D printer (Brůna et al., 2014). An extensive background search in regional archives and libraries was undertaken that focused on photographs, postcards and structural drawings describing the vanished water-mills. The background research was very successful and old postcards and structural drawings fully describing the water-mills were discovered. The structural drawings are used for precise 3D reconstruction in the selected software and the postcards were further used for texture determination.

Several software products were tested (Blender, ArcGIS City Engine, Maya, SketchUp, ArcGIS ArcScene) to produce a comprehensive 3D model containing all the gathered data. The 3D building/structure reconstruction was finally made using SketchUP software. The resulting visualizations were processed in the ArcScene environment, where the combination of DSM and orthophoto derived from the aircraft survey was used as the basemap. The 3D models presented within the SketchUp environment are presented in Figure 5. The final visualization of the processed 3D models in the ArcScene environment is shown in Figure 6.

4 CONCLUSION

The methods for reconstructing a vanished settlement in the abandoned Czech-German borderland are described in this paper. The Bezruè valley was chosen as there used to be an original system of water-mills until the 1950s. Several data sources were processed in order to identify the settlement residuals and to make a 3D visualization of the former state. A high quality orthophoto and digital terrain model is required to identify the settlement residuals and to create a 3D visualization. The data offered for public use by the Czech Office for Surveying, Mapping and Cadastre (orthophoto with spatial resolution 25 cm/pixel and LIDAR data with spatial resolution 5 m/pixel) were not sufficient to meet our visualization requirements. A new method using a small aircraft as a carrier for Small Format Aerial Photography developed at J. E. Purkynì University was used to collect aerial imagery in the area of interest. The data were processed into an orthophoto and raster DSM with a spatial resolution of 5 cm/pixel serving as a quality data source for visualization and analysis. The final 3D reconstruction of the watermills was performed in SketchUp software and visualized in ArcScene. The created 3D model is accessible online at https://skfb.ly/BUY7.

This article is part of a grant project for the Czech Ministry of Culture – NAKI DF12P01OVV043 Reconstruction landscape and databases of vanished settlements in the Ústí Region for the preservation of cultural heritage.

REFERENCES

Aber, J. S. & Marzolff, I. & Ries, J. B. 2010, Small-Format Aerial Photography, *Principles, Techniques and Geoscience Applications*, Amsterdam – London: Elsevier Science.
Bruna, V. & Pacina, Jan & Pacina, Jakub & Vajsova, E.. 2014 Città e Storia, *Modelling the extinct landscape and settlement for preservation of cultural heritage*, IX, č. 1, 131–153. ISSN: 1828–6364
Cajthaml, J. 2012 Czech technique – publishing CTU, *Analysis of old maps in digital environment on the example of Müller's map of Bohemia and Moravia*. Prague, Czech Republic. 172 s. ISBN 978-80-01-05010-1. (in Czech).
Cardenala, J. & Mataa, E. & Castroa, P. & Delgadoa, J. & Hernandeza, M. A. & Pereza, J. L. & Ramos, M. & Torresa, M. 2004. Altan, Orhan (ed.), *Evaluation of a digital non metric camera (Canon D30) for the photogrammetric recording of historical buildings*. ISPRS Congress Istanbul. Vol. XXXV, Part B5, s 455–460.
Chandler, J. H. & Fryer, J. G. & Jack, A. 2005. Photogrammetric Record, *Metric capabilities of low-cost digital cameras for close range surface measurement*.Vol. 20, no. 109, s. 12–26.
Duchnova R. 2015. *The reconstruction of the vanished town of Pressnitz*. [online] URL: http://prisecnice.eu/.
Kraus K. 2007. Walter de Gruyter, *Photogrammetry: Geometry from Images and Laser Scans*, Berlin.
Pacina, J. & Hola, J. 2014 14th International Multidisciplinary Scientific GeoConference, SGEM 2014, *Settlement identification in abandoned borderland*. Conference Proceedings, Book 2, Volume I, s. 769–776, ISBN 978-619-7105-10-0, ISSN 1314-2704. DOI:10.5593/sgem2014B21.
Pacina, J. & Cajthaml, J. 2014. Modern Trends inCartography, *Historical Data Processing, Modelling, Reconstruction, Analysis and Visualization of Historical Landscape in the Region of North-West Bohemia*. DOI: 10.1007/978-3-319-07926-4_36, Edition: Selected Papers of CARTOCON, Chapter: 477, Publisher: Springer International Publishing, Editors: Jan Brus, Alena Vondrakova, Vit Vozenilek, pp. 477–488.
Quan L. 2010. *Image-based Modelling*. Springer, New York.

Advances and Trends in Engineering Sciences and Technologies – Al Ali & Platko (Eds)
© 2016 Taylor & Francis Group, London, ISBN: 978-1-138-02907-1

Quantitative and qualitative differences in selected Prague cemeteries

K.F. Palánová, J. Kovář, I. Dlábiková, J.P. Janda, T. Babor & O. Juračka
VŠB-TUO, Ostrava, Czech Republic

ABSTRACT: When comparing old and newly established cemeteries it seems the value of a man is degraded. The purpose of this article is to compare two biggest cemeteries in Prague. The Olšany cemetery progressed mostly during 18th and 19th century, unlike the cemetery in Ïáblice which was founded in the early 20th century. Both are architecturally very important. In the Olšany cemetery many bodies of outstanding personalities were laid to rest and we can feel harmony here. On the other hand the Ïáblice cemetery gives the impression of disharmony and confusion, in this place it seems as if the bodies were put off rather than laid to rest. Seeing these differences we can start to search the burial adaptability for social changes and the positive and negative influence of these changes on cemeteries as well.

1 INTRODUCTION

Originally there was a settlement Olšany with a farm and vineyard in the place of today's cemetery. In 1679 the settlement was sold to Old city, management of which founded here the burial place here because of the plague epidemic. The original purpose of the burial site is evident in the reconstructed chapel St. Roche, St. Šebestian and St. Rozalia (1682), the patrons against plague. Under the reforms of Josef II., it was ordered to relocate the cemeteries outside of the city, the plague cemetery was declared as a public cemetery assigned to the locality in the right side of Vltava River. During 19th century and the beginning of 20th century the Olšany cemetery was enlarged by adding plots up to 12 cemeteries marked with Roman numerals. Today's the area expands behind the Jana Želivského Street which cuts it in two parts. Till the beginning of 20th century the inhumation prevails here. The part of the cemetery, from 18th and mainly the first half of the 19th century, has a lot of valuable, figural tombstones and so it documents the Czech funeral plastic. Many outstanding personalities in the field of politic, cultural or industry, such as Franz Kafka, Jiří Orten, businessman Rudolf Petschek, teacher Ladislav Blum, writer Ota Pavel and others, are buried there. Maybe thanks to the sense of value we feel in the Olšany cemetery some kind of a resonance between the cemetery and us, which for example is apparent in the gallery between a visitor and an artwork.

The current dense structure of the cemetery, together with evolution of funeral services in the Czech Republic during the last century, doesn't allow for inhumation on a large scale. Nowadays the cremation dominates and therefore only a little space for a quality tombstones remains. The urns are deposited into columbarium, to the ground or ash is dispersed on scattering meadows of smaller size which were established later on.

Due to the influence of Prague expanding, the Olšany cemetery situated in the urban area now, in the city part Žižkov and it's a part of Prague (Pražská informační služba, 2007). The Olšany cemetery is accessible by three entrances from Vinohradská Street: the lower gate near IV. cemetery was built in the 19th century, the main gate from the year 1928 is found in the middle and on the left side behind this gate the Middle ritual hall, which was built in 1894 and renovated in 1928, is situated. The upper gate from Vinohradská Street is near Želivského Street and it leads to a new ritual hall, the previous old crematorium, built in 1898 and converted in 1921 by architect František Nevola.

On the contrary, the Ďáblice cemetery was established only in the years from 1912 to 1914 in a cubist style in line with a competition project by architect Vlastislav Hofman. It's situated in the north Prague periphery on the boundary of quarters Ďáblice and Střížkov. The ordinary graves and shaft graves were present from the beginning, the vaste scaterring meadows for a burial by cremation were established only later. In the north part of the cemetery shaft graves of the tortured and executed victims from the 50's of the communist era are located (for example tortured priest Josef Toufar). Graves of children of imprisoned women from the same era are found here too. The Ďáblice cemetery is accessible by two gates from Ďáblická Street. The original main gate, designed by architect Hofman, became the side entrance. Kiosks with the expressively created windows and with small stepwise roofs are empty today, because of the Offices of Administration of Prague cemeteries moved to the new gate next to a tram stop.

2   THE TYPES AND METHODS OF BURIAL CEREMONY

The area of today's Olšany cemetery is 50,17 ha large. It keeps 112 000 of buried bodies according to the evidence, but over the whole period of its existence the number of buried people is estimated at two millions. There are 110 000 burial places in total, out of which are 200 chapel graves, 25 000 other graves, 65 000 tombs, 20 000 urn tombs, 6 columbarium along walls and 2 scattering meadows.

With the area of 29 hectares the Ďáblice cemetery is the second largest cemetery in Prague. There are over 20 000 tombs in evidence and extensive scattering meadows here.

The traditional disposition of the Olšany cemetery is represented by the rectangular net of service streets and paths with a strict position of tombs. This fact makes the cemetery a respectful place for pieta and a place where death is perceived as inseparable part of life. The high artistic quality and craftmanship of tombstones refines the place and the settlement of visitors and expresses the respect for the deceased. The artistic experience uniqueness becomes even more profound. The high boosting vegetation of high trees and ubiquitous cover of ivy unites tombs and enhances the feeling of serenity, that is needed for a visit of this garden.

The organic disposition of the Olšany cemetery, its location in the Prague periphery, improved by the large scattering meadows area and the area of free forested or grassed space, have turned the cemetery into a landscape. The cremation trend, which started growing only after the constitution of Czechoslovakia as an independent state, had a direct influence on the founding and development of the cemetery. The organic network of paths and large areas of scattering meadows give the cemetery natural character, which is supported by missing tombstones, absence of the names of deceased and anonymous urn tombs as well. The casual visit is thus distracted only by common tombs and hole tombs in the north part of the cemetery at the former main gate; the distraction being even deeper when one realizes these are the tombs of victims of Communist era. These tombs are gently followed by the lines of common and urn tombs, positioned in dense lines imitating a traditional look of a cemetery. Their placement doesn't respect the free composition of the cemetery. At present, the natural look of Ďáblice cemetery, that was created and grew along with the boom of cremation, reflects new trends in establishing The Forest of memory (for saving ash at roots of grown trees). However, the new trends of burying bodies and storing ash do not always allow space for placing flowers and candles. Thus a visitor looses „the contact"with the deceased, which is needed in the following stages of mourning and is recognized as a traditional form of honoring the dead.

3   CEMETERY BUILDINGS

The development of funeral services is reflected in cemetery buildings. The Olšany cemeteries were established in the surroundings of the chapel St. Roche, the patron against plague (1682, J. B. Mathey). Another building, the central ritual hall, was built near the main gate only in 1894. In 1898 the other ritual hall was constructed, but later it was converted into the crematorium.

It was in use until 1971, then returned to its previous function (Svobodová, 2013, p. 35). In 1928 the buildings for administration of Prague cemeteries were built near the main gate. The Russian Orthodox Church was constructed in the years 1924–1925 and it is surrounded by graves of Russian emigrants who left their country after the revolution. Some of the other important buildings are the chapel graves (200); they bloomed especially in the 19th century. In the 20th century the interest in the chapel graves decreased, nowadays they are rarely used for the burials. The reason for this *was a fundamental change of attitude towards the death, which became almost a taboo topic, divested of old myths and religious ideas…the architecture of death created a new formal language at the period, also new types of building, such as a crematorium, were formed, ritual halls were built, the formerly unknown phenomenon of urn forest and of a military burial site occurred* (Kašpinská, 2013, p. 23).

The director of the administration of the Prague cemeteries is considering the conversion of some the buildings on the Olšany cemetery border in order to make the objects not just for the dead, but also for the living people, which might be seen as an effort to eliminate the taboo of the death and make it a part of the course of life again.

The Ďáblice cemetery was built according to plans of the architect Vlastislav Hofman in the cubist style. The important shape element is the cemetery wall. The new entrance was rebuilt in cubist style in the last years. On the border of scattering meadows there is a cubist pavilion standing and the water pumps are designed in a cubistic style too. The ritual hall is situated near the second gate.

Only in 2004 Hofman's ideas were followed by architect Marie Švábová, who created the new ritual hall in a cubist style near the scattering meadows. The pavilion has an octagonal shape, similar to Hofman's kiosks near the entrance gate, and the roof is constructed in the same way as well.

The new type of building, the columbarium, occurred in the last century. It's a wall divided to racks, which are designed for one or more urns with ash. The opening is usually fixed with a stone ledger, with the name of the deceased, the date of birth and death marked. [4] Or it's closed with a glass door and then the urn is stored in another decorated urn, on which the information about the deceased are marked sometimes.

There are multiple types of columbarium differing according to period of their creation, in the Olšany cemetery. In general they are parts of the cemetery wall, high up to five racks with the glass doors, which makes it impossible to reach higher racks levels. Sometimes the racks are provided with a firm ledger, bearing the name of deceased. Occasionally there is a subborn effort to make the columbarium cosy by placing the seasonal decoration in it. The Absence of the space for candles and flowers is compensated by putting them on the pedestal. The urn graves with an excessive ledger gravestones are situated close to the columbarium. They bear the dates of birth and death of the deceased, sometimes only the family name. In case of the older inhumation graves the urns are stored into the gravestones niches, where the small safety copper doors are often missing. The numerous figural plastics, tombstones made in Art nouveau style with floral ornaments, elevated stone sarcophagus on pedestals or complicated portals with a peristyl show the early origin, the status and the wealth of the renter. Graves are bordered with a stone frame and often covered with soil. Tombs are covered with a stone ledger or metal sheet with the hitches for manipulation. The soil graves occurred rarely, they are mainly in the old cemetery part. Plenty of funeral plastics and quality artworks make the Olšany cemetery an open air art gallery.

Although the concept of the Ďáblice cemetery was compact with quality design of disposition and of the ritual hall, the main gate and kiosks, it cannot be considered as a good example of sculptural and craft artworks. It seems, like this trend was stopped at the beginning of 20th century and the further increase of the secularisation and the tabuization dragged away the attention from the dead dwelling. The tombstones, including the tombs of the victims of communism, have more informative character. The moderation or the indifference responds to the periphery location. This concept doesn't correspond with existing shaft graves with the bodies of politic victims and surrounding meadow with the epitaph ledgers for the deceased children of imprisoned women, who we should not forget. It makes the impression as if the dead were "put aside" (the political prisoner were treated similar in the totalitarian system) without any subsequent care.

4 CEMETERY GREENERY

In Olšany we can notice the continuous ground cover of ivy and grass crossed by asphalt paths, which is complemented by uninterrupted growth of tall trees and there are bird boxes and benches in the shadows of huge (trees) crowns.

In Ďáblice the trees create an alley passing along the path of granite cobblestones towards the honorable burial place. There are wide meadows – the scattering meadows for cremation adjacent to the alley. The entrance is dominated by the tall spruce tree. The area is densly plated by trees, without strict composition In the back of the cemetery. In comparison with the Olšany Cemetery it becomes evident, that the population density (settlement) is lower here. On the contrary this density seems to be much higher when assessing the columbarium dungeons (which are) decorated with glass doors. The demeaning decoration around the urns express inadequate piety and disrespect for the deceased (dead).

The location within urban area and the good access from the town center, but also the grown greenery makes the cemetery look like an urban public garden or park, the visit rate of Olšany cemetery is high. The Ďáblice cemetery, located in the outskirts of the urban area with nearby housing estate, despite its natural character, has quite low visit rate. Anyway, it must be mentioned, that the visit rate of cemeteries in general is a seasonal business. On All Saints' Day, or Christmas and Easter many people meet here and the cemetery becomes a meeting place.

5 THE PROBLEMS

The capacity of the Olšany Cemetery is quite full and its downtown location doesn't allow any further territorial expansion. Therefore it is the number of urns in the columbarium that increases rather than the number of tombstones. This fact, according to Mgr. Červený (present director of the Administration of the Prague Cemeteries "SPH") "kills" the tombstones and their creation. Since the beginning of the last century the significant downgrade of the art and craft level of tombstones and monuments has been noticed, these ones were regarded rather as "*a stone mound*" (Almer, 1928, pp. 7–8).

Also the chapel graves, the number of which is estimated at 200, are no longer used as the burial places at the moment. No new ones are built, although the chapel graves represent the art and architecture as such. Many of them are abandoned now and SPH management intends to use them as a columbarium. In the cemetery the small scattering meadows were added, but they require much larger area because of their purpose and use. Therefore their use is only seasonal The lack of the space in the cemetery is not a new issue. It was already mentioned in the article "*For the need of new treatment of The Prague Cemetery*" written by Ing. Jan Almer in 1928: "*The basic defect and the severe deficit of all our cemeteries …is that they miss the adequate space for a car parking and gathered audiences in large quantities …Similarly, there is no provision for audience shelter in case of sudden bad weather, especially rain….*" (Almer, 1928, pp. 5–6).

In the Ďáblice Cemetery there was newly established The Forest of memories, where the ashes will be stored to the roots of growth trees. These days the cemetery follows European trends having the scattering meadows and the free composition this is due to its founding at the beginning of the20th century, during the onset of the cremation expansion and thus enough free space was retained there.

The problem of the death taboo results in the increasing number of the social funerals. One of the factors influencing this fact is certainly a "*society inability to cope with the death of beloved ones, the own dead*" (Frolíková Palánová, Kovář, 2014, p. 42). While in the 19th century "*the cult of death was widespread together with the increase of the bourgeoisie economic power, during the 20th century the interest of this cult is decreasing…*" (Kašpinská, 2007, pp. 23).

The ISSP survey conducted in 2008 let us know how people solve the dead departure of this world and that means how they cope with the death. It shows the proportion of ceremonies or burials without ceremony in the total number of burials. The selection varies by age. The respondents

Table 1. Preference of funeral ceremony (in %).

	Total	By age			
		18–29	30–44	45–59	60+
Religious ceremony	19,8	8,2	13,3	21,4	34,4
Secular ceremony	25,7	27,0	22,1	27,2	27,1
Burial without ceremony	18,9	13,2	20,6	21,4	19,0
I do not care	30,4	43,1	37,6	26,2	16,7
I cannot choose	5,1	8,6	6,3	3,7	2,9

Origin: ISSP 2008.

who don't care about the kind of a ceremony comprise nearly one third of the surveyed. The people selecting a secular ceremony present one quarter of the survivors. The survivors choosing a religious ceremony less than 20% and those ones, who want no ceremony, comprise almost the same proportion of the total burial amount. The age indicates, that the opinion on a burial ceremony is not clearly formed especially with representatives of the younger generation.

6   THE CURRENT DIRECTION OF THE CEMETERY

The fundamental problem of cemeteries is that there is no place for the living people in their area. The death leaves us cold and unconcerned, because the death (in the cemeteries) takes up all the necessary space. Would there be a solution to watch the cemetery as a public space? A space for living people, where the dead are buried too. And not as it is at the present - a space for the dead, where sometimes someone alive appears. The Management of the Administration of the Prague Cemetery (SPH) informed about their plan to build in the Olšany Cemetery a café bar so that the visitors have the opportunity to enjoy refreshments during a long stay in the urban garden, which a cemetery can be; to have picnic, if a suitable area would be given, to stay in the company of the living, but even of the dead. In calm, under the roof of the giant tree crowns, in the company of quality funeral sculptures and surrounded by the names of the greatest personages of our history. "*Especially older cemeteries, such as Malostranský, l. and ll. and the requiem cemetery in Olšany nowadays become almost like museums and for intelligent man like a kind of a place to rest and think, even a place to study the changing public life. As well to study about prominent or even unknown individuals, who often are remembered just at their graves without frames and at a nearby built monument.*" (Almer, 1928, p. 13). Commemorating their legacy and thus being aware of the cultural development of the nation. Also an intended gift shop or store where one could choose a nice artistic tombstone or a bust (a sort of funeral gallery). the valuable and tasteful piece of artwork should not be replaced by a simple range of prefabricated objects. The best example of living connection with the world of the dead is a bench at the graveside. It's a symbol suggesting that we are welcomed there and that we're not just passers-by. There were several such benches In the Olšany cemetery, especially one is worth mentioning it was joint to the tombstone and whole ensemble was from one piece of stone. Chapel graves on their own represent the model of honors for the dead, as they have an interior space and therefore they truly constitute the abode of the dead. Also the stairs, which we climb up, represent an example of the path to spirituality, but expressed in the language of the living. We meet with other kind of art too. The respect and the more pleasant environment for visitors can be represented by a nice cemetery wall. The media once described the cubistic gate in Ďáblice as representing the crystalline structure of atoms, which humans are formed of, but maybe it's just a coincidence with the favorite crystalline form of the Cubist movement in general (Oulický, 2011).

The statement of Tomas Bata, who founded the Forest Cemetery in Zlín with the aim to create a place to live, presents the idea of future development of cemeteries at its best: "*We used to regard cemeteries as a place, where one retires to mourn. But a cemetery as all other things and places should serve the living. This is why it should not be frightening place so that the living people can spend nice and peaceful moments."* (Dobešová, 2012, p. 24).

ACKNOWLEDGEMENT

This work was supported by means of the VŠB-TUO Student Grant Competition. Project registration number is SP2015/62.

REFERENCES

Almer, J. 1928. *O potřebě nové úpravy pražských hřbitovů*. Praha: Důchody obce hlavního města Prahy.

Dobešová B. 2012. *Environmentální a společenské přínosy přírodního pohřebnictví*. Brno: Masarykova Univerzita, Fakulta sociálních studií, Katedra environmentálních studií. Head master's thesis Hana Librová.

Frolíková Palánová, K., Kovář J. a kol. 2014. Architecture in perspective VI. *Burying as Part of Life in the European Context*. 39–43. Germany: TTP Ltd

Kapišinská, V. 2007. Architektúra a urbanizmus. *Formovanie moderny vo funerálnom kontexte*. 41 (1–2): 21–29.

Oulický, J. 2011. *Ďáblice nemají hřbitov s kostelíkem, ale fyzikální rozbor smrti*. Ihned.cz [Internet]. Publicated 2 September 2011 [cited 23 February 2015]. Available from: http://life.ihned.cz/cestovani/ c1-52738000-dablice-nemaji-hrbitov-s-kostelikem-ale-fyzikalni-rozbor- smrti.

Pohaničová J. 2013. Architektúra a urbanizmus. *Štýlové variácie v komornom prevedení.*, 47 (1–2):107–121.

Pražská informační služba. 2007. *Památky – Olšanské hřbitovy*. Pražská informační služba [Internet]. Publicated 29 June 2007 [cited 20 November 2014]. Available from: http://web.archive.org/web/ 20071023062816/http://www.pis.cz/ cz/praha/pamatky/olsanske_hrbitovy.

Stejskal D., Šejvl J. a kol. 2011. *Pohřbívání a hřbitovy*: 464. Praha: Wolters Kluwer ČR, a.s.

Svobodová M. 2013. *Krematorium v procesu sekularizace českých zemí 20. století*: 182. Praha: Artefactum, nakladatelství Ústavu dějin umění Akademie věd České republiky, v.v.i.

The Czech Republic. 2001. *256/2001 Sb. Zákon o pohřebnictví a o změně některých zákonů, §5 (1)*

Advances and Trends in Engineering Sciences and Technologies – Al Ali & Platko (Eds)
© 2016 Taylor & Francis Group, London, ISBN: 978-1-138-02907-1

Non-destructive testing of settling tank

M. Pitonak & M. Cangar
Faculty of Civil Engineering, Universtiy of Žilina, Žilina, Slovakia

ABSTRACT: The paper deals with the rehabilitation technology settling tanks, which had a problem with significant leaking of water through concrete wall of the tank. Possible causes of this disorder were surveyed by using a non-destructive method – thermal camera vision. Based on the findings was selected appropriate to rehabilitation. Similar tanks were built in several manufacturing plants as part of the technology, so the chosen procedure can be implemented on a wider scale.

1 INTRODUCTION

Article deals with the possible rehabilitation technology the settling tank, which is located at the site Metsä Tissue Slovakia Ltd. in the administrative area of the city. The round structure is currently placed as settling tanks in papermaking factory which is being use for cleaning industrial water and then is released to public sewage system. The authors searched for the cause of overflowing through the construction structure the water settling tanks, during the full run of tank and without possibility emptying the tank.

1.1 *Existing construction parameters and condition of structure*

Settling tank is construction structure circular floor plan with a radius of 23.1 m, which is designed as a reinforced concrete structure. Tank height is 2.75 m above the surrounding terrain. The tank bottom is sloped towards its center at 4° inclination.

The heart of the tank is rotating lattice structure whose axis of rotation is located at the middle of the tank (Figure 1). The tank is filled with industrial water from the middle part. At the same time the pickup device, which wheeled vehicle moves around the perimeter of the tank on the concrete "wall head" is collecting waste from the top and from the lowest part of the tank (from the center of the tank). The waste is push to outside of the tank. Along the perimeter of the tank is a tank overflows, through which is flows industrial water cleaned from industrial part of the production (Figure 3).

Industrial water temperature is constantly during the year, between 25–30°C. During the operating time (age of construction is unknown – assumption 60 years), the tank has started to penetrate thru the wall of the tank. The visual inspection identified unambiguously that the outside wall is visibly wet and are visible the drops off the parts of the facade (Figure 4).

In places which are wet in several places on facade of the tank is visible reinforced of the concrete structure. The largest areas of wets from industrial water leak out throw the wall (Figure 5).

Another problem is the concrete facade, along which the vehicle pantograph runs. During the time influence and the wheel load affects is upper part of the concrete "wall head" (Figure 6).

1.2 *Water inside of the tank*

Water analysis was provided by the operator Metsa group tank. Analysis of unpurified waste water, which is located in the tank showed that the pH of the water is from 6.5 to max. 8. Even when

Figure 1.   Settling tanks with overflow.

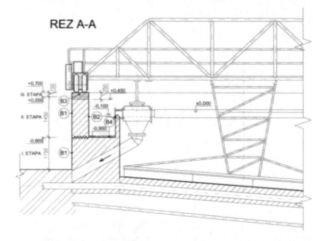

Figure 2.   Profile of tank.

Figure 3.   Detail overflow tank.

Figure 4.   Wet outside wall with drop off the parts of the facade.

Figure 5.   Details the exposed reinforcement.

Figure 6.   Damages of the upper part by wheel load.

Figure 7.   Pullout testing concrete.

the water is acid nor excessively basic, is aggressive. This can be seen from the high levels mainly COD, soluble substances from soluble inorganic and organic substances.

2   STRUCTURE DIAGNOSIS

After visual assessment of the existing situation, we started with possible the diagnosis of structure without emptying tank: *Pullout testing concrete* and *Thermography inspection*.

2.1   *Pullout testing concrete*

For the location of the pullout concrete tests were selected 4 points around the circumference the tank (Figure 7). The measurements were performed pullout device, on 04/22/2013 in the morning.

Measurement values ranged was from 1.5 to 2.7 MPa. Before was expecting the value at 0.5. The reasoning was based on a visual inspection, since it is not known special technical documentation. For control measurements can be concluded that the tested concrete locations manifests as healthy as measured and there is no need to interfere with the static part of the structure, but must be addressed is the facade remediation.

2.2   *Thermography inspection*

To use the thermography inspection diagnostics, was the purpose to identify areas of wet and detect leakage of water through the sfacade and detect any anomalies that can't be detected by the visual inspection. The use of thermal imaging has been useful in the moment, since the waste water has a temperature of about 25–30°C and the outside temperature is in the morning was about 2°C (relative humidity 75%), detection of critical areas were relatively simple, as is represented on next figures.

The measurements were performed with a thermal imager Thermotencer 7500 all around the tank on 8.4.2013 5:30 am. Images (Figure 8) are seen massive horizontal wet area in high approximately 1.1 m from outside above the ground, from this point by gravity is spreading downwards.

3   CONCLUSIONS

The main task was to find out the causes of leaking the tank structure and so can be identified rehabilitation technology to protect the tank against leaking. The find out the causes was necessary detect in time by full operation of the tank, beacau So over of maintenance shutdowns (five days

Figure 8.   Images from thermography inspection.

per year), it was necessary to know which rehabilitation technology has to be use for fixed the leaking problem.

Going through by visual inspection on settling tanks were found significant wetting of reinforced concrete structures on facade. In several places even water leaking through structure so it created small streams of water, which then flowed over the facade. Of wet as well as massive leaks were located at a height about 1.6 m from the upper edge of the facade traction. This amount corresponds to the bottom of the internal overflow canal which 1.15 m above the ground.

Thermography inspection has been confirmed that there is a horizontal of wet (leakage) Figure 8 above 1.1 m above the ground.

On the basis of measurements, which exclude the possibility of weakening the concrete structure by aggressive water from tank and provided that the joint is connected by an uninterrupted

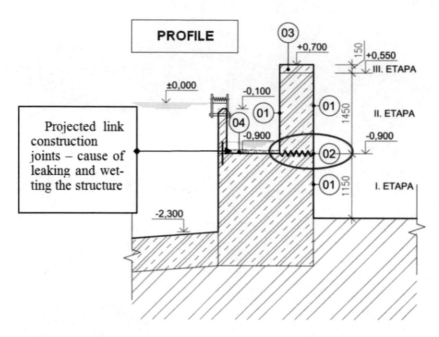

Figure 9. Place of cause the leaking tank.

concreting Figure 3, we come to conclusion the leak causes are at the bottom of the internal overflow canal were was insufficient interconnection completed. Probably during the construction works there was work break on whole structure, which caused not interconnected horizontal interface (construction joints) Figure 9. Long-term use of tank and the flow of the water around internal overflow canal, this plane gradually locally violated, and consequently was leaking visible at the periphery of the tank facade.

The aim of the article was pointing out the possibility of direct use applications of non-destructive methods, which are based on sophisticated scientific methods.

ACKNOWLEDGEMENTS

This contribution is the result of the project implementation: "Support of Research and Development for Centre of Excellence in Transport Engineering" (ITMS: 26220120031) supported by the Research & Development Operational Programme funded by the European Regional Development Fund.

REFERENCES

Decký, M. – Drusa, M. – Pepucha, Ľ. – Zgútová, K.: Earth Structures of Transport Constructions. Pearson Education Limited 2013, Edinburg Gate, Harlow, Essex CM20 2JE. Edited by Martin Decký, p. 180, ISBN 978-1-78399-925-5.
Manning, G.P. – Concrete reservoirs and tanks, (Concrete series) 1967 ISBN:6628408146
Zich, M. – Bažant, Z. Concrete structures, tanks and reservoirs: Akademické nakladatelství CERM, 2010, ISBN: 9788072046935
Zivčák, J. and coauthors – Thermography diagnostic: Technická univerzita v Košiciach, 2010 ISBN: 9788055305332

Advances and Trends in Engineering Sciences and Technologies – Al Ali & Platko (Eds)
© 2016 Taylor & Francis Group, London, ISBN: 978-1-138-02907-1

The application of the Electre I method for the selection of the floor solution variant

E. Radziszewska-Zielina
Department of Construction Technology and Organization, Institute of Management in Construction and Transport, Faculty of Civil Engineering, Cracow University of Technology, Poland

M. Gleń
Cracow University of Technology graduate, Poland

ABSTRACT: The Electre methods are relatively rarely applied in civil engineering; however, there are several interesting publications on the subject. The ELECTRE I method was first applied in civil engineering in 1985. The aim of the present paper is to demonstrate the possibility to apply the multi-criteria Electre I method for the analysis of selected floors in terms of selection of the best solution variant in accordance with the adopted criteria of analysis.

1 INTRODUCTION

The Electre methods are relatively rarely applied in civil engineering; however, there are several interesting publications on the subject. The ELECTRE I method was first applied in civil engineering in 1985. That research was then briefly presented in (Zavadskas 1987a), (Zavadskas 1987b). The application of the ELECTRE III method may be found in papers by numerous authors. Among them, Azar et al. (2001) presented the building optimization tools at the stage of development of an architectural sketch while Cavallo and Norese (2001) applied multi-criteria analysis for the evaluation of erosion and landslide risk. The multi-criteria decision support method was applied by Thiel and Mróz (2001) in the design of heating systems in museum buildings whereas Tam et al. (2004) used it in the evaluation of construction machine operation, presenting a case study of concrete vibrators. Multi-criteria analysis was applied by Zavadskas et al. (2004) in commercial construction projects for investment purposes while Mróz and Thiel (2005) used that analysis to evaluate the heating system in buildings. The preferences of participants in the decision support process were analysed by Thiel (2006), who also proposed a procedure for determination of the relative significance of criteria when the number of those who assess constitute a small sample (Thiel 2008). In a paper by Ulubeyli and Kazaz (2009) the multi-criteria method supports the decisions concerning the choice of concrete pumps. The Electre 4 method was applied, among others, by Ustinovichius, Zavadskas et al. (2006) for supporting the evaluation of projects of real estate revitalization in rural areas. Another application of the Electre III method may be found in Radziszewska-Zielina (2010), where it is used to select a construction enterprise for cooperation in construction projects.

The analytical foundations and the basic concepts of the Electre family methods are similar.

The threshold of equivalence expresses the maximum difference in the evaluation of two variants, for which it is assumed that the variants are equivalent in relation to the criterion.

The threshold of preference expresses the minimum difference in the evaluation of two variants, for which it is assumed that one variant is strongly preferred to the other. If the difference in evaluations is smaller than the threshold of preference, the preference can only be assumed.

Table 1. The ascribed evaluations of floor solution variants in accordance with the criteria.

Floor variants	Criteria					
	k_1 Execution cost [zł/m^2]	k_2 Labor intensity [r-g/m^2]	k_3 Construction lightness (weight) [kg/m^2]	k_4 Easy assembly [1–10]	k_5 Acoustics [1–10]	k_6 Esthetics [1–10]
w_1 Wooden	130	4.28	38	10	4	9
w_2 Ribbed	200	2.40	300	9	8	7
w_3 Reinforced concrete, monolithic	260	2.50	250	5	10	10
w_4 Prefabricated (of the filigree type)	250	2.33	230	4	10	9

Table 2. Assigned veto thresholds and criterion weights.

	k_1	k_2	k_3	k_4	k_5	k_6
max	260.00	4.28	300.00	10.00	10.00	10.00
min	130.00	2.33	38.00	4.00	4.00	7.00
Veto threshold	130.00	1.95	262.00	6.00	6.00	3.00
Criterion weights	0.30	0.20	0.05	0.10	0.20	0.15

The veto threshold determines the conditions which are the consequence of the existence of clear prerequisites that make exceeding impossible. The selection of veto threshold values depends on the scatter of a given value for the examined variants and preferences of the decision maker. A threshold value that is too low may turn out to be contrary to the reality and disturb the quality of the ordering. However, the adoption of a value that is too high may mean the non-existence of such a strong dominance for a given criterion. The veto threshold can be determined mathematically for every criterion according to the following formula: $Vk_i = \max k_i(w) - \min k_i(w)$, where $\max k_i(w)$ and $\min k_i(w)$ are, respectlively, the maximum and the minimum measure among the variants for the i-th criterion.

The aim of the present paper is to demonstrate the possibility of application of the multicriteria Electre I method for an analysis of sample, selected floors in terms of selection of the best solution variant in accordance with the adopted criteria and other assumptions.

2 METHOD ASSUMPTIONS AND THE RESEARCH PROCEDURE

Selected for analysis were 4 types of floors (decision variants) and 6 evaluation criteria. Weights ω are ascribed to each criterion k and weights ω_k fulfil the conditions of being non-negative and they total to units, i.e. $\Sigma_{m=1}^{6}\omega_k = 1$ (Table 2). The evaluation criteria and criterion weights were determined on the basis of an expert opinion survey. $W = \{w_1, w_2, w_3, w_4\}$ denotes a set of decision variants, $K = \{k_1, k_2, k_3, k_4, k_5, k_6\}$ a set of criteria (Table 1). For each criterion the veto threshold is defined that points to prerequisites that make exceeding impossible (Table 2). The $\phi_k(w_i, w_j)$ denotes variant evaluation $i, j = 1, \ldots, 4$ in relation to criterion $k = 1, \ldots, 6$. This is exemplified in Table 3, which shows evaluation analysis results for criteria k_1 and k_2.

The research procedure consists of 4 steps.

Step 1. Assignment of the consistency set

Table 3. Evaluation of floor variants in relation to criteria k_1 and k_2.

$\phi_k(w_i, w_j)$	$\phi_1(w_i, w_j)$ for k_1 execution cost				$\phi_2(w_i, w_j)$ for k_2 execution cost			
Floor variants	w_1	w_2	w_3	w_4	w_1	w_2	w_3	w_4
w_1 Wooden	1	1	1	1	1	0	0	0
w_2 Ribbed	0	1	1	1	1	1	1	0
w_3 Reinforced concrete, monolithic	0	0	1	0	1	0	1	0
w_4 Prefabricated	0	0	1	1	1	1	1	1

Table 4. Values of the consistency coefficient $c(w_i, w_j)$ of the floor variants for individual criteria.

Consistency matrix	$c(w_i, w_j)$ for k_1 execution cost				for k_2 execution cost			
Floor variants	w_1	w_2	w_3	w_4	w_1	w_2	w_3	w_4
w_1 Wooden	0.30	0.30	0.30	0.30	0.20	0.00	0.00	0.00
w_2 Ribbed	0.00	0.30	0.30	0.30	0.20	0.20	0.20	0.00
w_3 Reinforced concrete, monolithic	0.00	0.00	0.30	0.00	0.20	0.00	0.20	0.00
w_4 Prefabricated	0.00	0.00	0.30	0.30	0.20	0.20	0.20	0.20

Table 5. Values of consistency coefficients of floor variants for all criteria.

Consistency coefficients $c_{1-6}(w_i, w_j)$	w_1	w_2	w_3	w_4
w_1 Wooden	1.00	0.60	0.45	0.60
w_2 Ribbed	0.40	1.00	0.60	0.40
w_3 Reinforced concrete, monolithic	0.55	0.40	1.00	0.45
w_4 Prefabricated	0.55	0.60	0.75	1.00

For each pair of variants (w_i, w_j) the value of the consistency coefficient $c(wi, wj)$ is calculated using the formula (1) (Trzaskalik 2006):

$$c(w_i, w_j) = \sum_{k=1}^{ml} w_k \cdot \varphi_k(w_i, w_j) \tag{1}$$

where $\phi_k(w_i, w_j) = 1$ when w_i is not worse than w_j in relation to criterion k; $\phi_k(w_i, w_j) = 0$ in the opposite case.

Table 4 presents the values of the consistency coefficients for each pair of variants in relation to two criteria: k_1 and k_2. For the remaining criteria k_3, k_4, k_5, k_6 adopted in the study, the values of coefficients are calculated following the same rule.

Step 2. Determination of the consistency set $C(s)$ for the adopted consistency threshold value (s)

Juxtaposition of the total of consistency coefficients of particular variant solutions was done on the basis of the scope of the consistency threshold $s \in [0.5; 1.0]$, in accordance with the rule that the higher the consistency threshold value s, the more justified the exceeding (preference).

As a result of totalling the values of consistency coefficients of all criteria $k_1 - k_6$ in each relation of variant solutions, the values presented in Table 5 were obtained.

Table 6. Consistency coefficients of floor variants with the consistency threshold s = 0.6, arranged according to exceeding.

Consistency coefficients $c_{1-6}(w_i, w_j)$	s = 0.5	s = 0.6	HIERARCHY according to consistency coefficients
w_1 Wooden	2.20	2.20	w_4 Prefabricated
w_2 Ribbed	1.60	1.60	w_1 Wooden
w_3 Reinforced concrete, monolithic	1.55	1.00	w_2 Ribbed
w_4 Prefabricated	2.90	2.35	w_3 Reinforced concrete, monolithic

Table 7. Inconsistency matrices of the floor solution variants in relation to criteria $k_1 \cdot k_2$.

Inconsistency matrix $d(w_i, w_j)$ Floor variants	for k_1				for k_2			
	w_1	w_2	w_3	w_4	w_1	w_2	w_3	w_4
w_1 Wooden	*	0	*	0	*	0	*	0
w_2 Ribbed	*	*	0	*	*	*	0	*
w_3 Reinforced concrete, monolithic	*	*	*	*	*	*	*	*
w_4 Prefabricated	*	0	0	*	*	0	0	*

The highest coefficient values in Table 5 are 0.75, 0.60 and 0.55.

The adoption of the highest threshold s = 0.75 would equal one relation of exceeding, namely $C(0.75) = \{(w_4 \cdot w_3)\}$. However, the adoption of the lowest threshold s = 0.55 would be inconsistent with the rule of the exceeding test. For this reason the consistency threshold s = 0.60 was adopted. The conditional interval of the consistency threshold $s \in [0.6; 1.0]$ will include the coefficient set {0.60; 0.75}. These coefficients are obtained in 5 relations $(w_i \cdot w_j)$ forming the following consistency set: $C(s = 0.60) = [(w_1 \cdot w_2); (w_1 \cdot w_4); (w_2 \cdot w_3); (w_4 \cdot w_2); (w_4 \cdot w_3)]$.

The values of the consistency coefficients of each variant were totalled (Table 6). The obtained numerical data allowed for ordering the variants from the best solution to the worst. The values were compared by adding all values within the range of the consistency threshold according to the condition $s \in [0.5; 1]$ as well as the values within the limits of the adopted threshold, i.e. $s \in [0.6; 1]$. The obtained results show that the arrangements are consistent with one another. The highest numerical value was reached by variant w_4, i.e. the prefabricated floor of the filigree type, and then the wooden floor (w_1).

Step 3. Determination of the inconsistency set $D(v)$ based on the condition of a lack of inconsistency

The condition of a lack of inconsistency is related to the assumed veto threshold v_k. Exceeding this threshold means that a variant dominates over another one in terms of a given criterion so strongly that the remaining criteria cannot influence the change of this relation. The inconsistency set $D(v)$ is built out of pairs $(w_i.w_j)$ fulfilling the condition that is contrary to the condition of a lack of inconsistency. Using the inconsistency condition: $k_m(w_i) + v_m[k_m(w_i)] \geq k_m(w_i)$, the inconsistency matrix of the variants was developed (Table 7), in which the following symbols were applied:

0 – lack of inconsistency; 1 – occurrence of inconsistency; * – verification is unnecessary (pairs are beyond the inconsistency set). The table presents a fragment of the matrix, i.e. for criteria k_1 and k_2. Floor variants undergo the same analysis in relation to the remaining criteria adopted in the study.

Relations of exceeding $(w_i \cdot w_j)$ occur when the conditions of consistency and inconsistency are fulfilled simultaneously. In Table 8 the color denotes the relations in the consistency set (as well as the same ones in the inconsistency set). The 0 value occurs in each section, which means a lack of inconsistency. Therefore the relations of exceeding marked with symbol 1 occur in all pairs $(w_i \cdot w_j)$ contained in the consistency set.

Table 8.   Relations of the inconsistency matrix and of exceeding of floor variants.

Relations	Relations of the inconsistency matrix				Relations of exceeding			
Floor variants	w_1	w_2	w_3	w_4	w_1	w_2	w_3	w_4
w_1 Wooden	0	0	0	0	0	1	0	1
w_2 Ribbed	0	0	0	0	0	0	1	0
w_3 Reinforced concrete, monolithic	0	0	0	0	0	0	0	0
w_4 Prefabricated	0	0	0	0	0	1	1	0

Table 9.   Inconsistency coefficients of floor variants, arranged according to exceeding.

Inconsistency coefficients (d, k_{1-6}), *total*		HIERARCHY according to inconsistency coefficients
w_1 Wooden	2	w_4 Prefabricated
w_2 Ribbed	1	w_1 Wooden
w_3 Reinforced concrete, monolithic	0	w_2 Ribbed
w_4 Prefabricated	2	w_3 Reinforced concrete, monolithic

As in the case of analysis of the consistency relation, the values of the inconsistency coefficients of each variant were totalled (Table 9). The obtained numerical data allowed for the ordering of the variants from the best to the worst. The highest numerical value is reached by variants: w_4, i.e. the prefabricated floor of the filigree type and the wooden floor (w_1). The results of the arrangement of floor variant solutions, obtained on the basis of the consistency condition and the inconsistency condition, are unequivocal, i.e. the arrangement is the same.

It must be noted that the results in the inconsistency analysis are based on whole numbers; therefore the ordering of variants from the best to the worst solution is less precise. In this case two floor variants have the same position in the hierarchy of inconsistency. There is no contradiction in the analysis of consistency and inconsistency.

Step 4. Construction of the diagram of preferential relations

Subsequent stages of the diagram construction are presented in Figure 1 below, beginning from the highest value of the consistency threshold. The number of stages equals the number of consistency thresholds occurring between the top limit value $s = 1$ and the adopted bottom threshold s. The lowest consistency threshold adopted in the present analysis is $s = 0.60$; therefore the construction of the diagram consists of 2 stages for the adopted $s = \{0.75; 0.60\}$. According to the consistency coefficients determined in Table 6, the highest consistency threshold amounts to $s = 0.75$ and was obtained in the variant relation ($w_4.w_3$); therefore the set of the consistency threshold is a single-element one (one pair of preferences), i.e. $C(0.75) = \{(w_4 \cdot w_3)\}$. The arrow connecting the variants symbolises the occurrence of a relation while the arrowhead indicates the exceeding variant. The next limit threshold reached is $s = 0.60$; it occurs in several relations, forming the last set $C(0.60) = \{(w_1 \cdot w_2); (w_1 \cdot w_4); (w_2 \cdot w_3); (w_4 \cdot w_2); (w_4; w_3)\}$. In order to check whether there would occur changes in the preference if the threshold was lowered more, an additional diagram was made which determines the relations for threshold $s = 0.55$. Set C for the adopted threshold is as follows: $C(0.55) = \{(w_1 \cdot w_2); (w_1 \cdot w_4); (w_2 \cdot w_3); (w_4 \cdot w_2); (w_4; w_3); (w_3 \cdot w_1)\}$.

3   RESEARCH RESULTS AND CONCLUSIONS

As results from the diagram developed for $s = 0.75$, there occurs one relation between variants w_3 and w_4., i.e. the prefabricated floor and the monolithic reinforced-concrete floor. The arrowhead points towards w_3; therefore it indicates the supremacy of the prefabricated floor (w_4) as

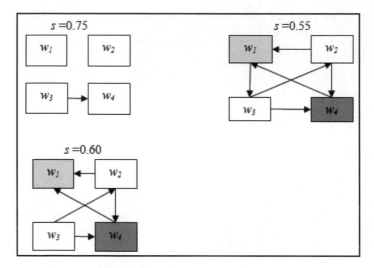

Figure 1.   Construction of the diagram of preferential relations of the floors.

compared to the monolithic reinforced-concrete one (w_3). Lowering the threshold of preference to $s = 0.60$ causes a considerable increase of the number of relations to 5. Due to the same number of exceedings, the prefabricated floor (w_4) and the wooden floor (w_1) become mutually competitve. The most interesting fact is that the wooden floor exceeds the prefabricated one, which with threshold s > 0.60 was beyond competition. Considering the validity of the thresholds, and in this case s = 0.75 > s = 0.60, variant w_4 – the prefabricated floor was adopted as the most favourable choice. Further lowering of the consistency threshold to s=0.55 does not bring significant preference changes. What is additionally obtained is only the information on relations w_3 and w_1 (the supremacy of the monolithic reinforced-concrete floor over the wooden floor). In order to verify the results, the authors also performed an analysis by means of another method using the von Neumann-Morgenstern encoding. The obtained results were comparable. It is worth noting that multi-criteria methods only support the decision-making system of the enterpreneur concerning the choice of the decision variant. The result may depend on the adopted assumptions and evaluation criteria. Although the Electre methods model the decision-makers' preferences, the final decision must always be taken by the decision-makers themselves.

REFERENCES

Azar S. Azar J.-M. Hauglustaine J.-M. 2001. Multicriteria and Multiple Actors Tool Aiding to Optimise Building Envelope at the Architectural Sketch Design. *Informatica* Vol. 12. No. 1: 3–24.
Cavallo A. Norese M. F. 2001. GIS and Multicriteria Analysis to Evaluate and Map Erosion and Landslide Hazards. *Informatica* Vol. 12. No. 1: 25–44.
Mróz T. M. Thiel T. 2005. Evaluation of a heating system for buildings using multiple criteria decision analysis. *Archives of Civil Engineering* Vol. 51. No. 2: 281–298.
Radziszewska-Zielina E. 2010. Methods for selecting the best partner construction enterprise in terms of partnering relations. *Journal of Civil Engineering and Management* Vol. 16. No. 4: 510–520.
Tam C. M. Tong T. K. L. Lau C. T. 2003. ELECTRE III in evaluating performance of construction plants: case study on concrete vibrators. *Construction Innovation* Vol. 3. No. 1: 45–61.
Thiel T. Mróz. T. 2001. Application of Multi-Criterion Decision Aid Method in Designing Heating Systems for Museum Buildings. *Informatica* Vol. 12. No. 1: 133–146.
Thiel T. 2006. A proposal of defining participant preferences in a decision aiding process with the participant representing a collective body. *Technological and Economic Development of Economy* Vol. 12. No. 3: 257–262.

Thiel T. 2008. Determination of the relative importance of criteria when the number of people judging is a small sample. *Technological and Economic Development of Economy* Vol. 14. No. 4: 566–577.

Trzaskalik T. 2006. *Metody wielokryterialne na polskim rynku finansowym*. Warszawa: PWE

Ulubeyli S. Kazaz A. 2009. A Multiple Criteria Decision-Making Approach to the Selection of Concrete Pumps. *Journal of Civil Engineering and Management* Vol. 15. No. 4: 369–37.

Ustinovichius L. Zavadskas E. K. Lunkevichius S. Shevchenko G. 2006. Multiple Criteria Analysis for Assessing the Investment Projects in Rural Property Revitalization. *International Conference on Operational Research: Simulation and Optimization in Business and Industry.* May 17–20. Tallinn. Estonia: 194–195.

Zavadskas E. K. 1987a. *Complex Evaluation and Selection of Resource-Saving Decisions in Construction*. Vilnius: Mokslas.

Zavadskas E. K. 1987b. The grounds of rational choice of the upper floor construction on taking into account some criteria. *Zeszyty Naukowe Politechniki Poznańskiej. Budownictwo Lądowe* Nr 30: 141–149.

Zavadskas E. K. Ustinovichius L. Stasiulionis A. 2004. Multicriteria Evaluation of Commercial Construction Projects for Investment Processes. *Journal of Civil Engineering and Management* Vol. 3. 10. No. 2: 151–166.

Advances and Trends in Engineering Sciences and Technologies – Al Ali & Platko (Eds)
© 2016 Taylor & Francis Group, London, ISBN: 978-1-138-02907-1

Development of a weighting system for the assessment of economic sustainability of buildings

D.A. Ribas
School of Technology and Management of Polytechnic Institute of Viana do Castelo, Portugal

M.M. Morais, A.L. Velosa & P.B. Cachim
RISCO and Department of Civil Engineering of University of Aveiro, Aveiro, Portugal

ABSTRACT: Assessment systems and sustainability certification of buildings plays an important role in the design, construction, operation, maintenance and decommissioning of a building. They promote the integration of economic, social and environmental aspects in building sustainability. A Methodology of Assessment of Economic Performance of Residential Buildings (MAEP-RB) which is an innovative approach to systematic evaluation of economic performance of a building within the concept of sustainability, based on the Life Cycle Analysis (LCA) according to prEN 16627:2013 has been developed. The purpose of this article is to present the results of the application of Analytic Hierarchical Process (AHP) to determine the hierarchical structure of the weighting system of MAEP-RB in order to achieve a consistent and rational model of analysis.

1 INTRODUCTION

The European Standards set developed by CEN/TC 350 "Sustainability of construction works", proposes a system for assessing the sustainability of buildings based on life cycle analysis (LCA). The assessment of the sustainability of buildings quantifies impacts and aspects to assess the environmental performance, social and economic buildings, using quantitative and qualitative indicators, measured without value judgment. Based on LCA in the *before use* phase as defined in EN 15643-4:2012 and prEN 16627:2013, a new approach has been developed for systematically assess the economic performance of a building within the concept of sustainability (Methodology of Assessment of Economic Performance – Residential Buildings: MAEP-RB). The assessment of economic performance of a building expressed in monetary terms and at the same time the allocation of economic sustainability levels for a building, needs the definition of a system of weights based on the hierarchical structure of attributes of MAEP-RB. This system of weights will allow the possibility of integration of MAEP-RB in a global evaluation of sustainability, in which environmental, social and economic dimensions are simultaneously considered. In MAEP-RB, the assessment of the level of economic performance and economic sustainability of a building are achieved by aggregating the attributes at each of the level of the hierarchical structure of the method: parameters, indicators, modules and stages of the life cycle. Although there is no doubt that there are some attributes that are more important to the economic sustainability than others, there is currently no universally accepted method for the definition of the relative weight of each attribute. The weight system adopted depends, among other factors, on the context and local priorities and on the different opinions of different stakeholders in the life cycle of buildings. This paper aims to present the methodology and calculation of a system of weights for MAEP-RB. The mathematical formulation of the AHP is presented as well as the developed tool for its application to the hierarchical structure of MAEP-RB. The weights calculated for the different levels of the hierarchical structure of MAEP-RB are presented.

Currently, many European countries, the United States, Canada, Australia, Japan and Hong Kong, have a system for the assessment of environmental performance of buildings. An overview of these different methods shows that the system of weights of each system is different and also that there is no unified methodology for the definition of these weights. The Hierarchical Analysis Process (AHP) is a multi-attribute decision analysis technique that belongs to the group of simple additive methods. It can be applied to problems where the decision-maker must sort or choose one out of a finite number of alternatives, which are measured by two or more relevant attributes. Overview of the main methods can be found in the vast literature available on the topic, such as Hwang (1981) and Chen (1992). In the absence of an objective basis for considering sustainability parameters, the AHP is suggested to determine the weighting of the attributes of the hierarchy by a number of authors (Norris, 1995; Chapman, 1998; Cole, 2002). The AHP was originally developed and applied by Saaty in the 1970s (Saaty, 1980). It is a fully compensatory analytical tool, applicable to multi-attribute decision problems that can be formulated as a decision tree, where each hierarchical level involves several types of attributes. This method was among the first to be developed in the context of multicriteria decisions and is probably the most used in the world (Gomes, 2004).

2 ANALYTIC HIERARCHY PROCESS

The decision problem in AHP is to compare the relative importance of attributes in a systematic and quantitative way. Mathematically, the objective is to determine the non-negative weights w_i of attributes (for i varying from 1 to n, where n is the number of attributes). If the weight vector $w = (w_1, \ldots, w_n)$ is known, the relative importance of attribute A_i compared to attribute A_j, a_{ij}, would be given by the ratio w_i/w_j (Haurie, 2001). All possible n column vectors can be combined in a matrix A (Saaty, 1980). The AHP approach assumes that the comparison matrix A has three special properties: (i) Identity: all diagonal elements of the matrix are equal to 1, because an attribute is always as important as itself; (ii) Reciprocity: if an attribute A is x times more important than an attribute B, then, the attribute B is $1/x$ times more important than attribute A. According to this principle, the terms of matrix A read $a_{ij} = 1/a_{ji}$, for all i and j; and (iii) Consistency: for consistency it is meant that, if there are three attributes A, B and C, if A is x times more important than B, and B is y times more important than B, then, A must be xy times more important that C. In this case, by linear algebra, the vector of weights, w, is the eigenvector of the matrix A, with eigenvalues λ, such that $Aw = \lambda w$. With this particular procedure, the AHP formalizes the conversion of a problem of assigning weights to attributes in a more tangible problem of comparing the relative importance of pairs of attributes competing attributes. This series of comparisons is summarized in square matrices, containing the relations between the compared attributes, as presented in Table 1.

For the comparison of attributes, a 5-point scale was adopted (4-much more important; 2-much important; 1-equally important; 1/2-less important; 1/4-much less important) to avoid incompatibility problems that might occur when three or more attributes are compared (Haurie, 2001). The calculation of the eigenvector of the matrix A, can be performed using a simplified approach that calculates the geometric mean of each row (Golony, 1993; Oliveira, 2008). In this case, the

Table 1. General form of comparison matrix (Silva, 2003).

	Attribute 1	Attribute 2	Attribute 3
Attribute 1	1	Importance of Attribute 1 compared to Attribute 2	Importance of Attribute 1 compared to Attribute 3
Attribute 2	Importance of Attribute 2 compared to Attribute 1	1	Importance of Attribute 2 compared to Attribute 3
Attribute 3	Importance of Attribute 3 compared to Attribute 1	Importance of Attribute 3 compared to Attribute 2	1

normalized weights w_i corresponding to each attribute can be obtained by:

$$w_i = \frac{\left(\prod_{j=1}^{n} a_{ij}\right)^{\frac{1}{n}}}{\sum_{i=1}^{n}\left(\left(\prod_{j=1}^{n} a_{ij}\right)^{\frac{1}{n}}\right)}, \text{ for } i=1,n, \text{ with } \sum_{i=1}^{n} w_i = 1. \tag{1}$$

The calculation of the maximum eigenvalue, $\lambda_{máx}$, associated with the calculated vector can be calculated by:

$$\lambda_{máx} = \frac{1}{n}\sum_{i=1}^{n} \frac{(Aw)_i}{w_i}. \tag{2}$$

The decision matrix is reciprocal, positive and consistent and has only one nonzero eigenvalue $\lambda_{máx} \geq n$ (Saaty, 2008). The consistency index (CI) is defined by $CI = (\lambda_{máx} - n)/(n-1)$. The consistency index calculated for the decision matrix is compared with the value of the Random Index (RI) to provide the consistency ratio (CR) so that $CR = CI/RI$. If the CR is less than 0.10 then the judgments used to build the decision matrix are considered to be consistent. Otherwise, there is some inconsistency in the judgments and the values of the decision matrix should be revised. The RI for the matrix of order n can be found in Saaty, 1987.

3 METHODOLOGY MAEP-RB

The MAEP-RB methodology was developed in order to allow the assessment of economic performance and the level of economic sustainability of a residential building during the design phase, based on the expected behaviour for the entire building life cycle. It is a modular approach for compiling information throughout the building's life cycle including the four phases of the life cycle of a building: *before use* phase, *use* phase, *end of lifecycle* phase and *beyond life cycle* phase. Each phase of the life cycle is divided into stages, modules, indicators and parameters. For the

Table 2. Stages, modules, indicators and parameters of the MAEP-RB Methodology (Ribas, 2014).

Level 1 Stages	Level 2 Modules	Level 3 Indicators	Level 4 Parameters
Pre-construction Stage	A0: Site and associated fees and counselling	A0.1: Cost of purchase and rental incurred for the site or any existing building.	P1: Costs with the site P2: IMT – Municipal tax on onerous transfer of property P3: IS – Stamp tax
		A0.2: Professional fees related to the acquisition of land.	P4: Costs related to real estate P5: Costs of viability studies P6: Costs of legal support P7: Costs related to the notary fees P8: Costs related to the land registry fees
Product Stage	A1: Supply of raw materials	A1.1: Cost of raw materials.	P9: Percentage cost of each type of material used
	A2: Transport of raw materials	A2.1: Cost of transportation of raw materials.	P10: Percentage cost of each type of material used
	A3: Manufacturing	A3.1: Cost of transformation raw materials.	P11: Percentage cost of each type of material used
Construction process Stage	A4: Transport	A4.1: Cost of transport of materials and products from the factory gate to the building site	P12: Percentage cost of each type of material used

Table 2. (*continued*)

Level 1 Stages	Level 2 Modules	Level 3 Indicators	Level 4 Parameters
		A4.2: Cost of transport of construction equipment such as site accommodation, access equipment and cranes to and from the site.	P13: Percentage of the cost of the building site
	A5: Construction-installation process	A5.1: Costs with exterior works and landscaping works.	P14: Cost for the earthmoving work P15: Cost of support structures and sealing P16: Cost concerning pavements P17: Cost relative to hydraulic networks P18: Cost related to outdoor lighting P19: Cost related to recreational equipment P20: Cost of sowing and planting
		A5.2: Cost of storing products including the prevision of heating, cooling, humidity etc.	P21: Percentage of cost for each type of material used
		A5.3: Cost of transportation of materials, products, waste and equipment within the site.	P22: Cost of equipment related to the achievement of the subcomponents of the building
		A5.4: Cost of temporary works including temporary works off-site as necessary for the construction.	P23: Cost construction site percentage of the total value of direct costs
		A5.5: Cost on site production and transformation of a product.	P24: Cost of hand labor P25: Cost of equipment P26: Cost of fuel P27: Cost of water
		A5.6: Cost of heating, cooling, ventilation, humidity control, etc. during the construction process.	P28: Cost of equipment P29: Cost of electricity
		A5.7: Cost of installation of the products into the building including ancillary materials.	P30: Cost of hand labor P31: Cost of equipment P32: Cost of auxiliary materials
		A5.8: Cost of water used for cooling, of the construction machinery or on-site cleaning.	P33: The cost of cooling water and cleaning
		A5.9: Cost of waste managing processes of other wasters generated on the construction site (RCD).	P34: Cost of the screening process of RCD P35: Cost of packaging of RCD P36: Tax amount
		A5.10: Transportation cost of waste RCD.	P37: Cost of transporting the RCD
		A5.11: Costs of commissioning and handover related costs.	P38: Cost of the extension of domestic wastewater sanitation. P39: Cost of the extension of sanitation storm water P40: Cost of extension of water supply

Table 2. (*continued*)

Level 1 Stages	Level 2 Modules	Level 3 Indicators	Level 4 Parameters
			P41: Cost of extension of electricity
			P42: Cost of extension of gas supply
			P43: Cost of extension of telecommunication
			P44: Cleaning cost
		A5.12: Cost for professional fees related to work on de project.	P45: Fees of the project team
			P46: Fees of the inspection team
			P47: Fees the technical director
			P48: Fees of the health and safety at work team
		A5.13: Costs of the taxes and other costs related to the permission to build and inspection or approval of works.	P49: Value of the license fee projects
			P50: Value of building permit fee
			P51: Exchange certifications gas project
			P52: Certification fee thermal design
			P53: Certification fee of electrical design
			P54: Rate design verification of fire safety
			P55: Certification fee of telecommunications project
			P56: National health service project certification fee
			P57: Certification fee of the gas network
			P58: Rate of energy certification
			P59: Certification fee electricity grid
			P60: Certification fee telecommunications network
			P61: Rate survey of municipal services
			P62: Rate survey of the firefighters
			P63: Survey national health service fee
			P64: VAT rate
		A5.14: Incentives or subsidies related to the installation.	P65: Value of the incentive

moment, MAEP-RB is developed only for the *before use* phase. The object of assessment is the building, including its foundations and landscaping within the building perimeter (Ribas, 2014). Table 2 shows the hierarchical structure of the method (stages, modules, indicators and parameters) that correspond to the *before use* phase. At each level, information is obtained by aggregating information at the lower level. For example, each of the twenty-one economic indicators (Level 3, indicators A0.1 to A0.2, A1.1, A2.1, A3.1, A4.1 to A4.2 and A5.1 to A5.14) is estimated by aggregating the results of one or more parameters (Level 4, parameters P1 to P65), following the hierarchical structure presented in Table 2. The assessment of the economic performance of the *before use phase* is obtained by aggregating of the results of each stage of the building's life cycle (Level 1).

4 RESULTS AND DISCUSSION

The AHP was used to define the weight system of the hierarchical structure of MAEP-RB, consisting of the four previously defined levels of attributes (level 1: stages; level 2: modules; level 3:

Table 3. Relative weights.

Weights for the level 1 attributes: before use phase

Weights	0.286	0.143	0.571
Stages	Pre-construction	Product	Construction process

Weights for the Level 2 attributes: stages

Stages	Pre-construction	Product			Construction process	
Weights	1.000	0.286	0.143	0.571	0.200	0.800
Modules	A0	A1	A2	A3	A4	A5

Weights for the Level 3 attributes: modules (indicators A0.1 to A5.14)

Modules	A0		A1	A2	A3	A4	
Weights	0.80	0.20	1.00	1.00	1.00	0.67	0.33
Indicators	A0.1	A0.2	A1.1	A2.1	A3.1	A4.1	A4.2

Modules	A5													
Weights	0.13	0.03	0.04	0.09	0.12	0.03	0.05	0.02	0.03	0.03	0.07	0.07	0.09	0.03
Indicators	A5.1	A5.2	A5.3	A5.4	A5.5	A5.6	A5.7	A5.8	A5.9	A5.10	A5.11	A5.12	A5.13	A5.14

Weights for the Level 4 attributes: indicators (parameters P1 to P65)

Indicators	A0.1			A0.2					A1.1	A2.1	A3.1	A4.1	A4.2
Weights	0.66	0.21	0.13	0.48	0.21	0.12	0.12	0.07	1.00	1.00	1.00	1.00	1.00
Parameters	P1	P2	P3	P4	P5	P6	P7	P8	P9	P10	P11	P12	P13

Indicators	A5.1							A5.2	A5.3	A5.4	A5.5			
Weights	0.12	0.25	0.18	0.15	0.10	0.10	0.10	1.00	1.00	1.00	0.56	0.20	0.14	0.10
Parameters	P14	P15	P16	P17	P18	P19	P20	P21	P22	P23	P24	P25	P26	P27

Indicators	A5.6		A5.7			A5.8	A5.9				A5.10	A5.11					
Weights	0.67	0.33	0.55	0.35	0.10	1.00	0.49	0.31	0.20	1.00	0.17	0.16	0.17	0.18	0.15	0.09	0.08
Parameters	P28	P29	P30	P31	P32	P33	P34	P35	P36	P37	P38	P39	P40	P41	P42	P43	P44

Indicators	A5.12				A5.13						
Weights	0.45	0.23	0.13	0.19	0.03	0.16	0.02	0.03	0.03	0.02	0.03
Parameters	P45	P46	P47	P48	P49	P50	P51	P52	P53	P54	P55

Indicators	A5.13 (Continue)									A5.14
Weights	0.03	0.05	0.12	0.09	0.06	0.05	0.05	0.05	0.18	1.00
Parameters	P56	P57	P58	P59	P60	P61	P62	P63	P64	P65

indicators; and level 4: parameters). For the definition of the system of weights, a AHP tool was developed, using Microsoft Excel with Visual Basic macros. At the lowest level, parameters, a total of thirty-one comparison matrices of various orders were defined: $15(1 \times 1)$; $4(2 \times 2)$; $5(3 \times 3)$; $2(4 \times 4)$; $1(5 \times 5)$; $2(7 \times 7)$; $1(14 \times 14)$; $1(16 \times 16)$. Systematic comparisons of successive elements pairs were performed for each matrix, and the respective weights calculated. In this paper a system of weights has been defined for the assessment of economic sustainability of buildings using the MAEP-RB methodology that is based on European Standards. These weights have been calculated using AHP structure is shown in Table 3. The results obtained indicate that the construction process stage is the most relevant, accounting for 57.1% of the economic building sustainability in the before use phase. The *pre-construction stage* (28.6%) has twice the importance of the *product stage* (14.3%). At the modules level, and within the economic indicators, the economic indicator A5.1 (*Costs with exterior works and landscaping works*) and A5.5 (*Costs on site production and transformation of a product*) are those that have a bigger impact in A5 module (*Construction installation process*), with 13% and 12%, respectively. The values of Consistency Ratio determined for this set of 31 matrices vary between 0.00 and 0.05, the maximum value corresponding to the matrix of order (14×14).

5 CONCLUSIONS

The definition of the relative importance of the different attributes of the hierarchical structure, a system of weights, is one of the most critical points in the development of an assessment methodology such as MAEP-RB. Although there is no consensus regarding the definition of a method for determining the relative weights, the AHP tool has been used worldwide with satisfactory results. In this paper a system of relative weights has been defined for the assessment of economic sustainability of buildings using the MAEP-RB methodology that is based on European Standards. These weights have been calculated using AHP structure. Overall weights can be very useful in comparative studies of performance and economic sustainability of buildings, because they allow the identification of the most relevant aspects towards sustainability. This study is part of a broader research project that aims to develop the MAEP-RB methodology of performance evaluation and economic sustainability of buildings, still running.

REFERENCES

Chapman, R. E. & Marshall, H. E. 1998. Forman E. H. User's Guide to AHP/Expert Choice for ASTM Building Evaluation. Manual 29 Software to Support ASTM E 1765: Standard Practice for applying Analytical Hierarchy Process (AHP). Related to Buildings and Building Systems.630 pp.
Chen, S. & Hwang, C. 1992. Multiple Attribute Decision Making. Lecture Notes in Economics and Mathematical System. Springer-Verlag, Berlin/Heidelberg/New York.
EN 15643-4: 2012 – Sustainability of construction works – Assessment of buildings Part 4: Framework for the assessment of economic performance.
Golony, B. & Kress, M. 1993. "A multicriteria evaluation of the methods for obtaining weights from ratio-scale matrices". European Journal Operation Research – Vol. 69 – Pags. 210–220.
Gomes L. M. & Araya, M. G. & Carignano, C. 2004. Tomada de Decisões em Cenários Complexos.Thomson Learning, 168p.São Paulo.
Hwang, C. & Yoon K. 1981. Multiple Attribute Decision Making. Lecture Notes in Economics and Mathematical System. Springer-Verlag, Berlin/Heidelberg/New York.
Haurie, A. 2001. The AnalyticalHierarchyProcess: Université de Genève/Centre Universitaire d'écologie humaine et des sciences de l'environnement. Topic 1.3. Genève.
Norris, G.A. & Marchall, H.E. 1995. Multiattribute decision analysis method for evaluating buildings and building systems. National Institute of Standards and Technology – NIST, Gaithersburg. 77 pp.
Oliveira, C.A. & Belderrain, M.C. 2008. Considerações Sobre a Obtenção de Vetores de Prioridades no AHP. I Erabio – XXI ENDIO – XIX EPIO. "Sistemas Boscosos y Tecnologia". Posadas. Argentina.
prEN 16627: 2013 – Sustainability of construction works – Assessment of economic performance of buildings – Calculation method.
Ribas, D., Morais, M. & Cachim, P. 2014. Economic Performance of Buildings: Development of Assessment Methodology for Second prEN 16627:2013. In: 40th IAHS World Congress on Housing – Sustainable Housing Construction, Funchal. p.10. Portugal.
Saaty, T.L. 1980. The Analytic Hierarchy Process. McGraw Hill. New York.
Saaty, R.W. 1987. The Analytic Hierarchy Process: What it is and how it is used? Mathematical Modelling. V.1, n.1, p. 310–316. Great Britain.
Saaty, T.L. 2008. Decision Making With the Analytic Hierarchy Process. Int. J. Services Sciences, v.1, n.1, p. 83–98. Pittsburgh
Silva V. 2003. Avaliação da Sustentabilidade de Edifícios de Escritórios Brasileiros: Diretrizes e Base Metodológicas". Tese de Doutoramento – Engenharia de Construção Civil e Urbana, São Paulo.

Advances and Trends in Engineering Sciences and Technologies – Al Ali & Platko (Eds)
© 2016 Taylor & Francis Group, London, ISBN: 978-1-138-02907-1

Smart city – the concept for Prague

M. Rohlena & J. Frkova
Czech Technical University in Prague, Prague, Czech Republic

ABSTRACT: The aim of this paper is to develop a concept of a Smart City, considering also the approach to environment. The concept is developed further for the capital city of Prague. The capital city of Prague has put forward its own concept of development called SMART PRAGUE 2014–2020. This concerns a long-term concept of economic, technologically effective and sustainable development supported by the use of sophisticated and integrated data. In order to designate targets in the area of SMART infrastructure, a case-study was carried out, entitled Smart Buildings, involving a sample of 20 public buildings. This enabled proposals for, and evaluation of, measures leading not only to reduction of energy demand but also to improvement the quality of the internal environment. This part of the paper has the purpose of making known the methodology and some of the results of this case-study in which the Technical University participated.

1 WHAT IS THE SMART CITY CONCEPT

In the Czech Republic the urbanization level is 73%, and the majority of inhabitants can suffer some negative impact from this process. A more conceptual and coordinated approach to solving problems on any territory with a high concentration of population, transport, energy consumption, waste production etc. can be very effective. Integrated projects which can be implemented by means of various coordinated instruments have just the kind of substantial potential required for needed creative synergies. An integrated combination of support across an energy efficiency area, ICT and transport – so called Smart Cities – or indeed other interventions put in place on the basis of a coherent set of goals in an area, can provide such an example of significant synergy. In 2014 the capital city Prague, in harmony with the European strategy and the Partnership Agreement, started preparation of the concept SMART Prague 2014–2020, the essential vision of which is to support effectively and systematically within Prague the development and interconnection of a quality energy, telecommunication, transport and environmental infrastructure, including education, culture and entrepreneurship to give a high added value.

1.1 *Definitions of the term Smart Cities and their ranking*

Today *smart cities* has great current topicality and is widely discussed among the professional community and there are many publications on this matter. (Townsend 2013, Thoumi 2013, Kumar 2014, Araya 2014, Brook 2013, Hollins 2013, Montgomery 2013, Smith 2012).

In the specialized literature we can find multiple definitions of the term *Smart Cities*. Let us introduce two here, the first one representing a more academic view and the second one more recognizable to the general public:

- "A city is considered smart when investments in human and social capital and traditional transport and modern (ICT) communication infrastructure generate sustainable economic growth and a high quality of life, overseen with a wise management of natural resources, through participatory governance." (Caragliu 2009)
- The second definition is presented by the climate strategist B. Cohen: "Smart cities use information and communication technologies in order to be more intelligent and efficient in the use of resources, resulting in cost and energy savings, improved service delivery and quality of life,

and a reduced environmental footprint–all supporting innovation and a low-carbon economy." (Cohen 2012)

Globally there exist several rankings of cities with different approaches to evaluating them from the viewpoint of the environment e.g.

- *The City innovation economy classifications and rankings* is the largest city classification and global rankings with 445 benchmark cities involved. Prague was placed 62nd among those evaluated, the first place going to San Francisco – San Jose, USA. (Global Innovation Agency 2014)
- *The Quality of Life Index*. Prague according to this rating is placed 45th out of 113 world cities evaluated, first place being given to Canberra, Australia. (Database NUMBEO 2015)
- Rankings of green cities. According to *the Green city ranking* Prague is placed 24th in a European rating, the first place taken by Copenhagen. (Siemens global 2015)
- *The Top 10 Smart Cities in Europe*. (Cohen 2014) B. Cohen developed a set of ideal indicators (28) to be used for benchmarking and ranking smart cities. Copenhagen is again first of Ten Smart Cities in Europe. Copenhagen is a very accomplished city. For example according to the ratings of The Green City Index 2014 Copenhagen's residential buildings consume almost 40% less energy than the Index average. (Siemens global 2015). Prague did not appear among the 10 Smart Cities.

One important indicator for a Smart City is environment and in this context mainly CO_2 emissions. Carbon dioxide emissions, the main contributor to global warming, are set to rise again in 2014 – reaching a record high of 40 billion tonnes. The Global Carbon Budget 2014 shows that global CO_2 emissions from burning fossil fuel and cement production grew 2.3% to a record high of 36 billion tonnes of CO_2 in 2013. Emissions are projected to increase by a further 2.5% in 2014. In 2013, the ocean and land carbon sinks respectively removed 27% and 23% of total CO_2 leaving 50% of emissions in the atmosphere. (Earth System Science Data 2014).

2 CONCEPT SMART PRAGUE

2.1 *Why just Prague*

The capital city of Prague is economically by far the most developed region of the Czech Republic and a center beyond simply regional significance. As such, it contributes in a significant sense to the development of competitiveness and economic growth in the Central European region. It provides a dense network of public services not only for the citizens of Prague but also for all citizens of the Czech Republic. It is also a center of education because it is the headquarters of two-thirds of all science and research institutions in the country and within its own boundaries it accounts for 40% of all expenditures in the Czech Republic.

The concept SMART Prague is a modern urban concept the basic vision of which is effective and systematic support for the development of a high quality urban infrastructure, enterprising activity with a high added value and the development of education and culture in the city, all processed at *the Prague Institute of Planning and Development* (IPD).

Concept SMART Prague 2014 – 2020 defines three main axes:

- *SMART Infrastructure*
- *SMART Specialization*
- *SMART Creativity.*

2.2 *Smart Infrastructure*

A connecting framework of energy savings in buildings introducing the concept of intelligent buildings, which will be effective in economic and energy terms and in future will also enable full integration into SMART Grid providing a big potential for the concept SMART Infrastructure. Regarding the extent of the UNESCO conservation zone in the town center it is rather difficult to achieve energy savings in historical buildings, and not least any implementation of renewable energy resources and indeed many other ecologically focused measures. Nevertheless, in total perspective it is necessary to state that Prague has reserve options in this area. For verifying the effectiveness of economical measures on buildings within the framework of the axis SMART Infrastructure there has been prepared a project Smart Buildings for the revitalization of buildings held in ownership by the city itself. The Czech Technical University participated in this project. Selected buildings will be transformed into intelligent buildings, the goal of which is to maximize savings on natural resources or to improve the internal environment of buildings (Clements-Croome 2010) In order to achieve this goal there have been allocated means from Operational Programme Prague – Growth Pole of the Czech Republic, and within the framework of energy savings and buildings transformation into intelligent buildings, these allocations amount to CZK 2 billion.

2.3   The Pilot Project SMART Buildings

In cooperation with *the Prague Institute of Planning and Development*, the company *ECOTEN Ltd.* and the *Czech Technical University* there was carried out a study of potential for the transformation of selected buildings into intelligent buildings. 20 buildings were selected for the pilot project: 8 basic and secondary schools, 11 buildings for social services and 1 administrative building.

There was analyzed the current energy consumption and energy sources for these buildings. On the basis of requests from users and building caretakers there were proposed measures leading not only to reducing the nature of energy demand but also to improve the quality of the internal environment. Owner requests were directed to introducing intelligent buildings principles, i.e. for automatic heating management, heating for hot water, lighting, and ventilation.

The following actions were proposed for all buildings regarding their technical condition and use such as:

– Insulation of the building envelope
– Replacement of windows and doors
– Installation of forced ventilation - recuperation
– Renewable sources of energy
– Energy efficient lighting
– Shielding objects
– Use of rainwater
– Dealing with drinking water
– Waste management
– Integration of Building Management Systems
– Security
– Building Information Modeling
– SBToolCZ – Czech certification tool in accordance with principles of sustainable construction.

All measures were individually priced for each building and the energy consumption savings and operational costs were calculated. The resulting investment value amounted to CZK 1.28 billion.

Given the high number of buildings, it was necessary to set up some criteria for their selection for the pilot project: savings in primary energy, savings in operational costs, investment return potential, level of effectiveness and many others.

Below, some results from the pilot project are presented.

The total anticipated investment costs for the implementation of economical measures and the implementation of systems of intelligent buildings reached the amount of CZK 1.28 billion for all buildings. On this investment there will be on average a reduction in primary energy consumption of more than 50% from that of the current level.

The primary energy savings in absolute terms is obviously the greatest with buildings in a bad technical condition, with the largest flooring areas which so far had not undergone any fundamental

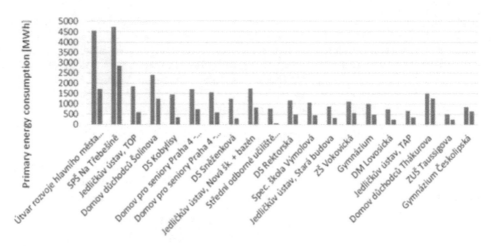

Figure 1. The scale of achieved primary energy savings in comparison with current consumption.

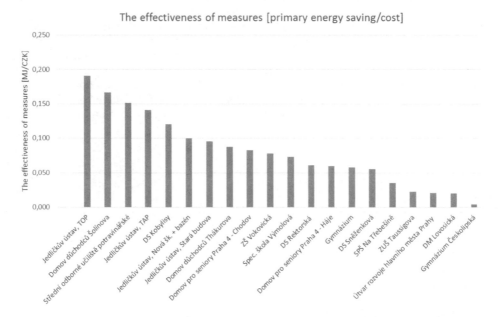

Figure 2. The effectiveness of measures (primary energy saving in the conversion per CZK 1 of the investment).

reconstruction, thus bringing the highest savings. Because of their technical condition, it was desirable to carry out reconstruction regarding these buildings, see Figure 1. On the basis of this primary recommendation the technical requirements were investigated in more details and real possibilities of implementation and financing were evaluated.

One further criterion for the particular selection of a building appropriate for such implementation is the effectiveness of the measures. This is the ratio between the achieved primary energy saving and investment costs. This criterion objectively evaluates imposed costs on 1 MJ of the primary energy. This proportional indicator is not distorted by the size of a building, as it is in the previous case – see Figure 2.

The Jedlièka Institute building is, from the viewpoint of effectiveness, seen as the most beneficial, i.e. it achieved energy savings per 1 investment crown.

2.4 Buildings certification

The evaluation of the complex quality of buildings, considering the wide spectrum of criteria of sustainability in many countries, is becoming a common part of the project and implementation process of buildings construction. Different participants in this process have different motivations for using the evaluation results. The intention of the state administration is to make savings in strategic raw material resources and to reduce the ecological burden and users will expect increasing quality of the internal environment of the building and its surroundings and all this while reducing total costs and impacts on the living environment.

The pilot buildings will be, both before and after the implementation of economical measures, evaluated and certified by the localized methods of the complex certification of buildings SBToolCZ which reflects national standards, or in other words those specific to the Czech Republic. (Vonka et al. 2013) Thus there will be guaranteed a quality performance of the implemented measures, which will bring not only energy consumption savings, but also minimization of the building impact on the living environment, improving the quality of the internal environment and using certified materials.

The buildings will be certified on the basis of three considerations:

– Environmental criteria (living environment). Environmental criteria evaluate consumption of energy and emissions, and they are evaluated in harmony with the principles of *Life Cycle Assessment* (LCA). It means that in the algorithm of evaluation there is covered not only the operational impact of the construction (e.g. operational energy consumption), but also energy consumption during the production of used materials and structures from which the building was itself built (so-called tied-up energy consumption, sometimes also as grey, or built-in energy). Social criteria (or also socio- cultural). Social criteria evaluate inner environment quality from the viewpoint of the user of the building. The evaluating criteria will be e.g. level of heat contentment, acoustic comfort, safety, wheel chair access, the use of materials certified as harmless to health and there are other criteria.
– Economic and managerial criteria. These criteria will encompass the evaluation of *Life Cycle Costs* (LCC), ensuring facility management, processing project documentation according to *Building Information Modeling* (BIM) and accessibility of the project documentation and further data including operational data from a central repository.

On the basis of the methodological process and the achieved results from the pilot study Smart Buildings, there are being prepared other fundamental materials for the selection of particular buildings for implementation in 2014–2020. Regarding the fact that the concept of intelligent buildings is to a high extent still a very innovative branch of interest , the academic sector will also be engaged in the preparation and implementation of further pilot projects. Thus there will be achieved synergies in harmony with the concept "triple helix", i.e. cooperation between the public administration (project owner), the academic sector (project co-author) and industry (project contractor).

3 CONCLUSION

The Czech Republic is among those countries with an above-average energy demand economy when placed in comparison with other EU countries. Regarding energy consumption and the accompanying pollution, the conceptual solution of these issues is therefore considered beneficial through the concept "smart city", i.e. interconnected measures for energy savings at municipal level. The capital Prague, took up the project SMART Cities, within the updating of the strategic plan of the city, a territorial energy concept and other development strategies. *The concept SMART Prague* is a modern urban concept, the essential vision of which is an effective and systematic support for the development of a high quality urban infrastructure, enterprise with a high added value and the development of education and culture in Prague. Within the framework of the priority axis *SMART*

Infrastructure there has been implemented a project *SMART Building*, where the buildings are transformed into intelligent buildings, such as schools, homes for social care and administrative buildings. Prague can use for financing all this resources available from the *Operational Programme Prague – Growth Pole of the Czech Republic*, where there is allocated the amount CZK 2 billion. This pilot project will serve for gaining experience regarding the implementation, not only in Prague but also in other towns in the Czech Republic. The next project development will focus on the evaluation of the economic benefits of carrying out particular types of transformation into intelligent buildings. Within both, the already existing and future cooperation of the P*rague Institute of Planning and Development* and the *Czech Technical University in Prague* there is provision for evaluating particular implementation, additionally to creating a common methodology for the selection of appropriate buildings, which can be transformed later into intelligent buildings. Specialists from the *IPD, CTU – University Centre of Energy Effective Buildings (UCEEB)* are working on further research regarding the pilot project *SMART Building*. Further research will be focused on a prospective methodology regarding building suitability in assessing potential for transformation into the intelligent buildings.

ACKNOWLEDGEMENT

This work was supported by the Grant Agency of the Czech Technical University in Prague, grant No. SGS15/020/OHK1/1T/11.

REFERENCES

Araya, D. 2014. *Smart Cities as Democratic Ecologies*. Palgrave Macmillan.
Brook, D. 2013. *A History of Future Cities*. W.W. Norton & Company.
Caragliu, A. et al. 2009. Smart cities in Europe. In *3rd Central European Conference in Regional Science – CERS, 2009*. Faculty of Economics, Technical University of Košice (ed.), *Proc. Europ. Conf.* Available from <http://www.inta-aivn.org/images/cc/Urbanism/background%20documents/01_03_Nijkamp.pdf> [27 February 2015].
Clements-Croome, D. 2010. *Intelligent Buildings*. ICE Publishing.
Cohen, B. 2012. *The Top 10 Smart Cities On The Planet 2012*. Available from <http://www.fastcoexist.com/1679127/the-top-10-smart-cities-on-the-planet> [5 January 2015].
Cohen, B. 2014. *The Top 10 Smart Cities in Europe 2014*. Available from <http://www.fastcoexist.com/3024721/the-10-smartest-cities-in-europe> [28 January 2015].
Database NUMBEO, 2015. *Quality of Life Index*. Available from <http://www.numbeo.com/quality-of-life/rankings.jsp> [10 February 2015].
Earth System Science Data 2014. *Global carbon budget*. An annual update of the global carbon budget and trends. Available from <http://www.globalcarbonproject.org/carbonbudget/> [15 January 2015].
Global innovation Agency 2014. *City innovation economy classifications and rankings 2014*. Available from <http://www.innovation-cities.com/innovation-cities-index-2014-global/8889> [12 Jan 2015].
Hollins, L. 2013. *Cities Are Good for You: The Genius of the Metropolis*. Bloomsbury Press, New York.
Kumar, T. M. V. 2014 . *E-Governance for Smart Cities*. Springer.
Montgomery, Ch. 2013. *Happy City: Transforming Our Lives Through Urban Design*. Farrar, Straus & Giroux.
Siemens global. *Green City Index*. [online]. 2015 [cit. 2015-01-28]. Accessible from http://www.siemens.com/entry/cc/features/greencityindex_international/all/en/pdf/gci_report_summary.pdf
Sinopoli 2009. *Smart Buildings Systems for Architects*. Butterworth-Heinemann USA.
Smith, P.D. 2012. City: *A Guidebook for the Urban Age*. Bloomsbury Publishing. Plc.
Thoumi, G. 2013. *From Intelligent to Smart Cities*. Taylor & Francis.
Townsend, A. M. 2013. *Smart Cities: Big Data, Civic Hackers, and the Quest for a New Utopia*. W.W. Norton & Company, Inc.
Vonka, M.; Tencar, J.; Hodková, J.; Schorsch, P.; Havlík, F. 2013. *SBToolCZ for apartment buildings*. Prague: Czech technical university in Prague, Faculty Civil engineering.

Advances and Trends in Engineering Sciences and Technologies – Al Ali & Platko (Eds)
© 2016 Taylor & Francis Group, London, ISBN: 978-1-138-02907-1

Water savings and use of grey water in the office building

M. Rysulová, D. Káposztásová & F. Vranay
Technical University of Košice, Civil Engineering Faculty, Košice, Slovakia

ABSTRACT: The modern decentralized water infrastructure can include site-collected rainwater, grey water, storm water, and black water systems. These alternative water sources may never totally replace the centralized system, but they can improve the treatment of a valuable natural source – potable water. With this example of grey water system application we can consider, that a system is efficient in terms of saving these water sources.

1 INTRODUCTION

The reuse of waste water is an entirely new concept in Slovakia. The main goal is to point out the existence of another source of water, suitable for non-potable purposes. Equally important is the need for standardization and to protect the public and to ensure that reliable systems are designed, installed and maintained (Markovic, 2015; Markovic et al., 2014). It is necessary to define regulation and set standards for designing hybrid systems for example according to foreign national standards and performed experiments in Slovak conditions. Grey water system application and its utilization as a part of building's water cycle, can definitely bring some saving potential, especially in water savings. The main topic of this article is to describe, how we can treat this source of water, and demonstrate its potential utilization, which means saving a particular source of potable water and in parallel to water savings, bring financial savings.

1.1 *What is a grey water system?*

A grey water system can be described as a system which is oriented on capturing waste water before it's discharged from a building. If we want to apply this system, the waste water has to be separated according to grey water and black water.

There are a lot of descriptions, of grey water implies, and according to British Standard (BS 8525-1, 2010), we can consider grey water as domestic wastewater excluding fecal matter and urine. This characteristic specifies using waste water for sanitary appliances, where one would otherwise not expect a high rate of water pollution. This usually includes sinks, baths and showers and washing machines, but only after a sufficient cleaning process, which ensures the required quality of water for its designated use. The primary role of the system is to provide the appropriate treatment for grey water, which depends on the types of contaminants, removed and required quality of white water (treated grey water) that returns back to the building (Figure 1).

2 CASE STUDY OF GREY WATER SYSTEM APPLICATION

2.1 *Designing the system*

The first important thing, when we are designing the grey water system is to define grey water system parts for the proposed building. The design phase requires that we determine what white water can be efficiently reused as grey water. Depending on this specification and according to the chosen types of sanitary appliances we can estimate the amount of water pollution, and required water quality for the intended use, which will be reused in the building water cycle (Vrana et al., 2013, Chaitidak et al., 2013, Vranayova et al., 2005).

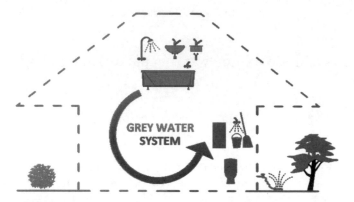

Figure 1.   The grey water system application.

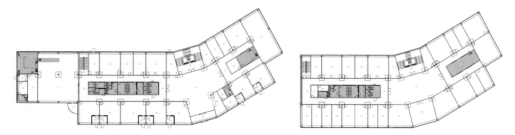

Figure 2.   The grey water system application in an office building, with marked parts of grey water system application.

2.2   *Building characteristics*

An office building was selected for the introduction of a grey water system. The building has four aboveground floors and one underground floor (Figure 2), where house the garages and technical room. The first floor consists of retail stores, offices and a restaurant. The rest of the floors are approximately identical, but with offices, kitchens, and sanitary facilities. There is only one difference – the second floor, which in contrast with other floors accommodate three meeting rooms.

2.3   *Grey water system in office building*

To ensure that the building's existing water system, uses minimal additional energy, a self-gravity grey water discharge from sanitary appliances was designed. Since the technical room is located at the lowest point of the building, where the wastewater treatment plant and water tanks will be placed, we can ensure this requirement.

Sanitary appliances which will produce grey water are sinks, kitchen sinks and shower. Cleaned white water will be used for toilets and urinal flushing and cleaning.

2.4   *Amount of water in system*

For the calculation it is important to quantify the daily production of grey water and daily demand of white water, and then compare these two amounts with an effort to fulfill the following condition:

$$Q_{prod} \geq Q_{24} \tag{1}$$

Q_{prod} – volume of produced grey water per day (l/day)
Q_{24} – volume of white water demand per day (l/day)

Table 1. Amount of sanitary appliances which produce grey water.

Sanitary appliance	Amount of sanitary appliance	Grey water production q_{prod} (l/day)
Sinks	40	12
Shower	1	2
Kitchen sinks	4	5

Table 2. Amount of sanitary appliances utilizing white water.

Sanitary appliance	Amount of sanitary appliance	White water production q_{prod} (l/day)
Toilets men	8	6
Toilets women	19	24
Urinals	8	9

Table 3. White water demand for cleaning.

Type of water demand	Area A (m^2)	White water demand q (l/m^2)
Cleaning	4 261.04	0.1

When this condition is fulfilled, we can consider that system has preliminary potential for grey water utilization.

Daily production of grey water. Determination of daily production of grey water per day was based on a number of sanitary appliances which produce grey water. We selected 45 sanitary appliances – 1 shower, 40 sinks and 4 kitchen sinks (Table 1).

$$Q_{prod} = \Sigma(q_{prod,i} \cdot n_{mj,i}) \tag{2}$$
$$Q_{prod} = (q_{prod,sink} \cdot n_{person} + q_{prod,shower} \cdot n_{person} + q_{prod,kitchen,sink} \cdot n_{person})$$
$$Q_{prod} = (12 . 375 + 2 . 10 + 5 . 100)$$
$$Q_{prod} = 5\ 020\ \text{l/day}$$

Q_{prod} – volume of produced grey water per day (l/day)
$q_{prod,i}$ – grey water production per unit or day (l/day)
$n_{mj,i}$ – amount of the same measuring units

According to the calculation the daily production of grey water in the office building averages 5020 l/day.

Daily white water demand (Table 2 and 3). White water will be used for flushing 27 toilets, 8 urinals and for cleaning 4 261,04 m^2 of floor area.

$$Q_{24} = (q_{toilet} \cdot n_{men} + q_{toilet} \cdot n_{women} + q_{urinal} \cdot n_{men} + q_{clean} \cdot A_{clean}) \tag{3}$$

Q_{24} – volume of white water demand per day (l/day)
$q_{n,i}$ – white water demand per unit or day (l/day)
$n_{mj,i}$ – amount of the same measuring units
A_{clean} – cleaned area (m^2)
q_0 – flushing volume (l)
p – number of utilization by one person per day

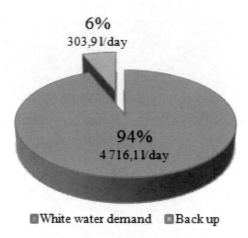

6%
303,9 l/day

94%
4 716,1 l/day

☐White water demand   ☐Back up

Figure 3.   Daily white water demand covered by grey water production.

Water demand for flushing toilets and urinals are determined by following relation:

$$q_{toilets} = q_0 \cdot p \tag{4}$$
$$q_{toilets-men} = 6.\ 1 = 6\ l/day$$
$$q_{toilets-women} = 6.\ 4 = 24\ l/day$$
$$q_{urinal} = q_0 \cdot p \tag{5}$$
$$q_{urinal} = 3.\ 3 = 9\ l/day$$

$$Q_{24} = (6.70 + 24.\ 135 + 9.70 + 0{,}1.\ 4\ 263{,}04) \tag{3}$$
$$Q_{24} = 4\ 716{,}1\ l/day$$

Daily amount of required white water for an office building is 4 716,1 l/day. According to calculated amounts of water in grey water system, we can estimate the preliminary potential of grey water system utilization for this building (Figure 3):

$$Q_{prod} \geq Q_{24} \tag{1}$$

5 020 > 4 716,1 l/day.

This calculation proves that the condition of preliminary system potential is fulfilled and the production of grey water covers the daily white water demand. Grey water production also creates a certain reserve, which can serve in unexpected water demands. Since the reserves are not always depleted, it is important to design safety discharges from the tanks.

2.5   *Amount of water according to real condition*

The calculation was assumed for a 100 percent occupied building (Table 4), which rarely occurs in reality. A realistic annual usage is provided in the following example for comparison purposes. According to the building type, we can consider, that during holidays and weekends, there will be less users, and also the water production and demand will undergo a different operational regime.

We can consider that this example of grey water application proves that system has some saving potential, especially for potable water. If we are considering the amount of water in the system mentioned above, our annual saving will approximate 1 392, 2 m³ of potable water. It is understood that with water savings, we save some sort of financial sources. For this specific case it represent 1 520 € savings per year, calculated from an actual price for water supply 1, 0922 €/m³ in Liptov region, and 1 849 € per year for discharging water to public sewer. The actual price for water discharging in Liptov region is 1, 3282 €/m³. Overall savings of water supply and discharging water will be 3 369 €.

Table 4.   Annual grey water production.

Month	Days	Grey water production Daily (l/day)	Grey water production Monthly (l/month)	White water demand Daily (l/day)	White water demand Monthly (l/month)
January	20	5020	100 400	4 716,1	94 322
	11	2008	22 088	1 886,4	20 750
February	19	5020	95 380	4 716,1	86 606
	9	2008	18 072	1 886,4	16 978
March	24	5020	120 480	4 716,1	113 186
	7	2008	14 056	1 886,4	13 205
April	20	5020	100 400	4 716,1	94 322
	10	2008	20 080	1 886,4	18 864
May	19	5020	95 380	4 716,1	89 606
	12	2008	24 096	1 886,4	22 637
June	22	5020	110 440	4 716,1	103 754
	8	2008	16 064	1 886,4	15 091
July	22	5020	110 440	4 716,1	103 754
	9	2008	18 072	1 886,4	16 978
August	21	5020	105 420	4 716,1	99 038
	10	2008	20 080	1 886,4	18 864
September	20	5020	100 400	4 716,1	94 322
	10	2008	20 080	1 886,4	18 864
October	22	5020	110 440	4 716,1	103 754
	9	2008	18 072	1 886,4	16 978
November	19	5020	95 380	4 716,1	94 322
	12	2008	24 096	1 886,4	20 750
December	19	5020	95 380	4 716,1	89 606
	12	2008	24 096	1 886,4	22 637
Annual amount of water			1 481,9 m³/year		1 392,2 m³/year

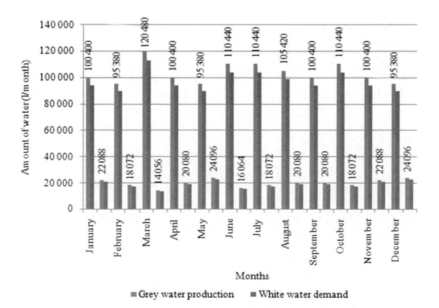

Figure 4.   Comparison of the grey water production and white water demand.

385

3   CONCLUSION

According to the results we can consider that grey water systems exhibit certain potential for economic savings, particularly in relation to operation costs. For clients, return on investment costs could be a deterring factor, but with ever increasing prices for water supply and discharging sewerage, the solution of using alternative water sources is a viable one.

It is clear that financial savings are often the most important factors for users, but we can say that nowadays when the world is also trying to think ecologically, it may be definitely considered as an effective solution to save resources, in this case – potable water. In our conditions we have enough sources of potable water; therefore it is common to not use alternative water supply systems. However, in the context of sustainable thinking we should take responsibility for the environment and ultimately our future.

ACKNOWLEDGEMENT

This work was supported by projects VEGA n. 1/0202/15: Sustainable and Safe Water Management in Buildings of the 3rd Millennium.

This work was supported by The APVV – SK-CZ-2013-0188 Lets Talk about the Water – An Essential Dimension of Sustainable Society of the 21. Century.

REFERENCES

BS 8525-1:2010 Greywater systems PArt 1: Code of practise. UK: BSI,2010

Chaitidak D. M., Yadav K.D., Characteristics and treatment of greywater-review. Anviron Sci Pollut Res (2013) 20:2795–2809. DOI 10.1007/s11356-013-1533-0

Markovič G., Zeleňáková M., Measurements of quality and quantity of rainwater runoff from roof in experimental conditions – 2014. In: ICITSEM 2014 : International conference on innovative trends in science, engineering and managment 2014 : 12th and 13th February 2014, Dubaj, UAE.

Markovič, G. Analýza potenciálneho využívania zrážkových vôd z povrchového odtoku v budove areálu TUKE In: Plynár. Vodár. Kúrenár+Klimatizácia. Roč. 13, č. 1 (2015), s. 29–32. – ISSN 1335–9614

Srážkové a šedé vody aneb "colors of water". Conference proceedings.2013. www.asio.sk

Vrána, J.; Raček, J.; Raclavský, J.; Bartoník, A. Preparation a new Czech Standard ČSN 75 6780 Reuse of greywater and rainwater in buildings and adjoining grounds, príspevok na konferencii CLIMA 2013 – 11th REHVA World Congress and the 8th International Conference on Indoor Air Quality, Ventilation and Energy Conservation in Buildings. STP. Praha. 2013. ISBN: 978-80-260-4001-9.

Vranayova, et al.: Evaluation of feasibility and benefits of a storm water systems for housing supply in slovak conditions, In: International Journal for Housing and Its Applications, vol. 29, 2005, no. 1, p. 3344.

Use of traditional and non-traditional materials for thermal insulation of walls

A Sedlakova & L. Tazky
Technical University of Košice, Faculty of Civil Engineering,
Institute of Architectural Engineering, Košice, Slovakia

S. Vilcekova & E. Kridlova Burdova
Technical University of Košice, Faculty of Civil Engineering,
Institute of Environmental Engineering, Košice, Slovakia

ABSTRACT: In world with limited amount of energy sources and with serious environmental pollution, an interest in environmental embodied impacts of buildings with different structure systems and alternative building materials will be increased. The selection of building materials used in the constructions (floors, walls, roofs, windows, doors, etc.) belongs to one of the most important roles in the phase of building design. This decision has impact on the performance of the building with respect to the criteria of sustainability. The energy needed for extraction, processing and transportation of materials used in building structures can be significant part of the total energy used over the life cycle of building, particularly for nearly-zero energy buildings. The aim of this paper is to identify the environmental quality of material compositions of proposed alternatives for exterior wall and to compare them by using methods of multi-criteria decision analysis.

1 INTRODUCTION

The construction sector accounts for at least a third of all resource consumption globally, including 12% of all fresh water use. Some 25 40% of produced energy is consumed in the construction and operation of buildings, which accounts for approximately 30–40% of all carbon dioxide (CO_2) emissions. And 30–40% of solid waste comes from construction. In terms of the economy, the sector produces approximately 10% of gross world product, and buildings represent a massive share of public and private assets (Taipale 2012). The majority of materials are created from primary raw materials. However, extracting those raw materials have impact on the environment (Godfaurd et al. 2005). Study (Arena et al. 2003) state that using insulated external walls produces 60% of the total energy savings; double glazing amounts for 24% and the remaining 16% is saved by the reduction in infiltrations. According to Worldwatch Institute data, the building constructions consume annually 40% of stone, sand and gravel, 25% of wood and 16% of water globally (Arena et al. 2003). Construction industry in England consumes over 420 million tons of a wide range of raw materials and generates about 94 million tons of waste – 13 million tons of which is estimated to be due to over specifications (Plank 2008). The greatly increased utilization of recycled building products and materials, together with the comprehensive utilization of agricultural and manufacturing wastes as raw materials for the production of building materials, combined with a systematic effort to privilege the use of renewable materials and a determined effort to minimize the use of virgin, non-renewable materials would also have a very big impact on resource depletion (Storey et al. 2001). If sustainable construction is to be achieved, it has to adopt more long-term sustainable strategies at the feasibility stage of a building project to promote environmental protection and conservation. These strategies must focus on continual improvement through the consideration of sustainable development in the decision process. Therefore, construction has to place a higher priority on sustainability considerations in building projects and ensure that the concept of sustainability is valued and rewarded as well as practiced at all levels throughout the project's entire life span (Akadiri 2011). Efforts for sustainable environment bring challenges leading to improvement and elimination of unwanted effects of noise in buildings. Therefore, it is necessary also to pay the attention to the need for vibroacoustic building certification (Flimel 2014). While the studied wall systems (mass, insulation and finish materials) represent a significant portion of the initial EE of the building, the concrete structure (columns, beams, floor and ceiling slabs) on average constitutes about 50% of the building's pre-use phase energy. This proportion could diminish to some extent if recurrent EE were included in the analysis (given that walls generally need maintenance and the structure is assumed to last for the duration of the building's life), but there can be no doubt that the energy-efficiency of structural systems is an issue which has not been sufficiently addressed in the LCA literature (Huberman et al. 2008). Growth of energy efficiency in buildings represents priority in EU. Directive 2010/31/EC on the energy performance of buildings requires implementing better law provisions of energy performance in buildings. Directive seeks to minimize energy consumptions and eliminate emissions production through sustainable energy strategies (EU Directive). The Member States of the European Union have made a unilateral commitment to reduce the overall greenhouse gas emissions by 20%, to increase the share of renewable energy sources up to 20%, and to reach 20% streamline of energy balance. In February 2011, the European Council re-confirmed a goal until year 2050 to lower the greenhouse gas emissions by 80–95% compared to year 1990. This process is in accordance with opinion approved by world powers at Copenhagen and Cancun agreement. These agreements contain a promise of reaching a long term low-carbon development strategy. Some of the EU members have already taken steps, or are planning to, toward this goal. Those political actions are targeted to support an effective environmental protection, to ease climate problems, and to create sustainability for the future generations. Buildings are large consumers of energy in all countries. In regions with harsh climatic conditions, a substantial share of energy goes to heat and cool buildings. This heating and air-conditioning load can be reduced through many means; notable among them is the proper design and selection of building envelope and its components. The proper use of thermal insulation in buildings does not only contribute in reducing the required air-conditioning system size but also in reducing the annual energy cost.

2 ENVIRONMENTAL CONTEXT – SELECTION OF BUILDING MATERIALS

Selection of building materials has significant effect on energy consumption and following emission production in a variety of its life cycles, with possible impact effects in contrary, because. For instance, increasing an insulation thickness can contribute to energy savings during the operation of building, but it can also increase embodied energy. Embodied effects on the environment are not yet considered in today's applicable requirements for new building constructions. However, securing a balance of those factors is important. Correct selection of building materials in the process of designing a building play a significant role in the whole building life cycle and it can have an important impact on the principles of sustainable development (Akadiri 2011; Huberman et al. 2008).

2.1 *Methods of research*

Environmental indicators are calculated by the Life Cycle Assessment method. The analysis investigates the role of different building material compositions in terms of the embodied energy from non-renewable resources and the embodied equivalent emissions of CO_2 and SO_2 in nearly zero energy buildings. Embodied energy (EE) is the energy utilized during manufacturing stage of building materials and represents the energy used to acquire raw materials (excavation), manufacture and transport. Similarly, CO_2 emissions (ECO_2 – global warming potential GWP) and SO_2 emissions (ESO_2 – acidification potential AP) represent the equivalent emissions within the LCA boundary – Cradle to Gate. The input data of these indicators are extracted from the LCA database – IBO. In the figure (Figure 1) is show environmental evaluation of wall construction layers. In this study, it

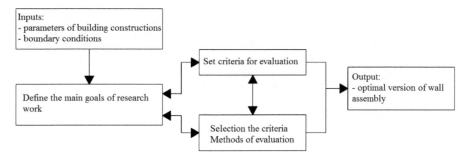

Figure 1.   Environmental evaluation of wall construction layers.

is also calculated environmental indicator Δ OI3 which describes impact of building material in given structure layer and is calculated according to equation (1) (Sedláková et al. 2015).

$$\Delta OI3 = \frac{1}{3} \cdot \left[\frac{1}{10} \cdot (EE_{BM}) + \frac{1}{2} \cdot (ECO_{2BM}) + \frac{100}{0,25} \cdot (ESO_{2BM}) \right] \tag{1}$$

Where:
EE_{BM} – embodied energy of one structure layer – building material [MJ/m^2];
ECO_{2BM} – embodied emissions CO_2 of one structure layer – building material [kg CO_{2eq}/m^2];
ESO_{2BM} embodied emissions SO_2 of one structure layer – building material [kg SO_{2eq}/m^2].

2.2   Thermo-physical evaluation of layers

For purpose of reduction of future energy demand, these versions of wall assemblies are designed to meet requirements for nearly zero energy houses (U = 0.15 W/(m^2.K)). The thermal-physical parameters are calculated for Slovak climatic conditions (STN EN 730540): θ_e – outdoor air temperature (−13°C); θ_i – indoor air temperature (20°C); Rh$_e$ – relative air humidity in outdoor (84%); Rh$_i$ – relative air humidity in indoor (50%); U – heat transfer coefficient (W/(m^2.K)); R – thermal resistance (m^2.K/W); μ – diffusion resistance factor; λ – thermal conductivity coefficient (W/(m.K)); ρ – density (kg/m^3) and c – specific heat capacity (J/(kg.K)). In the table (Table 1) is shown basic physical parameters of seven wall assemblies.

3   RESULTS AND DISCUSSION

3.1   Thermo-physical results

In the figure (Figure 4) is shown comparison of thicknesses of wall. In the figure (Figure 2) is shown thickness of layer in seven evaluated wall assemblies.

3.2   Environmental results

This study uses life cycle analysis in system boundary from Cradle to Gate and focuses on environmental indicators such as embodied energy and emissions of CO_{2eq}. and SO_{2eq}.. The selection and combination of materials influence the amount of energy consumption and associated production of emissions during phase of building operation. In the Figure 3, a), there are shown results of embodied energy of seven evaluated wall assemblies. Variant 3 with EPS thermal insulation with graphite is the best variant with value of 816.225 MJ. Variant 5 with mineral wool insulation is the worst variant with value of 1485.99 MJ. In the Figure 3, b), there are shown results of CO_2 emissions of evaluated wall assemblies. Variant 3 with EPS thermal insulation with graphite is again the best variant with result of 267.9625 kgCO_{2eq}. The worst is Variant 5 with mineral wool insulation with result of 598.022 kgCO_{2eq}. In the Figure 4, a), there are shown results of SO_2 emissions of evaluated wall assemblies. Variant 6 with foam glass thermal insulation is the best

389

Table 1. Basic physical parameters of wall assemblies.

No.		Wall assemblies	Thickness d [mm]	Density ρ [kg/m³]	Thermal conductivity coefficient λ [W/(m.K)]	Specific heat capacity c [J/(kg.K)]	Diffusion resistance factor μ [–]
1	1	Silicate plaster	5	1800	0.86	920	19
	2	Thermal insulation – EPS 20	220	20	0.038	1270	40
	3	Adhesive mortar	10	350	0.8	920	18
	4	Ceramic brick	175	800	0.22	960	5
	5	Lime cement plaster	25	2000	0.88	790	19
2	1	Silicate plaster	5	1800	0.86	920	19
	2	Thermal insulation – mineral wool	250	130	0.041	1030	1
	3	Adhesive mortar	10	350	0.8	920	18
	4	Ceramic brick	175	800	0.22	960	5
	5	Lime cement plaster	25	2000	0.88	790	19
3	1	Silicate plaster	5	1800	0.86	920	19
	2	Thermal insulation – EPS with graphite	190	15	0.038	1450	40
	3	Adhesive mortar	10	350	0.8	920	18
	4	Ceramic brick	175	800	0.22	960	5
	5	Lime cement plaster	25	2000	0.88	790	19
4	1	Silicate plaster	5	1800	0.86	920	19
	2	Thermal insulation – mineral wool	30	108	0.036	1020	1
		Thermal insulation – EPS with graphite	190	15	0.038	1450	40
	3	Adhesive mortar	10	350	0.8	920	18
	4	Ceramic brick	175	800	0.22	960	5
	5	Lime cement plaster	25	2000	0.88	790	19
5	1	Ceramic veener	100	–	–	–	–
	2	Ventilated air cavity	70	–	–	–	–
	3	Thermal insulation – mineral wool	210	60	0.035	1020	1
	4	Adhesive mortar	10	350	0.8	920	18
		Ceramic brick	175	800	0.22	1000	5
	5	Lime cement plaster	25	2000	0.88	790	19
6	1	Silicate plaster	5	1800	0.86	920	19
	2	Thermal insulation – foam glass	220	105	0.038	840	–
	3	Adhesive mortar	10	350	0.8	920	18
	4	Ceramic brick	175	800	0.22	960	5
	5	Lime cement plaster	25	2000	0.88	790	19
7	1	Silicate plaster	5	1800	0.86	920	19
	2	Thermal insulation – hemp	230	100	0.04	1500	6
	3	Adhesive mortar	10	350	0.8	920	18
	4	Ceramic brick	175	800	0.22	960	5
	5	Lime cement plaster	25	2000	0.88	790	19

with result of 0.19652 kgSO$_{2eq}$ and Variant 5 with mineral wool insulation is the worst with result of 0.48515 kgSO$_{2eq}$. In the Figure 4, b), there are shown results of OI3. In the Table 3, there are shown summary results of LCA assessment for seven evaluated wall assemblies.

3.3 MCDA results

The aim of analysis is to identify the thermo-physical and environmental quality of material compositions for proposed alternatives of seven exterior walls. The final values of assessments are compared by using methods of multi-criteria decision analysis. In the Table 3, there are shown results of MCDA for alternatives of wall assemblies from overall – thermo-physical and environmental impact. Variant 3 is the best according to CDA and TOPSIS and Variant 6 is the best according to IPA and WSA method. Variant 5 with thermal insulation from mineral wool is the worst according to MCDA from overall – thermo-physical and environmental impact. Variant 3 is the best and variant 7 is the worst from the MCDA analysis of environmental impact.

Table 2. Thermal performance results of versions 1–7 – wall construction.

Variant	Thickness of wall d [mm]	Surface temperature θ si [° C]	Heat transfer coefficient U [W/(m².K)]	Heat transfer resistance R [(m².K)/W]	Moisture resistance factor μ [10^9 m/s]
1	435	19.34	0.15	6.31	0.50
2	465	19.34	0.15	6.36	0.50
3	405	19.36	0.15	6.56	0.50
4	405	19.35	0.15	6.42	0.50
5	590	19.34	0.15	6.32	1.12
6	445	19.37	0.15	6.59	0.50
7	445	19.37	0.15	6.59	0.50

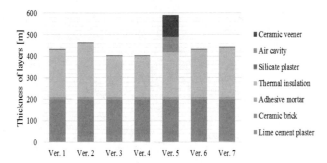

Figure 2. Thickness of layers of wall assemblies.

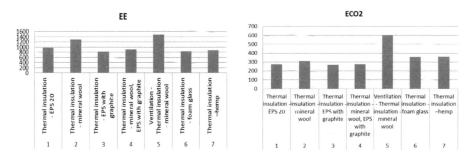

Figure 3. a) Embodied energy, b) CO$_2$ emissions for evaluated wall assemblies.

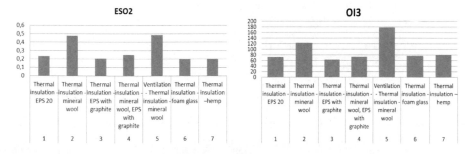

Figure 4. a) SO$_2$ emissions, b) OI3 for evaluated wall assemblies.

Table 3. Results of MCDA for alternatives of wall assemblies from overall impact.

Order	V	CDA	V	IPA	V	WSA	V	TOPSIS
1	V3	1.5926	V6	0.0889	V6	0.9111	V3	0.991
2	V6	1.9954	V3	0.0934	V3	0.9066	V4	0.9193
3	V7	2.3458	V7	0.0982	V7	0.9018	V1	0.8407
4	V4	4.0623	V4	0.3061	V4	0.6939	V6	0.8114
5	V1	7.025	V1	0.5017	V1	0.4983	V7	0.8018
6	V2	7.4733	V2	0.6487	V2	0.3513	V2	0.5626
7	V5	10.4267	V5	0.992	V5	0.008	V5	0.0009

4 CONCLUSION

The results from environmental evaluation as well as environmental profiles of variants of wall assemblies show that Variant 3 achieves the lowest values of EE, ECO_2 and ESO_2. Exterior wall – Variant 3 can assure the highest reduction of EE by 13%–45%, CO_2 by 39.5%–55%, SO_2 by approximately 2%–59.5% in comparison with other variants. Variant 3 with EPS insulation with graphite is the best from overall as well as from environmental impact according to MCDA analysis.

ACKNOWLEDGEMENTS

This article was written as a project solution entitled "The use of the virtual laboratory for designing energy-efficient buildings" Project no. 052TUKE-4/2013 and project no. 1/0405/13.

REFERENCES

A Roadmap for moving to a competitive low carbon economy in 2050, COM (2011) [cit 2015-02-11]. http://eurlex. europa.eu/LexUriServ/LexUriServ.do?uri=COM:2011:0112:FIN:en:PDF

Akadiri, O. P. (2011). ICT Development of a multi-criteria approach for the selection of sustainable materials for building projects, dissertation thesis, University of Wolverhampton, United Kingdom.

Arena, A.P. and De Rosa, C. (2003). "Life cycle assessment of energy and environmental implications of the implementation of conservation technologies in school buildings in Mendoza–Argentina," *Building and Environment*, 38(2), 359–368.

European Commission. 2005. Green Paper on Energy Efficiency: Doing More with Less. [cit 2015-03-11]. http://ec.europa.eu/energy/efficiency/doc/2005_06_green_paper_book_en.pdf

Flimel, F. (2014): "Vibro-accoustic Comfort Assessment Methodology of Residential Buildings in Urban Environment", *Advanced Materials Research*, 1041, 432–435.

Godfaurd, J., Clements-Croome, D. and Jeronimidis, G. (2005). "Sustainable building solutions: a review of lessons from the natural world," *Building and Environment*, 40, 3, 319–328.

Huberman, N. and Pearlmutter, D. (2008). "A life-cycle energy analysis of building materials in the Negev desert." *Energy and Building*, 40, 837–848.

OI3-indicator: Guidelines to calculating the ecological indexes for buildings, 2011. [cit 2015-02-11]. www.ibo.at/documents/ OI3_Berechnungsleitfaden_V3.pdf

Plank, R. 2008. The principles of sustainable construction. *The IES Journal Part A: Civil & Structural Engineering*, 1(4), 301–307.

Sedláková, A., Vilčeková, S., and Krídlová Burdová, E. (2015): "Analysis of material solutions for design of construction details of foundation, wall and floor for energy and environmental impacts," *Clean Technologies and Environmental Policy*, in press

Storey, J.B. and Baird, G. (2001): "Sustainable cities need sustainable buildings," *CIB World Building Congress*, Wellington: New Zealand.

Taipale, K. (2012). "*State of the World 2012*," Chapter 10, From Light Green to Sustainable Buildings. *Worldwatch Institute*, 13, 266.

Advances and Trends in Engineering Sciences and Technologies – Al Ali & Platko (Eds)
© 2016 Taylor & Francis Group, London, ISBN: 978-1-138-02907-1

Fibre concrete durability in relation to the development of permeability

M. Stehlík, T. Stavař & I. Rozsypalová
Department of Building Testing, Faculty of Civil Engineering, Technical University Brno, Czech Republic

ABSTRACT: The main goal of research is to demonstrate a positive or a negative effect of fibres in the cement composite on the development of permeability and subsequently aggressive liquid and gas chemical ingress and thus durability of the material in the structural elements. Durability depends on the properties of the surface layer of concrete but also on its internal structure. A number of testing methods has been described which assess the quality of capillary-porous system of concrete on the principle of measuring gas and liquid permeability. The surface permeability tests were carried out on concrete slabs of four concrete formulas. The determination of permeability of and chloride migration in the surface layer of fibre and reference concrete slabs by means of TORRENT, GWT and NORDTEST methods proved the objectivity of all the three methods with a view to the comparable relative results of air and water permeability and chloride migration.

1   INTRODUCTION

Durability of concrete, a composite with cement matrix, is a topic which is under great interest of technical and scientific community (CEP-FIP model Code 2010). According to the Guidance paper F to the European Construction Products Directive of December 2004, the durability of concrete can be defined as "the ability of a product to maintain its required performance over a given or long time, under the influence of foreseeable actions". The concrete structure durability is determined both by the cover layer of concrete (up to 50 mm), and also by the internal structure of concrete. A favourable capillary-pore system of the *"covercrete"* as well as the deeper layers of concrete protect the reinforcement of structures (Adámek & Juránková 2010a,b). Original methods developed in the last decades describe permeability of the surface layer of concrete as a physical property (Neville 1997).

Fibres distributed in the composite structure bring about important changes in the properties of composite materials (Adámek *et al.* 2009; Hronová & Adámek 2011). The fibres have substantial effect on mechanical properties of the composite during the stage without loading (Akers 2010), but mainly under loading. It is possible to say that short and also long structural fibres decrease the probability of the occurrence of microcracks during the early shrinkage, and the long structural fibres later increase the influence on the tensile strength during the occurrence of cracks and on the character of behaviour after their occurrence (post-crack behaviour), and influence also the compression strength, elasticity modulus and volume changes. It is supposed, that short fibres added in concrete increase its durability by preventing the occurrence of microcracks at various stages of concrete curing, of course with an appropriately determined water ratio and an optimum combination of admixtures. However, poor adhesion of fibres to the cement matrix, shear overstress or corrosion of fibres already in process can lead to increasing the permeability and subsequently decreasing the durability (Kim *et al.* 2011; Stehlík *et al.* 2014; Stehlík & Stavař 2014).

The research presented in this article focuses primarily on long polymer structural fibre composites with cement matrix. Appropriate type of polymer fibres with correct dosage has usually the maximum effect on the enhancement of mechanical properties of the composite. It is inevitable to deal with the durability topic as it has an important influence on the reliability of structures. If we accept the fact that the durability of concrete is determined, to a considerable extent, by the permeability of its surface layers, it seems to be adequate to use the modern non-destructive methods TORRENT and GWT to determine the permeability of the surface of the tested concretes, and the

Table 1. Four basic formulations of concrete.

Type of concrete	CEM II/B-S 32.5R [kg/m³]	Aggregate			Water coefficient –	Plasticizer [%] of w_c	PP fibres (% of concrete volume)	
		0–4 mm [kg/m³]	4–8 mm [kg/m³]	8–16 mm [kg/m³]			[kg/m³]	[%]
O	490	890	100	745	0.34	0.7	0.00	0.00
B	490	890	100	745	0.36	1.0	9.10	1.00
HV	490	890	100	745	0.35	0.8	1.37	0.15
C	490	890	100	RA-633	0.43	1.0	9.10	1.00

method NORDTEST to determine the non-steady-state migration coefficient of chlorides through concretes. The proportional comparison of the determined values of permeability and migration coefficients of tested concretes for various media (TORRENT – air, GWT – water, NORDTEST – chloride solution) will show whether the TORRENT, GWT and the NORDTEST methods and their combinations are suitable for a general estimation of the durability of fibre concretes or not.

2 MATERIALS AND METHODS

2.1 Concrete types and formulations

The presented research focuses on the determination or rather estimate of the durability of tested fibre concretes according to their actual permeability and chloride migration. The permeability and migration tests were carried out on one reference concrete, two fibre concretes made from dense aggregate with different additions of long polymer fibres (0.15% and 1% of concrete volume), and one fibre concrete made from concrete recyclate with 1.0% of fibres. The composition of tested concretes of the four formulas can be found in Table 1. Concrete containing concrete recyclate (C) was prepared by substituting the coarse fraction of natural aggregate 8–16 with raw concrete recyclate (RA) fraction 0–16 mm with 19% volume of fine particles. The consistence of fresh concrete in all formulas was unifiedby anindividual dose of CHRYSOPLAST 760 plasticizer to the degree of F2 in flow table test according to EN 12350-5. PP fibres FORTA FERRO 54 mm long were added to formulas B, HV and C.

2.2 Testing apparatuses and methods

From the methods for determining the permeability of concrete for liquids and gases which describe permeability as a physical property of the surface layer, we chose the TORRENT method (permeability of the surface layer for air) and GWT method (pressure absorption of water by the surface). Both methods were used for testing concrete slabs of 300 × 80 × 300 mm. The TORRENT method was used to test 6 testing points in each formula, the GWT method was used for 3 testing points. The third, slightly different method NORDTEST, evaluated the non-steady-state chloride migration in each formula through 2 concrete segments with a diameter of 100 mm and a height of 50 mm, cut from core drills.

The first testing method, TORRENT, is suitable for determining air permeability of the surface layer of concrete into the maximum depth of 50 mm. The principle of determining air permeability consists in measuring the decrease in vacuum (1000 mbar) due to the passage of air through the capillaries of concrete under the attached bell in the course of the defined time interval. The quality, or rather quality index, can be read from the table supplied by the producer (Torrent & Fernández 2007), the set of TORRENT vacuum apparatus is apparent from Figure 1.

The second testing method, GWT (Fig. 2), is designated for measuring pressure water permeability of the cover layer on the basis of a direct contact with the surface of concrete specimen. The

Figure 1.   Set of TORRENT apparatus.

Figure 2.   Set of GWT apparatus.

Figure 3.   AgCl indicates penetration of NaCl.

sealed pressure chamber is put on the concrete surface, filled with previously boiled water, and the given water pressure acts upon the concrete surface. The water pressure is kept at a chosen constant level by an attached piston of a micrometer gauge which compensates for the loss of water, and thus the volume of water penetrated into the concrete is measured. The reference quantity is the flow of water passing through the layer of concrete (Adámek *et al.* 2012).

The third testing method, NORDTEST, determines the chloride migration coefficient from the non-steady-state migration experiments. Electrical potential is applied from the outside in the direction of the specimen axis, due to the action of which outer ions of chloride are forced to pass inside the specimen. When the determined time interval for the test expires, the specimen is split in half along its axis, and the freshly split surfaces are covered with silver nitrate $AgNO_3$. In the areas of penetration (Fig. 3), it reacts into silver chloride $AgCl$, on which it is easy to see and measure the depth of penetration (Fig. 4). The migration coefficient is subsequently determined from this measured depth (Nordtest method 1999; Collepardi *et al.* 1972).

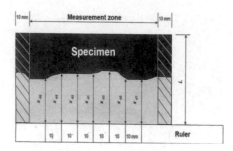

Figure 4. Measurement of NaCl penetration.

Table 2. Classification of the quality of concrete surface layer by TORRENT vacuum method.

Type of concrete	Moisture content w average [%]	Correction of value K_T for w = 3% average [$*10^{-16}$ m^2]	[%]	Depth of vacuum penetration L average [mm]	Quality of covercrete for w = 3% average
O	2.92	0.003	100.0	4.1	very good
B	3.25	0.007	233.3	4.2	very good
HV	3.42	0.011	366.6	4.5	good
C	3.68	0.035	1166.6	6.3	good

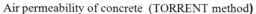
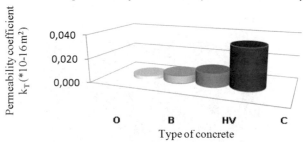

Figure 5. Dependence of the permeability coefficient for air on the type of concrete.

3 EVALUATION OF RESULTS AND DISCUSSION

The tests of permeability of and chloride migration in the surface layer of concretes using the methods of TORRENT, GWT and NORDTEST were carried out on the specimens of concrete that were three months old (1 month storage in water + 2 months in the laboratory conditions). Table 2 and next Figure 5 evaluate the quality of covercrete according to the defined values of air permeability coefficients using the TORRENT method recalculated to the contractual 3% of moisture content of concrete.

Table 3 and graphically Figure 6 show the determined flow of water passing through the surface layer of concrete using GWT method, after 10 and 60 minutes of measurement.

The procedure for determination of the chloride migration coefficient in concrete is relatively difficult and the results from Figure 7 cannot be directly compared with chloride diffusion coefficients obtained from the other test methods.

It is apparent from Figures 5–7 that polymer fibres slightly increase the permeability of concrete for air and for chlorides, inversely proportionally to their amount. It is possible to observe

Table 3. Flow of water into the concrete surface layer by GWT method.

Type of concrete	Actual moisture content w average (%)	Flow of water q after 10 minutes individual (mm/s)	Flow of water q after 10 minutes average (mm/s)	Flow of water q after 60 minutes individual (mm/s)	Flow of water q after 60 minutes average (mm/s)
O	2.92	1.73E-04 1.92E-04 1.69E-04	0.000178	5.69E-05 5.92E-05 5.99E-05	0.0000587
B	3.25	2.22E-04 2.38E-04 2.30E-04	0.000230	6.23E-05 6.38E-05 6.32E-05	0.0000631
HV	3.42	3.00E-04 3.11E-04 3.25E-04	0.000312	1.03E-04 1.04E-04 1.02E-04	0.000103
C	3.68	4.16E-04 4.03E-04 4.11E-04	0.000410	1.20E-04 1.25E-04 1.24E-04	0.000123

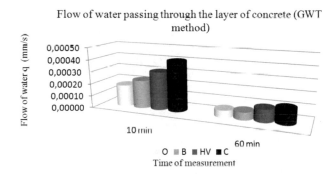

Figure 6. Dependence of the flow of water on the type of concrete.

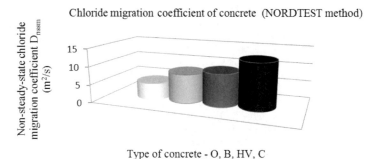

Figure 7. Dependence of the chloride migration coefficient on the type of concrete.

arelativelyclose correlation dependence between the results of TORRENT, GWT and NORDTEST methods. In the concrete made from concrete recyclate, the results of permeability characteristics according to the three presented methods are the highest, which is caused by a higher porosity of concrete recyclate in comparison with natural aggregate. Higher permeability is characteristic of concretes with 0.15% of fibres compared to concretes with 1% of fibres. This effect is caused by an insufficient amount of fibres in the volume of concrete, which cannot efficiently suppress the occurrence of microcracks during hydration and subsequently the shrinkage of concrete.

4 CONCLUSION

The permeability coefficients determined by means of the TORRENT method, the flow of water per time determined by means of the GWT method, and the coefficients of non-steady-state chloride migration determined by means of the NORDTEST method are in close correlation dependence and can be used separately or in combination to make a rough estimate of durability properties of fibre concretes made from natural aggregate as well as from concrete recyclate. If we compare the durability of the surface layer of concrete with fibres and that without fibres, it is possible to say that the fibres contribute to increasing the permeability of the surface layer for air, water and chloride solutions, but the increase is not too distinct. A higher permeability, especially for gaseous media, has been detected in the concrete made from concrete recyclate with fibres, probably due to a higher porosity of concrete recyclate in comparison with natural aggregate. For the flow of water through the surface of concrete and the migration of chlorides this difference is not too significant.

In conclusion it is possible to say that concrete made from concrete recyclate shows slightly worse results of permeability characteristics in the tested formulations, but it can certainly be used even for heavy constructions with low aggressiveness of the environment of grade X0 and XC1.

ACKNOWLEDGEMENTS

This paper was prepared under the support of project No. GAČR 13-18870S "Evaluation and prediction of durability of the surface layer of concrete".

REFERENCES

Adámek, J. & Juránková, V. 2010a. Durability of the concrete as a function of properties of concrete layer. *Transactions on Transport Sciences* 2 (4): 188–195.
Adámek, J. & Juránková, V. 2010b. Evaluation of durability of concrete by measurement of permeability for air and water. *In the 10th International ConferenceModern Building Materials, Structures and Techniques –* Selected papers. Vol. 1, Lithuania: 1–5.
Adámek, J., Juránková, V., Kucharczyková, B. 2009. Fibre concrete and its air permeability. *In Proceedings of 5th International Conference Fibre Concrete.* CTU Prague: 9–14.
Adámek, J., Juránková, V., Kadlecová, Z., Stehlík, M. 2012. Three NDT methods for the assessment of concrete permeability as a measure of durability. *In Nondestructive testing of materials and structures. Rilem Bookseries.* Istanbul, Turecko. Springer in RILEM Bookseries: 732–738.
Akers, S.A.S. 2010. Cracking in fiber cement products. *Construction and Building Materials* 24(2): 202–207.
CEP-FIP model Code 2010. 2010. Final Draft, Section 5.1.13.: *"Properties related to durability".* Ernst und Sohn, Germany: 106–110.
Collepardi, M., Marcialis, A.,Turriziani R. 1972. Penetration of chloride ions into cement pastes and concrete. *Journal of American Ceramic Society* 55(10): 534–535.
Hronová, P. &Adámek, J. 2011. Effect of the use of dispersive fibres on the size of cracks in the early stage of setting and hardening of concrete. *In Collection of Abstracts.* Praha. Faculty of Civil Engineering: 43–44.
Kim, B., Boyd, A.J., Lee, J.Y. 2011.Durability performance of fiber-reinforced concrete in severe environments. *Journal of Composite Materials* 45(23): 2379–2389.
Neville, A.M. 1997. *Properties of concrete / 4th ed.* John Wiley, New York, USA.
Nordtest method.1999. P.O. Box 116, FIN-02151 Espoo, Finland.
Stehlík, M., Heřmánková, V., Vítek, L. 2014. Opening of microcracks and air permeability in concrete.*Journal of Civil Engineering and Management, 21(2): 177–184.*
Stehlík, M. & Stavař, T. 2014.Carbonation depth vs. physical properties of concretes with alternative additions of concrete recyclate and silicate admixtures*Advanced Materials Research* 897(1): 290–296.
Torrent, R. & Fernández, L. 2007.*Non-destructive evaluation of the penetrability and thickness of the concrete cover.* State of the art report of RILEM Technical Committee 189-NEC. 157 rue des Blains, France: RILEM Publications S.A.R.L., RILEM Report 40.

Advances and Trends in Engineering Sciences and Technologies – Al Ali & Platko (Eds)
© 2016 Taylor & Francis Group, London, ISBN: 978-1-138-02907-1

Construction design of high-rise module for low-cost observation tower

M. Sviták & K. Krontorád
Mendel University in Brno, Faculty of Forestry and Wood Technology, Brno, Czech Republic

T. Svoboda
Czech University of Life Sciences in Prague, Faculty of Forestry and Wood Sciences, Prague, Czech Republic

ABSTRACT: The article deals with the drafting process of an observation tower's height module composed of low-cost wood-based materials. The wooden observation tower has a proposed height of 12.5 m with ground plan dimensions of 3.5 × 3.5 m. The construction of the tower is made up of modules of a specific height. The height of one module is 1 440 mm. The total height of the tower is formed by stacking the modules on top of each other with successive vertical 90° rotations around a central 2 steel rod which forms the axis. The emphasis of the design was on structural simplicity, low material consumption, the simplicity and speed of construction, as well as the realization cost of the desired structural characteristics, which can be achieved thanks to the wood-based materials and the construction possibilities. The construction design was verified by calculating the structural assessment according to the method of limit states and the structure is designed for climatic loads (wind, snow) for the selected site of the observation tower in the westernmost corner of the Czech Republic.

1   INTRODUCTION

The article deals with a topic that has recently been on the rise, namely the construction of wooden buildings or wooden structures. According to Kuklik (2005), it has, after a long pause, increased in popularity in Europe but more precisely in the Czech Republic. The paper focuses on the draft of a wooden high rise module observation tower, which should, despite its distinct character defined previously, meet the standards of 21st century wooden buildings. This fact is based on historical ties, because previously it was also possible to observe the development of such constructions (Miškovský, 2008). It is a specific type of construction, which differs from the proposal itself because this type of construction can be used for mass construction. Nouza (2014) stated that nowadays Czech cities and towns are trying to promote their surroundings with the construction of observation towers. For this purpose, structures made from wood are often selected, especially for a positive impact on the environment, the speed of construction, the economic aspect and aesthetic value. Wood is environmentally friendliest construction material because it is renewable source and whose reserves are practically unlimited in the case of rationality. Positive environmental impact is as well in the form of preservation carbon dioxide in the form of carbon.

   The design of the observation tower's shape structure is a task that defines various solutions. A concept is looked for that will further mingle with the entire building. It is all about how the object is used, how it should be oriented and situated on the land and, of course, what are the financial possibilities for realization. The financial part requires a lot of time and a lot of discussion. At this stage, principles can be set that will specify the entire structure significantly. Most important for the proposed model of the wooden observation tower is a versatile construction for the usage in multiple locations but still fulfillment with structural analysis requirements which is a necessary part of any such construction. There is, however, the loss of originality and uniqueness for construction. If this unification has more pros or cons is a matter of opinion, but the paper shows one of the solution.

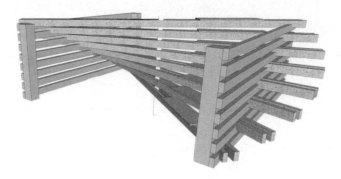

Figure 1. The high-rise module of observation tower

2 MATERIALS AND METHODS

The proposed concept allows complex utilization and tries to bring flexibility into the design so that the tower could be applicable to multiple locations thanks to its unique system of modules of which it is composed. The construction proposal analyzes and verifies the possibility of carrying out the architectural design of the whole structure from a technical point of view. The technical proposal is governed by Act no. 268/2009 Coll., on the technical requirements for buildings. From this perspective, the proposal mostly addressed the staircase because of the connection to the carrier part with the construction of individual step treads. The staircase proposal complies with the standard ČSN 73 4130. The staircase is the skeleton of the whole tower and can be regarded as the staircase construction. The envelope is the load bearing structure. In order to simplify construction, the stringers and elements of the supporting envelope are designed from planks with dimensions of 60×120 mm. It is creating one high-rise module, in Figure 1.

Not only technically, but also aesthetically, the tower construction is very simple. The simplicity lies both in the embodiment of the structure, as well as in the visual impression from the tower. The building should be an example for that simplicity is power. Exaggeration, one could say that it is a purposeful structure of the staircase, which was created in order to climb up above the treetops.

The observation tower proposal went according to ČSN EN 1995-1-1, and its guide (Koželouh, 2004). All countries of the European Union have incorporated Eurocodes as European standards into their technical standards; therefore they should be identical rules for designing building structures in Europe. ČSN EN 1998-6 was used, when assessing the impact of oscillation and vibration on the observation tower's structure. The observation tower is designed in the Aš headland at the western tip of the Czech Republic, at an optimum prospective location with an altitude of 560 m above sea level. The main purpose is to measure and monitor the deformation of structural elements and interactions of wooden structural element joints with the observation tower's steel elements. According to the authors Drdácký and Kasal (2007), the best suited implementation is the identification and implementation of non-destructive or sparingly destructive testing methods. Another goal is that it could be used by the general public, and thus use the building for a completely normal purpose. When designing the observation tower, the emphasis was laid on an easy and quick feasibility of the construction and design variability, depending on the overall cost and material consumption.

3 RESULTS AND DISCUSSION

The structure is made up of panels of larch wood (*Larix decidua*), with a central steel rod in the middle for spatial stiffness. The foundation is done with the help of KRINNER ground screws (Pšenička and Jebavý, 2010). This method may be carried out without excavation and concreting, irrespective of subsoil, with a capacity up to 72 kN per screw (Hrubý, 2011).

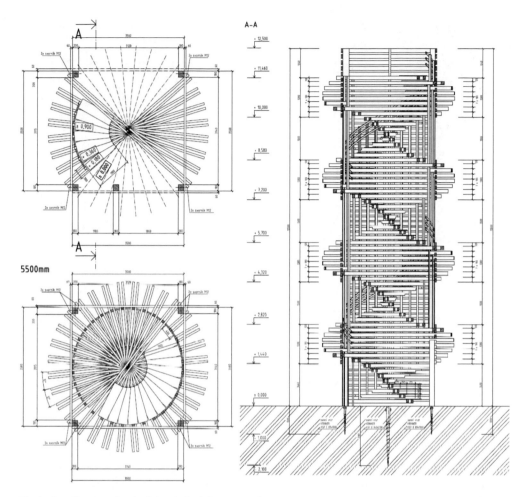

Figure 2.   Construction in the technical documentation.

Connections of timber elements are solved using bolts. The observation tower has a base of 3.5 × 3.5 meters; the height of the observation tower reaches 12.5 meters. The tower's base consists of one tweeter module with a height of 1 440 mm. It is made of four pillars on which planks are fastened that form the supporting elements for the stringer and also fulfill the function as a curtain wall. This module can be placed vertically on top of each other to achieve the chosen height, for various reasons, e.g. height of the trees and the like.

The carrier envelope is made from horizontal wooden planks with dimensions of 60 × 140 mm. Two associated wooden planks with a dimension of 60 × 140 mm are used for the stringer, which extend beyond the outline of the structure. The carrier corner columns are made of Trio glued wooden beams with dimensions of 200 × 200 mm; there are four in the entire construction. In contrast to that, the railing is made from a wooden prism with dimensions of 80 × 80 mm. The railing is made from wooden slats 60 × 20 mm and the central reinforcing steel pillar, to which all the wooden stringers are mounted, has a diameter of 80 mm. All used materials were selected to fulfill their stress functions, but also to be least demanding on price and consumption, and thus to guarantee a low cost of the whole building. To completely fulfill the observation tower's function, its viewing platform has to be higher than the nearest trees and obstacles that might obstruct the view (Shicker and Wagener, 2002). Overall technical layout of construction (Figure 3.) and structural analysis (Figure 2.) are part of the project documentation in accordance by Act no. 62/2013 Coll.

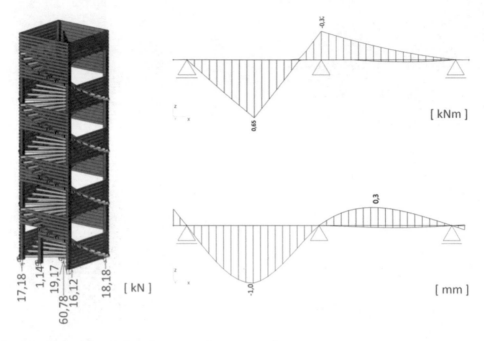

Figure 3.   Structural analysis of the construction.

If the natural greenery outgrows the height of the observation tower or approaches it, it should be removed. It should not be a problem because there is only such natural greenery in the area of the designated site for the observation tower construction. Another requirement for the functionality of the observation tower, respectively to keep the observation tower functional and accessible, is to treat the wood of the tower with oil paint. For construction reasons, this should be done at least once every two years.

The observation tower's proposal should respect the urban and historical landscape. For the place the tower is designed for, namely the cadastral area Kopaniny in the Karlovy Vary region, there are no construction restrictions. The observation tower is designed on the border of three republics, the Czech and two federal – Saxony and Bavaria. After climbing to the top platform of the tower, there will be an outlook in each of these republics. The observation tower should deepen interactions between the Czech and German borderland.

The establishment of the Schengen area virtually abolished borders, at least as tourism is concerned. Therefore, it is not a problem to cross the border from Germany, a few hundred meters from the observation tower located in the Aš headland. One of the main objectives for the tower construction is to bring visitors. These are, therefore, both Czech and German visitors and it is necessary to build bilingual signage nearby the observation tower and a media campaign with visualizations, such as in Figure 4.

The observation tower will be based on KRINNER ground screws for economic and technological reasons. The biggest advantage of screws is the absence of drying time, without which a foundation block of plain concrete cannot function. Important parts of the observation tower's supporting structure are joints. These will consist of galvanized steel studs. The joints of wooden structures are solved in detail by author (Krämer, 2011), with the standard ČSN 73 1702.

The construction of the tower is made from larch timber. Vertical supporting columns are made from Trio glued beams. Fasteners are galvanized because of corrosion resistance. The observation tower will be based on KRINNER ground screws for reasons of construction speed and the simplicity of implementation. The amount of material is directly related to the observation tower's height, which is important especially in the case of mass production. Visualization of the viewing platform and observation tower is shown in Figure 5.

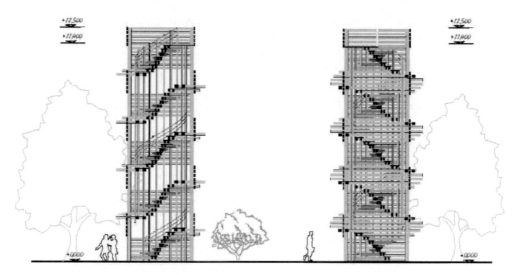

Figure 4.   The observation tower views.

Figure 5.   Visualization of the observation tower structure and viewing platform.

4   CONCLUSIONS

The overall design concept of the wooden observation tower's modules is based on the principle of supporting the envelope, which carries the staircase's step treads, circling around the upper observation platform. Thanks to this, the versatility of the observation tower was achieved in the form of module composition. Thus, there is the possibility to build an arbitrarily high observation tower based on predetermined priorities, such as exceeding the height of the surrounding trees, etc.

The cost of the observation tower, without unforeseen construction problems, should not exceed half a million crowns, including transportation and the amount spent on implementation. This is mainly due to efforts to as much as possible eliminate the use of laminated structural timber, which

403

is used only in the main supporting pillars. All other wooden elements are made from solid wood. This makes the observation tower, in relation to comparable buildings, a low-budget building.

REFERENCES

ČSN 73 1702. "Designing. Calculation and assessment of wooden constructions – General rules and rules for civil engineering," Prague: Czech Standards Institute, 2007, str. 174.

ČSN 73 4130. "Stairs and inclined ramps – basic requirements," Prague: Office for Standards, Metrology and Testing, 2010, str. 28.

Czech. Ministry for Regional Development. Decree no. 268 dated 20 August 2009, as revised 20/2012 on technical requirements for buildings. In Collection of Laws of the Czech Republic, 2009. The amount of 81, p. 3702.

Czech. Ministry for Regional Development. Decree no. 499 dated 10 November 2006 as amended by Amendment No. 62/2013 for the technical documentations. In Collection of Laws of the Czech Republic. 2006, part 163, p. 6872.

Drdácký, M. & Kasal, B. 2007. In-situ evaluation. Luigia Binda, Prague: ÚTAM AV ČR. 253 p. ISBN 978-80-86246-36-9.

EN 1995-1-1. Eurocode 5: Design of timber structures – Part 1-1: General – Common rules and rules for buildings. Czech Standards Institute, 2006.

EN 1998-6. Eurocode 8: Design of structures for earthquake resistance – Part 6: Towers, masts and chimneys. Czech Standards Institute, 2007.

Hrubý, P. 2011. "Ground screws, without concrete foundations," [online]. [cit. 2014-04-17]. Available from: http://www.zemnivruty.cz/.

Krämer, V. 2011. "Wooden constructions. Examples and solutions by ČSN 73 1702," Prague: the Czech Chamber of Chartered Engineers and Technicians in Construction has issued Information center ČKAIT. 316 p. 978-80-87438-16-9.

Kuklík, P. 2005. "Wooden constructions," 1st edition, Prague: Information center ČKAIT. 171 p. 80-867-6972-0.

Miškovský, P. 2008. "Our view towers," Prague: Dokořán, s.r.o. 326 p. 978-80-7363-189-5.

Nouza, J. 2014. "Pictures from history and present observation tower in the Czech Republic and abroad," [Online]. [cit. 2004-03-15]. Available from: http://itakura.kes.tul.cz/jan/rozhledny/rozhled.html.

Pšenička, F. & Jebavý, M. 2010. "Pergolas and shelters," Prague: Grada. 112 p. ISBN 978-80-247-2812-4.

Shicker, R. & Wagener, G. 2002. Drehmoment richtig messen. Darmstadt: Hottinger Baldwin Messtechnik. 316 p. 978-300-0090-158.

RPAS as a tool for mapping and other special work

J. Šedina, K. Pavelka & E. Housarová
Czech Technical University in Prague, Prague, Czech Republic

ABSTRACT: RPAS has become a very useful tool in many areas. At the Czech Technical University in Prague, RPAS is used for mapping, archaeological surveying, and monitoring of the biosphere. In this paper, the different projects are presented in which RPAS was used as the main data source. For mapping small areas RPAS is a very useful tool. The GSD of the orthophoto is usually within a few cm, which is sufficient for accurate mapping. The Digital surface model is used to determine the volume of refuse dumps, or their changes. For archaeological survey, orthophotos are usually used and digital surface models to detect debris objects. Utilizing infrared cameras together with visible spectral range cameras, it can monitor the biosphere (NDVI index). Special flights with a video camera were performed, heaps of Mininng museum were documented. These projects demonstrate the wide use RPAS as a tool for data collection and its advantages over classical geodetic measurements.

1   INTRODUCTION

RPAS (remote piloted aircraft systems), UAV (unmanned aircraft vehicle) or UAS (unmanned aircraft systems) in the U.S. are growing in importance due to their versatility. RPAS is equipped with many sophisticated micro-instruments such as IMU, GNSS receiver, wireless control, automatic stabilization flight planners, etc. RPAS provides not only photographic data, but also multispectral data and thermal data. Bigger RPAS are capable of carrying much more expensive and more accurate devices such as laser scanners, hyperspectral cameras and accurate GNSS receivers (Jon, et al. 2013). RPAS combines close range photogrammetry with aerial photogrammetry and has the ability to produce accurate outputs. However, it is necessary to consider many criteria affecting the accuracy of RPAS outputs, such as the GSD (Ground Sampling Distance), image overlapping, methods of flight over the area, number and distribution of GCPs (Ground Control Points), types of cameras, etc. RPAS has become a universal tool for data collection due to its ability to be equipped with a variety of devices. It can be used in many areas, such as archaeological surveying (Saleri, et al. 2013, Casana, et al. 2014), the monitoring of the biosphere, mapping of natural disasters, detection of the changes in river beds, determining the volume of an object (Wang, et al. 2014), the protection of cultural heritage, precision agriculture (Vega, et al. 2015), and mapping (Rijsdijk, et al. 2013).

2   ARCHAEOLOGICAL SURVEY

The biggest benefit of RPAS for archaeological surveying is its rapid deployment in the field and the possibility of multiple sensing in the area of interest. With the miniaturization of devices such as hyperspectral cameras and thermal imaging cameras, RPAS has become very popular for its versatility in field measurement. It allows us to obtain data from the various devices, such as cameras with various spectrum sensing, hyperspectral cameras or laser scanners. It can be data from the visible spectrum, near infrared spectrum, thermal data, hyperspectral data or the point cloud from laser scanning from the surveyed area. Thermal data is of great importance to the archaeological survey, as it is capable of detecting artificial objects on and below the ground surface (Casana,

Figure 1.   DSM of defunct field fortification.

Figure 2.   Orthophoto of defunct field fortification.

et al. 2014). Heat absorbs, emits and reflects from each material differently. Based on the different material properties, such as thermal conductivity and thermal capacity, we are able to detect these objects near or just under the surface. Acquisition of thermal images in time series, allows you to detect different material lying below surface. Using near infrared cameras and camera sensing in the visible spectrum, we are able to eliminate the influence of vegetation on the thermal data. The product of laser scanning or image-based modeling is a point cloud, which contains information about an object's surface (Saleri, et al. 2013). We can detect different information there, such as artificial edges, mounds or waves. This information can be poorly visible to the naked eye on the ground, see Figure 1.

Our project of this archaeological survey deals with the defunct field fortifications of the fortress Terezín, located near the city of Litoměřice. The fortifications were equipped with artillery and infantry weapons and belonged to the system of fortifications called Labské předmostí (Elbe bridgehead).

The area was imaged by the visible spectral range and near-infrared camera. GSD was 4 cm and 55 visible spectral range images and 54 infrared images were taken by RPAS eBee. The image resolution was approximately 16 MPix. It was measured 5 GCPs by RTK GPS Leica Viva using online correction CZEPOS net. Software Agisoft PhotoScan was used for creating a DSM (Digital Surface Model), see Figure 1, and for the creation of an orthophoto, see in Figure 2.

3   MONITORING OF THE BIOSPHERE

For the monitoring of the biosphere, RPAS used the hyperspectral camera, or a combination of cameras sensing in different bands. For the monitoring of the biosphere, the most commonly used

Figure 3.   Map of NDVI index of NNR.

bands are the red and infrared band. Comparing these bands, it is possible to create a vegetation index RVI (Ratio Vegetation Index), NDVI (Normalized Difference Vegetation Index) and TVI (Transformed Vegetation Index). The most commonly used vegetation index is NDVI, for its calculation, see Eq. 1. Other areas of use in RPAS in the monitoring of the biosphere is the detection of the changes in riverbeds and revealing their original riverbeds, monitoring natural disasters (e.g. Floods) or for precision agriculture (Vega, et al. 2015). Based on these requirements for equipment, there are created specialized RPAS and specialized cameras for the monitoring of the biosphere, such as cBee Ag, designed for precision agriculture. The main output is a map of the vegetation index of the area.

$$NDVI = \frac{NIR - RED}{NIR + RED} \tag{1}$$

Where NIR = near infrared band; and RED = red band.

For the creation of vegetation index NDVI, the locality of National Nature Reserve (NNR) "Božídarské rašeliništ", lying in the Karlovy Vary region, was selected. The subject of the protection are peat bogs as well as mountain peaty meadows, which merge into heathlands.

226 images from the visible spectral range camera and 226 images from the near-infrared camera were captured by RPAS eBee. Images were taken in two flights (one for each camera). GSD was about 5 cm and 5 GCPs was measured by RTK GPS Leica Viva using online correction CZEPOS net. Images were processed by Agisoft PhotoScan software, and all images were processed simultaneously. A set of images of each spectrum for the creation of different spectral orthophotos was chosen. The map of NDVI index is shown in Figure 3.

4   DETERMINING THE VOLUME AND THEIR CHANGES

For creating a 3D model of the object, laser scanning or image-based modeling can be used. Bigger RPAS are capable of carrying more sophisticated, heavier and more expensive equipment, such as a laser scanner or hyperspectral cameras. These bigger RPAS are usually equipped with more accurate GNSS receivers and IMU (Jon, et al. 2013). The accuracy of a specific point in the point cloud is only a few cm. Image-based modeling is used for smaller, cheaper and much more affordable RPAS (Wang, et al. 2014). The equipment of cheaper RPAS usually consists of GPS (m accuracy), a less precise IMU and a commonly available camera. The output is a point cloud generated by image correlation or a point cloud measured by laser scanner.

Our project is focused on the visualization of the Mining museum of Jan Šverma mine located in Zácleø, see Figure 4. The beginnings of coal mining are back to r. 1570 when the first written mention of the issue of permits for prospecting and mining of coal was recorded.

Figure 4.   Mining museum of Jan Šverma mine.

Figure 5.   Placement of signalized points.

For capturing images a video camera mounted on the wing of a small plane was used. The camera resolution was 1280x720. For visualization purposes GCPs haven't been used. For determining the volume and their changes by RPAS, it would be appropriate to equip RPAS with a camera with much better resolution and use the GCPs (to determine the coordinate system or model scale). The visualization of the point cloud of heap Jan Šverma is shown in Figure 5.

5   MAPPING

Recently, the use of RPAS for mapping of the cadastre of real estates (Rijsdijk, et al. 2013) has been considered. RPAS appears to be a useful tool for mapping work of a lesser extent due to its rapid data collection. The main problem of aerial photogrammetry is the overhang of the roof. The cadastre of real estates registers in the intersection of buildings with terrain or for unusual buildings with a vertical projection of the outer perimeter of the building to the terrain. For mapping, orthophoto is considered, but because it captures the roof of the building, it is necessary to determine the roof overhang by field measuring. Another problem is the second independent measurement of a detailed point. The question is whether the same technology should be used for the second independent measurement of the detailed point, or a completely different method (eg. GPS measurements).

Testing RPAS for mapping purposes was carried out in a gardening colony near Litoměřice city. In the area (600 m × 250 m), 21 signalized points were placed, which were measured by RTK GPS Leica Viva using online correction CZEPOS net. The size of the signalized points were about 30 × 30 cm, in a checkerboard pattern. Their arrangement is shown in Figure 6. The area was

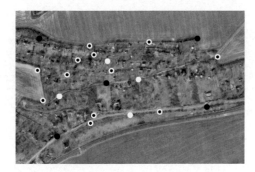

Figure 6.   Placement of signalized points in a gardening colony.

Table 1.   Influence of image forward overlap and perpendicular flight lines on GCPs and Checkpoints.

Flight		Forward		Forward + Perpendicular	
		Δp [m]	SD Δp [m]	Δp [m]	SD Δp [m]
Forward	80%	*0,04*	*0,08*	**0,03**	**0,04**
overlap	60%	*0,01*	*0,06*	**0,02**	**0,04**

Δp – average positional error; SD Δp – average standard deviation of positional error

captured at different altitudes with GSD 3 cm, 6 cm and 12 cm using perpendicular flight lines. The overlap of images was 80/60% and image resolution was 16 MPix. For taking images, the RPAS eBee was used, equipped by visible spectra range camera. The project was processed by software Agisoft PhotoScan.

6   ACCURACY TESTING

The influence of different effects was tested on the achieved accuracy of GCPs and Checkpoints (GCPs and Checkpoints were processed together) and on the stability of the image projection center (4 images were tested, one with 3 visible GCPs, one with one visible GCP in the center and two without visible GCPs). The gardening colony was used for accuracy testing.

The effect of added perpendicular flight lines was tested on the achieved accuracy of GCPs and of Checkpoints. Table 1 indicates that better results are obtained by adding perpendicular flight lines; the influence of the image forward overlap of 80% or 60% is negligible. Table 2 confirms this assertion that the standard deviation of the coordinate differences of image projection centers are significantly smaller after adding perpendicular flight lines.

Also, the effect of images taken at several altitudes and their processing together in one model was tested. Table 3 shows that there is no significant difference in the flight at one altitude (GSD 3 cm), or at two altitudes (GSD 3 cm and 6 cm). By adding a third altitude, the model deteriorates, it is probably due to its GSD (12 cm). Table 4 confirms that the best results are obtained when using two altitudes (GSD 3 cm and 6 cm).

Finally, the effect of GCPs placement and influence of tie points filtration was tested. From Table 5 it is apparent that for such a large area, 5 GCPs (Figure 6 black marks) is insufficient. 9 GCPs (Figure 6 black marks and white marks) gives us satisfactory results, comparable with 21 GCPs (Figure 6 black, white and black/white marks). From the point of view of tie points filtration, optimal results are achieved when filtering into the reprojection error 3 pixels.

Table 2. Influence of image forward overlap and perpendicular flight lines on image projection centers.

Flight Forward overlap GCPs	Coordinate	Forward 60 [%] Δ [m]	SD Δ [m]	Add perpendicular 60 [%] Δ [m]	SD Δ [m]	Forward 80 [%] Δ [m]	SD Δ [m]	Add perpendicular 80 [%] Δ [m]	SD Δ [m]
3	x	0,10	0,12	0,10	0,01	0,19	0,02	−0,10	0,01
	y	−0,02	0,03	0,03	0,02	0,08	0,01	−0,02	0,02
	z	−0,12	0,02	0,04	0,02	0,03	0,04	−0,02	0,03
0	x	−0,10	0,03	0,05	0,01	0,06	0,02	0,01	0,01
	y	0,01	0,03	0,02	0,01	0,15	0,01	−0,03	0,01
	z	−0,08	0,01	−0,01	0,03	0,04	0,03	−0,04	0,03
0	x	0,08	0,08	0,01	0,01	0,02	0,01	−0,01	0,01
	y	−0,04	0,06	0,03	0,00	0,00	0,01	0,01	0,01
	z	−0,10	0,06	0,01	0,02	0,08	0,05	−0,05	0,03
1	x	−0,07	0,04	0,00	0,00	0,02	0,02	0,01	0,00
	y	0,02	0,02	0,02	0,01	0,05	0,01	−0,03	0,01
	z	−0,13	0,02	0,00	0,03	0,07	0,04	−0,04	0,04
SD	[m]	0,07		0,03		0,05		0,03	
Mean	[m]		0,05		0,02		0,03		0,02

Δ – coordinate difference between averages of different altitudes, image forward overlap, tie point filtration and average with fixed different altitudes and forward overlap; SD Δ – standard deviation of image projection center for changing tie point filtration; SD – standard deviation of Δ; Mean – mean of SD Δ

Table 3. Influence of used altitudes on GCPs and Checkpoints in processing.

Altitudes	3	2	1
Δp [m]	0,06	0,03	0,03
SD Δp [m]	0,07	0,04	0,04

Table 4. Influence of used altitudes on image projection centers.

Used altitudes GCPs	Coordinate	1 Δ [m]	SD Δ [m]	2 Δ [m]	SD Δ [m]	3 Δ [m]	SD Δ [m]
3	x	−0,10	0,01	−0,02	0,00	−0,02	0,00
	y	−0,02	0,02	0,00	0,01	0,01	0,01
	z	−0,02	0,03	0,00	0,02	0,10	0,04
0	x	0,01	0,01	0,00	0,01	0,03	0,01
	y	−0,03	0,01	−0,05	0,01	−0,06	0,01
	z	−0,04	0,03	−0,01	0,02	0,05	0,04
	x	−0,01	0,01	−0,02	0,01	−0,05	0,00
0	y	0,01	0,01	0,02	0,01	0,03	0,01
	z	−0,05	0,03	−0,03	0,03	0,06	0,04
1	x	0,01	0,00	0,01	0,00	0,00	0,01
	y	−0,03	0,01	−0,02	0,01	0,00	0,01
	z	−0,04	0,04	−0,02	0,03	0,07	0,04
Mean [m]		−0,02	0,02	−0,01	0,01	0,02	0,02
SD [m]		0,03		0,02		0,05	

410

Table 5. Influence of used GCPs in bundle adjustment and tie points filtering.

Altitudes Used Filtering	Used GCPs	3 ΣΔp	2 ΣΔp	1 ΣΔp	Used Filtering	3 ΣΔp	2 ΣΔp	1 ΣΔp
No filtering	21	2,21	1,79	1,75				
	21	1,99	1,12	0,94	RE = 3 pix,	2,12	1,22	0,97
RE = 3 pix	5	2,86	1,66	1,64	VI = 3 img.	3,05	1,86	1,87
	9	2,18	1,25	1,19		2,33	1,39	1,29

ΣΔp – sum of positional errors of GCPs and Checkpoints; RE – maximal reprojection error for tie point, VI – tie point visibility at least on the number of images

7 CONCLUSIONS

Four areas for the use of RPAS and accuracy testing were presented. From the perspective of archaeological surveying, RPAS has great use for creating a detailed DSM, the thermal imaging of an area and the creation of orthophotos. In our project, RPAS was used for an archaeological survey of a defunct fortification. Another major application of RPAS is the monitoring of the biosphere. Maps of the vegetation index for smaller localities is formed in a very short time. Our RPAS was used to create NDVI index for a part of NNR "Božídarské rašeliniště". Bigger RPAS are capable of carrying devices like laser scanners. These bigger RPAS can be used for determining the volume, or to determine the change in the volume, e.g. heap. In our project, we dealt with the visualization of the mining museum of Jan Šverma mine. RPAS is of great importance in terms of mapping for cadastral purposes, where the actual measurements in the field can be done in a short time. Finally, different influences have been tested for the accuracy achieved on GCPs and Checkpoints such as image overlap, using perpendicular flight lines, the use of different altitudes, the influence of tie points filtration and influence of GCPs placement.

ACKNOWLEDGEMENTS

This paper project was supported by a grant of the Ministry of Culture of the Czech Republic NAKI DF13P01OVV002 (New modern non-invasive methods of cultural heritage objects exploration).

REFERENCES

Casana, J., et al. 2014, "Archaeological aerial thermography: A case study at the Chaco-era Blue J community, New Mexico", *Journal of Archaeological Science*, vol. 45, no. 1, pp. 207–219.

Jon, J., et al. 2013, "Autonomous airship equipped by multi-sensor mapping platform", *International Archives of the Photogrammetry, Remote Sensing and Spatial Information Sciences – ISPRS Archives*, pp. 119.

Rijsdijk, M., et al. 2013, "Unmanned Aerial Systems in the process of juridical verification of cadastral border", *International Archives of the Photogrammetry, Remote Sensing and Spatial Information Sciences – ISPRS Archives*, pp. 325.

Saleri, R., et al. 2013, "UAV photogrammetry for archaeological survey: The Theaters area of Pompeii", *Proceedings of the DigitalHeritage 2013 – Federating the 19th Int'l VSMM, 10th Eurographics GCH, and 2nd UNESCO Memory of the World Conferences, Plus Special Sessions fromCAA, Arqueologica 2.0 et al.*, pp. 497.

Vega, F.A., et al. 2015, "Multi-temporal imaging using an unmanned aerial vehicle for monitoring a sunflower crop", *Biosystems Engineering*, vol. 132, pp. 19–27.

Wang, Q., et al. 2014, "Accuracy evaluation of 3D geometry from low-attitude UAV images: A case study at Zijin Mine", *International Archives of the Photogrammetry, Remote Sensing and Spatial Information Sciences – ISPRS Archives*, pp. 297.

Advances and Trends in Engineering Sciences and Technologies – Al Ali & Platko (Eds)
© 2016 Taylor & Francis Group, London, ISBN: 978-1-138-02907-1

Renewable energy sources: analysis of two different alternatives of the heat pumps

M. Štefanco & D. Košičanová
Technical University of Košice, Civil Engineering Faculty, Košice, Slovakia

ABSTRACT: As the energy prices are constantly rising, the energy management is currently a very important issue. One of the main reasons is that obtaining the primary energy sources like oil is much more difficult than in the past. This may lead to the exhaustion of worldwide resources. The use of different sources e.g. solar and geothermal energy etc. could be seen as a potential way of improving thermal energy management. The implementation of heat pumps as the primary source of heating and cooling in buildings, compared to installations using fossil fuels has become increasingly economical. This is due to increasing prices of fossil fuels, which is directly influenced by their accessibility and limited quantity. In addition, heat pumps contribute to the energy objectives of the European Union set for 2020. It specifically deals with achieving energy savings, reducing CO_2 emissions and increasing the share of renewable energy by 20%.

1 INTRODUCTION

1.1 *Building description*

The main purpose of this article is the analysis of two different alternatives of the heat pumps. To be more specific, article describes the energetic analysis of the ground/water and air/water heat pumps in service building. The building is intended for the seniors 'accommodation as the heat source for heating and hot water heating-up. In order to achieve the best possible value of the seasonal performance factor – SPF, the heating system is divided into several separate heating circuits. The building is shown in Figure 1 (Štefanco, 2012).

There are two floors in the service building which is mainly intended for the seniors' rehabilitation. Ground plan for the first and second floor is shown in Figure 2.

Figure 1.   Building visualization (Štefanco, 2012).

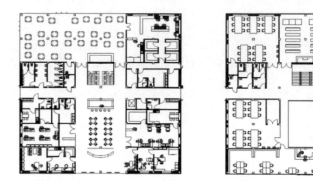

Figure 2. Ground plans of the service building (Štefanco, 2012).

Table 1. Technical parameters for heat pump ground/water specified by the manufacturer.

Primary medium temperature (°C)	Secondary medium temperature (°C)	Heat pump power (kW)	Heat pump imput (kW)	Coefficient of Performance (−)
0	35	30,8	6,8	4,5
0	50	29,5	8,9	3,3

Table 2. Technical parameters for heat pump air/water specified by the manufacturer.

Primary medium temperature (°C)	Secondary medium temperature 50°C (°C)	Secondary medium temperature 35°C (°C)	Heat pump power at gradient		
			55/45°C (kW)	33/28°C (kW)	55°C (kW)
−15	50	8,9	8,77	8,91	8,77
−10	8,8	10,4	10	10,5	9,87
−5	10	11,4	11,47	11,36	11,53
0	11,5	13,2	13,1	13,35	12,8
5	12,9	15,4	15,4	16,22	13,67
10	14,1	18	18,37	18,38	16,53
15	16,9	20	20,4	20,53	19,2
20	20,8	22,4	24	24	20,27
25	22,9	24,8	27,33	26,94	22,27

2 IMPUT PARAMETERS

2.1 *Parameters for the design source*

Input parameters include the heat loss of the building, energy consumption in terms of heating, power and energy consumption in terms of hot water heating, heat load of the assessed building and technical parameters of the heat pumps. Technical parameters of the heat pumps are shown in Table 1. and Table 2.

Each input for the design and assessment of the heat pump has been calculated via computing program PROTECH in accordance with STN EN 12831.

2.2 *Design and assessment of two different heat pumps*

Heat pump ground/water has been chosen as the first alternative source for the heating of the discussed building. The source itself consists of two heat pumps which are connected to the cascade

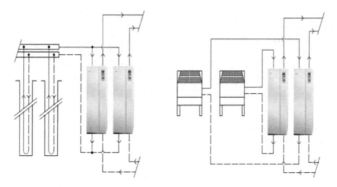

Figure 3.   Heat pump alternatives ground/water and air/water.

with the power of 30 kW. In the primary section of the system there are deep water boreholes which are 100 meters long. Active borehole length is 1500 meters. Active length and frequency of the boreholes have been calculated on the basis of theoretical profit from the borehole which was at 50 W/m (Karlík, 2009). Boreholes are located near the assessed building. When realizing the project, it is necessary to do the so-called TRT – borehole thermal response test. This test determines the exact value of profits from a borehole as well as the required active borehole length (Trs, 2008). The second alternative heating source is in the form of the heat pump air/water. The source similarly contains two heat pumps with the power of 30 kW. The primary section of the system consists of the heat pumps and their outdoor units which are located in the close proximity to the building. Power of this alternative depends on the outdoor temperature. Heat pump alternatives are shown in Figure 3.

3   ENERGETIC ANALISIS OF THE ENERGY SOURCE

3.1   Four modes of operation

Total energy needed for heating is covered in four modes of operation. Energy needed in individual modes is covered by the source which consists of two heat pumps with the same power in both alternatives. In case one of the heat pumps is out of order the other is able to cover the part of the required energy and prevent the shutdown in the building. Following is the description of the modes:

The first operating mode – heat pump 1 with its power covers the total energy for capillary mats at temperature gradient 33/28°C. Energy needed for capillary mats is about 1/4 from the total energy needed for heating.

The second operating mode – heat pump 1 covers with its reserve power part of the energy needed for board heating elements and fancoils with higher temperature gradient 55/45°C. Energy needed for board heating elements and fancoils is about 3/4 from the total energy needed for heating.

The third operating mode – heat pump 2 covers with its power the remaining part of the energy needed for board heating elements and fancoils. Rest of the energy is covered by bivalent source in the form of electric heating body.

The fourth operating mode – heat pump 2 covers with its power part of the energy needed for the hot water heating up during the heating season. Remaining energy is covered by bivalent source in the form of electric heating body. During summer, when the demand is only for hot water heating up, heat pump 2 covers the total energy needed for the hot water.

3.2   Values of the energetic use for heat pumps

Values of the energy used including bivalent sources for heating and values of the total energy used in individual operating modes of the heat pumps – alternative ground/water are shown in Table 3.

Table 3. Total energy used alternative ground/water (Štefanco, 2012).

Heat pump 1 1.mode (kWh)	Heat pump1 2.mode (kWh)	Heat pump 2 3.mode (kWh)	Heat pump 2 4.mode (kWh)	Bivalent source heating (kWh)	Bivalent source water heating (kWh)
40963	90877	52539	37590	1160	2363

Table 4. Total energy used alternative air/water (Štefanco, 2012).

Heat pump 1 1.mode (kWh)	Heat pump 1 2.mode (kWh)	Heat pump 2 3.mode (kWh)	Heat pump 2 4.mode (kWh)	Bivalent source heating (kWh)	Bivalent source water heating (kWh)
40963	79217	55927	36946	9432	3007

1. mode heat pump 1 33/28 °C with its power uses 18% from the total energy,
2. mode heat pump 1 55/45°C with its power uses 40% from the total energy,
3. mode heat pump 2 55/45°C with its power uses 23% from the total energy,
4. mode heat pump 2 50°C with its power uses 17% from the total energy,

Bivalent source for heating uses 0,5 % from the total energy,
Bivalent source for hot water heating up uses 1% from the total energy (Štefanco, 2012).
Values of the energy used including bivalent sources for heating and values of the total energy used in individual operating modes of the heat pumps – alternative air /water are shown in Table 4.

1. mode heat pump 1 33/28°C with its power uses 18% from the total energy,
2. mode heat pump 1 55/45°C with its power uses 35% from the total energy,
3. mode heat pump 2 55/45°C with its power uses 25% from the total energy,
4. mode heat pump 2 50°C with its power uses 17% from the total energy,

Bivalent source for heating uses 4% from the total energy,
Bivalent source for hot water heating up uses 1% from the total energy (Štefanco, 2012).

3.3 *Comparison of COP and SPF*

Heat pumps reach the highest heating factor – COP and seasonal performance factor – SPF at the highest outdoor temperature. The reason is that at higher outdoor temperatures lower energy as well as lower power and input of the heat pump are required (Quaschning, 2010). On the contrary at low temperatures, there is an increasing energy needed and higher power and input of the heat pump. Course of heating factors in separate operating modes – alternative ground/water is shown in Figure 4.

Operating modes – alternatives ground/water:

1. mode TČ1 33/28°C reaches total COP 5,18,
2. mode TČ1 55/45°C reaches total COP 4,34,
3. mode TČ2 55/45°C reaches total COP 3,98,
4. mode TČ2 TV50°C reaches total COP 2,91,

System ground/water total COP is 4,09,
System ground/water total SPF is 3,5 (Štefanco, 2012).
Course of heating factors in separate operating modes – alternative air / water is shown in Figure 5.

Operating modes – alternatives air/water:

1. mode TČ 33/28°C reaches total COP 3,44,
2. mode TČ1 55/45°C reaches total COP 2,79,

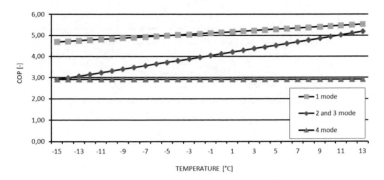

Figure 4. COP – coefficient of performance of the heat pump ground/water (Štefanco, 2012).

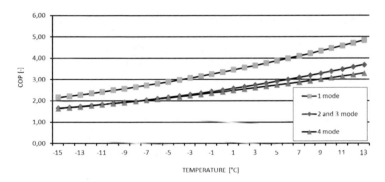

Figure 5. COP – coefficient of performance of the heat pump air/water (Štefanco, 2012).

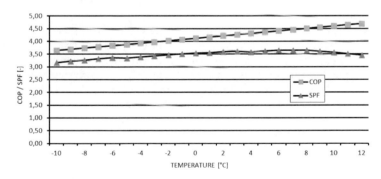

Figure 6. Course comparison of the total COP and SPF – heat pump ground/water (Štefanco, 2012).

3. mode TČ2 55/45°C reaches total COP 2,41,
4. mode TČ2 TV50°C reaches total COP 2,79,

    System ground/water total COP is 2,49
    System ground/water total SPF is 2,3 (Štefanco, 2012).
    Main aim of deliberate division into two heating circuits was to reach the highest possible value of the heating factor COP and seasonal performance factor SPF at the maximum load of the heat pumps. Course comparison of the total COP and SPF for heat pump ground/water can be seen in Figure 5.
    Course comparison of the total COP and SPF for heat pump ground/water can be seen in Figure 6.

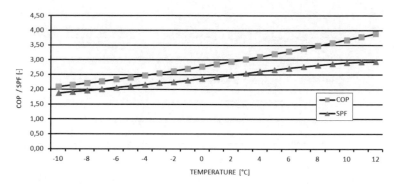

Figure 7.   Course comparison of the total COP and SPF – heat pump air/water (Štefanco, 2012).

Heating factor has a tendency to rise from the lowest to the highest outdoor air temperature. Seasonal performance factor reaches its maximum value in the first alternative ground/water at 7°C and at 13°C in the second alternative air/water. Seasonal performance factor has a tendency to fall beyond this temperature, which is caused by lower power of the heat pump and constant input of the system circulating pumps (Štefanco, 2012).

4   CONCLUSIONS

The article is focused on the design of the heat source for the operational section of the building which is in terms of renewable resources intended for old people's accommodation.

Two alternatives of the heat pumps ground/water and air/water were proposed and assessed. Heating system was deliberately divided into heating circuit with low temperature gradient for capillary mats heating and heating circuit with higher temperature gradient for heating by board elements and fancoils.

Main aim was to achieve high value of the heating factor COP and seasonal performance factor SPF at the maximum load of the heat pumps. Share of the energy needed for capillary mats is only 1/4 from the total energy needed for heating. Therefore, by dividing the heating into two heating circuits, there is minimum value improvement of the heating factor and seasonal performance factor. The first alternative ground/water achieve better overall value of COP and SPF than the second alternative air/water. However, it is more suitable to use alternative air/water than ground/water for this type of the building. The reason is that the alternative ground/water is much more costly in terms of investment costs than alternative air/water.

ACKNOWLEDGEMENTS

This work is directly related to the research project KEGA 052TUKE-4/2013: "The implementation of a virtual laboratory for designing energy-efficient buildings".

REFERENCES

Karlík, R. 2009. Tepelné čerpadlo pro váš dům. Praha 7: GRADA publishing, a.s., pp. 109–110.
Quaschning, V. 2010. Renewable Energy and Climate Chang. West Sussex: Wiley, Ltd., pp. 225–226.
Štefanco, M. 2012. Senior citizens home. Diploma thesis. Košice: TU, SvF.
Trs, M. 2008. Thermal response test. Liberec: Gerotop.

Advances and Trends in Engineering Sciences and Technologies – Al Ali & Platko (Eds)
© 2016 Taylor & Francis Group, London, ISBN: 978-1-138-02907-1

Comparison of ventilated air channels for historical buildings with moisture problems

L. Tazky & A. Sedlakova
Technical University of Košice, Faculty of Civil Engineering, Košice, Slovakia

ABSTRACT: The purpose of the research is to find adequate solution against rising damp in historical buildings. The improvement of this case is with ventilated air channel. Special part of historical buildings are churches because not use routinely this channels as for example museums, galleries. We designed 4 differently construction of ventilated air channel. These are: open or closed, with overpressure and underpressure. We use simulation software ANSYS CFX for verification relevance these construction. We modeled equal segments of the different construction. This research should be offer for architects the solution against problems of rising damp in historical buildings. For find the right solution is necessary to research everyone case. The aim of the research is to prepare the design process of the ventilated air channels for the architects without simulation every building, cross section or geometry of these construction.

1 INTRODUCTION

One of the special categories of historical buildings are churches. The reconstruction of these buildings is difficult and extraordinary. Historical buildings are sensitive to moisture, temperatures, especially churches. Villagers repaired the small errors on the walls, but they didn't use the correct methods and materials in this case. Because in the past the builders used traditional materials, nowadays they use cement and other adhesive mortar based on the cement. This materials must not be used in historical buildings with moisture problems because we worsen the state instead repairing it. Therefore we can use many construction materials to repair or relieve problems with moisture. The best and the most popular in the past is the ventilated air channel around the perimeter of the wall. Many types of these systems exists, for example: overpressure or underpressure channels, open or closed systems.

2 TYPES OF THE SYSTEMS

2.1 Open system of the ventilated air channels

One of the most popular system is the open system without inlet and outlet pipes. The construction of the channel consists of the wall and base of the channel, and the top of it is vacant or with iron bars. This system is the best for churches but not esthetic because around the building there is an empty hole.

2.2 Closed system of the ventilated air channels with overpressure

The second type is the closed channel with overpressure air ventilation. The outdoor dry air enters the channel through the inlet pipes and the damp air is evaporated through the outlet openings. The advantage of this type is that the channel is hidden, only the inlets and outlets are visible on the façade or on the ground. The prefabricated cover plates are covered with gravel.

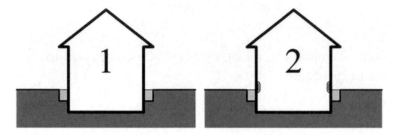

Figure 1.   Open system (left) and closed system with overpressure (right).

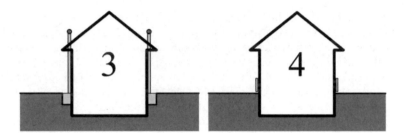

Figure 2.   Closed system with underpressure (left) and ventilated plinth system (right).

2.3   *Closed system of the ventilated air channels with underpressure*

The third type is the closed system with underpressure air ventilation. The underpressure in the channel is ensured with ventilation heads above the roof. The inlet openings are on the top of the channel or in the wall. The outlets are situated on the façade, this is a tall pipe, we can hide it in roof drain or on the façade only.

2.4   *System of ventilated plinth*

The last one is the ventilated plinth. This type is the soft method of the ventilation, because between the plinth and the wall is a thin layer of air. The inlets are situated at the bottom of the plinth and the outlets on the top of the plinth.

3   NUMERICAL MODEL OF THE VERSIONS

3.1   *Geometry of the domain*

These numerical applications present the aerodynamic analysis of the building sections and all versions of ventilated air channels. We designed the 5 m wide section in the software ANSYS CFX surrounded with air boundary. From the 2D model of the section we transformed a 3D version with the function "extrude". The sizes of the environments are designed according the principles of the air flow. The present simulation is considered the turbulence fluctuation in the inflow boundary conditions. The Figure 3 shows the geometry of the air environment.

The general domain size for the numerical model was set at 113.2 × 5.0 × 35.0 m3. The distance between the inlet section and the center of the building was 74.1 m. The full height of the building is 10.96 m, respectively the height of the wall was 4.7 m; the width was 8.2 m and the roof angle was 54.64°.

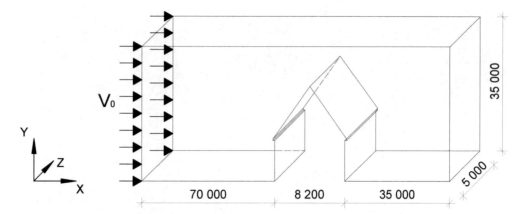

Figure 3.   Geometrical characteristics of the computational domain for a section model.

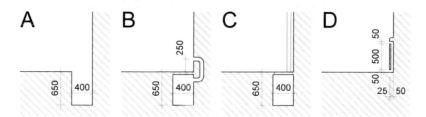

Figure 4.   Dimensions of versions: A – open channel, B – closed channel with overpressure, C – closed channel with underpressure, D – ventilated plinth.

3.2   Geometry of the versions

The open system air channels are the simplest solutions. The channel is situated at the perimeter of the building, near the base. The width of the channel was 0.4 m and the height was 0.65 m. The channel isn't closed with iron mesh or with other cover, the air is flow free in the channel Figure.

The geometry of closed system with overpressure air ventilation is similar. The width of the channel was 0.4 m and the height was 0.6 m, and the cover plate was 0.05 m. The air channel is connected to the exterior air with pipes (inlets)with a diameter of 0.075 m. The height of the inlets is 0.25 m on terrain. The outlets are situated on the terrain,the length of the outlets was 0.35 m and the width was 0.025 m.

The next version is similar, and the dimensions are the same. The inlet pipes are replaced with pipes at the full height of the wall. The end of the pipes are above the roof with 0.5 m. The outlets in this case function as inlets, and the pipe above the roof as the outlet. The diameter of the pipe is 0.075 m.

The last system is the ventilated plinth. In this case we used the basic dimensions for the simulation. The height of the prefabricated plates is 0.5 m and the thickness is 0.025 m. The air channel under the plates is 0.6 × 0.05 m. On the base there are the inlets with a height of 0.05 m and on the top of the plates are outlets with a height of 0.05 m.

3.3   Mesh generation of the models

The overall model is meshing to the maximum size of the elements of 0.3 m. Surfaces or edges are condensed to the element size of 0.25–0.05 m, depending on the versions. Figure 4 shows the mesh of the model, especially the ventilated plinth. This model is constituted by 6,495,987 elements and 1,146,303 nodes. Time period of generating is 5–7 min.

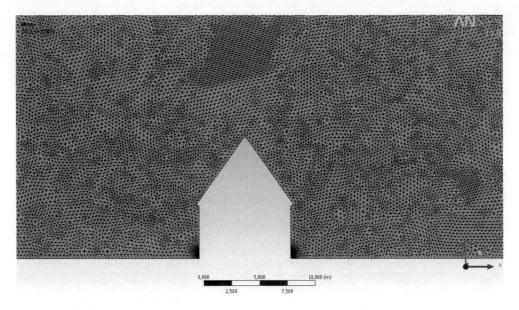

Figure 5. Mesh characteristics of the computational domain for a section model.

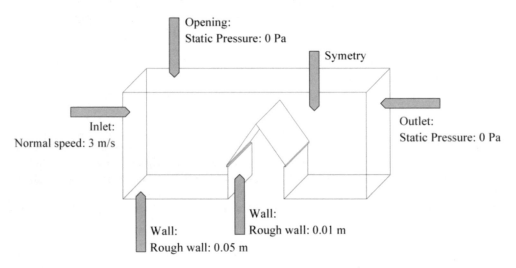

Figure 6. Simulation model and input parameters of the conditions.

3.4 *Boundary conditions of the models*

In this case we used a SST (Shear Streas Transport) numerical model. Temperature of the overall model is 25°C (summer temperature). Apparent density of the air is 1.1845 kg/m3 and the dynamic viscosity of the air is $1.86159.10^{-5}$ kg/m.s. All section models of the channels are simulated by a wind flow of 3 m/s. In Table 1 we show the boundary condition parameters for the model.

3.5 *Results of the simulation*

The figure 7 displayed contour of air in the open air channel around the church. The streamlines air is coloured according to air velocity. The tested samples have been collected from the bottom

Table 1.   Parameters of the setup model.

Name of surface	Boundary type	Parameters
Inlet	Inlet	Normal speed 3 m/s
Outlet	Outlet	Relative pressure 0 Pa
Sky	Opening	Relative pressure 0 Pa
Sides	Symetri	–
Terrain	Wall	Roug wall = 0,05 m
Building	Wall	Roug wall = 0,01 m

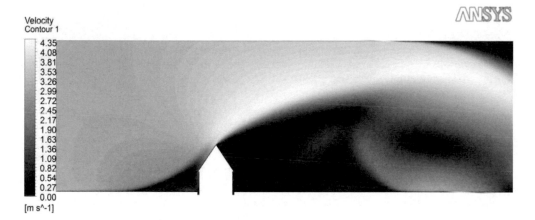

Figure 7.   Air velocity contour of open air channel model.

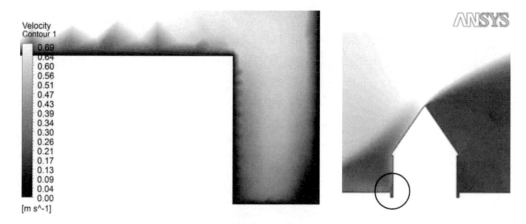

Figure 8.   The air contour of the open channel system.

part of the masonry wall. Results obtained from the numerical simulation in the software ANSYS CFX are satisfactory in the complicated parts of the church as well as at the apse.

In the Figure 8 is displayed air velocity contour in the channel. Therefore in the final numerical simulation we used lower depth of the air channel with the enlarged width. Final cross-section dimensions of the air channel are 650×400 mm.

4 CONCLUSION

Many construction solutions of air channels exist for historical buildings with damp problems in walls. The designed versions are the most widely used solutions as the: open air channel, closed air channel with overpressure, closed air channel with underpressure, ventilated plinth. We simulated these 4 different versions, because these solutions are practiced in the reconstruction of historical buildings. The versions are designed and set in accordance with regulations. Many literatures write about designing the right dimensions of air channels for example: Balik M. in the book "Dehumidification of buildings". As'a result of the simulations we will compare and find the best solution. Obviously, not all versions or models are right for all environments. Modeling is necessary in every case and verification of its function.

ACKNOWLEDGEMENTS

This article was written as a project solution entitled "The use of the virtual laboratory for designing energy-efficient buildings" Project code: 052TUKE-4/2013.

REFERENCES

Alexander, L.B., Armando, M.A. 2007. *Aerodynamic analysis of buildings using numerical tools from computational wind engineering.* Córdoba: AAdeMC.

Balik, M. et.al. 2008. *Dehumidification of buildings Vol. 2.* Praha: Grada.

Balik, M., Solar, J. 100. 2011. *Traditional details of buildings – rising damp.* Prague: Grada Publishing.

Brestovic, T., Jasminska, N., Kubik, M. 2013. *Development of software support for ANSYS CFX.* Zilina, Slovakia.

Horanska, E., Dvorakova, V. 2011. *National cultural heritage, importance and reconstruction a financial support.* Bratislava: Jeka Studio.

Rencko, T., Sedlakova, A. *Assesment of underfloor ventilation of historic buildings using Ansys CFX.* Krakow.

Vlcek, M., Benes, P. 2006. *Disturbances and reconstruction of buildings – Modul1.* Brno, Czech Republik.

Young, D. 2008. *Salt attack and rising damp – A guide to salt damp in historic and older buildings.* Melbourne: Red Rover.

Advances and Trends in Engineering Sciences and Technologies – Al Ali & Platko (Eds)
© 2016 Taylor & Francis Group, London, ISBN: 978-1-138-02907-1

Two and three-dimensional web presentation of Czech chateaux and manors

P. Tobias, J. Krejci & J. Cajthaml
CTU in Prague, Faculty of Civil Engineering, Prague, Czech Republic

ABSTRACT: Chateaux and castles are an important part of cultural heritage in the Czech Republic. The aim of this project is to develop a public web mapping application capturing the changes of manors and surrounding landscape during the last two centuries. This mapping application will be based on digitized and georeferenced old maps and various plans as well as both historical and up to date photographs. The application will allow users to compare various rasters, vector layers and photos from different time periods. Two-dimensional web mapping application will be also supplemented with a 3D web scene based on procedural modelling rules. The procedural modelling approach provides a quick way to create 3D models of buildings within a manor from 2D vector layers. Rules can be then continuously refined or even high detailed non-procedural models can be added to depict the most significant buildings (i.e. chateaux, castles or churches).

1 INTRODUCTION

The project is focused on historical photographs and old maps of castles and chateaux and their publication together within the web mapping application. The project deals with 60 chateaux and castles formerly in property of noble families and at present time owned by the Czech Republic and administrated by the National Heritage Institute (NPU). Old maps along with historic photographs, views and text documents can offer us new views into the history of castles, manors and aristocratic families using online technologies, web mapping and 3D modelling.

The focus is put on the chateau itself, its subsidiary buildings, gardens, parks, but also wider surroundings integrating economic and cultural background of the whole domain. Various old maps and plans are collected and processed. Floor and building plans show the castle interior. The coloured Imperial Imprints of maps of the Stable Cadastre, other cadastral maps and The State Derived Map provide a continuous base map depicting the castle area and the closest surroundings. Maps of domains and Maps of Military Mapping Surveys illustrate complete manor and its economy. Much information recorded only in text documents is collected via an archival survey. Important objects on maps as well as objects recognized by the archival survey are entered into the vector data model. The chateau area and its close surroundings are vectorised fully allowing a comparison of land-use and area development in different time.

Historic photographs of chateau, subsidiary buildings, gardens, etc. are collected, localized and possibly completed by present photo for comparison of state. Above mentioned maps, plans and vector data compose a frame for localized photographs publication within the web mapping application.

New approach for old maps and photo presentations brings the 3D modelling and visualization. The reconstruction of the historic landscape and the look of a chateau and its close surroundings can be created using old maps, the present digital terrain model and available historic photographs.

2 TWO-DIMENSIONAL DATA PROCESSING

2.1 *Georeferencing*

Selected old maps and plans suitable for the 2D processing should be digitized first. Raster digital data as an output of scanning must be georeferenced to be properly used in geographical information systems. There are a lot of methods of georeferencing, which can be divided into two groups (Cajthaml 2013). Global transformation methods use a set of ground control points to compute the transformation key. In the case of old maps affine or second order polynomial transformations are the most usable. On the other hand local transformation methods transform the image to be non-residual on the ground control points. There is no global transformation key and the image can be distorted to fit onto these points. For different data sources different transformation methods were chosen during our project. For older mappings either second order polynomials or local methods were used; for high-quality maps like 2nd or 3rd Military Mapping Surveys or State Derivated Maps affine or projective transformation using corner points were used.

A very important part of the whole georeferencing process is the estimation of the quality of the outputs. Within global transformation methods standard errors of position can be computed. For other methods the statistical testing of ground control points is highly recommended for the quality assessment.

2.2 *Vectorization*

The project is focused on buildings nearby castles' or chateaux' main buildings or buildings and areas somehow connected with them. Therefore, close surroundings of castles on selected maps are vectorised completely allowing a comparison of the land-use in different periods of time and castle area development. Vector models of maps of the Stable Cadastre, other old cadastral maps, State Derivated Maps and present cadastral data bring information about situation from the mid 19th century up to now. Vectorised polygons representing buildings then serve as a basis for procedural modelling.

Important economic and cultural objects on maps of domains, as well as objects recognized by archival survey are entered into the vector data model also. This allows broader view to the domain area and its economy. Old and present photographs are localized and stored as point layers including attributes of photograph description, owner, orientation angle, etc.

2.3 *2D web mapping application*

The common way of communicating spatial relationships is maps and plans, which represent two dimensional surface. Modern technologies enable interaction with maps in an online environment in the form of web mapping applications. The application created in this project serves three main purposes.

First, it offers spatial bookmarks (filters) of all studied objects enabling quick access to specific chateau or castle. The left panel with castle thumbnails gallery is connected with the map and the spatial extent of the map limits the display of thumbnails.

Second, it presents and compares various layers (maps and plans). The application allows overlaying of various rasters as well as vector layers from different times.

Third, it serves as a photograph gallery, which is linked to the map in the same way as castle thumbnails. There are historical photographs compared to up to date photographs. Each photograph is represented by a special cartographic point symbol, which next to the location also shows the point of view of the camera.

The application will be complemented by text information about chateaux, their surroundings, gardens, economic background of domains and their former owners, aristocratic families.

Example of the web mapping application can be seen in the Figure 1.

Figure 1.   Web mapping application of the Hradek u Nechanic Chateau.

3   3D SCENE AS AN ENHANCEMENT OF THE 2D WEB PRESENTATION

3.1   *Procedural and non-procedural modelling*

Two-dimensional web mapping application is surely a very efficient way to make the result of this project available to the wide audience. However, the benefit of 3D visualization for presentation purposes is unexceptionable. Basically, there are two ways to model a chateau/castle and its surroundings in 3D: classical modelling in CAD and procedural modelling in suitable software. There exist a lot of efforts dealing with 3D modelling of heritage buildings in CAD software. Description of these can be found, for example, in (Dore & Murphy 2012), (Jedlicka et al. 2013) or (Jedlicka & Hajek 2014).

The latter, namely, the procedural modelling is also well established at present but it is used mainly in the gaming industry and for quick creation of extensive models of modern cities. Basis for these models are prepared sets of suitable rules and two-dimensional GIS layers. Polygon layers with footprints can serve as a foundation for buildings, line layers as centerlines for streets growing algorithms; point layers denote the placement of trees or street furniture. For examples of real landscape modelling see (Edvardsson 2013) or (ESRI 2014).

The goal of the described project is to depict the state of a manor in several different time periods. The same applies to the 3D visualization. Considering available data, it was decided to choose a combination of the two afore mentioned approaches. Because, in most cases, there is available documentation of a castle or a chateau in a particular time period e.g. in the form of floor plans, these can (and should) be modelled in greater detail using standard CAD software. In the same way should be created models of all significant buildings within the chateau/castle complex. On the other hand it is not possible to model all buildings in the surroundings in the same way. This would be highly time consuming, and also there is no existing historical documentation of all objects inside the manor. A lot of buildings do not exist anymore or they were significantly renovated. Some information can be acquired on the basis of old photographs, but mostly the only data source is the old map. Therefore, the surrounding buildings could be produced based on vectorised old maps using procedural modelling approach. Missing information (e. g. the height of buildings) can be chosen randomly or better considering the look and the characteristics of buildings in a given time.

3.2   *Technical platform*

Because we use the ESRI platform for the 2D data processing (namely ArcMap), it is a logical step to utilize another application of this vendor for the procedural modelling – City Engine. This program was released in 2008 by the software company Procedural. This company was later acquired by

ESRI and a lot of effort was spent to ensure its compatibility with other ESRI products (Edvardsson 2013). City Engine enables to import 2D and 3D geometry in various formats, ESRI shapefile and feature classes of a file geodatabase are among them. Backwards compatibility is also ensured so after applying procedural rules, resulting models can be exported back to the geodatabase. It was also very important for us that 3D results can be exported to the 3WS file format which is suitable for sharing via ArcGIS Online with the use of HTML5 and WebGL.

There is no doubt that City Engine is a suitable tool for 3D procedural modelling. However, as mentioned in the previous section, models of heritage objects should be created with more detail using classical modelling approach. There exist some tools in City Engine which are designed for accurate polygonal modelling. Although with these tools 3D objects can be created in a very similar way to the SketchUp solution, it is clear that accurate modelling is not the greatest strength of City Engine.

Considering the mentioned the decision was made to use the SketchUp application for heritage buildings creation rather than City Engine. Resulting models can be then imported into a City Engine scene, for example, via OBJ or DAE file formats. SketchUp models are imported as static models. Static models can be further scaled, rotated and translated in the City Engine scene but they cannot be processed with procedural rules (ESRI 2015). On the contrary, two-dimensional building footprints of surrounding buildings originating in file geodatabase classes are imported as shapes so they can serve as a basis for procedural modelling.

At the time of writing this paper ESRI City Engine 2014.1 and Trimble SketchUp 2015 are the most up to date software versions so they are currently used in our project.

3.3 *Modelling process*

The first step in creating a 3D scene in City Engine is importing a digital terrain model. We use the aforementioned Digital Terrain Model of the Czech Republic of the 5th generation (DMR 5G). DMR5G is a bare-earth representation which describes terrain surface as heights of discrete points in TIN with total standard error of 0.18 m. This terrain model depicts, of course, the present state of a landscape. Because our goal is to capture historical state of chateau/castle surroundings, this would be unsuitable for areas which have changed significantly during the centuries (e.g. mining areas). However, for most areas this does not represent a serious problem.

The DMR 5G terrain model consists of points with X, Y planar coordinates and an H height value so it has to be interpolated to create a raster file. It is advisable to set a higher spatial granularity of the resulting raster because it will be further clipped together with other map layers. The point spacing of DMR 5G is approximately 3 meters so we use this value for the output cell size. This ensures correct clipping of the terrain and is still not excessively space consuming. The cell size can be also further increased directly in a City Engine scene so the visualization efficiency is also ensured.

The ArcMap application is used for the terrain interpolation and clipping. The terrain is saved to the georeferenced TIFF file and the IDW method seems to give the best interpolation results for our purpose. During the clipping process in ArcMap some pixels which are crossed by a clipping polygon are set to NoData. This would be misinterpreted as zero values in City Engine and the terrain would not be displayed correctly. So it is highly recommended to use rectangular non-rotated polygons and remove NoData values using Raster Calculator or some Matlab scripting.

Having imported the terrain we can also import an old map or other raster which will be used as a texture. This texture will be stretched across the extent of the DTM so it is important to use the same clipping polygon for the terrain and all used textures/maps. In a City Engine scene the georeferencing of a map is not necessary because it is ensured by the terrain model. However, all maps have to be georeferenced in ArcMap for the proper clipping.

Before we start programming procedural rules we have to import basic shapes. Building footprints can be either exported from a geodatabase to a shapefile in ArcMap, and then imported to a City Engine scene, or they can be imported directly from a geodatabase as a feature class. After

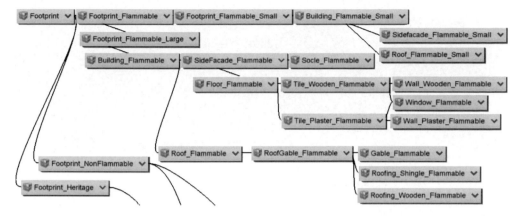

Figure 2.   Excerpt from a CGA rule file utilized for modelling based on the Imperial Imprints.

importing footprints aligning shapes to terrain and then terrain to shapes is performed to ensure them to fit together correctly.

Particular procedural modelling in City Engine consists in writing rules in a CGA rule file. Speaking about buildings, these rules usually first separate their footprints depending on building type, usage or material. Footprint attributes originating in corresponding feature class are here the basis. Then, each footprint is extruded to a certain building height. This height can be random or it can be described by another attribute. Furthermore buildings are split into roof surfaces and facades, facades are further split into floors etc. Finally actual roofs are created and textures are placed onto individual partial surfaces.

In the Figure 1 there is an excerpt from the visual representation of a sample CGA rule file. We use this rule file to create a semi-photorealistic 3D scene based on the Imperial Imprints of the Stable Cadastre. The hierarchical structure of the modelling process can be seen very clearly in this figure.

The Imperial Imprints contain information about the material and type of buildings – flammable, non-flammable and significant buildings are distinguished by colour. This information is stored in a geodatabase during vectorization so it can be used in the rule file to separate buildings. Moreover small buildings (with footprint area lesser then a certain set value) have to be filtered. These buildings can only have one storey (i.e. ground floor) and another texture will be placed on them.

Both flammable and non-flammable buildings are split in the similar manner that was mentioned above. The smallest part of any facade is a tile. All tiles are covered with textures as well as roofs and windows. Most buildings have a gable roof which was very common in the 19th century. Basically, there are so far two types of flammable (with and without plaster) and three types of non-flammable building facades. Buildings can have roofs covered by different types of shingles or roof tiles. Types of facades or roofing are chosen with a certain probability and when it is necessary there is also an interactive user input possible in the randomly generated scene. It is evident from the Figure 2 that there are also rules for heritage buildings. Resulting buildings modelled according to this rules are only provisional and they will be further replaced by non-procedural SketchUp models of chateaux or castles.

4   CASE STUDY AT THE CHATEAU OF MNICHOVO HRADISTE

Two-dimensional data collection and processing has already been performed in various manors. At this time the work is in various stages of completion. In some manors testing web mapping applications are already prepared for use. Two examples of this web application presenting the Hradek u Nechanic Chateau are depicted afore in the Figure 1.

Figure 3.   3D model of the surroundings of the Mnichovo Hradiste Chateau based on the Imperial Imprints of the Stable Cadastre and procedural modelling rules.

Procedural modelling of chateau/castle surroundings is currently being tested on the case of the chateau in Mnichovo Hradiste. The state owned Mnichovo Hradiste Chateau is located in the Central Bohemia Region. It is a baroque manor house which was rebuilt to its present state at the turn of the 18th century. The 3D procedural model of chateau surroundings based on the Imperial Imprints is shown in the Figure 3.

5   CONCLUSION

This article introduces a project focused on the web presentation of chateaux and manors in the Czech Republic. Besides the classic way of presenting 2D data via web mapping application, the article shows possibilities of 3D modelling and presentation of this data on the internet. The City Engine software and CGA rules are promising way of modelling 3D scenes. Procedural modelling was tested on the case of the chateau in Mnichovo Hradiste and this case study lead us to use this way in further development of our project. Meanwhile the web mapping application can be seen online at http://gis.fsv.cvut.cz/castles.

This work was supported by the Czech ministry of culture by the NAKI programme "Historical photographical material: identification, documentation, interpretation, record keeping, presentation, application, care and protection in the context of basic types of commemorative institutions" no. DF13P01OVV007.

REFERENCES

Cajthaml, J. 2013. Old Maps Georeferencing – Overview and a New Method for Map Series. In: *26th International Cartographic Conference Proceedings*. Dresden: International Cartographic Association.
Dore, C. & Murphy, M. 2012. Integration of Historic Building Information Modeling (HBIM) and 3D GIS for Recording and Managing Cultural Heritage Sites. In: *18th International Conference on Virtual Systems and Multimedia: "Virtual Systems in the Information Society", 2–5 September, 2012, Milan, Italy.*
Edvardsson, K.N. 2013. 3d GIS modeling using ESRI's CityEngine – A case study from the University Jaume I in Castellón de la Plana Spain.
ESRI 2014. Redlands Redevelopment 2014 Training. *2014 CityEngine Workshop.*
ESRI 2015. CityEngine Tutorial. Online: http://video.esri.com/series/62/cityengine/order/asc/
Jedlicka, K., Cerba, O., Hajek, P. 2013. Creation of Information-Rich 3D Model in Geographic Information System – Case Study at the Castle Kozel. In: *13th SGEM GeoConference on Informatics, Geoinformatics And Remote Sensing. 2013.*
Jedlicka, K. & Hajek, P. 2014. Large scale virtual geographic environment of the castle Kozel – best practice example

Advances and Trends in Engineering Sciences and Technologies – Al Ali & Platko (Eds)
© 2016 Taylor & Francis Group, London, ISBN: 978-1-138-02907-1

Analysis of effective remediation work in removing moisture masonry

S. Toth & J. Vojtus

Faculty of Civil Engineering, Technical University of Košice, Košice, Slovakia

ABSTRACT: The masonry moisture sanation process does not end, when the building job is over. This article contains results from long-term measurements of moisture levels in underground areas of several old buildings. These results can help in diagnostics and design of repairs in other buildings.

1 INTRODUCTION

The building must be understood as an organism with all its functions. It is exposed to a significant number of interior and exterior negative effects. The most common causes of damage of the building are particular long-term effects of weather conditions like temperature fluctuations, acidic rain water, the impact of air CO_2, road salt, frost and water with consequent formation of moisture and mold on the building surface. In examination the causes of failures – masonry moisture of walls should be systematically based on a wide range of causes. Their clarification is crucial for objective examination of the building condition, and it implies a proposal for the necessary remediation. Repeating chemical and physical processes, such as the conversion of water into ice, moisture capilar action, dissolution and crystallization of salts cause chemical reactions in construction processes that cause progressive disintegrating material. Due to the destructive phenomena and increased moisture of material, conditions for the growth and development of mold on interior and exterior surfaces are created [8]. Waterproof protection and moistureproof building protection is the most fundamental and the most important measure for ensuring the lifetime of the building and its operational capacity. When choosing a remediation method, it is necessary to review whole complex of factors that affect moisture regime of the building.

2 BUILDING DESCRIPTION

The historic building on Hlavna Street no. 8 in Presov, owned and used by Roman Catholic parish of St. Nicholas, is situated in a historic site of Historical Town Reserve of city Presov and is registered in the Central List of Monuments Fund in register of national cultural monuments no. ÚZPP – 3249/0. In the estate register it is registered as an object located on plot no. 44, in the tract area Presov. The first mention of the building is from 1223. The building on Hlavna Street no. 81 is a part of terraced houses on the west side of the square; it is situated opposite to the west entrance of the St. Nicholas Church. When reconstructing the older building, about half the size of the building in 1511, three tracts with storey and central passage originated. In 1983, the building layed on the gas, the central heating boiler with gas and the top hot water pipes in cast iron finned radiators. At that time, it seems using the fiddley stopped, whereby stove remained in original site. In 2002–2003 living rooms were designed in the attic.

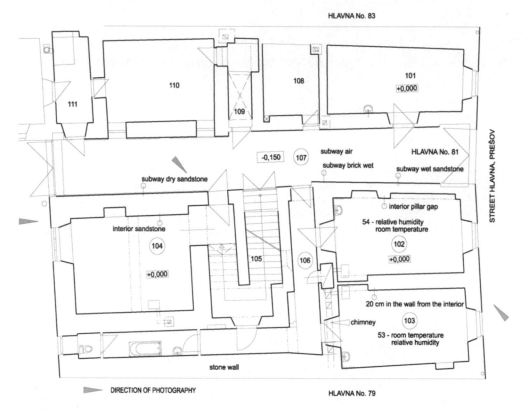

Figure 1. First floor plan of the parish building in Presov.

Figure 2. Exterior of the parish building in Presov.

3 FAULT DESCRIPTION AND ANALYSIS OF WETTING MANSONRY CAUSES

In accordance with the requirement to conduct a survey of the building in order to determine the causes of moisture of walls that was seen in almost every room, we performed more focused meterage of floor plans of the building and drawing of the factual floor plans including details, walls projections of each room on first floor where we also added a basement floor plan, on the basis of physical meterage that specified the image of the building.

Figure 3.   Interior of the parish building in Presov.

Table 1.   The results of laboratory evaluation of samples for moisture and salts from samples taken in January–February from boreholes.

No.	Height from floor	Material	Humidity [%] ±0.01%	pH	Chlorides [%]	Nitrates [%]	Sulphates quality
1a	1.6 m	sandstone	3.965	9.43	1.79	6.26	++
1b	1.0 m	sandstone, brick	2.976	9.20	1.01	5.3	+
2a	0.2 m	sandstone	1.778	9.73	0.285	0.498	++
3	0.3 m	sandstone	2.820	9.45	0.141	0.012	+++
4a	0.3 m	sandstone	8.536	9.29	0.250	0.437	+
4b	1.2 m	sandstone	6.485	9.78	0.245	0.498	+
4c	1.7 m	sandstone	4.542	9.32	1.027	6.20	+++
4d	2.1 m	sandstone	1.898	9.16	1.050	6.20	+++
5a	0.3 m	sandstone	4.43	9.53	0.331	0.405	+
5b	1.0 m	brick	5.687	9.35	1.650	5.79	++
6	1.8 m	brick	1.96	8.60	0.695	3.65	+
7	1.5 m	sandstone	2.32	9.48	0.106	0.185	+++
8	1.5 m	sandstone	2.917	9.67	3.14	0.488	+++
9	1.6 m	sandstone	1.388	9.20	0.678	3.56	+++
10	1.7 m	brick	16.49	8.40	0.831	3.48	+++
11	1.6 m	sandstone	19.69	10.20	1.72	4.2	+
12	0.3m	sandstone	2.441	9.18	2.51	5.64	++
13	1.5 m	brick	3.968	9.98	0.805	3.37	+
14	1.9 m	brick	1.734	9.74	2.43	5.46	+

Conserved project foundations situated in the parish from previous years did not have drew in basement floor plan. Likewise, position of fiddleys in each walls in the affected floors (first and second floor) was not drew in. After transferring the chimneys position from the attic (third floor), we investigated its presumable position in first floor, because their shape is dodging and they are expected, one of a range of different sources of moisture of walls.

In order to more accurately determine the rate of moisture of walls, we performed destructive sampling (drilling) in 14 locations on the first floor, in places with a clearly visible moisture on the surface of the plaster, respectively after removal of the plaster. Moisture in the walls is evaluated in % by weight. In the laboratory, the samples were dried and weighed, the chemical analysis of samples for salt content – nitrates, chlorides and sulphates was done.

From obtained results, increased to very high degree of salinity ensues. Based on laboratory results and analysis of moisture maps on the structures, we note that in the most parts of the

Table 2. Criteria of evaluation of the salinity in the sense of WTA 2-2-91 GB for remedial masonry systems.

Salinity degree	Cl^- %	NO_3^- %
Low	<0.075	<0.10
Increased	0.075–0.20	0.10–0.25
High	0.20–0.50	0.25–0.50
Very high	>0.50	>0.50

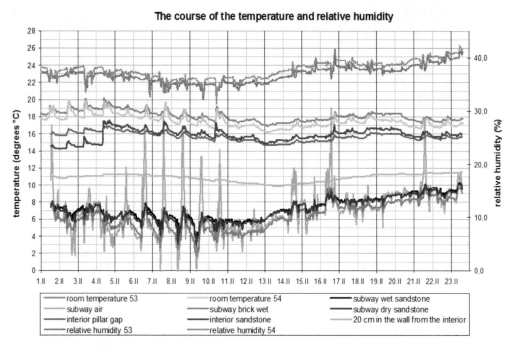

Figure 4. Measured temperatures and relative humidity in the building of the Roman Catholic parish in Presov.

walls on the first floor it is the condensation moisture (small thickness of masonry – niche, mixed masonry – brickwork interlaced with stone – sandstone) and partly humidity rising in the locations of the basement of the building, rooms no. 102 and 103. After the verification of the position of the vents of chimney bodies (principally in the room no. 102 and 103) shows up they are really dodged and they are also full of carbon black, probably since 1983, when new central heating was installed. Chimney vents are still ending above the roof, even they are covered with roofs, it does not leak in them directly, they are the cause of moisture of the surrounding masonry due to hygroscopic of material.

Laboratory analyzes discovered increased very high salt content in masonry and especially very high content of nitrates in stone walls, mainly in the tunnel area. In the past (18th century), in these places farm places may have been, what confirms the presence of the salts that are also significant manifestation of biocorrosive processes.

Semi-quantitative test for sulphate presence may be due to a large fire that was in Presov in 1887, and after it the houses were renewed. The presence of nitrates cannot be neutralized, their effect may after some time again occur and damage the structure of the remediation plasters (hygroscopic

Figure 5.   View on interior tunnel part of parish after the modification.

effect of increased salt content-the ability of congested salt to re-accumulate absorbing moisture from the air).

4   PROPOSITION OF CONSTRUCTION MEASURES FOR MOISTURE ELIMINATION

Projects proposed measures drew in the drawings, respect the requirements of Historic site of city of Presov, and can not 100% eliminate the effect of moistening even after a few years, but they try to eliminate demonstrations and thus extend the use of the rooms while maintaining hygiene requirements (minimum air temperature of 20°C, relative air humidity max. 50%). To improve the interior climate in the rooms on first floor would help balancing the heating on the first and on the second floor, or complement of convectors and in remote parts from the peripheral wall to warm the air in the room circulate better. Alternatively, the rooms during the time of use (the presence of people for more than 4 hours a day) may be heated by additional radiant heated electric radiators. In the corridor no. 106 would be appropriate to propose electric floor heating under the floor, which should separate switches, switching over timer. The situation in the underpass could significantly improve the creation of so-called interior space in exchange of glazed wall with steel profiles for wooden wall of Euro profiles with glazed double insulating glass and inserting a new glazed wall inside the front gate of Hlavna street. During the spring term it is necessary to clean with compressed manner purified downpipes connection to the public sewerage system in the facade of the Hlavna street.

5   CONCLUSION

When taking remedial actions, it is necessary to comply particular manufacturer's instructions, mostly technological remediation plaster mixtures, and it is necessary to observe the following procedures:

– Remediation systems is always applied on the wet surface (wet the surface before the application of the first layer, to ensure adequate adhesion of the plaster to the surface, otherwise there may be a later formation of cracks,
– Into the remediation mixture are not added any other ingredients than the specified ones by the manufacturer,
– Plaster mixture is being processed quickly (about 3 minutes), blending for longer period of time will decrease the strength of the plaster,
– Mortar that falls on the ground, is considered as devalued,

- The minimum air temperature and masonry temperature must be $+5°C$ during processing and during drying, should be avoided direct exposure to sun and rain,
- The next three to four days after the application the plaster is maintained moist – places e posed to air flow and direct sunlight are sprinkled,
- During the maturation it is necessary to create conditions for the uniform and regular air circulation at the treated site,
- On the plaster only recommended coating is applied,
- Technological breaks must be taken (for 1 mm thickness of plaster takes technological break one day) and required quality of implementation should be constantly monitored,
- Remediated walls cannot be wallpapered or tiled.

To the remediated walls it is not appropriate to just furniture, there must be preserved an air gap. Surface treatment cannot use acrylic or latex paint, which close pores, but the products that permeable water vapor, coatings of lime, silicate or silicone. Remediation works were made in accordance with design and financial options of the parish office.

ACKNOWLEDGEMENTS

This article was written as a project solution entitled "The use of the virtual laboratory for designing energy-efficient buildings" Project code: 052TUKE-4/2013.

REFERENCES

Chmúrny, I. 2003. Tepelná ochrana budov. Jaga group, Bratislava.
Flimel, F. 2014. Determining cumulative room temperature under in situ conditions using combined method. In: Advanced Materials Research. EnviBUILD. International Conference on Buildings and Environment. Bratislava. Vol. 899 (2014), p. 116–119. ISSN 1022-6680.
Guideline WTA CZ 2-2-91 Sanační omítkové systémy (German original translation).
Kmeť, S, Tomko, M., Demjan, I. 2014, Experimental diagnostics of steel structure after fire. In: Interdisciplinarity in theory and practice. No. 3, p. 1–7. ISSN 2344-2409.
Sedláková, A., Ťažký, L. 2014. Solution of the rising damp in the church in Gemerský Jablonec 2014. In: SGEM 2014: 14th international multidiscilinary scientific geoconference: Geoconference on Nano, Bio and Green-Technologies for Sustainable Future. Albena, Bulgaria. – Sofia. STEF92 Technology, p. 291–298. ISBN 978-619-7105-21-6. ISSN 1314-2704.
Terpáková, E. 2008. The study of building biocorrosion In: Chemické Listy. Vol. 102, no. Symposia, p. 919–920. – ISSN 0009-2770.
Tóth, S., Perečinský M., Vaškovičová A. 2006 Problémy historických budov v mestskej zástavbe – Problems of historical building in city town. In: Sanace a rekonstrukce staveb 2006. Praha, Česká stavební společnost ČSVTS, p. 200–205. – ISBN 8002018664.
Tóth, S., Vojtuš, J. 2014. Monitoring and analysis of fungal organisms in building structures. In: Advanced Materials Research. Vol. 969, p. 265–270. – ISSN 1022-6680.
Tóth, S., Vojtuš, J. 2014. Analysis of causes of mold growth on residential building envelopes in central city zone of Košice. In: Advanced Materials Research. Vol. 969, p. 28–32. – ISSN 1022-6680.
Urbanová, N. Prešov, pamiatková rezervácia z edície Pamiatky mestských rezervácií, vydavateľstvo Tatran ISBN 61-805-86, p. 62–63.
Vlček, M., Beneš, P. 2005. Poruchy a rekonstrukce staveb, Era, Brno.
Zákon NR SR číslo 272/1994 Z. z. O ochrane zdravia v znení neskorších predpisov, vyhláška č. 326/2002 MZ SR, ktorou sa ustanovujú najvyššie prípustné hodnoty zdraviu škodlivých faktorov vo vnútornom ovzduší budov.

Advances and Trends in Engineering Sciences and Technologies – Al Ali & Platko (Eds)
© 2016 Taylor & Francis Group, London, ISBN: 978-1-138-02907-1

An optimization of anchor of aluminum facade

K. Vokatá, M. Bajer, J. Barnat, J. Holomek & M. Vild
Faculty of Civil Engineering, Institute of Metal and Timber Structures, Brno University of Technology, Brno, Czech Republic

ABSTRACT: Nowadays, the glazed aluminum facades and ventilated facades are very often used types of facade single-storey or multi-storey buildings, e.g. Administration building, industry or logistics objects. This paper deals with optimization of angle supports which are the parts of aluminum grid of ventilated facade. The aim is an optimization of currently used types of supports – fixed and adjustable. Optimization is focused on effective usage of material. The result of this optimization will be the new shape for two types of anchoring supports (fixed and adjustable) with known bearing capacity. Bearing capacity will be obtain by numerical and experimental analysis. The results will be valid for the boundary conditions.

1 INTRODUCTION

In European trade are several systems of glazed aluminum facades offered by producers of facade systems (e.g. Reynaers, Schüco). Each producer has his own system solution. The base parts of each facade system are horizontal and vertical profiles. Each producer tries to make the most effective shape of each type of facade member. There are two important aspects – economic and reliability of using. One type of facade anchor member is chosen for our research. Because the producers often use similar shape of anchor for fixed anchor and adjustable anchor, both of them are optimized. These types of anchors are optimized with respect to higher reliability and suitable material usage.

1.1 Current system solution

Current system of ventilated facade is created by aluminum grid which is anchored by angle support to the structural part of building (e.g. floor slab). For example Alubond panel, HPL boards, ceramic boards, wooden boards or fiberglass concrete boards could be used as cladding.

1.2 Aim of solution

The aim of solution was optimization of two types of angle support (anchoring part for aluminum grid). The first type was fixed angle support, which is loaded by vertical load of weight of sheathing (e.g. Alubond panel) and by horizontal load of wind (suction or pressure of wind). The second type was adjustable angle support, which allows the vertical displacement. This type is loaded only by horizontal force from wind load (suction or pressure of wind). The aim of optimization was to find a new cross-section characteristic and dimensions of angle supports which satisfy a condition of effective usage of material.

1.3 Boundary conditions

The dead load of cladding is in vertical direction $g = 0.08\,\text{kN/m}^2$ and the value of horizontal load from pressure of wind is $w = 1.7\,\text{kN/m}^2$. The horizontal distance of grid between columns is 0.5 m and the vertical distance between angle supports is 1.35 m.

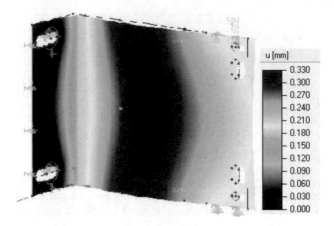

Figure 1.   Global deformation of fixed angle support.

2   SOLUTION

2.1   *Initial stage of experiment*

The initial stage of experiment was the first step in process of optimization. The aim of this experiment was to determine the load bearing capacity of fixed angle support and adjustable angle support. Special testing procedure was created for the purpose of this experiment. Basically, the angle support was mounted to the steel profile (simulation of mounting to the real structure). The load of 10 kg weight was suspended on the bottom of the angle (simulation of vertical load from sheathing). Pair of steel angles was mounted on the outstanding flange of angle. The horizontal load by hydraulic cylinder was brought through the steel plate mounted on the pair of angles.

There were tested 6 specimens for the tensile load and 3 specimens for the compression load for each type of angle supports in initial stage of experiment. The results of experimental analysis were used as validation of numerical model of angle supports. The validated numerical model of angle support was used for next step of optimization.

2.2   *Optimization of new shape anchor*

Numerical models were created according to the results of the experimental analysis. First model was created for fixed angle support; second one for adjustable angle support. Both of angle supports (anchors) were modeled in RFEM software by Dlubal Company.

The models were created using 2D elements. The loading of models was simulated the same way as in the case of experimental analysis. The course of von Misses stress was observed during process of optimization. The limit value of stress was set to 138 MPa due to fatigue. This was a key aspect for design of new shape of angle supports.

The areas with high value of stress were strengthened. The areas with low value of stress were attenuated. According to these results the outstanding flange of angle support was designed with gradation of thickness. The area around intersection of flanges was strengthened by rounding. These arrangements led to the decrease of the value of stress (see Fig. 2). The result of these adjustments was a new shape of fixed and adjustable angle support. The global deformation of new shape of angle support is satisfies the SLS criteria (see Fig. 1).

2.3   *Experimental verification of new angle support shape*

The new experimental analyses were realized according to design of new optimized types of angle supports. There were 10 specimens of fixed angle supports and 10 specimens of adjustable angle

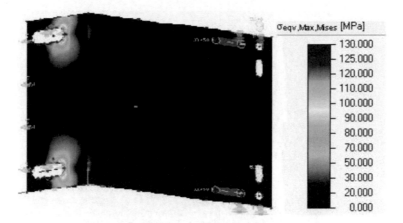

Figure 2. The von Mises stress results (MPa).

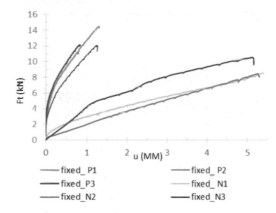

Figure 3. Tensile test – tension force versus horizontal displacement.

supports. The fixed angle supports were loaded by vertical and horizontal load. The adjustable angle supports were loaded only by horizontal load. Experimental analysis was divided to two parts for each type of angle support. One part of testing was transferring a compressive horizontal load and second part of testing was transferring tensile horizontal load to verify the behavior of the angle supports under compressive and combined tensile and bending loading conditions.

Six fixed angle supports were loaded by combination of tensile horizontal load and vertical load. Four fixed angle supports were loaded only by compressive horizontal load. Six adjustable angle supports were loaded by combination of tensile horizontal load and vertical load and 4 adjustable angle supports were loaded by compressive load. The aim of experimental analysis was to find the dependence of tension force on horizontal displacement, dependence of pressure force on horizontal displacement.

2.4 *Tensile test – verification of tensile resistance*

This experiment was focused on loading by combination of tensile horizontal load and vertical load. For this combination of load the influence of position of anchor bolts in slotted hole in flange on overall bearing capacity was monitored.

The specimen (angle support) was mounted on steel profile (simulation of mounting on real structure). The load of 10 kg weight was suspended on the bottom of outstanding flange (simulation

Figure 4.   Arrangement of tensile experiment.

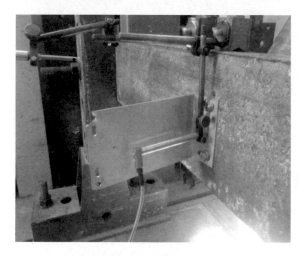

Figure 5.   Termination of experiment with specimen failure and position of LVDT.

of vertical load from cladding). The steel angle was mounted on the edge of outstanding flange. Tensile force was transferred to the specimen by rod through steel angle. The steel rod was connected to the hydraulic cylinder. Complete testing equipment is shown in Fig. 4.

During the testing the horizontal displacement was monitored by LVDT (see Fig. 5). The results of experiment are shown in Fig. 3. Termination of experiment with specimen failure is shown in Fig. 5.

2.4.1 *Compression test – verification of compressive resistance*
This experiment is focused on loading by compressive horizontal load. In this combination of load the influence of position of anchor bolts in slotted hole in flange on overall bearing capacity was monitored.

Complete testing equipment is shown in Fig. 6. The specimen (angle support) was mounted to steel base platform (simulation of mounting to real structure). The pair of steel angles was mounted

Figure 6.   Arrangement of compression test.

Figure 7.   Termination of the experiment by specimen failure – flexural buckling.

on the end of angle. The horizontal load by hydraulic cylinder was brought through the steel plate mounted to the pair of angles. The horizontal displacement was monitored by LVDT during the testing. The results of the experiment are shown in Fig. 8. Termination of experiment with specimen failure is shown in Fig. 7.

3   CONCLUSION

This paper deals with optimization of angle supports which are the parts of aluminum grid of ventilated facade. The aim was an optimization of currently used types of supports – fixed and adjustable. Optimization was focused on effective usage of material. The result of this optimization

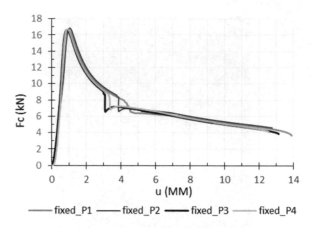

Figure 8.   Compression test – compressive force versus horizontal displacement.

is the new shape for two types of anchoring supports (fixed and adjustable) with known bearing capacity. Bearing capacity was obtained by numerical and experimental analysis. The results are valid only for the boundary conditions referred to in paragraph 1.3 in this paper.

ACKNOWLEDGMENT

This research has been supported by project MPO FR-TI4/332 and FAST-J-15-2859.

REFERENCES

ČSN EN 1991-1-4 Eurocode 1: *Actions on Structures – Part 1-4: General Actions – Wind Load*, Prague: Czech Standard Institute, 2011.
ČSN EN 1991-1-1 Eurocode 1: Actions on *Structures – Part 1-1: General Actions – Densities, self-weight, imposed loads for buildings*, Prague: Czech Standard Institute, 2004.
ČSN EN 1999-1-1 Eurocode 9: *Design of Aluminum Structure – Part 1-1: General Structural Rules*, Prague: Czech Standard Institute, 2009.
Kolář, V., Němec, I., Kanický, V. *FEM Principles and Practice of Method of Finite Number of Members*, Prague: Computer Press. 1997, ISBN 80-7226-021-9.
Radlbeck, Ch., Dienes, E., Kosteas, D., Sustainability of Aluminum in Buildings, *Structural Engineering International*, Volume 14, Number 3, 1 August 2004, pp. 221–224(4), ISSN 1016-8664.

Advances and Trends in Engineering Sciences and Technologies – Al Ali & Platko (Eds)
© 2016 Taylor & Francis Group, London, ISBN: 978-1-138-02907-1

Applying of dimensional analysis in rainwater management systems design

M. Zelenakova, I. Alkhalaf & V. Ondrejka Harbulakova
Institute of Environmental Engineering, Technical University of Košice, Košice, Slovakia

ABSTRACT: Design and use of infiltration facilities as a sustainable method of rainwater runoff disposal has become an integral part of drainage management and sewerage system projects for buildings or other paved surfaces. Infiltration facilities are devices designed for fluent and natural infiltration of rainwater from the roofs of buildings and paved surfaces. The basic principle and function of all types of infiltration facilities is to divert rainwater as quickly as possible to the infiltration zone, where it then infiltrates into the surrounding soil. One of parameters for rainwater systems design is emptying time in the percolation facility. The paper presents a model for prediction of emptying time in percolation shaft that was developed in the conditions of the campus of the Technical University in Košice.

1   INTRODUCTION

Design and use of infiltration facilities as a sustainable method of rainwater runoff disposal has become an integral part of drainage management and sewerage system projects for buildings or other paved surfaces.

The low impact development approach to rainwater management is an enormous change from conventional practices, which historically divert rainwater through engineered conduit systems to natural water bodies or costly treatment plants. In contrast, the low impact development approach carefully considers the rate, volume, frequency, duration, and quality of discharge so as to allow for groundwater and aquifer recharge and the overall health of ecological systems (Kibert, 2013). With a multifunctional landscape, it is possible to manage runoff, improve water quality, reduce power bills, increase property value, and save money and energy.

Disposal, or rather drainage of rainwater runoff is a problem for almost every new building in an urban area and in area with undersized sewage systems. Rainwater in built-up areas and other areas with closed surfaces can hardly find a natural path to return to the natural water cycle (Kibert, 2013; Ondrejka Harbul'áková & Homzová, 2013; Słyś, 2008; Krejči et al., 2002). This may result in gradual, long-lasting changes in soil structures and water regimes, entailing a reduction in the natural local groundwater replenishment, as well as impacts upon the chemical and biological conditions above and below ground (Hlavínek, 2007; Uhmannová et al., 2013; Słyś, Stec & Zeleňáková, 2012; Dziopak, 2001; Słyś, 2009).

The dimensioning of percolation facilities is worked out on the basis of the DWA Standard DWA-A 117E Dimensioning of Stormwater Retention Volumes. The following conditions are important: calculation of the inflow to the percolation facility, and calculation of the percolation rate (ATV-DVWK A 138E, 2005).

2   MATERIAL AND METHODS

The background data for infiltration facilities are maps, geological, hydrogeological data and hydrological conditions.

The main parameter affecting the intensity of the rainwater percolation process, and thereby the required geometry of the equipment and investment expenses, is the coefficient of soil filtration k_f. This characterizes the ability of soil to absorb and transmit water through a particular type of the soil, and its value depends mainly on the composition of soil, its particle size and porosity. The use of rainwater infiltration devices is especially recommended in areas which are characterized by values of soil filtration coefficient k_f in the range from 10^{-3} to 10^{-6} m/s. Long-term simulation also includes verification of emptying time. In the case of surface equipment covered by vegetation the maximum period of water detention should not exceed an average of 1–2 days (ATV-DVWK A 138E, 2005). The estimation of this emptying time is the objective of this paper.

For prediction of emptying time in percolation facilities using dimensional analysis, it is essential to state the parameters which characterize this phenomenon and which can be measured. We chose: rain intensity i [m.s^{-1}]; percolation area S [m^2]; coefficient of filtration k_f [m.s^{-1}]; density ρ [kg.m^{-3}]; emptying time T_e [s]. All the given variables are presented in basic dimensions, which is the condition for dimensional analysis application. The model for prediction of emptying time in percolation facilities is based on formation of non-dimension arguments π_i (Buckingham, 1914; Čarnogurská, 2000) from the stated variables influencing the emptying time.

The general relation among the selected variables, which can affect the pollutant concentration, can be put down in this form

$$\pi_i = i^{x_1} \cdot S^{x_2} \cdot \rho^{x_3} \cdot k_f^{x_4} \cdot T_e^{x_5} \tag{1}$$

We created a dimensional matrix-relation (2), which has the rank of matrix $m = 3$ and its lines are dimensionally independent on themselves. From $n = 5$ independent variables at the rank of the matrix m, we can set up $i = n - m$ of non-dimension arguments.

	i	S	ρ	k_f	T_e	
kg	0	0	1	0	0	
m	1	2	-3	1	1	(2)
(2)s	-1	0	0	-1	-1	

The matrix is modified for solution in the way that the determinant is not equal to zero. The matrix is changed to the system of four linear equations with five unknown parameters. Two independent vectors (3) and (4) are obtained by solution of the system of four linear equations

$$\pi_1 = S^{(1/2)} \cdot k_f^{1} \cdot T_e^{1} \tag{3}$$

$$\pi_2 = i^{1} \cdot k_f^{-1} \tag{4}$$

Non-dimension argument π_1 contains the unknown parameter T_e, so this argument can be expressed as a function of the argument π_2 in the form

$$\pi_1 = \varphi(\pi_2) \tag{5}$$

After modification the following equation is valid

$$T_e = A^{1} \cdot S^{-(1/2)} \cdot k_f^{(-1-B)} \cdot i^{B} \tag{6}$$

Relation (6) is the model for calculating the emptying time in percolation facility.

3   VERIFICATION OF THE MODEL

Prediction of emptying time in a percolation facility was performed for two percolation shafts located in the campus of the Technical University in Košice designed for rain with periodicity 0.2.

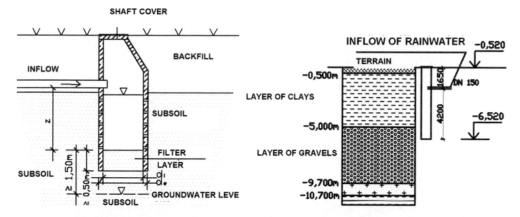

Figure 1. Percolation shaft A in the area of the Technical University of Košice.

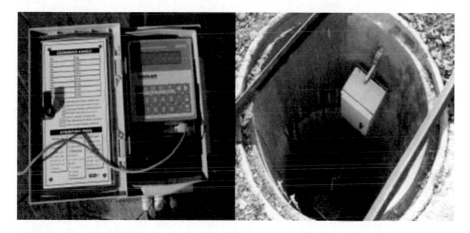

Figure 2. Comparison of measured and calculated emptying times in percolation shaft.

Shaft A (Figure 1) contains a universal data unit M4016 (Figure 2). All devices located in the shafts (A and B) are connected to this control unit.

In rain outlet pipe in the shaft there are measurement flumes for metering the inflow of rainwater from the roof of a nearby building. The area of the roof is 548 m² and the material is Ceberit roofing tile.

Pressure sensors are placed at the bottom of the shafts and are used for continuous measurements of the height of water levels in the shafts. The pressure sensor is connected with a data communication cable to control unit M4016 where the measured data are sent to the server (Ahmidat et al, 2014; Markovič et al, 2014a; Markovič et al, 2014b).

For modeling of emptying time, as already mentioned, it is essential to know the parameters that influence the percolation process. Values of filtration coefficient k_f are obtained from laboratory testing, with a percolation area of 0.785 m² in both shafts. We chose 12 rainfall events during 2011–2014 when we measured the relevant parameters (i and T_e), in Table 1.

According to the known relevant parameters from Table 1 the non-dimension arguments were stated from equations (3) and (4).

The regression equation between π_1 and π_2 is in the form

$$y = 1.0947.x^{-0.374}$$

where y is independent argument π_1, x is dependent argument π_2.

Table 1. Values of relevant parameters.

k_f [m.s^{-1}]	S [m^2]	i [m.s^{-1}]	T_e (measured) [s]
1.58E–03	0.785	7.31E–07	5580
1.58E–03	0.785	4.70E–07	9360
1.58E–03	0.785	4.32E–07	27840
1.58E–03	0.785	4.62E–07	26880
1.58E–03	0.785	4.20E–07	13800
1.58E–03	0.785	2.10E–07	12660
1.89E–04	0.785	7.31E–07	59580
1.89E–04	0.785	4.70E–07	55500
1.89E–04	0.785	4.32E–07	57780
1.89E–04	0.785	4.62E–07	31800
1.89E–04	0.785	4.20E–07	56700
1.89E–04	0.785	2.10E–07	35820

Table 2. Values of non-dimension arguments, measured and calculated values of emptying time.

π_2 [–]	π_1 [–]	T_e (calculated) [s]
0.0004627	9.95	10845.930
0.0002975	16.69	12793.120
0.0002737	49.65	13198.883
0.0002922	47.93	12879.408
0.0002657	24.61	13345.717
0.0001328	22.58	17299.900
0.0038677	12.71	40978.972
0.0024872	11.84	48336.002
0.002288	12.33	49869.088
0.0024429	6.78	48662.021
0.0022213	12.10	50423.867
0.0011099	7.64	65363.882

Emptying time was calculated according to equation (6) (Table 2).

Comparison of measured and calculated emptying time in shaft is shown in Figure 3.

The differences between measured and calculated emptying times can occur because the selection of relevant parameters did not involve all the factors which influence the process of percolation. In our future research, we should include storage coefficient to the model.

4 CONCLUSION

Rainwater in built-up areas and other areas with closed surfaces can hardly find a natural path to reach the natural water cycle.

Furthermore, harmless drainage of surface runoffs, in particular extreme runoffs during heavy rain events, calls for substantial technical and financial considerations during the stage of designing, construction and operation of sewer systems and wastewater treatment plants. However, despite all technical measures, some extreme runoffs do reach surface water. This may result in flood events, or increased pollution in small water courses in catchment areas with a major share of urban development.

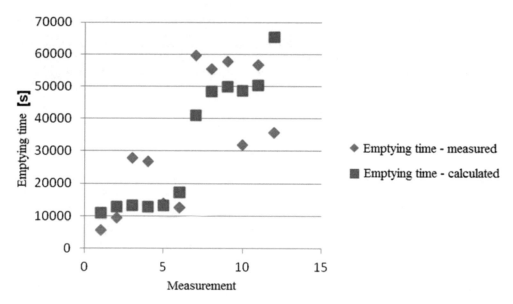

Figure 3. Comparison of measured and calculated emptying time in percolation shaft.

In inability of rainwater to return to the natural water cycle may result in gradual, long-lasting changes in soil structures and water regimes, entailing a reduction in the natural local groundwater replenishment, as well as impacts upon the chemical and biological conditions above and below ground (Hlavínek, 2007).

Rainwater infiltration systems aid rainwater runoff management and they also produce economic advantages, especially when prices of water rate and sewage charges are constantly rising (Słyś, 2012). The basic principle and function of all types of infiltration facilities is to divert rainwater as quickly as possible to the infiltration zone, where it then infiltrates into the surrounding soil. One of parameters for rainwater systems design is emptying time in the percolation facility.

The presented model that determines emptying time in a percolation shaft is based on dimensional analysis. Fundamentals of the modeling consist in derivation of function dependency from the expressed non-dimension arguments. Non-dimension arguments are stated from variables which influence the percolation process. From this function dependency it is possible to obtain values of emptying time in the shaft. Prediction of emptying time in a percolation shaft was performed on the campus of the Technical University in Košice.

ACKNOWLEDGEMENT

The Centre was supported by the Slovak Research and Development Agency under the contract No. SUSPP-0007-09.

REFERENCES

Ahmidat, M. K. M., Káposztásová, D., Markovič, G. & Vranayová, Z. 2014. The effect of roof material on rain water quality parameters in conditions of Slovak Republic In: *Advances in Environmental Sciences, Development and Chemistry: Proceedings of the 2014 Conference on Water Resources, Hydraulics and Hydrology*, Greece, Santorini: Europment, p. 275–280.
ATV-DVWK A 138E. 2005. *Planning, Construction and Operation of Facilities for the Percolation of Precipitation Water*.

Buckingham, E. 1914. On physically similar systems; illustrations of the use of dimensional equations. *Physical Review* 4 (4): 345–376, 1914.

Čarnogurská, M. 2000. *Basements of mathematical and physical modelling in fluid mechanics and thermodynamics* (in Slovak). Vienala, Košice.

Dziopak J. 2001. Co robić z nadmiarem ścieków. Warunki stosowania zbiorników retencyjnych w grawitacyjnych systemach kanalizacji. *Rynek Instalacyjny*, 12.

DWA-A 117E. 2005. Dimensioning of Stormwater Retention Volumes.

Hlavínek P. et al. 2007. *Rainwater management in urban areas* (in Czech). Brno.

Kibert, C. J. 2013. *Sustainable construction: green building design and delivery*, 3rd ed. John Wiley, 2013.

Krejči, V. et al. 2002. *Urban drainage area – conceptual approach*. Brno: NOEL 2000 Ltd. 2002.

Markovič, G., Zeleňáková, M., Káposztásová, D. & Hudáková, G. 2014a. Rainwater infiltration in the urban areas. In: *Environmental Impact 2*. Southampton: WIT press, Vol. 181, p. 313–320.

Markovič, G. Káposztásová, D. & Vranayová, Z. 2014b. The analysis of the possible use of harvested rainwater in real conditions at the university campus of Kosice In: *Recent Advances in Environmental Science and Geoscience: Proceeding of the 2014 Conference on Environmental Science and Geoscience*, Italy, Venice: Europment, p. 82–88.

Ondrejka Harbul'áková, V. & Homzová, E. 2013. Possibilities of capture and using rainwater from runoff (in Slovak). In: Proceeding from 3rd conference: Environment – Problems and possibilities of solutions: air-water-soil, p. 135–142, 2013.

Słyś, D. 2008. *Retention and infiltration of rainwater* (in Polish). Politechnika Rzeszowska, Rzeszów.

Słyś, D. 2009. Potential of rainwater utilization in residential housing in Poland. *Water and Environment Journal* 23(4).

Słyś, D., Stec, A. & Zeleňáková, M. 2012. A LCC analysis of rainwater management variants. *Ecological Chemistry and Engineering S* 19(3).

Zeleňáková, M. & Čarnogurská, M. 2013. A dimensional analysis-based model for the prediction of nitrogen concentrations in Laborec River, Slovakia. *Water and Environment Journal: Promoting Sustainable Solutions*. Vol. 27, no. 2, p. 284–291.

Zeleňáková, M., Purcz P., Ondrejka Harbul'áková, V. & Oravcová, A. 2015. Determination of pollutants concentrations in Krasny Brod river profile based on Buckingham theorem. *Desalination and Water Treatment*.

Author index

T - #0043 - 101024 - C466 - 246/174/25 [27] - CB - 9781138029071 - Gloss Lamination